THE ROUTLEDGE HANDBOOK OF ECO-PHENOMENOLOGY

The Routledge Handbook of Eco-Phenomenology is a comprehensive exploration of how we, as humans, interact with and perceive the natural world. Spanning seven thematic sections, this wide-ranging collection gathers contributions from leading scholars in philosophy, theology, anthropology, literature, music studies, and environmental thought. Together, over 30 chapters offer fresh perspectives on how we inhabit a more-than-human world.

At the heart of this volume is the dynamic intersection of phenomenology – a philosophical tradition attentive to embodied experience – and environmental ethics, inviting readers to rethink assumptions about activism, passivity, and ecological responsibility. Rather than advancing a unified doctrine, the chapters explore eco-phenomenology as a field of ongoing tension and dialogue, attentive to receptivity, perception, and the ethical demands of a world that exceeds us.

Key themes include re-evaluations of human embodiment in relation to nature, the role of perception in understanding our ecological entanglements, and cultural attitudes that challenge human exceptionalism. Literary explorations and reflections on music deepen the inquiry, showing how the arts open distinct pathways of attunement to the environment. The volume culminates in critical dialogues between eco-phenomenology and other intellectual traditions, including posthumanism, Indigenous cosmologies, and enactivism.

More than an academic survey, this handbook is an invitation to rethink environmental engagement as a mode of being rooted in care, vulnerability, and attentiveness to the Earth as our shared home. It is essential reading for scholars, researchers, and students in philosophy, environmental studies, anthropology, literature, and the arts.

Michael T. Heneise is Associate Professor of Religious Studies at UiT – The Arctic University of Norway. His work brings phenomenological and anthropological approaches to dreams, healing, and human–more-than-human relations in Highland Asia and the Arctic. He is President and co-founder of the Highland Institute, an independent research centre in Northeast India, and serves as Editor-in-Chief of *HIMALAYA: The Journal of the Association for Nepal and Himalayan Studies*.

Cassandra Falke is Professor of English Literature and leader of the Interdisciplinary Phenomenology group at UiT – The Arctic University of Norway. She has written two monographs and 50 articles and chapters, and has co/edited five collections. Her work has received support from Fulbright, the NEH, the NOS-HS, and Cornell University.

Espen Dahl is Professor of Systematic Theology at UiT – The Arctic University of Norway. He writes on phenomenological approaches to religious experience, holiness, aesthetics, evil, nature, and the body. His most recent book is *Incarnation, Pain, Theology: A Phenomenology of the Body* (2024).

Alice Sundman holds a PhD in English Literature from Stockholm University. Her current research focuses on literary imaginings of water in climate-changed future worlds. Her monograph *Toni Morrison and the Writing of Place* (Routledge 2022) explores the creation and presentation of Toni Morrison's literary places.

Edvard Lia is a PhD fellow at UiT – The Arctic University of Norway, working on a dissertation titled *The Phenomenon of Breath: An Existential Interpretation of Respiratory Facts* with Hans Jonas as the primary philosophical interlocutor. His other general research interests include the phenomenological tradition, Hegel's philosophy, and eco-Marxism.

ROUTLEDGE ENVIRONMENT AND SUSTAINABILITY HANDBOOKS

ROUTLEDGE HANDBOOK OF CLIMATE LAW AND GOVERNANCE
Courage, Contributions and Compliance
Edited by Marie-Claire Cordonier Segger and Christina Voigt

ROUTLEDGE HANDBOOK ON CLIMATE CRISIS COMMUNICATION
Edited by Alison Anderson and Candice Howarth

ROUTLEDGE HANDBOOK OF DEGROWTH
Edited by Anitra Nelson with Editorial Advisor Vincent Liegey

ROUTLEDGE HANDBOOK OF WETLANDS
Edited by Alan Dixon and Ian Maddock

ROUTLEDGE HANDBOOK OF WATER DIPLOMACY
Edited by Shafiqul Islam, Kevin Smith, Martina Klimes and Aaron Salzberg

ROUTLEDGE HANDBOOK OF WILDLIFE CRIME
Edited by Francis Massé, Annette Hübschle, Laura Gutiérrez, Rebecca W. Y. Wong, and Tanya Wyatt

ROUTLEDGE HANDBOOK OF GRASSLANDS
Edited by Heather Hager, David Gibson and Jonathan Newman

THE ROUTLEDGE HANDBOOK OF ECO-PHENOMENOLOGY
Edited by Michael T. Heneise, Cassandra Falke, Espen Dahl, Alice Sundman, and Edvard Lia

For more information about this series, please visit: www.routledge.com/Routledge-Environment-and-Sustainability-Handbooks/book-series/RESH

THE ROUTLEDGE HANDBOOK OF ECO-PHENOMENOLOGY

Edited by Michael T. Heneise, Cassandra Falke, Espen Dahl, Alice Sundman, and Edvard Lia

Routledge
Taylor & Francis Group
LONDON AND NEW YORK

earthscan
from Routledge

Designed cover image: Michael T. Heneise - Tromsø, Norway in 2022

First published 2026
by Routledge
4 Park Square, Milton Park, Abingdon, Oxon OX14 4RN

and by Routledge
605 Third Avenue, New York, NY 10158

Routledge is an imprint of the Taylor & Francis Group, an informa business

British Library Cataloguing-in-Publication Data
A catalogue record for this book is available from the British Library

ISBN: 978-1-032-88180-5 (hbk)
ISBN: 978-1-032-88498-1 (pbk)
ISBN: 978-1-003-53809-7 (ebk)

DOI: 10.4324/9781003538097

Typeset in Times New Roman
by SPi Technologies India Pvt Ltd (Straive)

CONTENTS

CONTRIBUTORS

Meera Baindur is Associate Professor of humanities at the T.A. Pai Management Institute, Bengaluru, and Co-ordinator at the Centre for Research – DWEEPA, Manipal Academy of Higher Education, Manipal, India. Her work spans various aspects of Environmental Humanities, nature in Indian philosophical and cultural traditions, cultural geography, and gender studies.

Bryan E. Bannon is Associate Professor of Humanities. His research applies phenomenological concepts to contemporary environmental issues. He is the author of *From Mastery to Mystery* (Ohio University Press) and has published in journals including *Environmental Ethics*; *Ethics, Place & Environment*; and *Research in Phenomenology*.

Forrest Clingerman was Professor of Religion and founding Director of the Center for the Advancement of Faculty Excellence at Ohio Northern University. He was a leader in environmental hermeneutics.

Thor Eirik Eriksen is Associate Professor at UiT and Researcher at UNN. His work explores the limits of objectivist epistemologies when confronted with lived, complex phenomena, drawing on phenomenology to reframe understanding and experience within interdisciplinary contexts. He co-leads the research group Interdisciplinary Phenomenology.

Zlatan Filipovic is an associate professor in English Literatures at the University of Gothenburg, Sweden. His research explores affect, deconstruction, and ethics in literature, including posthuman ecologies (Ecozona, 2025) and apocalyptic narratives (Broken Mirrors, Routledge, 2020). He is an editor of Moderna Språk and currently works within posthumanism and ecocriticism.

Marina Garcés is a philosopher and professor at the Open University of Catalonia, where she directs a master's programme and research group on contemporary challenges. Her work explores education, imagination, and collective life. Her recent book is *New Radical Enlightenment: Philosophy for a Common World* (Verso, 2024).

Louise Green is Professor in the English Department at Stellenbosch University, South Africa. Her research focuses on the diverse ways in which environmental damage enters stories being told about the world today. She is author of *Fragments from the History of Loss: The Nature Industry and the Postcolony* (2020).

Finn Thorbjørn Hansen is a full professor of Applied Philosophy, holds a PhD in Educational Philosophy, at the University of Aalborg (Denmark), and co-head of the research group TEN (Time, Existence & Nature connectedness). His research focuses on the phenomenology and ethics of wonder and methods in phenomenological action research through the so-called Wonder Labs.

Shannon Hayes is a senior lecturer at Vanderbilt University, researching continental philosophy and aesthetics with a focus on contemporary temporal experience and collective affect.

Maia Vige Helle is a PhD candidate at NTNU working on a dissertation called Enacting the Crisis, where she looks at the potential for using the enactive approach for developing an ethics fit for dealing with the environmental and climate crisis.

Magdalena Hoły-Łuczaj is an assistant professor at the University of Wrocław, Poland. Her research focuses on environmental ethics and philosophy of technology, with particular interest in their intersection. She has published in journals such as *Environmental Values*; *Environmental Ethics*; *Ethics and the Environment*; and *Philosophy and Technology*.

Jonas Jakobsen is associate professor of philosophy at UiT – The Arctic University of Norway. His research explores political philosophy, critical theory, and existential, contemplative traditions.

Hayden Kee is Assistant Professor of Philosophy at the Chinese University of Hong Kong. His research integrates phenomenology, cognitive science, and philosophy of language, focusing on embodied cognition and intersubjectivity. He also engages Indian philosophy, particularly the yoga tradition, in dialogue with Husserl, Merleau-Ponty, and 4E approaches to mind.

Simone Kotva is a scholar of theology and Continental philosophy. Her work explores mysticism, ecology, and vegetal life, with recent projects on decolonial theologies and Scandinavian Christianity. She is the author of *Effort and Grace and Ecologies of Ecstasy*, and co-translator of *Félix Ravaisson's Fragments on Philosophy and Religion*.

Charlotte Lindvang is Associate Professor in Music Therapy at Aalborg University, Denmark. She is a certified GIM therapist (EAMI) and psychotherapist (DPFO). Her research explores therapeutic training, existential health, music therapy interventions, phenomena of resonance and nature connectedness. She is author and co-editor of *Resonant Learning in Music Therapy* (2023).

Michael Marder is IKERBASQUE Research Professor in the Department of Philosophy at the University of the Basque Country (UPV-EHU), Vitoria-Gasteiz. His most recent books include *The Phoenix Complex* (2023); *Time Is a Plant* (2023); with Edward S. Casey, *Plants in Place* (2024); *Eco-Freud* (2025); and *Metamorphoses Reimagined* (2025).

Martin Mølholm is an associate professor at the University of Aalborg with a PhD in Philosophy of Language. Among his latest publications are the two essays *The Compulsion to Achieve as a Way to Immortality* and *The Sound of Wonder of Ontological Homecoming*, co-authored with Professor Finn T. Hansen.

David Morris is Professor of Philosophy at Concordia University, Montreal. His research draws on and advances phenomenology (especially through Merleau-Ponty), by focusing on the philosophy of the body, perception, mind, nature, and time, in relation to current sciences, especially biology, information science, and quantum mechanics and cosmology.

Svein Anders Noer Lie is Associate Professor of Philosophy at UiT – The Arctic University of Norway. His research spans environmental philosophy, philosophy of medicine and technology, and early engagements with ontology and phenomenology. Beyond academia, he is also a farmer, carpenter, and fisherman.

Petr Prášek is a postdoctoral researcher at the Institute of Philosophy, Czech Academy of Sciences. He has published *Le devenir-autre de l'existence. Étude sur la phénoménologie contemporaine en France* (Hermann, 2023) and co-edited *The Faces of Nature: Eco-Phenomenology and the Philosophy of Nature* (to be published by Lexington, 2025).

Kaja Jenssen Rathe is a PhD fellow at UiT – The Arctic University of Norway, where she recently completed her thesis "A Critical Phenomenology of Immigrant Indebtedness: Investigating First-Person Experiences of Permanent Guesthood."

Ida Solhaug is a clinical psychologist and an associate professor in the Department of Psychology, UiT – The Arctic University of Norway. Her research explores clinical and health psychology, particularly mindfulness practices, experimentally, qualitatively, and theoretically.

Patryk Szaj is an associate professor in literary studies at the University of the National Education Commission in Krakow, Poland. His research focuses on the environmental humanities, contemporary Polish poetry, and the Anthropocene discourse. He serves as the scientific editor of the "Humanistyka Środowiskowa" series at słowo/obraz terytoria press.

Calvin D. Ullrich is Senior Lecturer in Interdisciplinary Theology and Director of the Centre for Interdisciplinary Research in Religion, Community, and Ecologies (CIRRCE), University of the Free State. He is author of *Sovereignty and Event* (Mohr Siebeck, 2021) and publishes in the fields of continental philosophy and political theology.

Lisa Vargo is Distinguished Emerita Professor of English, University of Saskatchewan. Her research interests include Romantic women writers and textual editing. Her recent publications are Volume 2 of the *Collected Works of Anna Letitia Barbauld* with Paula Feldman and Volume 4 with William McCarthy and Scott Krawczyk (Oxford University Press).

Robert A. Winkler is Assistant Professor in the Department of English & American Studies at the University of Salzburg, Austria. His research explores the "cabin" in American cultural and literary history. He has published widely, for instance, on hardcore punk, whiteness studies, heavyweight boxing, Thoreau's Walden, the Unabomber, or Heidegger.

ACKNOWLEDGEMENTS

The editors express their deep gratitude to all those who made the *Routledge Handbook of Eco-Phenomenology* possible. We are especially indebted to UiT's publishing fund, administered through The University Library at UiT – The Arctic University of Norway, whose generous support enabled the full open-access publication of this volume and helped bridge the gap for contributors from institutions without dedicated OA funding.

We gratefully acknowledge the universities, foundations, and individual scholars whose contributions made this handbook freely available to readers worldwide: the Paris Lodron University of Salzburg (Austria); Concordia University and the University of Saskatchewan (Canada); the Institute of Philosophy, Czech Academy of Sciences (Czechia); the Department of Communication & Psychology, Aalborg University (Denmark); The Chinese University of Hong Kong (Hong Kong); UiT – The Arctic University of Norway and NTNU – The Norwegian University of Science and Technology, with additional support from affiliated research projects (Norway); the University of the National Education Commission in Krakow (Poland); the Centre for Interdisciplinary Research in Religion, Community, and Ecologies (CIRRCE), University of the Free State (South Africa); the Universitat Oberta de Catalunya (Spain); the University of Gothenburg and the Sven and Dagmar Salén Foundation (Sweden); and, Ohio Northern University (United States).

We extend our sincere thanks to contributors who personally covered their open-access costs, ensuring that the volume could reach a global readership. Their commitment to open scholarship reflects the collaborative spirit that underpins this work.

We also thank Routledge (Taylor & Francis) for its cooperation in facilitating multi-institutional open-access arrangements and for its careful attention throughout the production process. We are particularly grateful to Grace Harrison, our commissioning editor, and Morwenna Scott, who kindly oversaw the project during Grace's leave, for their professionalism and guidance, as well as to the wider Routledge team for their support.

Finally, we thank all contributors, whose collective effort, patience, and dedication have made this volume not only a shared intellectual achievement but also a testament to the collaborative ethos that sustains eco-phenomenology today.

—The Editors

1

INTRODUCTION TO *THE ROUTLEDGE HANDBOOK OF ECO-PHENOMENOLOGY*

Michael T. Heneise, Cassandra Falke, Espen Dahl, Alice Sundman, and Edvard Lia

Situating Eco-Phenomenology

In recent decades, *eco-phenomenology* has emerged as both a compelling and necessary intervention in environmental thought. The term gained prominence through the work of Charles Brown and Ted Toadvine (2003). Rather than functioning as a supplement to existing ecological discourse, eco-phenomenology presents a challenge to the conceptual architectures that undergird much of modern environmentalism. Prevailing approaches—often shaped by managerial or technocratic logics—tend to frame ecological crises in terms of sustainability, risk mitigation, or systems optimization. While these frameworks may offer tools for governance, they typically presuppose a worldview in which nature is external to human subjectivity: either as a totality in front of us, a passive backdrop to history, a system to be regulated, or a resource to be optimized. Eco-phenomenology interrupts this logic by returning us to the lived body and to the world as it is experienced. "Eco" does in our context denote the ecology understood as the natural scientific study of the dynamic and complex systems of living organisms and their environment. Such perspective is certainly appreciated, but phenomenology guides us to the more immediate ways in which these relations are lived. Accordingly, we invoke a broader and etymologically more original understanding of "eco," stemming from "oikos," that is, home or household. In this sense, eco-phenomenology aspires to call us back to our environments as our (often neglected) home—to return our attention to our original place in nature in, around, and beneath us. Thus, eco-phenomenology follows Husserl's slogan "back to the things themselves" in a specific direction, namely "back to the *earth* itself" (Brown, Toadvine 2003). We ask: *What does it mean to perceive, to be affected by, and to dwell within a world that is not only human? What difference does it make to our embodiment and to the world that they are both part of nature?*

Rooted in the phenomenological tradition and dynamically informed by ecological thought, eco-phenomenology insists that any meaningful engagement with the environment must begin not with abstraction, but with concrete, affective, and intersubjective dimensions of experience. It redirects attention away from objectified conceptions of "environment" and toward the ways in which the world is lived—in perception, memory, and bodily attunement—as vibrant, relational, and affectively charged. In contrast to a logic of mastery, it

 DOI: 10.4324/9781003538097-1

cultivates a posture of receptivity: a mode of being marked by attentiveness, exposure, and response, rather than control (Merleau-Ponty 1968, 2012 [1945]).

This volume brings together contributions from philosophy, theology, anthropology, literature, and cognitive science, each of which takes seriously the challenge of reimagining phenomenology in ecological terms. Rather than advancing a unified doctrine, these chapters treat eco-phenomenology as a field of productive tension: a site of ongoing negotiation between classical methods and the urgencies of contemporary ecological crisis. If phenomenology teaches us to return to "the things themselves" (Husserl 1982 [1913]), then this volume asks what those things become when the more-than-human world we return to is increasingly recognized as unstable and ethically demanding. Even as the "thing itself" remains the lodestar, approaching it, we suggest, is not reducible to a single disciplinary lens—it demands multiple, situated, and at times discordant modes of apprehension. Eco-phenomenology, as presented here, is not a fixed method but a sensibility, a mode of attunement to a world that affects and implicates us at every turn.

Origins and Philosophical Lineages

Rather than establishing a fixed system or a unified school of thought, the term designates a confluence of philosophical trajectories: from one direction, the phenomenological tradition inaugurated by Edmund Husserl; from the other, the broader field of ecological thought shaped by figures such as Arne Næss (1973), Val Plumwood (2002), David Abram (1996), and Isabelle Stengers (2011). Each of these thinkers approaches environmental questions in ways that challenge the assumptions of anthropocentrism and the logics of domination that have defined much of Western modernity. Brown and Toadvine's *Eco-Phenomenology: Back to the Earth Itself* (2003) provided a name and an integrated conceptual field for thinkers working at this convergence, where phenomenology's attentiveness to the structures of lived experience proves remarkably fertile for rethinking ecological relations.

From its inception, phenomenology has oriented itself toward what is given—what appears within the horizon of experience—and thus resists the notion that meaning arises solely through conceptual abstraction or instrumental calculation. In Husserl's early writings, this meant tracing how consciousness actively constitutes the world through intentional structures. But in his later works, particularly *Analyses Concerning Passive and Active Synthesis* (Husserl 2001 [1966]), Husserl complexifies this picture. He does not abandon the primacy of subjectivity, but expands it—revealing that perception is not simply the result of active synthesis but also of passive receptivity to temporality, affectivity, and sedimented lifeworld experience. Subjectivity, in this late Husserlian frame, is both generative and constituted: an open field through which the world impresses itself before any conscious act of reflection.

In parallel, ecological philosophers such as Næss, Plumwood, and Stengers call for a relational ontology that foregrounds interdependence and contingency, often in tension with modern assumptions of control, mastery, and separation. Their work helps prepare the conceptual ground for a phenomenology that no longer assumes a neutral subject appropriating a passive nature, but instead confronts the entangled and affective reality of ecological life. Eco-phenomenology thus begins to take shape as a site of mutual provocation: a dialogue between traditions that share a commitment to embodiment, situatedness, and the irreducibility of the more-than-human world.

Flesh, Passivity, and the Ecology of Perception

It is in the work of Maurice Merleau-Ponty that this reconfiguration of subjectivity becomes ecological. While Husserl's later writings gesture toward a passively constituted subject, Merleau-Ponty radicalizes this orientation in his late ontology of the *flesh*. In *The Visible and the Invisible*, he abandons any remaining vestiges of Cartesian dualism in favor of an account of perception as a reversible intertwining—where seer and seen, toucher and touched, are bound in a chiasmic relation (Merleau-Ponty 1968). The world, he writes, is not in front of us like a spectacle to be surveyed; it envelops and includes us. We do not look at things from a distance, but are *of the world*, made of the same flesh. This insight undoes any lingering notion of a disembodied subject and insists instead on the primacy of a corporeal and participatory perception (Merleau-Ponty 2012 [1945]).

Renaud Barbaras builds on this shift by pushing phenomenology toward what he calls a "desire for the world." Intentionality, he argues, does not begin with object-constitution but with a pre-objective movement of life—a desire that precedes and conditions cognition. Rather than defining perception as a structuring of givenness by the subject, Barbaras presents it as an exposure to alterity, a passivity that enables the very emergence of meaning (Barbaras 2006, 114). In this view, subjectivity is not the ground of experience but its effect—a crystallization of movements that originate outside it.

David Morris, in turn, extends these insights by situating perception within a thoroughly ecological frame. He introduces the notion of the "ecological flesh," expanding Merleau-Ponty's ontology beyond intersubjectivity to encompass planetary and environmental relations. Our embodied responsiveness, Morris shows, is shaped not only by other human beings but by historical conditions, biological rhythms, and the textured presence of place. Even the most intimate acts of perception—feeling the warmth of sun on skin, breathing mountain air, standing on shifting terrain—are shaped by a more-than-human world. Passivity here is not a negation of agency but its very condition: it is through being affected that one becomes capable of response (Morris 2008, 2017).

Passivity, Excess, and the Ethical Exposure to the More-Than-Human

The ethical dimensions of passivity are brought into sharp focus in the philosophy of Emmanuel Levinas, for whom subjectivity begins not with intentional activity but with exposure to an other who calls, commands, and disrupts. In his formulation, responsibility is not the outcome of rational deliberation or volitional choice, but something that comes before freedom—a demand that is felt, endured, and suffered. For Levinas, passivity is not a lack of action but a deeper responsiveness: "more passive than all passivity," as he famously writes (Levinas 1969, 110). While Levinas himself did not develop an environmental philosophy, his emphasis on unchosen responsibility—on being addressed by that which exceeds comprehension—offers eco-phenomenology a powerful ethical vocabulary. When the figure of the other is extended to include animals, rivers, trees, or future generations, the stakes of passivity become unmistakably ecological.

Yet if phenomenology is to remain attuned to ecological realities, it must also confront the limits of its own conceptual framework. Ted Toadvine takes up this challenge with urgency, arguing that phenomenology must not only describe nature as it appears to consciousness, but also acknowledge what in nature remains *inaccessible*—that which withdraws, eludes, or exceeds phenomenological thematization. Nature, in this view, is not simply a field of appearances or a horizon of lived significance; it is also that which resists being brought to presence (Toadvine 2009, 1–12). Toadvine calls for what he terms a "self-deconstructive phenomenology," one that retains its commitment to description while remaining open to its own failure—open to what refuses to be made fully intelligible.

This tension—between receptivity and resistance, between intertwining and alterity—is one of the most generative in contemporary eco-phenomenology. While many phenomenological approaches emphasize relation, attunement, or embodied reciprocity, there remains the persistent challenge of accounting for difference: for the irreducibility of other forms of life, of matter, and of worlds. In this context, passivity is not simply the ground of perception, but the threshold of ethics and of ontology—a reminder that to dwell ecologically is also to dwell incompletely, vulnerably, and with humility.

Dialogue and Transformation: Eco-Phenomenology Beyond Its Borders

If eco-phenomenology is to remain responsive to the world it seeks to think, it must risk transformation. Several of the contributions in this volume explore what happens when phenomenology enters into dialogue with other traditions of thought—posthumanism, agential realism, Indigenous cosmologies, enactivism, and critical phenomenology. These engagements do not necessarily dilute or sidestep phenomenology, nor do they merely apply phenomenological concepts to new domains. Instead, they can be taken to test phenomenology's assumptions and stretch its conceptual capacities, drawing it into relation with alternative modes of knowing, being, and responding.

Posthumanism, for instance, challenges the centrality of the human as a stable ontological or ethical category, insisting instead on networks of entanglement that include animals, technologies, microbes, and landscapes (Braidotti 2013; Wolfe 2010). Indigenous ontologies frequently reject the dichotomy between nature and culture altogether, presenting the world as inherently relational, agentic, and alive (Coulthard 2014; Todd 2016). These perspectives do not merely extend phenomenological thinking—they reconfigure it, opening it toward cosmologies in which perception is never disembodied, and where ethics is embedded in kinship relations with nonhuman others.

Such encounters require phenomenology to reflect on its own limitations: its Eurocentric roots, its anthropological presuppositions, and its traditional emphasis on the human lifeworld. But they also highlight phenomenology's potential for renewal. As Merleau-Ponty once suggested, phenomenology is not a doctrine but a style of thinking—one always open to interruption, displacement, and revision (Merleau-Ponty 1968, 139). In this spirit, the chapters that follow do not aim to defend a tradition, but to dwell within it, push against its contours, and discover what it might yet become in response to a world that exceeds it.

Several contributors also engage the later thought of Heidegger, especially *The Question Concerning Technology* (Heidegger 1977) and his reflections on *Gelassenheit*, or "releasement" (Heidegger 1966). These works help articulate a non-instrumental mode of thinking and dwelling—one that resists technological enframing and cultivates a receptive stance toward being. In this context, Heidegger's late writings offer eco-phenomenology a language of restraint, attentiveness, and letting-be that resonates deeply with themes of passivity, relationality, and ecological humility.

Intellectual Origins and Institutional Milieu

This encounter between phenomenology and ecological thought is not only conceptual but also institutional and communal, grounded in a shared scholarly milieu. The volume itself emerged from the ongoing work of the Interdisciplinary Phenomenology Research Group at UiT. The Arctic University of Norway—a collective of philosophers, theologians, anthropologists, and literary scholars committed to rethinking phenomenology across disciplinary boundaries. Through monthly reading seminars and collaborative workshops, the group has cultivated a space for philosophical reflection that is at once rigorous and hospitable: a place where classical texts and contemporary crises are taken up in dialogue, and where phenomenology is approached as a living, evolving tradition.

The project took further shape during the international conference *Eco-Phenomenology and Passivity*, held in Tromsø in September 2023. Organized by members of our research group, the conference gathered scholars from across the globe to reflect on passivity not as inertia or resignation but as a mode of ecological response-ability. Keynote speakers included Bryan Bannon, Simone Kotva, Michael Marder, and Arne Johan Vetlesen, each of whom brought a distinct voice to the question of how we receive and respond to a world that resists mastery. Rather than reinforcing the binary between activism and passivity, the event explored how phenomenological attention to passivity might illuminate the affective, ethical, and ontological dimensions of ecological life.

Many of the contributions in this volume were first presented at that conference, while others emerged through subsequent invitations that extended the conversation outward—across disciplines, lifeworlds, and intellectual traditions. The generosity and responsiveness of those who answered this call have been a source of encouragement throughout the editorial process. Their engagement affirms the relevance of eco-phenomenology today—not only as a theoretical intervention but also as a form of scholarly community grounded in care, attention, and collaboration.

We are especially honored to include a posthumous chapter by Forrest J. Clingerman, whose work on environmental hermeneutics was foundational to this project (Clingerman and O'Brien 2016). Forrest's unexpected passing in 2024 was a profound loss, both personally and intellectually. His contribution to this volume, and the warmth and generosity with which he participated in the conference, remain with us. We dedicate this volume to his memory. Our deep thanks also go to Professor Ray Person, Forrest's colleague and friend, who helped prepare Forrest's chapter for publication and ensured that his voice would be present in these pages.

Structure of the Volume and Closing Reflections

Although designated a *handbook*, this volume does not aim to codify eco-phenomenology into a singular methodology or fixed domain. Its purpose is not to define a field but to explore its range and to cultivate a sensibility—one that remains open to interruption, excess, and the call of the more-than-human. A running thread in both the handbook and the conference from which it emerged is the emphasis on passivity as a crucial concept of eco-phenomenology, which is explored in many of the individual chapters. Here, passivity is regarded not as a singular topic but as a mode of responsiveness: an orientation to the world as that which precedes and exceeds us (Barbaras 2006; Levinas 1969; Toadvine 2009). The chapters are grouped into six thematic sections, but the structure ought not to be taken as a dogmatic and externally imposed structure. Rather than discrete and all-encompassing categories attempting to exhaust the meaning of eco-phenomenology, these sections form constellations—regions of inquiry where ideas resonate, diverge, and converge anew. Affirming dialogue across traditions and disciplines, a core organizational principle has been to put individual chapters into dialogue with each other.

The first two sections, "Given Nature" and "Passivity and Perception," explore the distinctly *eco*-phenomenological dimension of many basic concepts of phenomenology, such as givenness, passivity, and perception to make them appear in a fresh light for future thinking. "Place and Cosmos" invites us to not take our place in the natural world for granted, but instead wonder about what it means to live and dwell in a world shared by more-than-human others. "Responsible Relating" attends to the ethics and responsibility of eco-phenomenology, wondering how we may lead our lives in a more responsible fashion toward what precedes and exceeds us, in contrast to the almost all-pervasive domination of nature in modernity. "Vegetal and Animal Life" follows an important trend in eco-phenomenology but also environmental philosophy in general, which is a general concern for other forms of life. Yet again highlighting the dialogic nature of the handbook, the final two sections exemplify this principle in different ways. "Literary Intersections" explores the intersecting lives of eco-phenomenology and literature in various ways. The literary texts in question appear not as illustrations but as sites of phenomenological insight, articulating what philosophy cannot do alone. Meanwhile "Eco-Phenomenology in Dialogue" puts eco-phenomenology into critical dialogue with other philosophical traditions and approaches to the issues taken up in the handbook.

To read this volume is to enter an unfolding conversation. The chapters do not merely catalogue ideas or describe concepts; they issue invitations—to pause, to listen, and to dwell differently in a world that is never encountered in solitude. They speak in many registers—scholarly and poetic, rigorous and speculative—and across these voices one hears a shared question: *How might eco-phenomenology respond to a world that does not merely await our gaze, but already touches, addresses, and implicates us?*

The hope is not to settle eco-phenomenology as a field, but to gesture toward a form of thinking that is always provisional, always responsive. To dwell with this volume is to begin, perhaps, to think and feel otherwise—to find, in passivity, not a failure of agency, but the beginning of care.

References

Abram, David. 1996. *The Spell of the Sensuous: Perception and Language in a More-than-Human World*. New York: Pantheon.

Bannon, Bryan. 2014. *From Mastery to Mystery: A Phenomenological Foundation for an Environmental Ethic*. Athens: Ohio University Press.

Barbaras, Renaud. 2006. *Desire and Distance: Introduction to a Phenomenology of Life*. Translated by Paul B. Milan. Stanford: Stanford University Press.

Braidotti, Rosi. 2013. *The Posthuman*. Cambridge: Polity Press.

Brown, Charles S., and Ted Toadvine, eds. 2003. *Eco-Phenomenology: Back to the Earth Itself*. Albany: State University of New York Press.

Clingerman, Forrest J., and Kevin J. O'Brien, eds. 2016. *Theological and Ethical Perspectives on Climate Engineering: Calming the Storm*. Lanham, MD: Lexington Books.

Coulthard, Glen Sean. 2014. *Red Skin, White Masks: Rejecting the Colonial Politics of Recognition*. Minneapolis: University of Minnesota Press.

Heidegger, Martin. 1966. *Discourse on Thinking*. Translated by John M. Anderson and E. Hans Freund. New York: Harper & Row.

Heidegger, Martin. 1977. *The Question Concerning Technology and Other Essays*. Translated by William Lovitt. New York: Harper & Row.

Husserl, Edmund. 1982 [1913]. *Ideas Pertaining to a Pure Phenomenology and to a Phenomenological Philosophy, First Book*. Translated by F. Kersten. Dordrecht: Springer.

Husserl, Edmund. 2001 [1966]. *Analyses Concerning Passive and Active Synthesis: Lectures on Transcendental Logic*. Translated by Anthony J. Steinbock. Dordrecht: Kluwer Academic Publishers.

Kotva, Simone. 2020. *Effort and Grace: On the Spiritual Exercise of Philosophy*. London: Bloomsbury.
Levinas, Emmanuel. 1969. *Totality and Infinity: An Essay on Exteriority*. Translated by Alphonso Lingis. Pittsburgh: Duquesne University Press.
Marder, Michael. 2013. *Plant-Thinking: A Philosophy of Vegetal Life*. New York: Columbia University Press.
Merleau-Ponty, Maurice. 1968. *The Visible and the Invisible*. Edited by Claude Lefort, translated by Alphonso Lingis. Evanston: Northwestern University Press.
Merleau-Ponty, Maurice. 2012 [1945]. *Phenomenology of Perception*. Translated by Donald A. Landes. London: Routledge.
Morris, David. 2008. "Bodies and Technologies: Dialectics of Emergence, Ecological Style." *Techné: Research in Philosophy and Technology* 12 (2): 69–92.
Morris, David. 2017. "The Enigma of Nature and the Flesh of the World: Reassessing the Ecological Scope of Merleau-Ponty's Philosophy." In *Phenomenology and the Natural Sciences*, edited by Joseph K. Schear, 205–228. Cambridge: Cambridge University Press.
Næss, Arne. 1973. "The Shallow and the Deep, Long-Range Ecology Movement." *Inquiry* 16 (1–4): 95–100.
Plumwood, Val. 2002. *Environmental Culture: The Ecological Crisis of Reason*. London: Routledge.
Stengers, Isabelle. 2011. *Cosmopolitics II*. Translated by Robert Bononno. Minneapolis: University of Minnesota Press.
Toadvine, Ted. 2009. *Merleau-Ponty's Philosophy of Nature*. Evanston: Northwestern University Press.
Todd, Zoe. 2016. "An Indigenous Feminist's Take on the Ontological Turn: 'Ontology' Is Just Another Word for Colonialism." *Journal of Historical Sociology* 29 (1): 4–22.
Vetlesen, Arne Johan. 2015. *The Denial of Nature: Environmental Philosophy in the Era of Global Capitalism*. London: Routledge.
Wolfe, Cary. 2010. *What Is Posthumanism?* Minneapolis: University of Minnesota Press.

SECTION 1

Given Nature

2

TRANSCENDING THE OBJECT HORIZON

Nature as Given and Saturated Phenomenon

Thor Eirik Eriksen

Introduction

To encounter nature in its diverse expressions often leaves an indelible mark on us. We might find our conversations quietly deepened, our restless thoughts quieted and burdens gently eased, or find ourselves overwhelmed, rendered silent in awe. What, then, transpires in these seemingly transformative encounters? What threads of connection are woven, drawing us into this interplay, as we stand face-to-face with natural phenomena? Initially, we can relate such events to the descriptions found in Heidegger's (1968 [1954]) work *What is Called Thinking?*

> We stand outside of science. Instead we stand before a tree in bloom, for example—and the tree stands before us. The tree faces us. The tree and we meet one another, as the tree stands there and we stand face to face with it. As we are in this relation of one to the other and before the other, the tree and we are. This face-to-face meeting is not, then, one of these "ideas" buzzing around in our heads.
>
> (41)

The encounter with the tree in bloom, "face-to-face," allows our conceptual constructions or abstract ideas to recede into the background. This event similarly enables our scientific or biological explanations, including philosophical and definitional determinations of the "entity" tree, to be set aside. As Heidegger suggests, we have "leapt out of the familiar realm of science and even…out of the realm of philosophy" (41). However, we are not left in a void or vacuum. Instead, following Heidegger, we find ourselves more properly on "firm soil," "that soil upon which we live and die," and "on which we really stand" (41).

This deep-rooted embeddedness stands in stark contrast to the dominant approach that has shaped the world over the past centuries: a period in which humanity has broken with nature, elevated itself above it, and subjected it to static, human-imposed laws that have confined it. In the epoch we designate as the Anthropocene, this approach has reinforced our notions of humanity *here* and nature *out there*.[1] In other words, as Foltz (2014) argues, we

 DOI: 10.4324/9781003538097-3

have come to regard nature as "pure exteriority, without face and without inner life" (abstract). Such ways of thinking have allowed us to treat nature as an accessible arsenal and, as Heidegger (1977, 17) reminds us, a standing reserve of materiality subject to our productive capacities. The results of such practices are now visible, and we are in the process of waking up to a hangover and the realization that humanity's thoughtless practices have brought us closer to an ecological crisis.

However, this problematization is not about opposing human agency, including the potential for a green transformation, but rather emphasizes the importance of *rediscovering "that" which has been downgraded, restricted*, and *diminished*. In a broader sense we can thus ask: How might nature and its phenomena unfold or reveal themselves if released from our habitual framing within the category of the object? Moreover, what approach might facilitate such a release, allowing these phenomena to reemerge as expansive, magnified, and "abundant"? More specifically, we will explore how the contributions of philosopher Jean-Luc Marion can shed light on these inquiries. This endeavor entails a radicalization of Marion's thought, aiming to deepen his core concern in resonance with the diverse and rich phenomena of nature.

Transcending the Object Horizon

Human beings exhibit a strong inclination to impose systematic order upon the world—res extensa—to the extent that it is intelligible to the intellect, graspable, and subject to mastery. This pertains to a form of rationality that primarily affirms what aligns with human interests and unfolds within what might be described as an *object or objectifying horizon*.[2] This painstaking reduction, from phenomena to objects we can master, represents a process of transcribing that can be discerned also in Heidegger's (1968) blossoming tree:

> The sciences do in fact decide what of the tree in bloom may or may not be considered valid reality. Whence do the sciences—which necessarily are always in the dark about the origin of their own nature—derive the authority to pronounce such verdicts? .... In truth, we are today rather inclined to favor a supposedly superior physical and physiological knowledge, and to drop the blooming tree.
>
> (43–44)

This scientific reduction thus culminates in what is tangible and factual: that which can be physically and physiologically stated as the truth about the blooming tree. Drawing explicit inspiration from Heidegger, Mugerauer (2009) emphasizes that the deeper problem here is that

> ...Occam's razor has shaved away the poetic, the gods, and the divine features of the human (such as the soul...), so that we are left with an intellectual consensus that what is given is a realm in which earth and heavens and humans are all the same—*material stuff* operating according to the same set of forces as described by universal laws.
>
> (82)

Similarly, Foltz (2014) is concerned with how such diminishments of nature, which in reality represent a long-standing assault on the integrity of nature, reflect how

> …nature gets wrested from its *givenness* to so many degrees, and in so many ways, that it is no longer nature at all, until at last it really is nothing more than *vorhanden*, 'just there', a lovely corpse in urgent need of 'preservation'.
>
> (52; emphasis added)

Extending Foltz's remarks, we will direct our attention to a thinker who directly engages with the challenge of a phenomenon being "wrested from its givenness"—namely Jean-Luc Marion. Although he very rarely addresses the topic of nature, the premise here is that his emphasis on revitalizing the phenomenological focus on *givenness* offers a promising foundation for a possible reinterpretation of it. A crucial point of departure for Marion (2002a) is that we ordinarily remain content with "common-law" phenomena—those phenomena that are "accomplished according to objectivity" and typified by "the objects of physics and the natural sciences" as well as by "the objects of technological production" (223). Therefore, we must turn to givenness, for "only a phenomenology of givenness can return to the things themselves. To return to them, it is necessary first to see them, to see them as they appear, and, ultimately, to bear their unpredictable arrival." In other words, phenomena that have been transcribed into objects must be re-transcribed and allowed free passage within a boundless realm of possibility—*givenness*.[3]

Releasing the Phenomenon

What possibilities, then, does the introduction of givenness unveil concerning our engagement with nature? A preliminary step in such an exploration might involve cultivating a clearer understanding of the very concept of the object itself. In other words, we must explore how phenomena are initially transformed into constrained, and reduced entities—*phenomena of the object type*. Consequently, we must further ask: What do we mean by object and why is it so often preferred, especially when we set out to engage with nature? In his book *Negative Certainties*, Marion (2015) provides a clarifying answer. The object "manifests itself as that which remains of the phenomenal flow once it has been regulated by 'the standards of reason', screened through the filters of the concept of under-standing, in short, judged by the norms of the apriori" (163). Marion describes a kind of "distillation process" whereby the phenomenon is filtered in such a way that it presents itself as *graspable*. This is a process of eliminating "*that* in those things which does not allow itself to be abstracted according to order and measure" (16).

How, then, might we comprehend such transformation more concretely, particularly in relation to nature? Consider, for instance, an example drawn from the realm of labor, where tangible *natural events* are exerting an ever-growing influence on workers and their health. Describing this challenge, prominent international organizations explain how workers are increasingly *exposed* (through identified cause–effect mechanisms) to a range of *hazards and pathogens* (e.g., high temperatures), which represent significant *health risks* (e.g., heatstroke) and ultimately demand comprehensive *risk mitigation* and *control strategies* (e.g., workplace regulations).[4] When the aim is to protect workers, it is not difficult to recognize the importance of explaining the mechanisms of exposure involved. However, it is somewhat disconcerting that authoritative sources continue to employ an object-oriented language that, in many ways, reinforces the notion of nature and its forces as "hazards and threats," "out there." If one believes that the world's largest group, workers, should only engage with manageable residual objects of natural events, one overlooks the necessity, as Vetlesen (2015) reminds us, of recognizing that these challenges also require "urgent action, based on a *multi-sensuous appreciation* of humans' dependence on nonhuman lifeforms" (cover). A receptivity that enables us to perceive the *cry of Nature*.[5]

Saturated Phenomena

If this cry from nature thus encompasses something beyond mere "things" (*Gegenstände*), how might we come to understand this "otherness," and how can we cultivate our receptivity to it? To use Marion's (2015) terminology, we might say that we are confronted here with phenomena that possess a *surplus or excess*. Consequently, the reduction—the "leading back"—must shift in character, moving from the identification of objects to preparing the ground for phenomena of the event type: *the saturated phenomena* (252). This entails clearing the way for phenomena to reveal themselves, on their own terms, as they give themselves, thereby determining their own degree of phenomenality (Marion 2002a, 15). This means that a saturated phenomenon is not something we intend, analyze, or rationalize (produce), but rather something that reveals itself to us through experience. Such phenomena are neither extraordinary nor unusual but, according to Marion, they are exceptionally common in our lives. They include love, birth, pain, death, happiness, trust, and, we might add, *a tree in bloom—or countless other phenomena of nature*.[6] These are phenomena, Marion emphasizes, where the initiative lies with the phenomena and *not* in the *observing gaze*. In line with such descriptions, let me return to Heidegger's (1968) tree in bloom:

> When we think through what this is, that a tree in bloom presents itself to us so that we can come and stand face-to-face with it, the thing that matters first and foremost, and finally, is not to drop the tree in bloom, but for once let it stand where it stands. Why do we say "finally"? Because to this day, thought has never let the tree stand where it stands. (44)

Heidegger's descriptions set the tone for Marion's (2002a) event, where the phenomenon takes the initiative. The subject in this case is a receiver, a witness or a screen on which the phenomenon is projected. The expression *"I think"* is replaced here by *"I am affected"* (255; (emphasis added). Instead of being the meaning-giver, the subject becomes the meaning-receiver. Marion elaborates: "At the center stands no 'subject,' but a gifted, he whose function consists in receiving what is immeasurably given to him, and whose privilege is confined to the fact that he is himself received from what he receives" (322).

Such descriptions indicate a profound reorientation, particularly concerning our relationship with nature. Marion scholar Gschwandtner (2014a) elaborates on this challenge as follows: "Identifying some natural phenomena as saturated might change not only our attitude toward them but it might also help us rethinking more deeply what it means to be human" (93). This very possibility and ambition serves as the central guiding thread for the exploration that follows. The critical point of reference here will be Marion's book *In Excess* (2002b). In it, he presents a framework of saturated phenomena: *the event* (e.g., a historical event), *the idol* (e.g., a painting), *the flesh* (*Leib*), and *the icon* (the face of the Other). These phenomena aim to transcend Kant's categories of quantity, quality, modality, and relation by overwhelming the subject's capacity to fully conceptualize or categorize them. The event saturates the category of quantity, the idol saturates quality by dazzling the subject, the flesh surpasses relational understanding, and the icon transcends modality by being a self-giving that cannot be objectified. Each type thus overflows cognitive boundaries, requiring a shift beyond traditional epistemic limits. Furthermore, what they also have in common is their *event-like* nature. In the following, we will explore how we might examine nature's diverse modes of giving in light of these different saturated phenomena. The overarching aim is to expand the scope of Marion's phenomenology to give the profoundly pluralistic experience of nature its due significance.

The Event: Nature Happens

Nature, in its multifaceted complexity and unpredictability, often presents itself to us as something that radically surpasses our efforts to fully understand it. We will explore here how Marion's (2002b) concept of the event can serve as a key to experiencing the deeply transcendent character of nature, where nature reveals itself beyond the frameworks we impose upon it.[7] What characterizes the event is its ability to *saturate our sense of quantity* by overwhelming us with an influx of data and information, presented with such richness that it exceeds our capacity to "grasp" or "manage" the phenomenon. It unfolds through an excess of intuition, without any concept having anticipated its possibility or conceivability. That which shows *itself* thus overrides our response-ability. An important aspect of this surplus is that it "attests to an unforeseeable origin, rising up from causes often unknown, even absent, at least not assignable, that one would not therefore any longer reproduce, because its constitution would not have any meaning" (31).

This implies that when we find ourselves before the tree in bloom, we must acknowledge our incapacity—even with the aid of scientific instruments—to fully comprehend the saturated prehistory and unfolding of this singular tree. It is coming from "an uncontrollable past" (33). Standing before this tree knowing that it was planted by my parents in the last century does not confine its story to a single moment of origin. Instead, it invites an unending series of questions that gesture toward an unfathomable depth of history. What kind of forest did this sapling emerge from? What ecological network nurtured its earliest roots? What weather-related dramas has it endured? This tree is not merely a fixed entity but a living archive, bearing within it an immeasurable confluence of events—its growth entwined with cycles of weather, soil, light, and the countless hands of time that brought it to stand here now. It reveals a history far beyond our comprehension, a story layered with moments beyond what our finite understanding can encompass.

This is not about disputing the tree's physical form—its trunk, bark, branches, and leaves—but about recognizing that its history surpasses my grasp and exceeds my understanding. Thus, we cannot fully articulate or account for all the dimensions of what we are confronted with in this singular encounter, unfolding *here and now*. The intricacies are too manifold and the phenomenon itself too complex. It means that "it escapes all constitution" (ibid). Furthermore, we are incapable of foreseeing the future trajectory of this specific tree. While we may, in a gesture of exerted control, cut it down, we still have no power over how, in an unforeseeable future, the tree stump and its remaining roots will become a habitat for fungi, bacteria, and other organisms. We are left with an infinite hermeneutic process.[8] Thus, when Heidegger (1968) asks, "Did the tree anticipate us and come before us? Did the tree come first to stand and face us, so that we might come forward face-to-face with it?"(42), we can, in this context, answer affirmatively. This understanding invites us into an ongoing dialogical relationship with nature, where every interpretative effort falls short of capturing its fullness and instead points toward an inexhaustible excess of meaning.

In the bigger picture, when the tree becomes part of a forest, the forest part of a landscape, and the landscape part of a vast ecosystem, we must also involve natural disasters like earthquakes and hurricanes.[9] These occurrences remind us of the inherent power and unpredictability of nature. An important aspect of this also concerns the effects of *climate change*, which resists presentation as a straightforward, linear phenomenon susceptible to reduction through simple explanatory frameworks. In a rare mention of this phenomenon, Marion asserts that "climate change, like peace, war, poverty, and social justice, cannot be reduced to objects; they come to us as events" (Arbib 2017, 168). It reconfigures our entire conception of nature's stability and future contingencies, functioning as an event within a global temporal schema—one not contained by singular occurrences but rather unfolding as a comprehensive progression that profoundly challenges our grasp of nature's boundaries.

Weather phenomena such as storms, floods, and tornadoes represent the manifest imprint of these changes. Their full consequences cannot be entirely conceptualized until after they have occurred. Consequently, this pertains not only to the beautiful and the sublime but also, as Mugerauer (2009), drawing inspiration from Marion, reminds us, to the overwhelming force of these natural phenomena, which "is so great that we may not be able to bear its impact. Battered by them physically or psychologically, we commonly find ourselves blankly uncomprehending, puzzled or frozen into inaction, or perhaps fainting" (75). Our challenge in the wake of such life-altering events is not only to take seriously the suffering of those affected but also to acknowledge it as a desperate cry of anguish emanating from nature itself.

The Idol: Telluric Manifestations and Qualitative Excess

In what follows, we will direct our attention to Marion's (2002b) concept of the idol. The idol is a concept within his theory of saturated phenomena that explores the limitations and overwhelming *qualities* of certain types of experiences or phenomena. It pertains to something that "rises up before us, silent, irresistible, adorable" (74). He uses a painting as an example of a phenomenon that, in this way, captures the gaze of the viewer, filling their vision with its overwhelming presence and captivating them entirely. The idol saturates the viewer's capacity for perception and interpretation, so much so that it seems to "bedazzle" or "fix" the viewer's gaze. This means that the phenomenon "presses urgently on the gaze more than the gaze presses toward it, and the gaze receives its impression from the phenomenon before any attempt at constituting it" (Marion 2002a, 159). The idol thus exposes the inadequacy of our concepts to fully encompass the richness of certain experiences.

Using Marion's concept of the idol as a gateway to nature initially involves stepping out of the Louvre Museum, leaving behind an encounter with the painting and entering a different kind of auditorium. Thus, when we direct our attention toward a landscape, a forest, or a plant, we may be overwhelmed by the visible that presents itself to us. When, for instance, after sustained effort, I ascend to a mountain peak and gaze upon the boundless expanse of the landscape, I might experience a moment of profound captivation, where the overwhelming intensity of what manifests to the senses evokes a sense of being "transfixed" or even "dazed" by the phenomenon's sheer qualitative excess. In this moment, I am, if only temporarily, unable to constitute or produce an assemblage of "natural objects." The abyss that manifests itself just a few meters away, the surrounding and towering mountain formations in the distance, and the passing cloud systems that give me a sense of being part of them, do not invite fixation, or explanation; instead, they coalesce into a totality of unbearable yet calming intensity.

That the mountain serves as a site for an elaboration of qualitative excess is of particular significance in this context. This is because Marion (2002b, 64), standing face-to-face with a painting—specifically, Cézanne's *Le Sainte-Victoire au grand pin*—offers the following description:

> I do not try to bypass the mountain in thought in order to imagine the other side, nor to reconstitute the complete trunk of the pine tree, nor to sketch the elevation of the house in the middle ground to the left. I have enough to look at with *what is presented to my look with such a brutal force—the telluric architecture of the mountain* so strangely sheltered by the oscillation of the plants and so embedded in it that the stone and the branch exchange their attributes—the first vibrates from heat, whereas the second is immobilized by calm.
>
> (64; emphasis added)

Marion illustrates how one can be enthralled by the grandeur of nature and the majesty of mountains, even within the confines of a museum gallery. The "telluric" quality—where the mountain originates and is anchored in the earthly, the ground, and the planet—effectively blurs the distinction between the "painting" and the "painterly spectacle" encountered atop the mountain.

What further exceeds the limits of our understanding in this encounter is how that which comes to the forefront on this stage—the idol—pertains, according to Marion, to "a visible that has never previously been seen by anyone" (68). That is, when we once again stand face-to-face with the painting in the museum or find ourselves as guests in nature's own gallery, we never see the same thing in the same way. This means that this singular encounter cannot be recreated or reproduced. It unfolds in the immediacy of the here and now. We are thus repeatedly confronted with an overwhelming tableau in which the "metaphysical" horizon of the object is bracketed or suspended. Nature gives itself to the "I," the receiver, the gifted, on its own terms.

The Flesh: The Earthing of Flesh and the Flesh of the Earth

What does it mean to exist as, or to possess, a body? This question lies at the heart of many disciplines, and it is on this terrain that Marion offers his reflections on *flesh* as a saturated phenomenon.[10] Inspired by the French philosopher Michel Henry, Marion argues that the flesh is not merely an object among other objects but is instead the medium through which the subject directly experiences its own body and existence. It signifies a fundamental self-awareness, unique in its dual character as a phenomenon that both gives itself and is given. It is at once active and passively receptive. Positioned as the paradigm of an *absolute phenomenon*, the flesh is defined by its immediate and overwhelming affectivity, wherein no ontological or phenomenological distance separates the self (*ego*) from the flesh (immanent auto-affection). This radical immediacy—preceding and transcending any intentionality directed toward an object—establishes the flesh as an absolute phenomenon in the domain of *relation*.

While Marion (2002b) occasionally acknowledges that bodily phenomena can relate to "objects of the world" or "other subjectivities," he grounds his contribution in concepts such as *my* flesh, "mineness," and "originally mine" (98–9). This conceptual framework encloses our embodied being-in-the-world within a paradigm that verges on solipsism. In one of his earlier works, *Being Given*, Marion (2002a) acknowledges this in his discussion of the flesh, "the flesh provokes and demands solipsism; for it remains by definition mine, unsubstitutable—nobody can enjoy or suffer for me" (232). Phenomena such as love, joy, anxiety, and grief thus "take flesh." that is,

> happens to *me*, affects *me*, and determines *me*, in short because flesh, of itself and always already, takes *me*. I do not give myself my flesh, it is it that gives me to myself. In receiving my flesh, I receive *me* myself—I am in this way gifted…to it.
>
> (Marion 2002b, 98)

This emphasis on the "mineness" of the human being presents a significant challenge when attempting to extend his ideas on flesh toward a revaluation of nature's phenomena.[11] This obstacle thus necessitates moving beyond the framework Marion establishes. One prominent figure advancing such efforts is MacKinlay, who challenges Marion's assertion that flesh is purely self-affective (auto-affection) and independent of relational frameworks. Consequently, he disputes Marion's characterization of flesh (*Leib*) as an "absolute phenomenon" (MacKinlay 2010, 130). He elaborates on this as follows:

> my analysis of flesh has shown that even a phenomenon such as flesh, whose immediate and immanent auto-affection precludes its being constituted as an object, must be understood as part of *an original relatedness between the world and me*. Marion may be correct to describe flesh as an experience of self-givenness, but this very self-givenness is always given with the givenness of the world.
>
> (158)

When Gschwandtner (2014a), along similar lines of thought, asks "is this account of the flesh [in Marion] really possible without some attention to nature and the earth?" (97), she also points to valid reasons why the answer must be negative.[12] We can even draw on some of Marion's (2002b) own formulations as a guide for broadening our perspective when he states, "as soon as the ego takes flesh, it finds itself stuck to itself as to its ground, to its phenomenological *earth…as earthed*." He continues in a similar vein: "Suffering rivets me to myself as one rivets something to the ground—*by earthing*" (92; emphasis added). Marion never develops these reflections on grounding—our connection to the earthly. This connection will, therefore, form the central focus of the exposition that follows.

The first of these concerns our own embodied existence, our *"fleshly" being-in-the-world*—which does not merely perceive nature as a visual or cognitive object but deeply experiences and lives it through the flesh itself. Nature does not appear as a distant or abstract reality but as an integral part of our being, *experienced through the flesh*, through the bodily grounding we have in the fabric of the world. Each breath connects us to the shared life source, the pulsing flow that sustains all living things on earth. The nourishment we consume depends on an interwoven ecosystem that sustains our existence through its interaction and balance. Further, and when we feel the touch of rain, it is our body that engages in a pre-reflective experience of nature's presence, an immediate and profound sense of its power. The feeling of grass underfoot or the scent of pine trees are all ways in which nature is experienced through and as flesh.[13]

In the next step, we must also explore how we might transfer this thinking to a "bodily" or "fleshly" dimension of nature. Gschwandtner (2014b) sets the tone once again, aptly asking "if an encounter with a painting or the experience of a historical event can individuate me, why not an encounter with a (nonhuman) animal or the experience of natural (rather than historical) phenomena?" (100). Can we thus envision a scenario in which we "experience ourselves envisioned (transformed and constituted) by the nonhuman other" (101)? An illustrative example might involve an encounter with the smallest, almost imperceptible, embodied beings inhabiting the natural world: insects. Consider the following scene: upon the windowsill, my gaze captures a bee, its body fragile and still, ensnared in a state of quiet captivity. A few decades ago, such a visitor might have been seen as a threat to the inhabitants of the house—an unwelcome intruder. Yet in light of a growing awareness of nature's precarious state, even the lives and destinies of insects have come under our care. The bee, once easily dismissed, now carries the weight of fragile ecosystems. The bee must be cared for, protected, and set free. Finally, we might propose—contrary to both Henry's and Marion's separation of human life and flesh from non-human beings, plants, or ecosystems—that bees, too, concern life and possess flesh.[14]

We can expand our perspective on these issues, even if it requires further extending Marion's theoretical framework beyond its original boundaries. This broader view encourages us to reflect on the planet we inhabit on a larger scale. Mitchell (2014) emphasizes that, according to Paul Shepard,

> the plant–insect symbiosis...gave birth to a true humus and soil, the most complex organic system in the world. *The soil became the flesh of the world—a skin, mediating the mineral and biological kingdom, regulating temperature and moisture levels.*
>
> (118; emphasis added)

These unforeseen and overflowing multitudes of interwoven organisms (quantity), manifesting in phenomena such as a vast, breathtaking meadow of wildflowers (quality), can thus also be regarded as flesh—the flesh of the Earth. In connection to this, Merleau-Ponty (1968) reminds us,

> that means that my body is made of the same flesh as the world (it is a perceived), and moreover that flesh of my body is shared by the world, the world reflects it, encroaches on it and it encroaches on the world.
>
> (248)

Finally, at some point, we can recognize that humans, animals, and plants share a common vulnerability of flesh.

The Icon: The Many Faces of Nature

So far, we have explored how Marion's philosophical framework can be activated and expanded when engaging with nature's *saturated phenomena*. Next, we will examine how Marion's (2002a) concept of the *icon* centers on an encounter with the Other and the face of the Other as a transformative experience that transcends mere representation. It is a phenomenon saturated according to modality; it cannot be directly observed, as it is irreducible and "irregardable" (214). This irreducible aspect prevents us from looking at the face as an object. Rather, it "reverses the order of visibility in that it no longer results from my intention but from its own *counter-intentionality*" (Marion 2002b, 113). On an interpersonal level, this encounter draws us into a space of humility and openness, where the self becomes "saturated" by the boundless depth of the Other's presence. The icon thus challenges us to engage relationally, allowing us to glimpse an alterity that surpasses comprehension and demands an ethical response.

With some familiarity with the field of phenomenology, it becomes evident that Marion draws inspiration from one of his teachers, Emmanuel Levinas. The focal point here is, naturally, the "face," which, like the event, the idol, and the flesh, confronts us in a manner that defies objectification. If we turn directly to Levinas (1974), a significant point in his work—which is also thoroughly examined by Marion—is the assertion that "to see a face is already to hear: 'Thou shalt not kill'" (22), an imperative which encapsulates an inescapable ethical demand. If we, even at this point and without reservation, introduce the concept of the "face of nature," this implicit call takes on a particularly profound significance.[15] Current reports highlight the escalating pollution of the world's oceans, leading to a form of *suffocation*: rising temperatures, decreasing oxygen levels, and widespread deoxygenation driven by global warming.[16] This warming is occurring on a scale that far exceeds our ability to grasp or fully comprehend. It involves an accelerating transformation, also referred to as *ecocide*. The imperative, "Thou shalt not kill," as interpreted here, rises from such events and thus appeals to our ethical responsibility, urging us not only to refrain from perpetuating environmental destruction but also to care for the countless individuals already suffering profoundly from the effects of ongoing climate change.[17]

Naturally, there is a risk that we may have, from the outset, extended the phenomenology of the Other too far, using it as a metaphorical framework that could ultimately devolve into more of the same: an anthropocentric appropriation of nature. Although merely an analogy, it still fulfills a crucial function, as we have heretofore treated nature and its gifts as sterile objects—what Foltz (2014) previously described as "pure exteriority" and "a surface" (43). What is thus brought forward here is the notion of nature as being alive, *"with face" and "with inner life."* This enlivening, according to Campbell et al. (2011), may signify that nature's face invites us to explore important "ways of reading the shapes, textures and colors of the land, its stone, waters, wind, and the other living creatures." Further, and in harmony with Heidegger and his blooming tree, they emphasize how "we're trying to stand face-to-face with the land so that we might read and understand its *character*" (x).

Through this, we align ourselves here with Prášek (2023, 158), who is also inspired by Levinas when he maintains "that sensible nature…does not have a human face, *but the encounter with its transcendence in its various facialities has a similar ethical force, from which an eco-phenomenological ethics of nature could grow"* (emphasis added). The crucial point here is that nature, in the event of giving and coming forward, advances a *call*, to which we can attune ourselves with receptivity. This is a crucial point also for Marion, because the call not only characterizes *all* saturated phenomena but the face in particular: "The face, saturated phenomenon according to modality, accomplishes the phenomenological operation of *the call* more, perhaps, than any other phenomenon (saturated or not)" (Marion 2002b, 118–19). That we are indeed face-to-face with nature, being called, can be illustrated with the following example: when we enter the mountains to behold a glacier, we may discover, especially if we undertake this journey regularly that it is gradually disappearing. It thus reveals itself, *its self*, through an urgent, paradoxical, and alarming act of disappearance.

Here, we are confronted with something that is more than a "simple object" of nature. Instead, it is something we must truly *see*—engaging with its phenomenological significance. In many ways, and especially if we closely follow Marion's (2002b) line of reasoning, we will find that "the face is imposed on [us]," a face which "speaks in silence" (115–16). We stand accused because "the face in its injunction obliges [us] to situate [ourselves] in relation to it,"

Mackinlay, Shane. 2010. *Interpreting Excess: Jean-Luc Marion, Saturated Phenomena, and Hermeneutics*. New York: Fordham University Press.
Magrini, James. 2019. *Ethical Responses to Nature's Call: Reticent Imperatives*. London and New York: Routledge.
Marion, Jean-Luc. 2002a. *Being Given: Toward a Phenomenology of Givenness*. Translated by Jeffrey L. Kosky. Palo Alto, CA: Stanford University Press.
Marion, Jean-Luc. 2002b. *In Excess: Studies of Saturated Phenomena*. Translated by R. Horner and V. Berraud. New York: Fordham University Press.
Marion, Jean-Luc. 2013. The Banality of Saturation. In *The Essential Writings*, edited by Kevin Hart, 135–55. New York: Fordham University Press. https://doi.org/10.1515/9780823292905-007
Marion, Jean-Luc. 2015. *Negative Certainties*. Translated by Stephen E. Lewis. Chicago: University of Chicago Press. Kindle Edition.
Marion, Jean-Luc. 2018. *On Descartes' Passive Thought: The Myth of Cartesian Dualism*. Translated by Christina M. Gschwandtner. Chicago: University of Chicago Press.
Merleau-Ponty, Maurice. 1968. *The Visible and the Invisible*. Translated by Alphonso Lingis. Evanston: Northwestern University Press.
Michener, Ronald T. 2018. A 'Value' Prolegomenon for Creation Care: Theological Appropriations from Marion and Levinas. In *Theological Ethics and Moral Value Phenomena*, edited by Steven Van den Heuvel, Patrick Nullens, and Angela Roothaan, 147–65. London: Routledge, Taylor & Francis.
Mitchell, Laura H. 2014. Nomadic Dimensions of Education with the Earth-in-Mind. In *Ecopsychology, Phenomenology, and the Environment: The Experience of Nature*, edited by Douglas A. Vakoch and Fernando Castrillón, 109–27. New York: Springer Science+Business Media.
Mugerauer, Robert. 2009. Call of the Earth: Endowment and (Delayed) Response. In *Heidegger and the Earth: Essays in Environmental Philosophy*, edited by Ladelle McWhorter and Gail Stenstad, 70–99. Toronto: University of Toronto Press.
Prášek, Petr. 2023. Levinas vs. Maldiney On the Face of Sensible Nature. *Phainomena* 32 (126–127): 157–84.
Schrijvers, Joeri. 2006. "Ontotheological Turnings? Marion, Lacoste and Levinas on the Decentering of Modern Subjectivity." *Modern Theology* 22, no. 2: 221–53. https://doi.org/10.1111/j.1468-0025.2006.00316.x
Svenaeus, Fredrik. 2023. The Phenomenology of Objectification in and through Medical Practice and Technology Development. *The Journal of Medicine and Philosophy: A Forum for Bioethics and Philosophy of Medicine* 48: 141–50.
Vetlesen, Arne J. 2015. *The Denial of Nature: Environmental Philosophy in the Era of Global Capitalism*. London: Routledge.
Vetlesen, Arne J. 2023. *Animal Lives and Why They Matter*. London: Routledge.
Wiseman, Wendy A. 2013. Nothing Is Without Reason: Climate Change and the Global Future as Saturated Phenomena. In *Phenomenology and the Human Positioning in the Cosmos. The Life-world, Nature, Earth: Book Two*, edited by Anna-Teresa Tymieniecka, 227–43. Dordrecht: Springer.

3

ECO-PHENOMENOLOGY AND THE THEOLOGICAL TURN?

Passivity, Activity, and Traumatic Being

Calvin D. Ullrich

Introduction

Given that the biblical legacy and subsequent theological conceptions of nature are ambiguous at best, it is a welcome development to observe the insights of the so-called "theological turn" in phenomenology being deployed to foster healthier relations with our natural environment. Indeed, as is the case with several contributions to this volume, the thought of figures like Jean-Luc Marion are offered as providing phenomenological resources to re-articulate human and non-human relations (Gschwandtner 2014, 79–94; Manolopoulos 2009).[1] This chapter contributes to this eco-phenomenological development, but by adopting a reverse track: In place of the primacy of passivity, givenness, and saturation, characteristic of the "theological turn", the notions of activity, chaos, and traumatic being are submitted as potentially more fruitful, or at least as an alternative way of conceiving the phenomenological relationship between the human and nature. Although the interest here is not in cataloguing all the movements within the theological turn which could serve an eco-phenomenology in general[2] – a contribution in this vein is still waiting for its articulation to this author's knowledge – we will, nonetheless, venture a step in this direction, namely by approaching the human–non-human relationship from within the "theological turn" itself, albeit subversively through the work of Emmanuel Falque. What will be implicitly at stake for any future of what we might call *a theological turn in eco-phenomenology*, however – which will become clearer through this engagement with Marion as representative – is that the very status of the phenomenological principle of "givenness in itself" (*Selbstgegebenheit*) (Husserl 1999, 42), its later articulation in Heidegger as "that which shows itself by itself" (Heidegger 1996, §7), and all later syntagms in the theological turn of the "event" and the "eventing of the event" (*l'événementialité*), are all called into question (Sebbah 2012).[3] We can formulate such a questioning in this way: is the choice for an eco-phenomenology within the theological turn, the choice between either the possibility of the excess of the event (here, the event of Nature) for which we would learn to accommodate (passively) and respond to ethically, or of a "non-manifest" – neither privation nor negation of phenomenality, but an "extra-phenomenality" constitutively outside but with the force (activity) of an exigency to be "otherwise", or what might be called "traumatic being"?

DOI: 10.4324/9781003538097-4

We will proceed in the following steps: In a first section, (theological) phenomenology's emphasis on the *appearance* or givenness (*Gegebenheit*) of phenomena will be challenged as it is found in the work of Jean-Luc Marion. Such "supra-phenomenality", even if it dramatically de-centers the subject, still presupposes a subject, albeit a passive one, to whom such a phenomenality makes its appearance. In a second section, contra-Marion, it will be argued that through Emmanuel Falque's unique method, the theological assumptions of Marion's phenomenology not only tend toward a dangerous irenicism with nature, which diminishes human responsibility, but also that a sacralization of nature through saturation, comes to occlude the "activity" of the subject read in terms of its organicity and material density. The third section will begin by further expanding on Falque's concept of the spread body (*corps épandu*), its chaos and activity. It does so through a brief detoured reading of Heidegger's Aristotle, where an account of material organicity construed as "activity" is given by emphasizing the natural movement of life (*Lebensbewegtheit*) characteristic of all natural living things. In this analysis, the deep ambivalence of active and passive forces in the drive-based life of living beings is described in terms of a fundamental *lack* or *sterésis* – an absent movement that is nonetheless not without force. In a final section, connecting to the latter, it is argued that the material density, activity, and force of our organic existence, which we share with nature, is fundamentally phenomenologically inaccessible (extra-phenomenal). This can be characterized as the "traumatic being" of our own material–organic existence, and it is precisely this which we share with nature: that is, we are a part of nature to the extent that we are not coincident with our own organic nature. In the final analysis it is argued that this non- or extra-phenomenological position is constitutive of a more responsible relation between humanity and nature, one that neither collapses the distinction nor renders a passive hierarchy to which we are subject, but rather necessitates a creative "solitude of solidarity" between both, in which we "stand by" nature just as nature "stands by" us.

The "Theological Turn" and the Sacralization of Nature: The Case of Jean-Luc Marion

In this section, the work of Jean-Luc Marion will be orientated accordingly as a kind of representative of the so-called "theological turn" in phenomenology (Courtine 1992; Janicaud 2000, 1–103). Rather than a full account of his work or the broader movement, the aim here is to clarify the theological structure of his phenomenology and its implications – particularly how the subject's passivity before excessive phenomena leads to an unproductive sacralization of nature and thus an impotent ethical response.

Jean-Luc Marion along with Emmanuel Levinas, Michel Henry, and Jean-Louis Chrétien have all been charged with abandoning the phenomenological method, or *betraying* a characteristic feature, namely, the exclusion of transcendence from the phenomenological inquiry. To understand Marion's counter-claim, namely, that he has *not* betrayed phenomenology but rather has only expanded and radicalized its field, we have to turn to his phenomenology of givenness (Marion 1998, 2002, 2008, 2016). For Marion, the principle of non-presuppositions or the "reduction" in Husserl should concern primarily the *appearing* of phenomena to consciousness. In a critical development of Husserl and then of Heidegger, where the reduction is considered too restrictive for Marion (in terms of "objectness" in Husserl and "beings" in the case of Heidegger), Marion proposes his infamous "third reduction" (Marion 1998, 192–198). Under the knife of the third reduction, the imposition of the constituting ego which provides intentionality is finally severed, allowing phenomena to "appear from themselves" as they show or "give" themselves more fully (Marion 2002, 39–31). In other words, Marion establishes a new principle of phenomenology, which is the principle of phenomenality *as such*: the phenomenon is not reduced to an object or in terms of the phenomenality *of beings* but rather what first appears in the phenomenon is necessarily *the appearing of appearing itself*. What is *given* under the reduction is *givenness*, or simply, "the given" (Henry 2015, 1–21).[4] One can see here why this would now exclude a "metaphysical" account of a Being which "gives", since Marion is insistent on the phenomenological immanence of his philosophy. Yet, this does not rule out for Marion, the possibility of a religious phenomenon that actually gives itself within the dimension of immanence, and it is to this phenomenon that a phenomenological analysis must also be able to give an account (Marion 2004, 24, 27). The latter becomes the (religious) phenomenological context for the "saturated phenomenon" – indeed, it follows that by establishing the principle of givenness one must take this to its "absolute" conclusion.

The saturated phenomenon is precisely the "absolute givenness" of intuition without the requirement of a limiting horizon or of a reduction to an I (Marion 2008, 13–14); this "impossible manifestation" according to classical phenomenology becomes the condition of possibility for saturation or what Marion calls "revelation" in terms of a phenomenology of religion. Revelation, in a certain sense, is not simply a particularly intensive degree of intuition, although it is also this, but more importantly is paradigmatic because it means thinking the possibility of the principle of absolute givenness itself: "the possibility of a phenomenon where intuition would give *more, indeed immeasurably more*, than intention would ever have aimed at" (Marion 2002, 193–94). This does not mean that all phenomena give themselves absolutely, but it does mean that they depend on *absolute givenness* or revelation as the governing phenomenological principle of all phenomenality (Marion 2002, 179–247). And yet, it is precisely because of this governing principle that questions are raised regarding the status of a "saturated phenomenon" and, in particular, it's "theological" character. For according to Marion, both of the basic conditions for the appearance of phenomena developed by Husserl and Heidegger become illegitimate: first, there is no horizontal character of experience under the saturated phenomenon, because per definition, it defies the requirement of something appearing *as* something, or *against* a horizon of expectation (Caputo 1999, 66–68). Thus, a phenomenon so intense with intuition is in *excess* of any possible intentional horizon – it "appears" only by paradox, only by making itself "felt" through the very disruption of an intentional horizon (Marion 2016, 55). Second, the saturated phenomenon refuses reduction of the phenomenon to an *I* who constitutes an experience of it, rather, it is the phenomenon which constitutes the *I*. We can hear in this language the theological resonances; the revelation *imposes* itself on a subject, it "happens" as an unexpected, surprising event, to which the I responds and bears witness: "Revelation reveals the *I* to itself" (Marion 2016, 52–53). Ostensibly, Marion is not concerned with the historical fact (actuality) of Revelation, but only with setting up the phenomenological conditions (its mode of givenness, i.e., saturation) for the *possibility* of such appearing, regardless of whether it is in fact given. It is here where the relationship between phenomenology and theology in Marion's work becomes blurred and where one begins to wonder whether theology has come to determine his phenomenology.

Indeed, Marion and many others who adopt his phenomenological approach, suggest that when the principle of a saturated phenomenon has been accepted, it can be *universally* recognized (e.g., in the Scriptures or, as we will see, in Nature) (Marion 2008, 2; Rivera 2018, 181–197). To put this another way: because Marion has deduced his concept of phenomenality (saturation) from (theophanic) Revelation, the *possibility* that all saturated phenomenon can subsequently be identified with Revelation in *actuality*, remains open. However, *what* is given cannot necessarily be deduced from *how* it is given, thus, the issue then becomes one of *faith*, as John Caputo correctly diagnoses (Caputo 2007, 67–94; Ullrich 2021, 154–160). Is it possible for Marion, then, as he consistently claims, to describe the possibility of revelation without needing to presuppose its actuality? Marion's critics suggest this is *not* possible, since how does one account for the fact that when Marion encounters saturation or revelation, others see something else – for example, hallucination, or indeed, simply nothing? (Benoist 2001, 102; Schrijvers 2011, 85; Falque 2018b, 124). To identify absolute givenness with God/Revelation is an imposition on the subject which has not been methodologically prepared. Perhaps the best one can do is to say that a phenomenon may have high degrees of intuition but that without faith there is nothing which can adjudicate its specifically religious character. This will have further implications, as we will see, when identifying nature with the saturated phenomenon.

If Marion's phenomenology finds its legitimacy in theology, whereby "absolute givenness" can be identified with God,[5] and that such givenness is universally available or publicly accessible in principle, then a curious and problematic dogmatism enters the equation. This dogmatism arises, as was memorably the case in his exchange with the atheist philosopher, Jocelyn Benoist, where the latter asks: "what will you say to me if I tell you that where you see God, I see nothing or something else?" (Benoist 2001, 102). Marion's response is simply that one "must learn to see otherwise" (Marion 2008, 124). Since the saturated phenomenon gives itself *unconditionally* to be seen, to *not* see it can only reside where there is no "will to see" (Marion 2002, 305–306). The latter can be understood as a kind of metaphysical introduction of an "I" which must precede the possibility of seeing, so that the fault of any non-phenomenalization is placed squarely on the side of a subject which is, as Emmanuel Falque writes, both "impotent as it is faulty" (Falque 2018b, 124). The moral consequence of this standardization and universalization of excess, is that the only way to convert what gives itself into what shows itself, is by way of overcoming the defective nature of human finitude in a strong metaphysical gesture that can only *then* pave the way for a passivity toward the other (Schrijvers 2011, 72–80).

With this brief exposition of Marion's phenomenology of givenness, saturation, and its theological underpinnings, let us now turn to "nature". As Christina Gschwandtner has demonstrated, "nature" or the non-human does not really appear in Marion's oeuvre under discussions of saturated phenomenon, and yet such a connection can be made, as already alluded (Gschwandtner 2014, ch. 3). Gschwandtner argues that there are elements of nature which equally result in counter-experiences that evade any control of a transcendental ego, and under their bedazzlement they may elicit "awe, wonder, maybe even worship" (Gschwandtner 2014, 79). Similar to Marion's insistence that the saturated phenomenon resists technological mastery of the concept and intentionality's priority of manifestation, Gschwandtner suggests that non-human elements from biological systems, natural history, geology, complex eco-systems, animals, etc. – are never fully understood or constituted by human consciousness. In short, they can exceed the Kantian categories necessary for transcendental understanding and, thus, are good examples of saturation under Marion's schema (Gschwandtner 2014, 80–82). Interestingly, and a point we shall return to below, on Gschwandtner's interpretation it is also the case that the sheer destructive potential of nature – storms, events like tsunamis and earthquakes – are also instances of saturated phenomenon; they are "chaos system[s]" of which we can only be "negatively certain" of their unknowability, since they involve countless interactions of relationality with other phenomena (Gschwandtner 2014, 82–83). On this reading, non-human life and nature more generally, cannot simply be reduced to "technical objects", indeed, it is the very *logos* of techno-science to actualize all possibility into mastery (Janicaud 1994). Understood as saturated phenomena, however, the elements of nature arrive at us with an ethical dimension toward which we must respond. Although phenomenologically distinct to the sheer transcendent otherness of Levinas' *visage*, Gschwandtner asks nevertheless whether saturation provokes the response of "I shall not kill" with its imposition of alterity? Or perhaps whether Marion's focus on the "love" of the saturated natural phenomena would be more appropriate to motivate "us to care about them and consider them worthy of protection…[being] 'devoted' to them, addicted to them, or even given over to them in way that forbids treating them as objects" (Gschwandtner 2014, 92).

This brings us to the essential point regarding the limitations of treating a project like Marion's for an eco-phenomenology. As the discussion above highlights, the phenomenology of givenness and saturation is made possible only by deducing phenomenality (saturation as *absolute givenness*) from the actuality of Revelation – or theophanic experience. Taken to this most intense degree of saturation then means that identifying it *as such* is a matter of preparation (faith), since contra-Marion, it cannot be the case that the experience of saturation is universally available to all. Gschwandtner's book, *Degrees of Givenness*, to which we have just referred, is in part precisely an attempt to circumvent this problem, that is, how to account for the fact that not all phenomena are saturated and moreover, that there must then surely be a *hermeneutic* dimension involved in constituting a phenomenon as excessive with intuition, and thus suggesting that there must also be "*degrees* of givenness" instead of just "poor" or "rich" phenomena (Gschwandtner 2014, 5). Gschwandtner's hermeneutic proposal is thus a response, both to the critique of passivity in Marion's phenomenology – a passivity that overwhelms the activity of the subject – and to what she detects as a present but under-developed hermeneutic operation. But by implication of this hermeneutic admittance, nature as saturated thus still requires a prepared subject, one *who is able to see* nature as saturated/sacred. Without situating the argument for saturation and thus the possibility of natural phenomena as saturated with respect to its theological rationale, Gschwandtner and others (Falke 2023, 268–72)[6] end up placing the moral responsibility for nature's sacrality on the subject, and whosoever would not see it as such are, in the words of Emmanuel Falque, condemned as "fallen…from a primitive state of saturation which is impossible to realise but that always constitutes the norm" (Falque 2018b, 123–24). While respecting such a hermeneutic approach, we have still not therefore dealt with the problem of saturation as the principle of all phenomenality; the subject is passively constituted such that the economy of the gift-given is predetermined, any *activity* (e.g., respecting, caring, and loving nature) always comes after-the-fact of saturation. To wrest ourselves from this exclusivist (theological) model, we turn now to Emmanuel Falque, who questions this pre-emption of the infinite on the finite, and argues via theology for a phenomenology of finitude, materiality, and organicity.

Falquean Philosophy: An Eco-Phenomenological Ally?

Falque, a medievalist, phenomenologist, and theologian, has gained recognition over the past decade with the English translation of his early works (Falque 2019b; 2016c; 2012). His key contribution lies in rejecting core tenets of the French theological turn, notably through a critical stance toward his mentor, Jean-Luc Marion (Falque 2007, 181–199, 2016b, 97–140). While scholars have attempted to summarize his method (Koci and Alvis 2020), this section will focus on two central aspects: the argument for finitude (*l'homme tout court* or *Endlichkeit*) and the theological emphasis on animality and organicity. These are closely tied to what some see as a radical shift in his recent work, which explores the "extra-phenomenon" and a "phenomenology of trauma", the latter to be discussed further below (Costa 2022, 279–290).

How might we begin to think of Falque as an ally for an eco-phenomenology? If Marion's program defended the coming into view of the existential conditions for givenness *only* to the extent that they are *exceeded by* givenness, Falque argues for the exact opposite: existential conditions form the horizon in which givenness is received for phenomenalization. In the case of nature, it would not be that the extremity of saturated natural phenomena reveal (gift) the conditions of our own existence (supposedly therefore establishing the ethical relation), but rather that existential conditions – "natural" or "material" horizons – are needed for any phenomena to appear at all. To elaborate this more clearly, Falque believes unlike much theological phenomenology, we must follow Heidegger and begin with the human as such; not just the "finite" human but a positive doctrine of "finitude" (Falque 2012, 13–19, 2016b, 122–128). This almost classically Heideggerian starting point for phenomenology (and theology) distinguishes him from most authors in the theological turn, whose approach, as we have seen in the description of Marion, can be characterized by a certain Cartesianism. Like Descartes, for whom the idea of an infinite (God) is prior to the perception of my own finite self, the logic of the theological turn (and Marion in particular, a Descartes scholar) is governed by what Falque, following Dominique Janicaud, calls "the pre-emption of the infinite on the finite", whereby the latter is always understood in light of the former (thus, usually in terms of "lack", "fault", "deficiency", etc.) (Falque 2012, 16–19). For Falque, this says nothing about or to the human, since this infinitude is foreign to the *experience* of the human being; at least *at first* we exist "within the blocked horizon that comes from the simple fact of existing" (Falque 2012, 17). This human being, understood first *without* reference to an infinite, must thus be taken as the starting point for any phenomenology or theology: "only the *positiveness of finitude*, understood as realised within temporality by the future (death)...can tell us what there is of the Being-there of man (*Dasein*)" (Falque 2012, 18).[7]

Crucially, for Falque – and here is the justification for his continued, albeit subversive, *inclusion* within the theological turn – it is precisely Christianity, particularly the Incarnation, which reveals to us not only the divinity of God but also something essential about our own *humanity*. In other words, Christ's incarnation serves first an *anthropological* function – it demonstrates to an extreme degree our own human condition. It is from this that Falque's infamous dictum is derived: "the more we theologize, the better we philosophize" (Falque 2016b, 147–52). The third instalment of Falque's *Tridiuum philosophique*, "The Wedding Feast of the Lamb" provides us then with the second methodological development which connects us directly with an eco-phenomenology through a new articulation of embodiment, as well as by Falque's own accounting, an early drafted radicalization of his later thinking on chaos, activity, and trauma by way of the extra-phenomenal (Costa 2022, 285; Falque 2016c, 11–17, 2021, 226). Taking his cue from the Eucharistic feast, whereby Christ is offered as the lamb to be consumed, Falque develops a new conception of embodiment that begins from finitude and animality, opposing the doctrines of phenomenological openness and an originary passivity for a type of active "thereness" or a resistance of the material weight and density that we also share with the non-human.[8]

On this reading, namely, that the divine kenotic movement involves a fully comprehended incorporation of all dimensions of our human existence by Christ – including our animality, organicity, and chaos – Falque attempts to displace the prioritization in phenomenology of the embodiment based on *flesh* (or *chair* in French). The latter, Falque believes, involves a spiritualism that is indebted to a problematic heritage of translation which glosses over an original theological meaning (Falque and Kearney 2019, 76–77). On this account, much of the phenomenological tradition from Husserl to Henry enacts a "swerve of the flesh" by renewing the modern dichotomization of soul and body into the phenomenological distinction between *Leib* (flesh) and *Körper* (body), and in this way provides no "access" to the body (*Körper*) through the flesh (Falque 2016c, 1–4, 2018b, 144, 2024a, 234–262; Ullrich 2024a, 171–194).[9] Therefore, instead of an experience of the body-as-flesh, whereby the self is radically distinguished from its material conditions in a primary auto-affection of the self by itself (e.g., Michel Henry), we must come to conceive of that which is outside-of-flesh (the body) (Romano 2016, 114–48). It is important to note that Falque does not deny a phenomenology of flesh (*Leib*) or its importance (Falque 2016c, 118) but argues instead that its progressive priority misses a structural condition that makes the very experience of oneself possible. Included within the scope of human embodiment, the animality of our bodies (*corps organique* as *hors-chair*) – our organic, material, visceral, and biological quality – offers not only a more complete picture of our condition as continuous with the non-human but is also the only way by which we can have any experience of our own flesh (*chair*) in the first place (Falque 2018b, 166). Falque's phenomenological inversion of the priority of flesh as depicting my originary ownness, is not simply to argue for its opposite as geometric extension (*Körper – objective body*), but to claim that this body must be "incorporated" or undergone, which is phenomenologically *impossible* but not without an experience. This phenomenological paradox – describing an experience which is necessarily outside of experience – leads Falque to hypothesize what he calls the *spread body* and where we can further derive the eco-phenomenological import of Falque's work (Falque 2019a, 91–116).

For now, we can conclude this section by rearticulating how far we have come from Marion and the passivity of the saturated phenomenon. An eco-phenomenology that integrates the theological turn within its repertoire should be weary of a Marionesque bait and switch: while harboring the possibility for saturation, the consistency of absolute givenness with respect to natural elements requires that it remains in principle unconditional and universal. Yet, this *structural sacralization of nature*, suggests that in order for its passive constitution to be effective – the consequence of Gschwandtner's hermeneutical supplement – a subject possessing the correct "will to see" must be (metaphysically) posited. This theological phenomenalization of *the given* makes of the subject a deficient and defiant will, whose "limit fades into the unlimited" (Falque 2018b, 124). By contrast – a different track located subversively within the theological turn – reveals via a phenomenological understanding of the Incarnation, as well as the Eucharist, a radical finitude according to which subjectivity is not a limit with respect to an unlimited (the pre-emption of the finite to the infinite), but simply limited by the sheer finitude of its inescapable *being-there*. This does not deny the saturated phenomenon for those with the intentionality of *faith*, but it supplants it with the limits by which all of humanity experiences themselves. In Falque's words: "a phenomenology from below precedes and grounds any theology from above" (Falque 2016b, 124). This "from below" is for Falque the distinctive feature of our embodiment that is shared by all, and surely a better universality from which to approach the human–non-human relation. We turn now more precisely to the nature of this body and its relation to Nature.

The Spread Body and "Activity" as *Lebensbewegtheit*

The distinctive features of the spread body (*corps épandu*) that comes to supplement (or to go beneath) the *Leib-Körper* dualism by accounting for the full corporeity of our material existence, emerges precisely to the extent that it recedes from phenomenalization. The latter is true for Falque because the lived-body of traditional phenomenology can never experience the organicity, viscerality, and material density of our corporeity; it only appears as an objective correlate (Falque 2016a, 159). This unsatisfactory situation of embodiment leads to this central concept: *corps épandu* registers the fact that the *thereness* of my body is not only an object "out there" but also in some sense *mine*. At this stage of Falque's conceptual development, the primary metaphor behind the spread body is the context of palliative care – a body that is in slumber or splayed out on the operating table "is me" yet also foreign or alien to me as it weighs "down on me" against my will (Nancy 2008, 129). This "in-between of the spread-out body" Falque writes, " – neither flesh nor body, neither purely subjective nor exclusively objective – is then, as I see it, what philosophy needs to recover and what theology needs to deal with" (Falque 2016c, 15; Dahl 2024, 41–45). In *Wedding Feast of the Lamb* this philosophical recovery issues in a still more expanded account of the spread body, one that now extends to animality, chaos, the underground of our being, and in direct contrast to the phenomenology of passivity, what he calls the forces, desires, and drives (Falque 2016c, 5–30). In the *active resistance* of a reduction to lived-experience, indicated by the non-sense of visceral physiological and psychological forces and drives – "the animal and the instinctual" (Falque 2016c, 24) – Falque claims to have overturned the primacy of passivity over activity in phenomenology.

It is evident through the selective synthesis of sources Falque deploys for the nature of the spread body, not only theological (e.g., Tertullian) but also philosophical, from Freud, Nietzsche, Deleuze, Derrida, early Levinas to late Merleau-Ponty – that he seeks to rethink the body. He aims to reconsider the very limits of phenomenology (Ullrich 2024b, 6–8). But these sources and their somewhat eclectic distribution mean that Falque leaves himself vulnerable to the criticism of an undifferentiated romantic subterranean excess, stripped of any inherent intelligibility, or ontological value: of what does this spread body consist? What *are* these forces, drives, desires? (Connelly 2020, 155–62). Are we to understand the body in the affective sense as a realm of pure becoming? In our view, a closer analysis of the "drive" concept and its attending notion of force (*dunamis*) is therefore necessary. In particular, we should consider how drives are related to the *active forces* of nature which are the common root of all living beings in their shared *Lebensbewegtheit* (movement of life). Here, a Falquean eco-phenomenology would not only challenge an eco-ethics based on saturation and question lingering angelism in phenomenology, but also, by radicalizing the spread body through its active forces, reveal the shared dynamics of the *common root of the forces of nature of all living beings*. The final section will argue that this common root is experienced as extra-phenomenal: a foundational traumatism marked by lack and distension, exhibiting a non-coincidence within us and nature.

The phenomenologist Rudolf Bernet offers a text which gives us the next step to assist in bridging this gap. *Force, Drive, Desire: A Philosophy of Psychoanalysis*[10] is a series of detailed studies which cover much ground, but the central thesis is that the drive concept in psychoanalysis can benefit from a history of its origins in philosophies of nature, regenerating and renewing its concepts of force, drive, and desire. At the same time, Bernet suggests that by attending to what psychoanalysis has brought to the fore in these concepts, the history of philosophy itself can understand why its heritage is more sensitive "to conflict than harmony, more attentive to negativity than positivity, and more preoccupied with the disorder of excess than with the certainties of reassuring final goals" (Bernet 2020, 5). Our limited focus here turns on Bernet's reading of Heidegger's interpretation of Aristotle's *Physics* and particularly the metaphysics of forces, marked not so much by *energeia* but by *dunamis* and *sterèsis* – the latter denoting a mode of being of the drive as "a non-actualized virtuality…restrained but effective power…pressing but provisionally inhibited force" (Bernet 2020, 5).

Drawing on a selection of texts that primarily focus on Heidegger's reading of *Aristotle's Physics*, Bernet suggests that Aristotle chooses to analyze the general structure of movement according to the movement involved in making (*poiein*) something, because (a) making reveals the common moments of movements in a more distinguishable way and (b) Aristotle has a marked preference for entities which have higher degrees of permanence or being-present (*ousia*). A chair thus has an origin (*archè*) and an end (*telos*) which reside outside of the work/movement (*ergon*) required for its realization. These characteristics throw into relief the type of movement of that which is *not* made, namely, that the being (*phusis*) of living beings must be characterized by a *natural* movement, whose movement of life is fundamentally in the process of actualizing itself toward an always incomplete totality (Bernet 2020, 22). The constant being-en-route of natural living things – their "movedness" as their life (*phusis*) itself or *Lebensbewegtheit* – means that this movement is never only an *energeia* or actualized power (as in *poiein*), it is rather accompanied by a *dunamis* or force that allows it to live and to pass it on to other living beings (Bernet 2020, 23) in the form of a "reciprocation" (*Wechselwirkung*) (Bernet 2020, 30). Crucially, if the movement of life of living things involves a force of *dunamis* that is per definition non-actualizable, it means that all the resources of this disposing force of *dunamis* will not have been exploited. Bernet summarizes: "in other words, this life is marked by a *lack* (*sterèsis*) of actualization (*energeia*) that is at the same time a *richness* of as yet unused or unrealized potentialities" (Bernet 2020, 24). What this amounts to saying is that the forces of the spread body as well as all living beings of nature share essentially three modes of movement: a not-yet actualized potentiality (*dunamis*), the actual movement that is in the course of taking place (*energeia*), and the mode of an absent movement, that is, a lack (*sterèsis*). *Sterèsis* as the category of negation or essential negativity that is not without force – for example, a being in the mode of *energeia* is only a "movedness" to the extent that it harbors the present–absence (*sterèsis*) of the *dunamis* which has either already been realized or still resides dormant – becomes the positive engine for all movement of life.

This connects to a second dynamic important for our purposes. Bernet follows closely the development of Heidegger's interpretation of the ontological understanding of *dunamis* as the *archè* of *kinesis* to the 1931 lecture course on "Aristotle's Metaphysics Theta 1–3: On the Essence and Actuality of Force". There he identifies in Heidegger a novel and more acute sensitivity to this "original *negativity* that inhabits all force, independently of its relation to a not yet realized *energeia*" (Bernet 2020, 37). This sensitivity is accounted for by two essential relations that force has with both "unforce" (*Unkraft*) and the force of "resistance" (*Widerständigkeit*). First, to the latter, in what we can interpret as a further problematization of the "passive" force of "being worked upon" (*pathein*), which we have been alluding to via Falque, there are involved here both a "force of receptivity" in the form of a "capacity to endure" (*Ertragsamkeit*) or "plasticity" (*Bildsamkeit*), as well as an intolerance or "attitude of resistance" (*hexis apatheias*) (Bernet 2020, 38). Such resistance to the force of *pathein* is precisely a negative mode of force essential to *dunamis* as the foundation of all life movement, thus, Heidegger can write: "That which *resists* is the first and most familiar form in which we experience a force" (Heidegger 1995, 76; quoted in Bernet 2020, 38). Second, to the former relation, "unforce", one can also approach "active force" (*poiein*) through its negative counterpart in *Unkraft* or impotence, where the negativity is located not so much in an external counterforce nor even an internal resistance, but in a force or power that imposes itself precisely as a withdrawal or lack (*Entzug; sterèsis*) of force. The nature of force (*dunamis*) is that it is constituted by an essential negativity, an "active" impediment within the very dynamics of the natural movement of life shared between all living things.

Traumatic Being and the Eco-Ethics of the Extra-Phenomenal?

While philosophical negativity has a rich heritage in the continental tradition as well as in theology, and in that sense is not necessarily a novel intervention, the virtue of this chapter as we have been developing, is to demonstrate its emergence from within the contemporary theological turn in phenomenology. It is therefore appropriate that the preceding argument can now be framed within the perspective of what Falque calls the "extra-phenomenal" as well as a generalizable trauma, or "traumatic being", which elicits a fundamental negativity of non-coincidence within us and nature, and where through such an experience of solitude an eco-ethics can emerge in terms of a "solitude of solidarity".

The spread body, as alluded, was an early drafting of Falque's more precise phenomenological position, one that was set to challenge the propensity for phenomenality (*apparoir* as appearance) as such (but which does not abandon the corporeal, it should be noted). A series of essays have now culminated in what we can call Falque's mature work with the publication of *Hors phénomène* (2021), *La chair de Dieu* (2023), and *Spiritualisme et phénoménologie* (2024b).[11] This discussion will be limited to a brief excursion into the generalities of *Hors phénomène* and a philosophy of trauma.

The extra-phenomenal is Falque's attempt to think through the possibility not of non-manifestation or over-manifestation, but the possibility of the non-possibility of manifestation. It will not be the non-appearance of the phenomenon but rather the full impossibility of the possibility of the phenomenon (Falque 2021, 63; 89–93). What occurs when one is faced with trauma – sickness, divorce, death of a child, a pandemic, or natural disaster – when the axioms of phenomenological or even hermeneutical meaning or sense-making are rendered insufficient? The subject of such an experience is not gone completely, but it certainly is no longer itself or at least is broken down in "the chaotic character of our lives" (Falque 2021, 151). In protest of the "cheerful phenomenology" (e.g., Marion) content with its stress on the Other, enchanted with its own conclusions, and tied in a straightjacket of its own etiquette that resists asking questions of its very own axioms, Falque proposes a "modest phenomenology" or the "night of phenomenology" (Falque 2021, 9, 55, 339, 379, 382). The latter, as the extra-phenomenon, does not promote a phenomenal preparation as in Husserl's passive synthesis, what Falque calls the "infra-phenomenal", nor is it interested in the overflowing of phenomenality, Marion's "supra-phenomenal", but rather registers the destruction of all possible categories of synthesis and horizons of phenomenalization. And yet, the burden of Falque's text is to demonstrate that the trauma, while lived "outside of lived experience" is nonetheless still an experience that shatters and changes the subject in the sheer fact of its being-there (Falque 2021, 18).

A fundamental takeaway for us from Falque's bold and one must admit, somewhat abstracted, phenomenological program is the suggestion that the extra-phenomenon, while revealed in exceptional cases, in fact reveals a more generalizable understanding of the human condition: this time, not in terms of an "openness" *to* experience, but in a transformation *by* experience. In the face of trauma and the "defeat of my categories [of experience]…I can only reinvent myself differently – not just in the expectation of eventuality or the so-called salvation of alterity, but also in the confidence of a 'power of the self'" (Falque 2021, 8). The final three chapters of *Hors phénomène* develop this philosophical anthropology of trauma, or what we are calling "traumatic being". Traumatic being reinscribes the formula: "so much exception, so much modification" (Falque 2021, 299). That is, the exceptional experience of the experience outside of experience (crisis or trauma as the extra-phenomenon), becomes the rule insofar as we are all "traumatizable" while not all necessarily being traumatized

(Falque 2021, 253, 262) and yet this undoing allows us to constitute ourselves again, differently (modification). Traumatic being is not, however, a resilience of the subject in the face of our own faults or flaws, it is rather, following Deleuze, the very ontological "cracks [*fêlures*] of our existence" (Falque 2021, 28–29) from which we continue to survive and become other: "'other *from* himself,' in fact, 'other *to* ourselves' but not 'other than oneself'" (Falque 2022b, 326).

One can surely question the phenomenological technicalities of Falque's argument as others have done – that is, whether he has succeeded in a phenomenological description outside phenomenology (Audi 2023; Barbaras 2023; Cassidy-Deketelaere 2023). Nevertheless, if Falque has at least opened a radical new field of experience, we can now ask whether this "'new field'– that of what is without 'field' or horizon" (Falque 2021, 89–90), this non-coincidence and traumatic being of the finitude of our own existence, in which we nonetheless persist, is not also a good description of nature? Does not nature also harbor, as we have indicated, its own moments of negativity in the "unforces" which constitute its very movement of life? Indeed, it would also be possible to suggest – by drawing a line from Schelling to Lacan and Žižek, for example (Schelling 1994; Žižek 1996) – that an arguably (deformed) theological depiction of nature in terms of a stable backdrop of Natural Law fails to account for its own generativity, underlying dynamics, and indeed irrational contagions – its own traumas, diseases, and natural disasters, from which natural life is interrupted while also constituting its own possibility. It is, thus, perhaps no surprise then that Falque would be drawn to Merleau-Ponty's late work and interest in "the barbarian principle" (Falque 2022c, 673–696). While he does not develop this, it is within this link of "traumatic being" with the "barbarian principle" that one can begin to sketch a vision for eco-phenomenology from within the theological turn. For Nature, in this view, like it is for Schelling and Merleau-Ponty, is certainly not a representation of a closed set of recursive laws. Nature is neither *before* us, in *front* of us, nor are we *in* it. But rather, as *of* Nature, entering "traumatic being" allows us to think the question of Nature today in terms of its *active* natality: out of an irreducible remainder and a traumatic "shadow" of nomadic force that not only resists settled modes of thinking, but which precedes the very distinction between representation and reality.

What sort of ethics, if any, emerges from this traumatic being? Here we suggest that the real stakes of Falque's proposal emerge, for the phenomenological consequence of the extra-phenomenal and traumatic being is a radical solitude which resists the "mysticism of the other" (Falque 2021, 257) in phenomenology. To be clear, Falque does not seek to wholly deny the intersubjectivity which is taken as *a priori* in phenomenology (Falque 2021, 403), but he does seem to wish to say that out of the context of the extra-phenomenon there can be, by definition, no premature intervention of the other which "gives to me" or what I "give to it". The traumatic being of the extra-phenomenon is not something I can share with the other precisely because it is not even really my own. This means that originary solitude – not solitariness (Rousseau), solipsism (Husserl, Descartes), or even isolation (Heidegger) – as a consequence of traumatic being, cannot rely on the intervention of the other as an "other possibility" (Marion and Levinas), for this would mean that the ego shutting itself away in the face of trauma and its inability to recover from it, resides in a fault of its own (in essence, what was argued contra-Marion above). If the other intervenes, it is not because they can share in the trauma: there is "the irrefragable 'nucleus of solitude' which constitutes each one of us, impossible to overcome, and even less to share" (Falque 2022b, 328). Rather, the other can at most "be there with me" (*cum meo*) in the extra-phenomenon, just as we can simply be "there" with them – "not to penetrate but to stand by his side" (Falque 2022b, 328). We have thus arrived at an *eco-ethics of solidarity*, one that does not presuppose by virtue of its inaccessibility, an equality with nature that would seek to penetrate it from a moral high ground, nor are we subject to a passive hierarchy whereby the responsibility to which we are held to account comes as a result of privative reasoning (Marion). Instead, what is shared with Nature is precisely and paradoxically a core of solitude (even if this is experienced differently than trauma), out of which a common inaccessibility forever knots us together in solidarity. To be knotted with Nature in this way is to think an (ethical) indivision, which is nonetheless not a reduction to the same. Perhaps herein lies the contribution of the theological turn in phenomenology to an eco-phenomenology: a radical understanding of our corporeal finitude that phenomenology can develop together with theology, and an account of how such finitude is shared with Nature so that, above all, a solidarity can be conceived that is at once unpresumptuous but also universally accessible.

Notes

1 For example, Gschwandtner (2014, 79–94), where she develops the possibility of saturation and "natural" phenomena, as well as Manolopoulos (2009), where he draws on Derrida and Marion's conception of "gift" to develop an "eco-ethos". The human starting point of experience and its relation to nature is the opening gesture of an eco-phenomenology, as argued by Toadvine and Brown (2003, xii) in their seminal volume.

2 In addition to the studies above who have engaged Marion and eco-phenomenology, we can also refer to studies which have attempted to incorporate Emmanuel Levinas, a key figure in the theological turn: see for example, Llewelyn (1991); Macauley (2011); Sallis (2000). One could also conceive of the work of Michel Henry contributing to this discussion, especially given his emphasis on "Life", materiality, and the phenomenology of the flesh which equates this "Life" with the divine. See for a critical attempt in this direction: Gschwandtner (2012, 116–138).

3 The syntagms referred to here can be broadly associated with thinkers in the theological turn as follows: Levinas (face), Marion (gift/givenness), Chrétien (word), Henry (flesh), and Lacoste (liturgy).

4 See Henry's (2015, 1–21) helpful discussion.

5 Caputo (2007, 85) calls this Marion's "theological magical realism."

6 Falke (2023), to an extent following Gschwandtner, and in an otherwise beautifully written piece, also neglects the theological dimension of Marion's thought in relation to the saturation of natural phenomena.

7 Later, Falque (2021, 9, 339, 379, 382) will complete this gesture of distancing himself from French phenomenology's stress on the Other, by introducing what he calls a "minimalist" or "modest" phenomenology.

8 While Michel Henry is more directly the foil for this development, Marion largely depends on Henry for his own account of *flesh*. For further discussion see, Ullrich (2024b).

9 See also Ullrich (2024b, 171–194).

10 Falque, in fact, cites Bernet (2020) (i.e., the French version) approvingly several times in his small but powerful text: Falque (2020). Here Falque is more concerned with the *somatic* origins of the drive concept in Freud associated with the Id, whose force is later lost in symbolic representations by Lacan.

11 See also Falque (2018a, 9–28); Falque (2022b, 315–330); Falque (2022a, 190–217).

References

Audi, Paul. 2023. "L'extra-phénoménalité à l'épreuve de la solitude." *Transversalités* 166: 73–94.

Barbaras, Renaud. 2023. "Hors Monde." *Transversalités* 166: 39–53.

Benoist, Jocelyn. 2001. *L'idée de phénoménologie*. Paris: Beauchesne.

Bernet, Rudolf. 2020. *Force, Drive, Desire: A Philosophy of Psychoanalysis*. Translated by Sarah Allen. Evanston, IL: Northwestern University Press.

Caputo, John D. 2007. "The Hyperbolization of Phenomenology: Two Possibilities in Recent Continental Philosophy of Religion." In *Counter-Experiences: Reading Jean-Luc Marion*, edited by Kevin Hart, 67–94. Notre Dame, IN: University of Notre Dame Press.

Caputo, John D., and Michael J. Scanlon, eds. 1999. *God, the Gift, and Postmodernism*. Bloomington: Indiana University Press.
Cassidy-Deketelaere, Nikolaas. "Towards a Critique of Phenomenological Normativity." *Phenomenological Reviews*, January 29, 2023. https://reviews.ophen.org/2023/01/29/emmanuel-falque-hors-phenomene-essai-au-confins-de-la-phenomenalite/
Connelly, William L. 2020. "The Intelligibility of the Flesh: A Response to Emmanuel Falque." *Crossing: The INPR Journal* 1: 155–162.
Costa, Joao Paulo. 2022. "Un tournant? Interview avec Emmanuel Falque." *Revista Filosofica de Coimbra* 31: 279–290.
Courtine, Jean-François, ed. 1992. *Phénoménologie et théologie*. Paris: Criterion.
Dahl, Espen. 2024. *Incarnation, Pain, Theology: A Phenomenology of the Body*. Evanston, IL: Northwestern University Press.
Falke, Cassandra. 2023. "Eco-Phenomenology in the Dark." In *Symbiotic Posthumanist Ecologies in Western Literature, Philosophy and Art: Towards Theory and Practice*, edited by Peggy Karpouzou and Nikoleta Zampaki, 268–272. Berlin: Peter Lang.
Falque, Emmanuel. 2007. "Larvatus pro Deo: Jean-Luc Marion's Phenomenology and Theology." Translated by Robyn Horner. In *Counter-Experiences: Reading Jean-Luc Marion*, edited by Kevin Hart, 181–199. Notre Dame: University of Notre Dame Press.
Falque, Emmanuel. 2012. *The Metamorphosis of Finitude: An Essay on Birth and Resurrection*. New York: Fordham University Press.
Falque, Emmanuel. 2016a. "Is There a Flesh without Body? A Debate with Michel Henry." *Journal of French and Francophone Philosophy* 24. 139–166.
Falque, Emmanuel. 2016b. *Crossing the Rubicon: The Borderlands of Philosophy and Theology*. New York: Fordham University Press.
Falque, Emmanuel. 2016c. *The Wedding Feast of the Lamb: Eros, the Body, and the Eucharist*. New York: Fordham University Press.
Falque, Emmanuel. 2018a. "The Extra-Phenomenal." Translated by Luke McCracken. *Diakrisis Yearbook of Theology and Philosophy* 1: 9–28.
Falque, Emmanuel. 2018b. *The Loving Struggle: Phenomenological and Theological Debates*. Translated by Bradley B. Onishi and Lucas McCracken. London: Rowman & Littlefield.
Falque, Emmanuel. 2019a. "Toward and Ethics of the Spread Body." In *Somatic Desire: Recovering Corporeality in Contemporary Thought*, edited by Sarah Horton, Stephen Mendelsohn, Christine Rojcewicz, and Richard Kearney, 91–116. Lanham, MD: Lexington Books.
Falque, Emmanuel. 2019b. *The Guide to Gethsemane: Anxiety, Suffering, and Death*. New York: Fordham University Press.
Falque, Emmanuel. 2020. *Nothing to It: Reading Freud as Philosopher*. Translated by Robert Vallier and William L. Connelly. Leuven: Leuven University Press.
Falque, Emmanuel. 2021. *Hors phénomène: Essai sur les confins de la phénoménalité*. Paris: Hermann.
Falque, Emmanuel. 2022a. "God Extra-Phenomenal: For a Phenomenology of Holy Saturday." Translated by Jan Juhani Steinmann. *Journal for Continental Philosophy of Religion* 4: 190–217.
Falque, Emmanuel. 2022b. "Outside Phenomenology?" Translated by Victor Emma-Adamah. *Open Theology* 8: 315–330.
Falque, Emmanuel. 2022c. "Princip barbare' et 'Il y a'. Lecture croisée: Merleau-Ponty et Levinas". *Revista Portuguesa de Filosofia* 78 (3): 673–696.
Falque, Emmanuel. 2023. *La chair de Dieu*. Paris: Cerf.
Falque, Emmanuel. 2024a. "The Turn of the Flesh." In *The Emmanuel Falque Reader: Key Writings in Phenomenology and Continental Philosophy of Religion*, edited by Nikolaas Cassidy-Deketelaere, 235–262. London: Bloomsbury.
Falque, Emmanuel. 2024b. *Spiritualisme et phénoménologie: Le cas de Main de Biran*. Paris: PUF.
Falque, Emmanuel, and Richard Kearney. 2019. "Embrace and Differentiation: A Phenomenology of Eros." In *Somatic Desire: Recovering Corporeality in Contemporary Thought*, edited by Sarah Horton, Stephen Mendelsohn, Christine Rojcewicz, and Richard Kearney, 76–77. Lanham, MD: Lexington Books.
Gschwandtner, Christina. 2012. "What about Non-Human Life? An 'Ecological' Reading of Michel Henry's Critique of Technology." *Journal of French and Francophone Philosophy – Revue de la philosophie* 20 (2): 116–138.

Gschwandtner, Christina M. 2014. *Degrees of Givenness: On Saturation in Jean-Luc Marion*. Bloomington: Indiana University Press.
Heidegger, Martin. 1995. *Aristotle's Metaphysics Theta 1–3: On the Essence and Actuality of Force*. Translated by Walter Brogan and Peter Warnek. Bloomington: Indiana University Press.
Heidegger, Martin. 1996. *Being and Time*. Translated by Joan Stambaugh. New York: SUNY Press.
Henry, Michel. 2015. "The Four Principles of Phenomenology." Translated by Joseph Rivera and George E. Faithful. *Continental Philosophy Review* 48 (1): 1–21.
Husserl, Edmund. 1999. *The Idea of Phenomenology*. Translated by Lee Hardy. Dordrecht: Kluwer.
Janicaud, Dominique. 1994. *The Powers of the Rational: Science, Technology, and the Future of Thought*. Translated by Peg Birmingham and Elizabeth Birmingham. Bloomington: Indiana University Press.
Janicaud, Dominique. 2000. "The Theological Turn in French Phenomenology." Translated by Bernard G. Prusak. In *Phenomenology and the 'Theological Turn: The French Debate*, 1–103. New York: Fordham University Press.
Ullrich, Calvin D. 2024a. "The Eclipse of the Body? Flesh and Materiality in French Phenomenology." In *The Unthinkable Body: Challenges of Embodiment in Religion, Politics, and Ethics*, edited by Rebekka A. Klein and Calvin D. Ullrich, 171–194. Tübingen: Mohr Siebeck.
Koci, Martin, and Jason Alvis, eds. 2020. *Transforming the Theological Turn: Phenomenology with Emmanuel Falque*. London: Rowman and Littlefield.
Llewelyn, John. 1991. *The Middle Voice of Ecological Conscience: A Chiasmic Reading of Responsibility in the Neighborhood of Levinas, Heidegger, and Others*. New York: Palgrave Macmillan.
Macauley, David. 2011. *Elemental Philosophy: Earth, Air, Fire, and Water as Environmental Ideas*. Albany, NY: SUNY Press.
Manolopoulos, Mark. 2009. *If Creation Is a Gift*. Albany, NY: SUNY Press.
Marion, Jean-Luc. 1998. *Reduction and Givenness: Investigations of Husserl, Heidegger, and Phenomenology*. Translated by Thomas A. Carlson. Evanston, IL: Northwestern University Press.
Marion, Jean-Luc. 2002. *Being Given: Toward a Phenomenology of Givenness*. Translated by Jeffrey L. Kosky. Stanford: Stanford University Press.
Marion, Jean-Luc. 2004. *In Excess: Studies in Saturated Phenomenon*. Translated by Robyn Horner and Vincent Berraud. New York: Fordham University Press.
Marion, Jean-Luc. 2008. *The Visible and the Revealed*. Translated by Christina M. Gschwandtner. New York: Fordham University Press.
Marion, Jean-Luc. 2016. *Givenness & Revelation*. Translated by Stephen E. Lewis. Oxford: Oxford University Press.
Nancy, Jean-Luc. 2008. *Corpus*. Translated by Richard A. Rand. New York: Fordham University Press.
Rivera, Joseph. Winter 2018. "The Myth of the Given? The Future of Phenomenology's Theological Turn." *Philosophy Today* 62 (1): 181–197.
Romano, Claude. 2016. *There Is: The Event and the Finitude of Appearing*. Translated by Michael B. Smith. New York: Fordham University Press.
Sallis, John. 2000. *Force of Imagination: The Sense of the Elemental*. Bloomington, IN: Indiana University Press.
Schelling, Friedrich W. J. 1994. "Treatise Explicatory of the Idealism in the Science of Knowledge." Translated and edited by Thomas Pfau. In *Idealism and the Endgame of Theory*. Albany, NY: SUNY Press.
Schrijvers, Joeri. 2011. *Ontotheological Turnings? The Decentering of the Modern Subject in Recent French Phenomenology*. Albany, NY: SUNY Press.
Sebbah, François-David. 2012. *Testing the Limit: Derrida, Henry, Levinas, and the Phenomenological Tradition*. Translated by Stephen Barker. Stanford, CA: Stanford University Press.
Toadvine, Ted, and Charles S. Brown, eds. 2003. *Eco-Phenomenology: Back to the Earth Itself*. Albany, NY: SUNY Press.
Ullrich, Calvin D. 2021. *Sovereignty and Event: The Political in John D. Caputo's Radical Theology*. Tübingen: Mohr Siebeck.
Ullrich, Calvin D. 2024b. "The Spread Body and the Affective Body: A Discussion with Emmanuel Falque." *Religions* 15 (30): 1–20. https://doi.org/10.3390/rel15010030
Žižek, Slavoj. 1996. *The Indivisible Remainder: An Essay on Schelling and Related Matters*. London: Verso.

4
BIRTH, NATURE, AND REPETITION
Arendt and Beyond

Espen Dahl

As long as we live, we are bound by the rhythms and movements of the surrounding nature for air and nourishment. Our own biological body follows the same cyclical rhythms as the nature around us: We grow, decay, and finally return to the earth from which we came. It seems that this ever-renewing cycle of nature goes on and on, as an eternal backdrop to the scene where human lives come into being, unfold for a while, and then perish. Yet something has changed in our age. It is almost impossible to rely on the everlasting renewals of nature. Faced with the environmental crisis that already can be felt around us, we must, for the first time in human history, realize that the constant birthing of new life might cease. Still, children are born into this world.

For the time that remains, birth is the emblem of the generosity of nature and the gift of life. Human birth is certainly no exception, even as it takes on special significance. We are not just a specimen of a general repetitive structure of nature, for we are also born as unique individuals, as if birth is the starting point of something new that has never existed before. Hardly anyone has stressed the latter point more than Hanna Arendt, suggesting that the new beginning, or more generally, our natality, is indeed emblematic of human life as such. However, emphasizing the new beginning in this way, Arendt is led to take birth as an exception to the repeated cycles of nature, as it transcends it with the unrepeatable start that puts our free actions into motion. Yet this is not the whole story, for Arendt also seems to suggest that birth and growth have their original home in nature. Indeed, natural birth inserts us into the processes of nature. To grow from itself, we learn, "is also the authentic meaning of our word 'nature,' whether we derive it from the Latin root *nasci*, to be born" (Arendt 1958, 159). Given this complexity, I want to raise this question: If birth and birthing lie at the heart of nature, is it possible to reconcile nature with the strong insistence on birth as the spontaneous new beginning that cuts across the repetitive structures of nature? Or conversely, if birth is spontaneous, is it then possible to conceive our natality as part of nature? And finally, can nature, conceived as the eternal recurrence of life after death, be retained in the face of the environmental crisis?

My initial discussion of Arendt's notion of birth has the purpose of bringing her implicit notion of nature into sharper focus. Since she trades on a sharp distinction between nature and birth, that is, between the ceaseless repetitions of nature and the spontaneous beginning of each individual, she is arguably unable to think of a positive relation between nature and birth.

DOI: 10.4324/9781003538097-5

My next step will be to seek out a more fruitful notion of nature that can potentially preserve Arendt's cherished insight into human birth without cutting it off from nature. I will proceed by revisiting the notion of nature, allowing for a richer notion of repetition that comprises difference and movement in a way that can encompass birth. Arendt's notion of miracles points in the latter direction.

Nature and Birth

Natality, our ontological condition of being born, is certainly among Arendt's most significant contributions to philosophy, countering the early Heidegger's one-sided emphasis on death. While it has ramifications for philosophical anthropology in general, it is important to notice that Arendt's concept of natality primarily evolved from a specific context, namely her analyses of actions as they appear in the human world. In defending human freedom, Arendt strongly emphasizes that however actions or speech acts are interwoven with the plurality of others, they can never be reduced to some causal structures or mechanical reactions of interpersonal life. Every single individual has the original and spontaneous capacity to start something new and thus interrupt the web of utterances and responses. Such capacity for new beginnings is, metaphorically, a birth, or better, a second birth: "With word and deed we insert ourselves into the human world, and this insertion is like a second birth; ... its impulse springs from the beginning which came into the world when we were born" (Arendt 1958, 176–77). It is from this second birth that Arendt regressively leads us back to the first birth, which serves as the ontological foundation of active life. By birth, we arrive in the world, yet in Arendt's discussion, there is no mention of the biological processes or the nature by which and into which we are born. The reason, I surmise, is that she restricts her scope to birth into the human world in contradistinction to the birth of nature.

The distinction between nature and the world is very much operative in *The Human Condition* and can first be gleaned in the way she breaks human forms of life down into three fundamental forms of activities. What she calls labor is the most basic activity, where our dependency on nature is obvious. Labor grows out of the necessities of life where ever-recurring needs must repeatedly be satisfied for our survival. As biological and needful creatures, we become subject to the cyclical movement of nature: we sustain life by participating in the metabolism with the earth and secure coming generations through procreation (Arendt 1958, 98). However, human beings are not only subjected to nature but can also fabricate enduring cultural artifacts and things that outlast immediate consumption (Chapman 2007, 435–37). By the second form of activity, work, we produce the cultural world of things which gives us permanence in time and space; we keep the forces of nature at bay and establish space between individuals that safeguards peaceful coexistence. By works, we transpose the natural earth to a cultural world, a world that finally can, at the third level, be the home of action, which includes speech and deeds, and which makes up the most important and distinct form of human activity, according to Arendt (Arendt 1958, 176). Action not only presupposes a world but also a plurality of human beings where each individual appears to others and participates in common affairs, as equals while remaining different. The realm of human affairs is where action belongs, a realm that in Arendt's broad sense of the term, is political (Arendt 1958, 7). In political life, we are not occupied either with needs or artifacts, but with common interests.

For Arendt, the human, political world is obviously not identical to the natural earth of labor and survival. Labor never ends, for as soon as food is consumed, new needs arise, in an endless repetitive rhythm. Arendt often describes the tension and difference between world and nature emphatically. She can say that outside the man-made world of work, we meet "the sublime indifference of an untouched nature, whose elementary force ... will compel them to swing relentlessly in the circle of their own biological movement, which fits so closely into the overall cyclic movement of nature's household" (Arendt 1958, 137). There is no romantic notion of nature here, neither religious gratefulness of its gifts; nature essentially does not care about us in its "sublime indifference." Indeed, Arendt at times invokes Nietzsche's teaching of the eternal recurrence: "A philosophy of life," Arendt contends, "that does not arrive, as did Nietzche, at the affirmation of 'eternal recurrence' (*ewige Wiederkehr*) as the highest principle of all being, simply does not know what it is talking about" (Arendt 1958, 97). But unlike Nietzsche, Arendt does not affirm the eternal recurrence as the guide to a specific human life. Rather, the recurrence of nature is precisely what must be held in check and transcended in order to resist being swallowed by its endless repetitions. Where the world allows for meaningfulness and free initiatives in the human realm, nature is an impersonal force that blindly and inexorably keeps repeating itself.

However, Arendt does not deny that we are also creatures of nature. Indeed, all three activities – labor, work, and action – are subject to the general natural preconditions of human existence that show forth in birth and death. Once again, labor seems to stand in the most intimate connection to our natural condition in so far as reproduction and birth are concerned. Yet things change when it comes to the level of action. Summing up labor's, work's, and action's relation to birth, Arendt writes that,

> of the three, action has the closest connection with the human condition of natality; the new beginning inherent in birth can make itself felt in the world only because the newcomer possesses the capacity of beginning something anew, that is of acting.
>
> (Arendt 1958, 9)

While Arendt can connect birth to both natural preconditions and to action, she does so by indicating their distinct phenomenology: in the former case, birth figures as the natural precondition for entering the world; in the latter case, birth is retroactively understood through actions in the world. Still, it is the latter sense of birth that is at the core of the human condition, that is, of our natality, as Arendt conceives it. Arendt does obviously not identify this sense of natality with natural processes. Whereas birth and death are events of nature, only humans relate to them as the beginning and end of their individual life stories. Put differently: whereas natural birth is chained into the endless reproduction inscribed in the rhythm of the earth, natality is transposed to the human world, so that "birth and death… are not natural occurrences" (Arendt 1958, 97).

If we solely look at human birth, one might expect Arendt to derive her understanding of it through the paradigmatic case of childbirth by which a new person enters the world. But, as already mentioned, she does not. The paradigm is rather human action. From the spontaneity of action that marks a new beginning in the web of human plurality, she regressively follows its condition back to childbirth. Authentic actions manifest the power of beginning something new, and that power, she claims, presupposes birth as its ontological basis. The priority of action, or as she calls it, "the second birth," is in keeping with her primary political ambitions of laying bare the *vita active*. Yet since it somehow relies on the "first birth," she must give convincing accounts of them both. The question is if she can adequately account for the first birth once she has severed it from its biological and natural dimensions.

> But is not the *pheuin* of *physis* first of all the donation of what gives birth, the originary productivity that engenders, causes to grow or increase, brings to light and flowering? Is it not what gives form and, by bringing things into the phenomenality of the light, unveils or develops the truth *of that which* it gives? Of the very thing it gives and of that it gives?
>
> (Derrida 1992, 128)

There is a richness to nature that continues to give birth in a manner that is at once productive and yet withholds itself from appropriation and mastery (Fritsch 2015, 19). It brings to light and still loves to hide. Heidegger's *On Time and Being* looms large in the background here (Derrida 1992, 21–22). For Heidegger, the point is to overcome the metaphysical notion that being (and nature) *is*, and instead proposes that nature appears as an event. Yet in giving and coming to light, the event also holds back and thus withdraws from our grasp or control (Heidegger 1972, 2). Derrida, with Heidegger, suggests a generous, giving nature that gives birth to life, that is, a sense in which giving and birthing are inscribed in the very event of nature. What then of Arendt's primary concern, the human birth: if the constant birthing is part of the repetitions of nature, could we regard human natality as part of it?

The Miracle of Birth and the Miracle of Nature

Arendt speaks of "changeless eternal recurrence" and "deathless repetition" (Arendt 1958, 96–97), "which is purposeless and turning within itself" (Arendt 1977, 42). If nature is cyclical in this sense, it becomes very hard to recognize any internal connection between natality and nature. Not only is the implied understanding of repetition here what Deleuze criticizes as falling under the Same and thus, according to him, no true repetition at all. But it also leaves unexplained what role the plain fact that human birth is inscribed in nature has for us. If we are constituted by natality, it seems strange to give no significance to the role of nature, say, such facts that we are conceived by procreation, develop during pregnancy, and make our appearance in the world through labor. Now, Arendt is certainly not completely blind to these facts, for according to human labor, birth has already from antiquity been connected with preserving all kinds of necessities of life, including birth.

According to the Greek city-state, the household takes care of our survival and procreation, which takes "the labor of man to provide nourishment and the labor of the woman in giving birth." (Arendt 1958, 30). Yet there is already here a strong distinction between labor and political life; political life presupposes labor, but is no longer concerned with biological maintenance. Of course, this relies on a historical division of labor that has changed. In some more normative passages, however, Arendt upholds the strong division, between labor and action. Labor is necessitated by our natural needs: "Cyclical, too, is the movement of the human organism, the human body not excluded… small, singular, cyclical, life processes, returns into the overall gigantic circle of nature herself, where no beginning and no end exist" (Arendt 1958, 96). In contradistinction, it is first as humans become agents in a public realm that this changes; first here is human life seen as one with beginning and end, with life stories and freedom, lifted above the eternal occurrence of nature (Arendt, 1958, 96–97).

But repetition does not stand still but moves ahead and creates something new (Caputo 2002, 13). Nature must somehow be inherently birthing, which, as Arendt mentions, is contained in the Latin root of nature (*nasci*), which means "to be born." (Arendt 1958, 150). Is there a way to see birth as the miracle of the beginning of something new without denying its dependence on the birthing of nature? Not immediately so, it seems. "Left to themselves," Arendt writes, "human affairs can only follow the law of mortality, which is the only certain and reliable law of a life spent between birth and death" (Arendt 1958, 246). But despite this destiny, the various human activities can give rise to a sense of wonder. There is more than labor. As one gradually climbs up the ladder, the form of life to which one aspire, appear like a miracle.

> From the viewpoint of *animal laborans*, it is like a miracle that it is also a being which knows of and inhabits a world; from the viewpoint of *homo faber*, it is like a miracle, like the revelation of divinity, that meaning should have a place in this world.
>
> (Arendt 1958, 236)

On the one hand, the quote strongly suggests the sense of something miraculous as we proceed upward toward action. But on the other hand, Arendt seems to imply that life according to nature is bereft of any sense of wonder. Would it also be possible to pursue the reversed direction, leading us back to nature – and find a miracle at the bottom of life in nature?

When Arendt at times comments on miracles, she considers the possibility of seeing miracles in ordinary reality and nature, even if miracles per definition must occur contrary to expectations. Seen from natural science, she claims, every beginning is an "infinite improbability." Transposed to the realm of the human world, Arendt holds that the miracles of human actions are "interruptions of some natural series of events, of some automatic process, in whose context they constitute the wholly unexpected" (Arendt 1977, 168). While miracles are unforeseen events breaking into the order of ordinary life, Arendt's invocation of miracles in this passage does not give us a richer notion of nature or birth. Yet there exist more inclusive notions of miracles, for instance as proposed by one of Arendt's favorites, namely Augustine. The wonders that he has in mind are not conceived as a clash but rather attest to the wider horizon of creation and nature. From a particular perspective, wonders can be seen to be at work in seeds growing, day following night, and in the fact that children are born (Augustine 2006, XV. 33). Whether this is taken, theologically, as God's ongoing creation or as nature's generous birthing, makes little difference here. The point is that the miracle is not pitted against the repetitions of nature but flows from those repetitions according to their power to create the new from the old.

Interestingly, Arendt can also employ a broad sense of miracles. This broad sense can, I believe, suggest a perspective from which birth can be embraced in its double sense, both as beginning and nature. Looking more deeply into what she says about the "infinite improbabilities," it becomes clear that there is a perspective from which reality as such appears as a miracle: "Our whole existence rests, after all, on a chain of miracles, as it were – the coming into being of the earth, the development of organic life on it, the evolution of mankind out of the animal species" (Arendt 1977, 169). If evolving nature as such must be regarded as resting on a chain of miracles, one might still wonder whether human actions fall outside the grand scope, as perhaps swallowed into the trivialities of human lives. But Arendt maintains that even if part of our ordinary life, actions are no less miraculous for all that. Miracles come to pass in the everyday, too: "The experience which tells us that events are miracles is neither arbitrary nor sophisticated; it is, on the contrary, most natural and, indeed, in ordinary life almost commonplace" (Arendt 1977, 168). While the recurring actions seldom occasion marvel, there is a sense in which it should; she claims that they are at once novelties and repetitions, or as Arendt adds at one place, they are: "infinite improbabilities which occur regularly" (Arendt 1958, 246).

That children are born is commonplace, but no less miraculous. From such a perspective, one might even say that we can turn Arendt against herself and thus gain a richer understanding of birth: the miracle conveys a new beginning that is no less natural. Augustine has put the point thus:

> A dead man rises again; men marvel: so many are born daily, and no one marvels. If we reflect more considerately, it is a matter of greater wonder for one to be who was not before, than for one who was to come to life again.
>
> (Augustine 2012, VIII.II.1)

So many are born daily, and yet, Arendt would agree, it is still a miracle. Here, I want to insist that the miracle is not to be conceived as a line that cuts into the circle, but as the wondrous repetition that always progresses toward the new. We are naturally born of mothers, and we grow gradually toward acting persons, without any leap that cuts us off from nature, even as we gradually acquire a richer notion according to the world as we grow.

Final Reflections

If nature and birth are implicated in one another, we need to adjust Arendt's account in two ways, one concerning the status of labor and the other concerning nature. To Arendt, labor concerns our biological constitution and needs, the sheer sustenance of human life and procreation. But Arendt argues that such activities are overcome and transformed as we reach activities of work and acts, where the significance of beginning and end replaces the eternal recurrence of the same. However, natality is always tied to those processes that we find in labor, and although natality might take on wider signification, it cannot undo labor and birth. They must somehow be preserved. To adjust Arendt's strict divisions of human activities, does not mean to reduce human life to labor (which is one of Arendt's criticisms of modernity), but to acknowledge that natural life is taken up, transformed and integrated with all other forms of life. Birth is never left behind but can reflectively be enriched in light of the stories of our lives.

If this is feasible, then it also means a clear shift in the conception of nature. Nature is no longer the eternal backdrop from which we depart, but the condition that will always remind us of our dependency: Dependent in the sense of coming into being through birth and the birthing of nature, and dependent on nature for upholding our lives. There is, I believe, no need to pit spontaneous birth against nature. There is room for new beginnings, but we need to regard these beginnings as made possible by the generous birthing of nature as such. Beginnings are not separated from the repetitions of nature. But since repetitions are not conceived as returning to the same, but as an onward movement, then each repetition has room for the coming of the singular, the unique, the new beginning. This is the miracle of birth: the new emerges from the old.

Arendt is apparently not very concerned about nature, and certainly not its future since it is both without beginning and end. And yet, in the prologue to *The Human Condition*, she is very much concerned with modern conceptions of the world, especially how those conceptions seek to overcome or deny the human condition. She notes how secularization has made us turn away from our Father in heaven, but then asks if such turning away will "end with an even more fateful repudiation of an Earth who was the Mother of all living creatures under the sky" (Arendt 1958, 2). What must be attended to is something like Earth's cycles and repetitions, moving from birth to death, and back to birth. Michael Marder calls this movement the Phoenix complex, which he takes to underlie much of Western thinking of nature. The complex denotes the deep-seated conviction that Phoenix, life itself, will always resurrect from the ashes. However, our climatic situations call this complex into question: It is far from certain that nature will keep returning forever due to our consumption, industrialization, depletion of fossil resources, and pollution of water and air. If we cannot cling to the Phoenix complex anymore, the answer is not, Marder holds, fatalism and nihilism. We should rather take the imminent limit as an opportunity to rid the philosophy of nature of all false beliefs in eternities. This opportunity can move us to appreciate nature's "absolute finitude [and] nonrenewable existence" (Marder 2023, 244).

This is, however, not the conclusion I will draw. Yes, we have become aware that there is no guarantee that nature's circle of life and death continues into eternity. But to reject nature's repetitive renewal and stick to its nonrenewable existence, seems to me tantamount to denying nature's inner movement. When there are dark horizons, hope emerges – the stubborn will to think that life, despite everything, must continue. It is precisely as we face that nature can come to an end, that hope arises – not as an escape, but as an ethical motivation. The child, Arendt reminds us, installs faith and hope. Even if its birth is not a pure spontaneous beginning, it nevertheless opens up a new future, a possibility, and hope. If birth is not possible to detach from nature, birth inspires hope also for nature. Hence also our care and hope for the future of our environment come to light in the new child, expressed in what Arendt holds as its most succinct expression: "A child has been born unto us" (Arendt 1958, 247).

Notes

1 The quote is taken from Augustine 2003, XII.21. Arendt repeats it on several occasions, such as Arendt 1973, 479, 1977, 167, 1978, 108, 1996, 55.

2 For its Greek origin, see Arendt 1977, 42–43. See also Marder 2023, 237–38.

References

Arendt, Hannah. 1958. *The Human Condition*. Chicago: The University of Chicago Press.

Arendt, Hannah. 1973. *The Origins of Totalitarianism*. New York: Harcourt Brace.

Arendt, Hannah. 1977. *Between Past and Future*. London: Penguin Books.

Arendt, Hannah. 1978. *The Life of the Mind*. New York: Harvest Press.

Arendt, Hannah. 1996. *Love and Saint Augustine*. Edited by J. Vecchiarelli and J. Chelius Stark. Chicago: The University of Chicago Press.

Augustine. 2003. *The City of God*. Translated by H. Bettenson. London: Penguin Classics.

Augustine. 2006. "The Usefulness of Belief." In *Augustine: Early Writings*. Edited and translated by B.H.S. Burleigh. London: Westminster Knox Press.

Augustine. 2012. *Homilies of the Gospel of St. John*. Edited by P.A. Böer. Dublin: Veritas Splendor Publications.

Caputo, John. 2002. "Looking the Impossible in the Eye." *Kierkegaard Studies Yearbook* 7:1–25.

Caravero, Adriana. 2014. "'A Child Has Been Born to Us': Arendt on Birth." *philoSOPHIA* 4: 123–140.

Chapman, Anne. 2007. "The Ways That Nature Matters: The World and the Earth in the Thought of Hannah Arendt." *Environmental Values* 16: 25–45.

Derrida, Jacques. 1973. "Difference." In *Speech and Phenomena and Other Essays*. Translated by D.B. Allison and N. Garver. Evanston, Ill.: Northwestern University Press.

Derrida, Jacques. 1988. *Limited Inc*. Translated by S. Wever. Illinois: Northwestern University Press.

Derrida, Jacques. 1992. *Given Time: I. Counterfeit Money*. Translated by P. Kamuf. Chicago: The University of Chicago Press.
Fritsch, Matthias. 2015. "The Gift of Nature in Mauss and Derrida." *The Oxford Literary Review* 37: 56–78.
Heidegger, Martin. 1972. *On Time and Being*. Translated by J. Stambaugh. New York: Harper and Row.
Kierkegaard, Søren. 1962a. *Gjentagelsen. Samlede Værker 5*. Copenhagen: Gyldendal.
Kierkegaard, Søren. 1962b. *Begrebet Angest. Samlede Værker 6*. Copenhagen: Gyldendal.
Marder, Michael. 2023. *The Phoenix Complex: A Philosophy of Nature*. Cambridge, Mass.: The MIT Press.
Williard, Mara. 2013. "Birth as Labor and Natality: Hannah Arendt and the Politics of Maternity." *Theology and Sexuality* 19: 76–90.

5

THE MIRROR-PLAY OF BEING USED AND NEEDED

Rereading *(Ge-)Brauchen* in Heidegger for Eco-Phenomenology

Magdalena Hoły-Łuczaj

Introduction

Environmental philosophy aims to eradicate two conflicting yet closely interconnected harmful beliefs: that nature is subordinate to humans and that we are alienated from it. This chapter explores how moving beyond these beliefs can help foster a sense of gratitude and belonging to the natural world. It seeks to achieve this through a phenomenological examination of the dynamics of inseparability between the phenomena of use and need, with a particular focus on their passive aspects—being needed and being used—in human interactions with natural beings.

This investigation is inspired by Martin Heidegger, a central figure in eco-phenomenology, as demonstrated by the seminal work of Charles. S. Brown and Ted Toadvine, along with contributions in this volume. I propose adding another theme: Heidegger's exploration of the ambiguity in the relationship between use and need, particularly evident in his later writings, including the recently published volumes of his collected works (GA 100–102), where he engages with the German verb *(ge-)brauchen*. The use of this verb in Heidegger's work is notoriously difficult to translate—and understand—especially in its passive forms, which can convey both "used" and "needed." This ambiguity presents a challenge, as these terms suggest opposing power dynamics. To say something is "needed" implies our dependence on it to fulfill a requirement, while "using" indicates a position of dominance by the agent. Likewise, "being used" typically implies subordination, whereas "being needed" underscores importance or relevance.

This chapter argues that Heidegger deliberately employed the ambiguity of the verb in question to challenge the binary opposition between use and need, expressing their non-reductive inseparability. To unpack this idea, these relations will be analyzed through the figure of chiasmus—a key concept in eco-phenomenology—focusing specifically on Heidegger's version of chiasmus, which he describes as the "mirror-play."

Eco-Phenomenology: Dismantling Environmental Prejudices

In exploring the significance of eco-phenomenology for environmental thought, we should emphasize its two overlapping focal points. The first concerns *what* it examines. Traditionally,

DOI: 10.4324/9781003538097-6

phenomenology has been regarded as a highly abstract, theoretical inquiry into consciousness, being, or the structure of objects (Toadvine 2003: 10). Eco-phenomenology recalibrates this focus, shifting it toward nature. This shift is precisely what makes it *eco*-phenomenology. The subtitle of the seminal book *Eco-Phenomenology* by Brown and Toadvine—*Back to the Earth Itself*—is particularly telling in this regard. As Iain Thomson puts it, this phrase cleverly twists the famous rallying cry of Husserlian phenomenology, "Back to the things themselves!" (*Zu den Sachen selbst!*), by replacing the central *Sache* ("the heart of the matter") with "earth" (Thomson 2004: 386). It is the earth and our relationship with it that we must rethink to find remedies for the current environmental crisis.

Here, "the earth" serves as a counterpart to nature in its primordial sense—not as an essence, but as an environment of non-human beings. However, its boundaries remain fluid, shaped by each person's perception of nature. In other words, our orientation toward the environment and the connection we believe we have with it define our understanding of nature. This leads to eco-phenomenology's second focal point: *how* it investigates this phenomenon.

This second focus is more closely tied to the "phenomenological" core of eco-*phenomenology*. It emphasizes the necessity of investigating the character of our attitude to the environment. Eco-phenomenology asks in what way and why we perceive nature in the way we do. Crucially, phenomenology—and, by extension, eco-phenomenology—takes experience as its starting point (Toadvine, 2017: 175–176). It aims to describe what is experienced—in the case of eco-phenomenology, this is nature—in an unprejudiced manner, or, upon closer examination, to revisit this experience.

One of the greatest achievements of phenomenology, utilized later in eco-phenomenology, is its description of reality as a value-laden and meaning-structured lifeworld (Kūle 2018: 54–55, Toadvine, 2017). Traditionally, phenomenology has distinguished itself from metaphysical naturalism, which is the theoretical view that existence is limited to physical reality governed by natural laws (Toadvine, 2017: 176). It has also differentiated itself from psychological perspectives by pointing out how experience is mediated by metaphysical categories, even when they are not explicitly recognized, and how these categories can result in objectifying illusions.

Eco-phenomenology engages both the descriptive and normative dimensions of experience, exploring how nature and its evaluation—its worth or moral significance—appear inseparably within our perception. A key task is to uncover the source of these values.

Supporters of eco-phenomenology argue that Platonism and Cartesianism create problematic dichotomies in this regard (Toadvine, 2003). These traditions often devalue material reality in favor of the spiritual or ideal and depict objects as either separate from or subordinate to the subject. Such divisions, eco-phenomenologists contend, are counterintuitive and harmful, reinforcing technological, economic, and managerial approaches to environmental issues rooted in instrumental rationality (Thomson 2004).

One response to this modernist view of nature is to "bracket" the (metaphysical) prejudices shaping human conceptions of it, critically reassessing their impact (Prášek 2023: 356; Toadvine, 2017: 178). But can we truly adopt a transcendental perspective and set aside all biases? Heidegger reshaped this challenge within phenomenology by introducing a hermeneutical dimension (Ahmed 2006; Blok 2020). In doing so, he arguably radicalized Husserl's notion of the lifeworld as an inescapable foundation.

This does not mean we are trapped in fixed imaginaries, powerless to change them or their effects. Rather, we can critically examine these attitudes, uncovering how prejudices or "fore-understandings" are constructed, their layers, and their impact. Instead of discarding them outright, we should think through them—dismantling and reshaping them into new understandings.

In summary, the first path ("transcendental") aligns with Husserl, while the second ("hermeneutical") follows Heidegger. As Sara Ahmed (2006: 184) notes, Heidegger offers a dynamic account of how fore-understanding shapes our engagement with new phenomena. A key aspect of this approach is the "as-structure" (*Als-Struktur*), which, as Lucilla Guidi (2021: 105) explains, is fundamental to all experience—things always appear "as-something" (a table, a stone) or "as-some" (available, broken, and needing care). Heidegger argues that both interpretation and assertion rely on this structure. Because entities are already understood in terms of their function and significance in our practical engagement, we can articulate assertions about them. He calls this potential "apophantical," referring to the way things manifest for understanding and become explicable (Guidi 2021: 105; Heidegger 1962: 188–191, 1999: 287–290).

For our considerations, the crucial issue is how nature reveals its meaning to us—specifically, through which "as-structures" we perceive and evaluate it. We will focus on two in particular: "as-needed" and "as-used." Inspired by Heidegger, this inquiry explores the tension between these perspectives. Rather than simply reflecting multiple viewpoints, they appear fundamentally contradictory. But perhaps their opposition is only apparent. We will explore this possibility to enrich our understanding of nature, revealing the complexity of our relationship with it.

Between Alienation and Domination

Before examining the significance of Heidegger's engagement with the dynamics of use and need for eco-phenomenology, it is important to outline the foundations of beliefs in both the subordination of nature to humans and our sense of alienation from it, as well as how these beliefs are related to the phenomena of use and need.

One of the main targets of environmental ethics, and consequently eco-phenomenology, is the demanding attitude toward nature that many people have held—and continue to hold. This attitude rests on the assumption that human beings are entitled to freely manage non-human entities to fulfill their needs. Such an approach is grounded in the perception of nature as subordinate to human beings, existing at our disposal. Environmental ethics, and eco-phenomenology as a part of it, criticize these beliefs on two fronts.

The first criticism concerns the idea of nature's subordination and human domination over it. Environmental ethics problematizes the concept of hierarchy and questions the supposed privileges that humans claim within it. The second criticism addresses the boundaries of human needs that would actually justify taking advantage of non-human entities. Where should we set the limit of needs, and where do they cross into mere whims? Are we, in fact, entitled to take advantage of non-human beings at all?

These dilemmas reflect two contradictory conceptions of the human–nature relationship that Toadvine characterizes: (1) humans are not exceptional but rather part of nature and (2) humans are simultaneously separated from nature, experiencing alienation—the most obvious consequence of which is the ecological crisis. This results in a paradox: how can we be part of nature and, at the same time, act against it? (Prášek 2023; Toadvine, 2017, 181–182).

This brings us to another crucial issue: the sense of human alienation from nature. This term has two basic meanings that are interdependent. The first is "estrangement," understood as a state of being separated or cut off from something. The second concerns property—a renunciation of ownership (Hailwood 2015, 16). As Simon Hailwood writes, the first sense often involves a lack of awareness of humanity's embeddedness within and dependence upon the broader natural world, as if that world either did not exist or we were somehow ecologically separate from it. However, he notes that this estrangement is not simply a matter of ignorance or disregard for our ecological relationships. There is also another dimension related to the "otherness" of nature, which should be endorsed. It suggests that nature is beyond human jurisdiction and disposal.

Viewed in this light, endorsing and living with some estrangement from non-human nature entails a positive endorsement of the alienation of non-human nature in terms of ownership (Hailwood 2015, 16–17). Thus, alienation as estrangement does not always need to be entirely overcome. Living with a degree of estrangement from non-human nature, rather than constantly seeking to overcome it, can be seen as a form of resistance to what some call the "domination" or "mastery" of nature (Hailwood 2015, 18).

Identifying this ambiguity pertaining to the phenomenon of alienation does not resolve the paradox posed by Toadvine. How can we coexist with the environment without being accused of harming it? Hailwood (2015, 18) offers a hint: we should live with *some* estrangement from nature. This suggests that our relationship with the environment should be understood in terms of degrees. The question then arises: at what point should we feel guilty for using nature?

The issue of guilt—specifically, environmental guilt—does not appear here by chance. Research on this topic is extensive, as studies show that guilt closely relates to a moral norm, defined as the sense of obligation to act in a pro-environmental manner. Guilt often functions as a key predictor of the felt obligation to repair environmental damage.

What is also of interest is the contrast between environmental guilt and environmental shame. Broadly speaking, guilt tends to encourage reparative behavior, while shame often leads to denial or avoidance (Nielsen et al., 2024). This distinction builds on the influential framework introduced by Helen Lewis, who argued that shame focuses on the self as the object of negative evaluation, while guilt focuses on the wrong action (Lewis, 1971). In addition, guilt is outwardly directed, rooted in a sense of caregiving and the desire to avoid harm to the external world, while shame is self-directed (Nielsen et al., 2024).

In terms of our relationship with the environment, eco-guilt related to harm can be triggered when individuals believe they have caused unnecessary damage and are aware that the environmental harm could have been avoided. For instance, in one study (interestingly employing a phenomenological approach for the empiric goal), a female participant who experienced eco-guilt for watering her plants explained: "I often think to myself: is it really necessary to use all this water just because you want some bloody flowers in your garden in summer?" (Nielsen et al., 2024).

It is noteworthy that the source of her guilt lies in the use of water, whose legitimacy she questions. She wonders whether it was justified to use water in this way. This reflection may arise when someone feels alienated from nature, of which water is a part. Both senses of alienation seem relevant here. First, the realization that water is not merely her property to be freely managed. Second, underlying this, is a deeper sense of separation or isolation from nature, as though she is not fully part of it and does not deserve to use it.

Significantly, the core of her doubt in the statement lies in the juxtaposition of use and necessity ("is it really necessary?"). This is crucial for the subject of our further analysis, as necessity relates to need, derived from the Latin *necesse*, meaning "to be needful." It seems that her sense of guilt stems from perceiving a discrepancy between use and need. But do we truly understand how this relationship should be defined? How do we grasp it? What enables us to speak of their relationship at all?

Ambiguity of *Brauchen*

We will explore these questions through Heidegger's analysis of the convergence between use and need. First, however, we will briefly assess the compatibility of his philosophy with environmental ethics. While its potential is significant, as shown in this volume and past works, challenges remain—particularly the charge of anthropocentric subjectivism in Heidegger's early thought. This critique centers on his strict human–animal distinction, transcendental view of time, and, most relevant here, his account of tools. Heidegger argues that human engagement with tools is the most fundamental way of relating to non-human entities, which reveals their Being.

Heidegger argues that metaphysics has incorrectly identified Being with the mere presence (*Vorhandenheit*) of objects, whereas the true essence of a thing is revealed in its handiness (*Zuhandenheit*). He explains that if we merely observe a tool without considering its functionality, its Being remains concealed. We might perceive its ontic features—such as material, color, and components—but it is only through its use that its true nature is disclosed. The ontological dimension of a chair, for instance, is revealed when we understand it as something to sit on (or occasionally to place clothes on, stand on, etc.). Thus, Heidegger describes the ontological structure of tools as something that exists "in-order-to" (*um-zu*) (Heidegger, 1962: 96–98).

This raises the concern of reducing beings to mere means, as the term "mere" itself suggests a depreciatory, reductionist view (Prauss, 1999: 35, 45; Zimmerman, 1990: 140). Extending the "in-order-to" structure (viewing things as existing for a purpose) to all non-human beings risks reinforcing a world entirely subordinate to human needs. However, Heidegger's exploration of the interdependence between use and need offers a compelling counterargument.

Two of Heidegger's most well-known works that address the question of use in relation to the complex semantics of *brauchen* are *The Anaximander Saying* (2002) and *What Is Called Thinking?* (1968) (Davis 2007; Merwin and Moore, 2023; Mitchell, 2023; Kleinberg-Levin 2005). Heidegger traces its etymology through its Greek and Latin roots, which allows him to argue that the original sense of "having at one's disposal" was grounded in "having the enjoyment of" the thing in question, eventually leading to a construal of this enjoyment as a "letting be" (Mitchell, 2023: 178). This forms the basis of Heidegger's elaboration on "proper use" (*Eigentlich Brauchen*), which he describes as that which "brings the thing to its essential nature" (Heidegger 1968: 187).

Using something, therefore, is not a depreciation of the thing but rather an affirmation, as it enables the thing's "presencing" (*Anwesen*). Usage is thus linked with the sphere of authenticity (*Eigentlichkeit*), as it is oriented toward what is proper (*eigentlich*) to the thing. This recalls Heidegger's comment in *Being and Time*, where he claims that, in its handiness, a thing "manifests itself in its own right" (Heidegger 1962: 98).

However, the concept of proper use dispels the anthropocentric undertone of this earlier idea: its basis lies in the act of being, not in the human being as *Dasein*. As Heidegger states, "Proper use… in general is not the business of mortals. Mortals are at best illumined by the radiance of use" (Heidegger 2002: 42; see Davis 2007: 227–230).

Heidegger's exploration of use (and need), expressed through *brauchen*-related terms, is scattered throughout his works. While he addresses this issue most intensively in the 1950s, it also appears in earlier texts like *Introduction to Metaphysics* and *Contributions to Philosophy* (1935–1936) (Merwin and Moore 2023: 104). However, these subtleties are often overlooked, especially in translations. A comparison of two English versions of *Contributions to Philosophy*—the older 1999 translation by Parvis Emad and Kenneth Maly (*From Enowning*) and the 2012 translation by Richard Rojcewicz and Daniela Vallega-Neu (*Of the Event*)—highlights this issue.

Both generally translate *brauchen* as "to need," though the appropriateness of this is debatable. What is especially notable are the differences in translating *gebrauchten*, the past participle of *brauchen*. For example, the original phrase: "Wie aber, muss dann das 'Nicht' (und das Ja) seine wesentliche Gestalt in dem vom Seyn gebrauchten Da-sein haben?" (*Beiträge zur Philosophie (Vom Ereignis)* (GA 65), p. 178) is translated by Emad and Maly as: "But how? Must not the 'no' (and the 'yes') have its essential form in the Da-sein that is used by Be-ing?" (*Contributions to Philosophy (From Enowning)*, p. 125).

Meanwhile, Rojcewicz and Vallega-Neu translate it as: "How is that possible, however, unless the essential form of the 'no' (and of the yes) resided in the Da-sein which is needed by Being?" (*Contributions to Philosophy (of the Event)*, p. 140).

Neither translation places the German verb in square brackets, allowing the subtle but significant ambiguity in this sentence to go unnoticed. This is problematic because the interpretive implications can vary greatly depending on whether we understand the phrase as "being used" or "being needed." Before delving into those implications, however, it is important to first explain the source of this discrepancy.

The genesis of the double meaning of *brauchen* as "to use" and "to need" in German—lacking a direct equivalent in many other languages (at least European ones), is rather complex. The most critical issue for us is the challenge of translating *gebraucht*, its derivative. On the one hand, *gebraucht* is a past participle (following the pattern *ge-brauch-t*), meaning "needed" (e.g., *Er war nicht da, als sie ihn gebraucht hat*—"He was not there when she needed him"). On the other hand, it functions as an adjective meaning "used" (e.g., *Dieses Auto ist gebraucht*—"This car is used," i.e., pre-owned). The dual meaning of *brauchen* is then most evident in its passive form, which creates the greatest ambiguity and will later be a crucial issue for our considerations.

Furthermore, a car that is "still in use" (though quite old) is described in German as *noch in Gebrauch*. Adding to the complexity, *der Brauch* means "custom," "convention," or "tradition" in the sense of repeated, meaningful practice rather than simply "need." Interestingly, English preserves a link between these meanings in the phrase "being used," which suggests the formation of a practice. Thus, in Heidegger's context, *Brauch* may also be rendered as "usage."

Without resolving the ongoing debates over translation (see Merwin and Moore 2023: 102–103; Mitchell 2023: 177; Polt 2023: 131), we should note that rather than employing distinct words for "use" and "need"—which German allows—Heidegger purposefully engaged with the polysemy of *brauchen*. He explored its meaning, stretched between "having need" (*nötig haben*) and "being taken into use" (*in Gebrauch nehmen*) (Marten 2018: 149), to the extent that even native German speakers sometimes find his writings ambiguous (Keiling & Moore 2022; Kettering 1987). He did so to reveal the underlying connection between these two phenomena.

The debate on *brauchen* in Heidegger's philosophy was reignited by the recent publication of his manuscript "The Argument Against Need (for the Being-in-Itself of Entities)" (*Das Argument gegen den Brauch (für das Ansichsein des Seienden)*, ca. 1963). As the title suggests, Heidegger examines the role of ecstatic Dasein in the disclosure of Being (Polt 2023: 125).

Scholars have linked his position to the broader debate on correlationism, influenced by Quentin Meillassoux (Keiling 2023), or more generally, to the issue of transcendental subjectivity in Heidegger's understanding of Being and time (Merwin and Moore 2023: 99).

These scholars agree that Heidegger rejects unidirectional dependence, instead proposing a multilayered interdependence between Being, entities, and human beings (Keiling 2023: 199). While he grants Dasein a special relationship with Being, this does not mean entities depend on humans for their existence (Merwin and Moore 2023: 103, 105). However, though Being is not created by Dasein, it does require human beings, in their ecstatic nature, for its disclosure—just as Dasein depends on Being to ground its essence.

This dependence would likely become even more complex if scholars were to engage more deeply with alternative meanings of *brauchen*-related phrases, as suggested in works already discussed in this paper, as well as in the recently published volumes of the *Gesamtausgabe*: *Vigiliae I, II, Winke I, II, Vorläufiges I-IV* (GA 100–102). In these texts, Heidegger leaves open to interpretation whether human beings are *used* in the Event of appropriation/adaptation or are *needed* for it (see GA100: 21; GA 101: 13; GA 102: 11).

The following analysis will not aim to resolve this tension, as doing so would go against Heidegger's insights. Instead, it seeks to explore their relationality, including the inherent tension and connection between need and use. The later section of this essay will shift these problematics to the realm of interactions between humans and nature. While overcoming subjectivism and anthropocentrism is crucial for eco-phenomenology, reexamining notions of inferiority and supremacy associated with need and use, along with activity and passivity in human–nature interactions, seems equally significant. In doing so, we will not reduce *Brauch* and *brauchen* in Heidegger to mere neediness or utility in their ordinary senses (see Polt 2023: 131). Instead, we will revisit the phenomena of need and use and highlight their broader significance.

Chiasmatic Dynamism of Use and Need

It has been noted that translating passages containing *brauchen*-related phrases can lead to significant controversies, as "need" and "use" suggest opposing power dynamics and vary greatly in their semantics. We will unpack these distinctions now and revisit them later.

At first glance, the meaning of "use" seems neutral ("to use a phone"), but upon deeper reflection, it takes on a more pejorative tone. This shift stems mainly from the assumption of a strict distribution of agency. Agency and activity are attributed to the user, whereas the object being used appears entirely passive. While we will complicate this view in the course of our interpretation, it becomes immediately clear when we consider the passive form of the verb—being "used." As the dictionary tells us, "what is used is often depreciated or diminished."

Moreover, when applied to persons, "use" often implies exploitation. This connotation is frequently linked to Immanuel Kant's imperative from *Groundwork for the Metaphysics of Morals*, which commands us to treat human beings as ends in themselves, not merely as means. However, a closer examination of the classical formulation reveals that the concept of *brauchen* lies at the core of the injunction itself, as expressed: "Act in such a way that you use humanity, whether in your own person or in the person of any other, always at the same time as an end, never merely as a means" (*Handle so, daß du die Menschheit sowohl in deiner Person, als auch in der Person eines jeden anderen jederzeit zugleich als Zweck, niemals bloß als Mittel brauchest*).

Interestingly, the translator chose to render the German word *brauchest* as "use," a choice that is often overlooked. Discussions of this Kantian dictum typically substitute the term "use" for "treat," likely due to the negative connotation of "use," which suggests the degradation of the used and the superiority of the user. As Pauline Kleingeld (2020: 399) astutely observes, Kant does not prohibit the use of persons *per se* but rather the act of using them *merely* as means, in a *purely* instrumental manner (Kerstein 2019; Kleingeld 2020). This distinction will later prove important when engaging with the concept of the intrinsic worth of natural beings, particularly when defined as non-instrumental value in environmental ethics.

Now let us underline that today the term "use" tends to be restricted to instrumental meanings, implying subordination and dominance, which traps it within a binary framework of inferiority (being used) and domination (using). Similarly, the semantics of "need" also fall prey to such dualistic conceptualization.

On the one hand, it is associated with a certain form of ennoblement. "Need" refers to a state of requiring something or someone because they are essential or very important, not merely because one desires them. We must distinguish needs from interests, wants, desires, or whims, which are less crucial. Need, then, is associated with the vital importance of something or someone, indicating that something or someone is necessary. It can also be added that someone is necessary, or needed, to accomplish or achieve something. In this case, there is an implicit instrumentality at play.

However, it's worth noting the inverse relationship between passivity and activity when compared to the phenomenon of use. In the case of need, being in the passive state of *being needed* carries a certain dignity, while the active state of *needing* points to vulnerability. While needing can be associated with demanding or requiring something, it fundamentally reflects a lack—an inability to do something without the help of someone or something. This paradoxical aspect of the active form of need is perhaps better captured in the phrase "being in need," which implies a condition that requires care.

Thus, needing tends to elicit compassion rather than admiration, in contrast to *being needed*, which often commands a higher regard. So in the case of need, passivity—typically viewed negatively—takes on a higher status than the active side of the phenomenon.

Observing this reversal of the status of activity and passivity in the contexts of use and need leads us to interpret Heidegger's juxtaposition of these terms through the figure of chiasmus, a concept whose significance has already been highlighted by Ted Toadvine. Chiasmus has broad appeal and much to offer eco-phenomenology, particularly in its attempts to rethink traditional binaries (such as mind and body, intelligible and sensible, human and animal, experience and nature) and overcome the dualistic heritage of Western thought for environmental thinkers (Toadvine 2017). For our purposes, this concept is key to revisiting the binarism of use and need in how human beings relate to non-human beings.

In brief, the term *chiasm* is a shortened form of two distinct words: one physiological (*chiasma*) and one rhetorical (*chiasmus*). A *chiasma* refers to the point where anatomical structures cross, such as the optic nerves meeting the brain, while *chiasmus* is a rhetorical figure involving inverted parallelism. Crucially, the figure of the chiasm exemplifies a general structure of mediation that operates across various relationships and levels of complexity, avoiding both simple opposition and synthesis (Toadvine 2017).

Chiasmus is often achieved by inverting parallel structures of assertions and negations, or activity and passivity. As such, it typically results in four elements (A+, A−, B+, B−). Heidegger's engagement with use and need, where he emphasizes their passive aspects, inspires us to reconstruct the following chiasmus: using, being used, needing, and being needed.

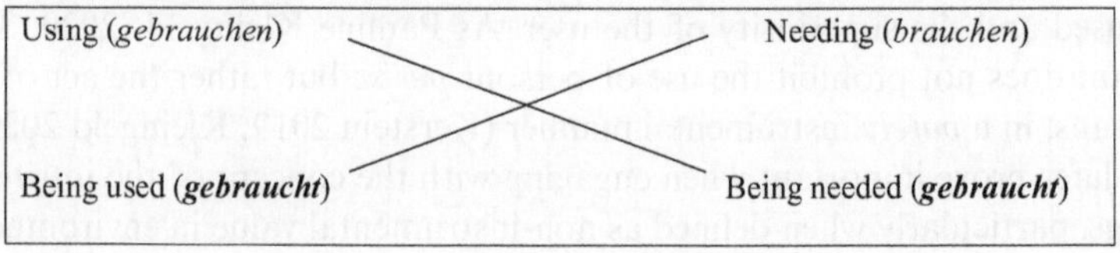

Sketching such a chiasmus scheme enables us to explore three types of relations: vertical, horizontal, and diagonal. While all are relevant, the specificity of both chiasmus and chiasma often lies in their focus on diagonal relations, as these tend to reveal the most intriguing dynamics. In our case, this involves examining the diagonal relationships between "needing" and "being used" as well as "using" and "being needed." However, Heidegger's engagement with the ambiguity of *brauchen*, culminating in a special focus on the word *gebraucht*, and his privileging of passivity in his ontology (see the other contributions in this volume), invites us to take the bottom axis—juxtaposing "being used" with "being needed"—as the starting point for reexamining the relations between use and need.

The role of chiasmus is not merely to pair phenomena but to collate or set them against each other. This is not limited to contrasting or opposing them, but rather involves comparing them and ultimately identifying the connections that span their differences. Such an endeavor can be traced in Heidegger's exploration of the ambiguity of *brauchen*, which, at first glance, refers to need and use as opposing phenomena. Interestingly, Heidegger allows us to start from the fundamental unity of these two sides, reflected in the simple fact that they are expressed by the same word as a specific form of *brauchen*, that is *gebraucht*. This allows us to interpret the axis in question as follows: "the one used is the one needed." Such a reading shifts the perspective on the subject (not the agent) of use, pointing to their indispensability and revealing a previously discussed layer of vital significance. It moves from a sense of depreciation to one of appreciation, reversing the standard account from a state typically seen as diminishing toward one of admiration.

One could, however, validly argue that the axis could also be read as: "the one needed is the one used," which might disclose a negative, instrumental undertone to need. Yet, I shall argue that while this reversal can be identified in Heidegger, its goal is different. It points toward the reversibility and interchangeability of need and use, revealing their originary inseparateness.

Heidegger hints at this reading in *What Is Called Thinking?*, where he writes:

> Proper use (*Eigentlich Brauchen*) is neither a mere utilizing (*bloßes Benützen*), nor a mere requiring (*noch aber auch nur ein Benötigen*). What we merely require, we utilize from the necessity of a demand (*Bedarfes*). Utilizing and requiring always fall short of proper use. (…) The essential nature of use can thus never be adequately clarified by merely contrasting it with utilization and requiring.
>
> (*Was heisst Denken?* (GA 8), p. 190; Heidegger 1968, 187)

The core of Heidegger's claim is that the relationship between use and need cannot be understood by considering them separately. Focusing solely on use, when it transforms into mere utilization, or on need, when it becomes an arrogant demand, misses the point. Nor can we reduce one to the other; they mutually belong to each other. The tension between them reveals their connection. In other words, not only are use and need highly relational in themselves (see Mitchell 2023: 177), but the inherent relationality between them also defines their essential character.

The challenge Heidegger sets for us, or the key to grasping this non-reductive, indivisible relationality, is to think and perceive both aspects of *Brauchen*—use and need—*simultaneously*. This presents a challenge for phenomenology, and consequently for its ecological application, as the science of understanding or seeing: how can we keep both aspects in view at once?

In addressing this question, we must first acknowledge the dynamism within the tension between use and need. Referring to the idea of "mirror-play" within the structure of the Fourfold could be helpful. This can be seen as a Heideggerian application of chiasmus, one more dynamic than traditional forms. Here, the focus is not on what the four elements (earth, sky, mortals, and divinities) symbolize in terms of the disclosure of being, but rather on the mirror-play through which they are tied together. This "mirroring" or "reflecting" (*Spiegeln*), as scholars explain, is Heidegger's term for the process through which each dimension becomes intelligible by reflecting the others. For instance, the sky is only intelligible as such through its interaction with the earth, striving to be the earth it is (or in terms of the mortal activities it blesses or restricts) (Wrathall 2021b).

The interpretive potential of mirroring as a dynamic interplay—suggesting that some matters can only be understood through ongoing confrontation rather than through static connections—sheds new light on the inherent relationship between use and need. This perspective invites us to reconsider their interdependence and the dual nature of our relationship with beings, necessitating a continuous movement back and forth between use and need. It calls for a circular approach to explore this dynamic. It is not merely a matter of recognizing that they are two sides of the same coin; rather, it emphasizes that one cannot be fully understood without the other, and their relationship is not unidirectional.

Examining the relationship from the perspective of use highlights the significance of the entity being used, revealing that it ultimately fulfills a need. However, moving from need to use does not necessarily imply derogation. Instead, it can serve to temper excessive desires, showing that what was once needed often becomes used—or even used up (Ahmed 2019: 53–54[1]). This reflection serves as a sober reminder of consequences, which might act as a limitation or restraint on our needs.

This brings us back to a previously raised question: should we feel guilty for watering the flowers? To address this, we must first consider how we frame our relationship to the problem. Is it a matter of *using* or *needing* water? The choice between these perspectives shapes the narrative. One could say, for instance, that we use water to shower, to drink, or to water flowers in the garden. Alternatively, one could say that we need water for these purposes. The next question then becomes: why do we use or need water? The answer is that we use it because we need it. But why do we need something? Because without it, we couldn't manage—reflecting the semantics of need as something vital.

Understanding needs as something that pertains to what remains outstanding (see Mitchell 2023) allows us to uncover the obscured dependencies between humans and nature. It highlights our lack of self-sufficiency and our reliance on the natural world. This resonates with Keith Peterson's (2020) argument, drawing on Val Plumwood's work, to move from a "logic of domination" to a "principle of dependence." This view emphasizes that humans are not superior to nature but are deeply reliant on it. Acknowledging this dependence is essential for transforming our relationship with the natural world.

In rejecting the idea of subordinating the Earth, this approach transforms our relationship with nature into one of admiration, where we recognize nature not as a resource to be controlled but as a source of life. We turn to nature because it offers something we lack—essential elements for our survival and well-being. This shift in perspective allows us to see that we are not separate from or above nature; rather, we are intrinsically bound to it. Our profound connection to nature is reflected in our deficiencies, revealing that what we often perceive as human limitations actually highlight our interdependence with the natural world.

But does need automatically justify use? Does every need warrant its corresponding use? To explore this, we might consider the interchangeability of need and use as a way of assessing whether a particular use is justified: Did I use something because I truly needed it? Was my need genuine? What kind of need was it satisfying, and what lack or deficiency was it intended to resolve? If my answers reveal a disconnect between use and need, this might indicate that my use was improper.

The passage above highlights the importance of using the active voice when discussing use. Grammatically, passive voice often serves to conceal the subject, and in doing so, it can obscure the agent responsible for the action. In other words, by relying on passive voice, we risk hiding our own role, making it unclear that it was we who used or consumed something. By contrast, using the active voice makes the relationship between the agent and the action more transparent, emphasizing that our choices have direct consequences.

Bouncing back to the passive voice, however, highlights another of its advantages, one that contrasts with the earlier claim. When we consider not only ourselves as users of natural beings but also reflect on the fact that these beings are *used by others*. While this may seem obvious or even naive, we often forget that non-human beings also make use of natural resources. In the example of watering flowers, for instance, the focus shifts to the flowers themselves, reminding us that they too are part of this dynamic, not just humans.

The point is not to lessen human responsibility for using or depleting resources, but rather to emphasize that we share these resources with other living beings. Natural resources are used because they *are needed by non-human beings* as well. Just as humans depend on nature, so do other creatures, relying on what the Earth provides.

This leads to a broader claim, offering yet another inversion of the typical passivity and activity associated with the phenomenon of use. Instead of viewing the user as the active agent, the focus shifts to the thing itself, which offers itself up for use. This interpretation resonates with Mark Wrathall's recent translation of *Bewandtnis* as "affordance," suggesting that objects present themselves in a way that allows them to be used.

Bewandtnis has traditionally been translated as "involvement," "relevance," "functionality," or "deployment." However, Wrathall notes that in its archaic usage, *bewenden* meant "to use" or "employ" something. The prefix *be*-likely conveyed the sense of "supplying" or "endowing." Thus, *bewenden* originally implied "to supply, endow, or offer something for use." The passive construction (*bewenden lassen*) suggests that it is the entities in the world themselves that offer or supply their use, creating opportunities for us to act and alter circumstances. Wrathall, therefore, proposes translating *Bewandtnis* as "affordance," borrowing from J.J. Gibson's concept (Wrathall 2021a, see Blok 2014).

An affordance refers to the meaning of an object or organism in the environment that is perceived by an observer, enabling specific actions—air affords breathing, a chair affords sitting, and so on. An affordance is neither a purely "subjective" quality of an object nor an entirely "objective" property detached from any user (Blok 2014). It transcends this subjective–objective divide, as well as dualism of activity and passivity, revealing the limitations of such dichotomies. Moreover, it does not carry the instrumental undertone of the phrase "in-order-to" (*um-zu*), shifting the focus toward giving or offering rather than applying something for specific goals.

Thus, if we combine reading *Bewandtnis* as "affordance" with interpreting the state of being used (*gebraucht*) as the confirmation of being needed (*ge-braucht*), or as "being used out of need" (and not out of demand) we arrive at a significantly less anthropocentric worldview. Human beings are not self-sufficient and exclusively active agents, but rather, dependent on what other beings offer, or afford them. This renders invalid the dichotomy of superiority and inferiority underlying the traditional account of relations between humans and non-humans. This enables us to think through the claim "the one used is the one needed" as a valid thought, rather than see it merely as a wordplay.

This encourages us to complicate the concept of intrinsic value and underscore its core aspects. While it is true that we cannot reduce beings to mere means, we must also acknowledge their significance, or vitality, for other beings. In other words, while we should respect their inherent value or telos, we cannot separate them from the intricate networks of affordances and needs in which they exist. Recognizing this interconnectedness invites us to appreciate the dynamic relationships that shape the lives of both human and non-human entities. Ultimately, a deeper understanding of intrinsic value requires us to engage with the complexities of these relationships, fostering a more holistic perspective on our place within the broader ecological web.

Conclusions

This chapter has explored eco-phenomenology as a lens through which we can examine our experiences with the natural world, emphasizing the inseparability of use and need. It draws upon Martin Heidegger's insights. Heidegger emphasizes the importance of being directly engaged with our surroundings and seeing them as useful, that is *zuhanden*) rather than merely contemplating them as present (*vorhanden*). This engagement compels us to recognize that we always relate to things "as-something," which highlights the complex interplay between our perceptions and experiences.

Contradictions arise when we attempt to reconcile different perspectives, for instance when we see other beings "as-used" or "as-needed." Heidegger's exploration of these two phenomena, particularly through the semantics of *brauchen*, encourages us to move beyond binary interpretations of these relationships. Instead, we are invited to appreciate how needing and using are mutually conditioned. Our interactions with the world reveal that while use is often accompanied by need, neither state can be fully understood in isolation. As Polt (2023) notes, neither phenomenon is solely active nor solely passive in these dynamics.

The relationship between need and use is multifaceted, resembling a chiasmus that reflects a more intricate reality than a simple coin analogy. This multidimensionality arises from the interplay of both passive and active elements of two phenomena that we too often tend to consider separately. As we engage with this complexity, we are reminded by both Heidegger and eco-phenomenology to reverse traditional logic by foregrounding passivity. Rather than focusing solely on what we need or use, we must also examine the beings we use or need. In doing so, we can reclaim the affirmative potential inherent in these relationships.

By shifting—or constantly circulating—between viewing nature through the lens of use or need, or as being used or needed, both human hubris and guilt can be tempered. In their place, we can cultivate a sense of gratitude toward nature and transform feelings of alienation into a sense of belonging. Recognizing our needs and vulnerabilities highlights our interconnectedness with the Earth, reinforcing the idea that we are an integral part of a greater ecological whole. This shift toward a less anthropocentric worldview reveals that humans are not entirely self-sufficient or solely active agents; rather, we rely on what other beings provide. This perspective challenges the conventional dichotomy of superiority and inferiority in human–non-human relations, fostering a sense of gratitude and belonging rather than alienation and subordination. By embracing this understanding, we can transform our interactions with nature, nurturing a more harmonious and reciprocal relationship that celebrates our shared existence within the natural world.

Note

1 Sara Ahmed's extremely interesting book on use (Ahmed 2019) and her engagement with Heidegger's scholarship (2006) might prompt one to wonder why I do not engage more extensively with her work. The main reason is that she does not explore the dynamics of use and need, which are central to my paper.

References

Agamben, Giorgio. 2016. *The Use of Bodies* (Homo Sacer IV, 2), trans. Adam Kotsko. Stanford: Stanford University Press.

Ahmed, Sara. 2006. *Queer Phenomenology: Orientations, Objects, Others*. Durham, NC: Duke University Press.

Ahmed, Sara. 2019. *What's the Use? On the Uses of Use*. Durham, NC: Duke University Press.

Blok, Vincent. 2014. Being-in-the-World as Being-in-Nature: An Ecological Perspective on Being and Time. *Studia Phænomenologica* XIV: 215–235.

Blok, Vincent. 2020. *Heidegger's Concept of Philosophical Method Innovating Philosophy in the Age of Global Warming*. New York: Routledge.

Davis, Bret. 2007. *Heidegger and the Will. On the Way to Gelassenheit*, Evanston, IL: Northwestern University Press.

Deutsches Wörterbuch von Jacob Grimm und Wilhelm Grimm n.d. online: https://woerterbuchnetz.de

Dreyfus, Hubert. 2002. Heidegger's History of the Being of Equipment. In *Heidegger: A Critical Reader*, ed. H. Dreyfus. Oxford: Hall, Basil Blackwell.

Guidi, Lucilla. 2021. As-structure (Als-Struktur). In *The Cambridge Heidegger Lexicon*, ed. M. Wrathall, 64–67. Cambridge: Cambridge University Press.

Hailwood, Simon. 2015. *Alienation and Nature in Environmental Philosophy*. Cambridge: Cambridge University Press.

Heidegger, Martin. 1962. *Being and Time*. Translated by John Macquarrie and Edward Robinson. New York: Harper and Row.

Heidegger, Martin. 1968. *What Is Called Thinking?*. Translated by F.D. Wieck and J.G. Gray. New York: Harper and Row.

Heidegger, Martin. 1999. *The Fundamental Concepts of Metaphysics. World, Finitude, Solitude*. Translated by William McNeill and Nicholas Walker. Bloomington, IN: Indiana, University Press.

Heidegger, Martin. 2002. Anaximander's Saying. In idem, *Off the Beaten Track* (pp. 242–281). Translated by Julian Young and Kenneth Haynes. Cambridge: Cambridge University Press.

Heidegger, Martin. 2012. *Contributions to Philosophy (Of the Event)*. Trans. Richard Rojcewicz and Daniella Vallega-Neu. Bloomington: Indiana University Press.

Heidegger, Martin. 2020a. *Vigiliae und Notturno (Schwarze Hefte 1952/53 bis 1957) (GA 100)*. Frankfurt am Main: Vittorio Klostermann.

Heidegger, Martin. 2020b. *Winke I und II (Schwarze Hefte 1957–1959) (GA 101)*. Frankfurt am Main: Vittorio Klostermann.

Heidegger, Martin. 2022. *Vorläufiges I–IV (Schwarze Hefte 1963–70) (GA 102)*. Frankfurt am Main: Vittorio Klostermann.

Keiling, Tobias. 2023. Being before Time? Heidegger on Original Time, Ontological Independence, and Beingless Entities. *Gatherings: The Heidegger Circle Annual* 13: 199–233.
Keiling, Tobias, Ian Alexander Moore. 2022. Heidegger on deep time and being-in-itself: introductory thoughts on "The Argument against Need". *British Journal for the History of Philosophy*, 30 (3): 508–518.
Kerstein, Samuel. 2019. "Treating Persons as Means", in: *Stanford Encyclopedia of Philosophy* (online: https://plato.stanford.edu/entries/persons-means/)
Kettering, Emil. 1987. *Nähe. Das Denken Martin Heideggers*. Pfullingen: Neske.
Kleinberg-Levin, David. 2005. *Gestures of Ethical Life: Reading Hölderlin's Question of Measure After Heidegger*. Stanford: Stanford University Press.
Kleingeld, Pauline. 2020. How to Use Someone 'Merely as a Means'. *Kantian Review*, 25 (3), 389–414.
Kūle, Maija.2018. Eco-Phenomenology: Philosophical Sources and Main Concepts. In *Eco-Phenomenology: Life, Human Life, Post-Human Life in the Harmony of the Cosmos*, eds. Daniela Verducci, Jadwiga Smith, William Smith. Cham: Springer Verlag, 43–58.
Lewis, Helen. 1971. Shame and guilt in neurosis. *Psychoanalytic Review*, *58*(3), 419–438.
Marten, Rainer. 2018. *Denkkunst: Kritik Der Ontologie*, Freiburg im Breisgau: Verlag Karl Albert.
McManus, Denis. 2013. *Heidegger and the Measure of Truth. Themes from his Early Philosophy*. Oxford: Oxford University Press.
McManus, Denis. 2014. *Introduction In Heidegger, Authenticity and the Self, Themes From Division Two of Being and Time*, red. D. McManus, New York: Routledge.
Merwin, Christopher D., Ian Alexander Moore. 2023. Heidegger's Legacy and the Need/Use of Being. *Gatherings: The Heidegger Circle Annual* 13: 98–123.
Mitchell, Andrew. 2023. On Need: Mortality and Heidegger's "Argument" with Science. *Gatherings: The Heidegger Circle Annual* 13: 173–198.
Nielsen, Rikke S., Christain Gamborg, B. Lund Thomas. 2024. Eco-guilt and Eco-shame in Everyday Life: An Exploratory Study of the Experiences, Triggers, and Reactions. *Frontiers in Sustainability*, 5: 1–15.
Peterson, Keith R. 2020. *A World Not Made for Us. Topics in Critical Environmental Philosophy*. Albany, NY: State University of New York Press.
Pfeife, Wolfgang et al., ed., n.d.. *Etymologisches Wörterbuch des Deutschen* online: https://www.dwds.de/wb/Gebrauch
Polt, Richard. 2012. *The review of Contributions to Philosophy (Of the Event)*. Notre Dame Philosophical Reviews. https://ndpr.nd.edu/reviews/contributions-to-philosophy-of-the-event/
Polt, Richard. 2023. On the Right to Kick a Rock: The Argumentum ad Lapidem and Heidegger's "The Argument against Need". *Gatherings: The Heidegger Circle Annual* 13: 124–145.
Prášek, Petr. 2023. Ecology, Eco-phenomenology, and the Immanent Ethics of Nature. *Meta: Research in Hermeneutics, Phenomenology, and Practical Philosophy*, XV (2): 342–366.
Prauss, G. 1999. *Knowing and Doing in Heidegger's Being and Time*, New York: Humanity Books.
Thomson I. 2004. Ontology and Ethics at the Intersection of Phenomenology and Environmental Philosophy, *Inquiry*, 47 (4): 380–412.
Toadvine T. 2003. Introduction. In *Eco-phenomenology. Back to the Earth Itself*, eds. C. Brown, T. Toadvine. Albany: State University of New Yok Press.
Toadvine T. 2017. Phenomenology and Environmental Ethics. In *Oxford Handbook of Environmental Ethics*, eds. S.M. Gardiner, A. Thompson. Oxford: Oxford University Press.
van der Heiden, Gert-Jan. 2020. On Use and Care: A Debate between Agamben and Heidegger. *International Journal of Philosophy and Theology*, 81 (3): 310–327.
Wrathall, Mark. 2021a. Affordance. In *The Cambridge Heidegger Lexicon*, ed. M. Wrathall, 31–33. Cambridge: Cambridge University Press.
Wrathall, Mark. 2021b. Fourfold (Geviert). In *The Cambridge Heidegger Lexicon*, ed. M. Wrathall, 335–341. Cambridge: Cambridge University Press.
Zimmerman, Michael. 1990. *Heidegger's Confrontation with Modernity: Technology, Politics, and Art*. Bloomington: Indiana University Press.

SECTION 2

Passivity and Perception

6

EARTH UN-EARTHED

Total Solar Eclipses as *Sur-Réflexion* on the Earthborne Topology of Flesh, with Remarks on Transcendental Geology

David Morris

6.1 Bringing Phenomenology and Science Down to Earth

Adam Frank, Marcelo Gleiser, and Evan Thompson's recent book *The Blind Spot: Why Science Cannot Ignore Human Experience* (2024) is inspired by Husserl's argument that science surreptitiously replaces experience with the mathematical abstractions of science. This places our experience in science's "blind spot" and obscures the experiential beginning points of science. This leads to various problems within science, including climate science, as they show in their penultimate chapter, which focuses on the methodological and conceptual consequences of placing our experience of Earth in science's blind spot. Their point about the Earth vanishing in science's blind spot is important for climate science and eco-phenomenology.

For eco-phenomenologists who draw on Merleau-Ponty to challenge this blind spot, the conceptual–methodological problem is even worse. Eco-phenomenology seeks *phenomenological* insights into our relations to Earth, to advance ecological science and responses to global heating. We must begin from within *human* experience of Earth and arrive at a *science* of Earth. Isn't this eco-phenomenological beginning point hopelessly anthropocentric? Wouldn't Earth remain hidden in the unsurpassable blind spot of our anthropocentric, egoic, beginnings? Eco-phenomenology has developed strategies for addressing this challenge. Ted Toadvine (2024) and David Abram (2011), for example, have shown how our human beginning point can turn into an anthropocentric blind-spot. They do this by showing what happen, if we acknowledge our evolved, animal being, and the ways that opens us to experiences of fossils, stones, places, animals, trees, which offer insights into our experience being interlaced with the planet, versus sealed up inside of us. As Merleau-Ponty might say, they bring philosophy and science down to Earth.

I attempt to bring philosophy and science down to Earth by the paradoxical move of looking *up*, going cosmic. But this was not *my* idea: the method struck me forcefully, from above, during a total solar eclipse. Admittedly, I was prepared for this by long study of Merleau-Ponty, including ever greater appreciation for how much his highly original reading of Husserl is crucial to his philosophy—and how Husserl's *Urmsturz* text, "Foundational Investigations of the Phenomenological Origin of the Spatiality of Nature,"[1] is at the heart of his reading.

 DOI: 10.4324/9781003538097-8

Following these methodological pointers, I seek to unearth a new sense of Earth, for phenomenology and science, by interweaving conceptual and methodological resources from Husserl, Merleau-Ponty, and solar eclipses.

6.2 Methodological Remarks on the Eco-Phenomenologist and Their Earth Shadow

This new sense of Earth is hard to glimpse, name, or conceptualize within the phenomenological tradition—and perhaps in any tradition. We will see that it calls for new methods and ontology. Merleau-Ponty foreshadows these issues in *The Visible and the Invisible* (*VI*), in the section "Science Presupposes the Perceptual Faith and Does Not Elucidate It." He writes that

> Philosophy is not science, because science believes it can soar above its object and holds the correlation of knowledge with being as established, whereas philosophy is the set of questions wherein he who questions is himself implicated by the question.
>
> (*VI* 47/27[2])

Husserl's *Urmsturz* illustrates this with the case of Earth: it appears that we can only get Earth in view by soaring far above it, or that the human world engulfs the Earth, subsuming it as the constitutive ground of our sense of movement. We keep substituting something else for Earth, as if *we* can say what it is from above, instead of grasping Earth as grounding and opening questions of being from within us.

Notably, Merleau-Ponty finds sciences that run counter to "soaring-science." "[A] physics that has learned to situate the physicist physically" (e.g., quantum mechanics) and "a psychology that has learned to situate the psychologists in the socio-historical world" (e.g., developmental psychology) "have lost the illusion of the absolute view from above: they do not only tolerate, they enjoin a radical examination of our belongingness [*appartenance*] to the world before all science" (*VI* 48/27). An astrophysics, cosmology, astrobiology or geology, that situates the scientist within the sphere they study (this should go without saying with these sciences!), would likewise open, within science, an approach to our belongness on Earth "before all science."

These points connect with Merleau-Ponty's remarkable essay "The Philosopher and His Shadow," which wrestles with Husserl's *Umsturz*. Toward the end, he writes that "The philosopher must bear his shadow, which is not simply the absence of future light" (1964b, 178). This "shadow" is an inborn pointer to a generative passivity that cannot be fully lit by active philosophical efforts. This echoes his indirect ontology, wherein the visible is always shadowed by a generative invisible in its depths. On this view,

> the transcendental field has ceased to be simply the field of our thought and has become the field of the whole of experience, and that Husserl trusts the truth that we are *in* from birth and which ought be able to contain both the truths of consciousness and the truths of Nature.
>
> (1964b, 177)

When the Moon's shadow sweeps over our bodies, it eclipses and brings to light our sense of belonging on Earth. The philosopher's shadow flares outward in the Moon's umbra, revealing a further penumbral shadow of philosophical activity, our "strange belonging" to Earth, as a generative passivity that "we are *in* from birth." This offers a peri-phenomenological[3] opening, through which eco-phenomenology can glimpse something beyond anthropocentric limits. But this also flares the transcendental field outward into what Merleau-Ponty calls a "transcendental geology" (*VI* 312/258-9). Within transcendental geology, blind spots are not failings of phenomenology but its inborn ontological condition: we are in the cosmos, we cannot get out of it. That is our most comprehensive transcendental, yet generative, blind spot. What is the most anthropocentric of all is a philosophy that would claim to surpass all blind spots and any anthropic beginnings.

I proceed by outlining the problem in relation to ongoing discussions in eco-phenomenology. This leads into an exposition of Husserl's *Umsturz* text, through Merleau-Ponty's engagement with it, then my phenomenological study of a total solar eclipse, and concluding, brief remarks on Merleau-Ponty's gnomic note on transcendental geology.

6.3 Earth and the Logic of Planar Immovable Regions

Phenomenology follows the phenomena in order to notice, analyze, and sometimes create conceptual distinctions adequate to phenomena. The conceptual distinctions I am noticing here were unnoticed and unimaginable to me until I experienced the eclipse, and seem to me unnamed and ill-conceived in the languages I know. But this is no accident. In *Creative Evolution* (2023), Bergson notes that evolution gears our perception and language to a "logic of solids," of manipulating blocks of delimited stuff, which reduces indivisible change and duration to mere successions of "immobiles." Similarly, languages I know gear us to a "logic of planar immovable regions": of navigating as finite creatures within relatively delimited regions experienced as flat planes. We move on earth, soil, or ground, as immovable stuff. We then use those terms to name our planet: earth turns into the proper name Earth, *terra* into Terra, a pervasive pattern over many languages Even the word "planet" is a problem: it names the wandering of *distant* celestial bodies that move around ours. But the planetary-body on which we stand no more wanders from us than does our own body.

Languages I know lack good names for our planet as an entire, lived sphere, that is more than just an immense spread of earth or ground beneath our feet. Hence my conceptual-poetic coinage of various words such as *toposphere* (place, *topos*, as a sphere surrounding us, in analogy with "atmosphere"); *Earthscape* (the toposphere as the Earth's unique landscape of *paysage*); *Ge* (as the name of our unique planetary body, studied by geology); and, finally, *Geosphere*. In geology, *geosphere* has divergent usages, one naming spheres of rock, water, air, ice, as a unity. I use *Geosphere* to name all of that as the unique orb of life in which we are enveloped, as a liminally experienced toposphere–Earthscape.

6.4 Radicalizing Phenomenological Reflection, from Deep Time to Deep Geosphere

In her book *Timefulness: How Thinking Like a Geologist Can Help Save the World* (2018), geologist Marcia Bjornerud notes that on our time-scale, it appears we are changing a climate that Earth *has*. But geological time—"deep time"—reveals that climate is a historically contingent phenomenon. The Earth didn't start with an atmosphere. That was generated by eons of meteoric bodies beating the hell out of our planet until eventually life emerged, ate the iron and rocks, and belched an atmosphere, which alone allows climate. Life has *always* been changing climate. The attitude that Bjornerud calls "timefulness," which attends to this deep timescale of Earth, thus offers an importantly different way to think about climate change.

Toadvine's *The Memory of the World: Deep Time, Animality, and Eschatology* (2024) develops an eco-phenomenological counterpart, in response to Quentin Meillasoux's (2010) charge that phenomenology remains sealed within its anthropic beginnings and can never get in touch with the "Great Outdoors." Meillasoux's charge famously appeals to what he calls the "archefossil," immensely old things, for example, Bjornerud's Earth before it had an atmosphere or humans. He argues that phenomenology subsumes the archefossil to human experience, here and *now*, engulfing deep time in the shallow time of phenomenological reflection. Toadvine upends Meillasoux by careful describing actually touching a real fossil, showing how this opens us to the deep time worked into, or indeed "lived through," by the fossil.

Toadvine advances his point through Merleau-Ponty's project of radical reflection, inaugurated in the *Phenomenology*, which doesn't simply reflect on objects from within experience, but reflects on the roots and *radix* of the givenness of reflection itself. In the *Visible and the Invisible* radical reflection deepens into what Merleau-Ponty calls *sur-réflexion* (*VI* 61/38, translated as "hyper-reflection"), which reflects over, back, under, and beyond the roots of reflection, onto its ontological condition of possibility. This requires an interrogation that operates within the very being on which *sur-réflexion* reflects. Radical reflection can only turn into *sur-réflexion* if it "keeps its feet in being," if it proceeds on top, against, toward, and over this very being (the very opposite of the soaring *pensée survol*!). (This "on/over-reflection" is well expressed by *sur-réflexion* in French, which might be translated as "over-reflection," to indicate it goes "over the top" with reflection, in the way that the *sur-réel* overtops the real into the hyperreal, surreal. But it's best to leave it untranslated).

Sur-réflexion is a method that admits that phenomenology is not a self-authorizing, self-activating, cogitative project: phenomenology is itself a *phenomenon*, an appearing, open to and contingent upon appearings and their history, to a generative passivity. As Merleau-Ponty puts it in the *Phenomenology of Perception* (*PhP* 424[4]), phenomenology must always turn into a "phenomenology of phenomenology."

Toadvine's openness to the fossil reveals that we are made here on this planet from and through the same stuff and process as the fossil. It is just that the fossil animal's *living* didn't make it through deep time into now. Only its fossilized life-remains did. We feel that deep-time affiliation open in our touch, now, because we are born within the same Geosphere. Contra Meillasoux, *our experience is open from within to deep time*. We could call this "*sur-temporality*": deep time as a near surreal experience.

Toadvine is tracking a further radicalization of Merleau-Ponty's continual emphasis on the openness of history and ontological openness, on an internal, indeterminate, open generativity, as invisibly inborn and key to being. I take that further by tracking this into an openness to geological and cosmological history. Phenomenology of phenomenology must open to the deep time in which phenomenology can be born and this opens into the Geosphere on which life evolves and on which we are born and borne.

6.5 Our Strange Belonging on Earth and the Shadow of Passivity

The problem is that our openness to Geosphere is engulfed in the Blind Spot. Many have noted how the "Blue Marble" photo, of Earth seen from space, offered a previously unavailable experience of our relation to our fragile planet.[5] The need for this outer space view of Earth implies that our standing within it as tiny creatures badly places us to experience our planet as such (like being a body badly places us to perceive our bodies as objects). This strangely echoes Meillasoux's correlationism charge. Call this the macroplanet problem: we are too small to experience Geosphere in its own deep bigness, in the way that we're too short-lived to experience the archefossil in its own deep time. I contend that this view is prejudiced by an inadequately radical reflection on what experience is—and that eclipses can counter this prejudice.

Meillasoux presumes phenomenologists are *obviously* sealed within themselves and have no affiliation with nature beyond us. This is what Toadvine challenges by touching the fossil. In related work, Toadvine studies what Merleau-Ponty calls our "strange kinship" with other animals. A Meillasoux might claim that our human conception of other animals remains hopelessly sealed "indoors" of human experience because we are so utterly different from animals. As Toadvine shows, this presumes what it claims to demonstrate. Indeed, Merleau-Ponty subtly and profoundly challenges such claims by arguing that the source of our difference from other animals just is our strange way of *being animal*. "Indoors of ourselves," we are, in fact animals! We are kin with them, but in a strange way, that can lead us to deny our animality—but also open to embrace it.

Meillasoux's shallow and presumptuous "indoor phenomenology," barricaded against time, animals, and Earth, flies in the face of Steinbock's argument (discussed below) that for Husserl our experience internally opens to what is beyond its limits. I am extending this openness to Geosphere. Our being bears a "strange belonging" in/on/over/against/toward Earth: we belong *sur terre*, in the strange mode of leaving it behind, forgetting it, moving away from it. We so much belong to Earth that we subsume it into an abstract, immovable ground of motion, or otherwise obscure its immensity. But it is precisely our long, evolutionary belonging here that, paradoxically, evolves ways of easily moving and navigating Earth… that obscures that it is an immense *Geosphere* on which we belong. It's like not noticing the glasses on your head because they are so integral to looking for anything at all.

Methodologically, the problem is noticing our strange belonging within Geosphere. Merleau-Ponty remarks something similar regarding our experience "up" versus "down": this so basic that "in the everyday course of life" the basis of up and down "is already concealed beneath its own acquisitions. We must look to some exceptional case in which it breaks down and rebuilds itself before our eyes." (*PhP* 291) In *The Sense of Space* this led me to study the way up and down break down and rebuild in the experience of weightlessness by astronauts and others. This revealed that the topology of our body schema opens us to and crosses over with Earth as a larger place, "excorporating" Earth as part of our bodies (Morris 2004, ch 5).

Below, I suggest how eclipses invite a revelatory *sur-réflexion* that breaks down and rebuilds our strange belonging on Geosphere in the first place. To illustrate by briefly anticipating a result: when you are jet-lagged, you feel you are not from the time-zone that you're now in. You're from some other time, to which you are passive, as an abiding temporal ghost fossilized in bodily rhythms you carry with you when jetting away from home. When you enter totality, flaring through the shadow of the Moon, the strange mutations of light that ensue visually rocket you through "planet-lag": things look strange, you strangely *feel* you are not from the lit-planet that you are now in. You don't belong *here-now*, you're from some other planet. But *that* home planet is *strangely felt* within the other, weird one you now inhabit. Geosphere thereby steps forward from "its own acquisitions," from the blind spot, revealing itself as that to which you are abidingly passive: an inbiding placial ghost, fossilized in your ways of seeing and looking, shadowing you always. (This is in analogy with Toadvine and Abrams attending to experiences that reveal the animality that is strangely and always felt within us).

6.6 Bringing Husserl's *Umsturtz* Text Back Down to Earth

I approach these phenomenological and ontological points returning to Husserl's "Foundational Investigations of the Phenomenological Origin of the Spatiality of Nature: The Originary Ark, the Earth, Does Not Move." This is called the *Umsturz* text because it "overthrows" the Copernican world-picture of the Earth as a planet moving around the Sun, and associated Galilean and Newtonian frameworks in which motion and rest are merely relative terms.[6] It's worth returning to because it is a rich text that sometimes is taken too much for granted. (Also, if you find Merleau-Ponty claiming something about Husserl that sounds unorthodox, I'm prepared to bet he's explicitly or implicitly referring to the *Umsturz*).

Husserl overthrows Copernicus by showing how Earth is the abiding ground and basis of our experience of concrete motion and spatiality. Husserl calls this *Erdboden*, which I'll call *grounding-earth*. Earth is usually our *Erdboden*. But Husserl conducts imaginative variations on experience, leading him to argue that when travelling on a spaceship moving away from Earth, that spaceship "ark" would serve as *Erdboden*. Our living on this ark would, however, retain a reference to Earth as the "originary ark" that more grandly orients our movement. The space-ark is the grounding-earth of my movement on it, even as I also feel myself moving away from my original planetary home. (Contrast this with "planetary-lag" in the eclipse: we feel transported from our Geosphere whilst *remaining* on it).

Contra Copernicus revealing Earth as a planetary body moving amongst the stars, grounding-earth "does not move." As constitutive ground of my sense of movement, it is entirely beyond motion and rest. It is what I called an "immovable," and what Merleau-Ponty (2011, 192–3), drawing on Henri Wallon, calls an "ultra-thing." Wallon is referring, for example, to the Moon as peculiar thing that, unlike distant buildings or trees, is not perceived as moving when we walk or drive around. For the child, who doesn't know the Moon is a moving thing, it may be perceived as an "ultra-thing," beyond all regular moving things. So too the dome of the sky. Optically, features of Moon and sky are too far away for us to perceive a change in visual angle, relative to us, when moving by foot or car—they are "angular immovables." But the moving planetary body of Copernicus is also an inaccessible ultra-thing. The Earth is too big to perceive while on it—is a "scalar immovable."

This is the macroplanet problem. Husserl puts it in terms of *bodies*. Like all bodies, the Earth as planetary body is a whole composed of parts, each of which are bodies. Husserl charmingly names the Earth-body as a whole "*großer Klotz*" usually translated as the "big block," but it has the lovely sense of "big lump" or "big clod." I'll call this block-earth. I am such a small speck-lump on the big-lump that I can't perceive it. These tiny lumpen-dice I am holding, *these* I can perceive as a whole, when I turn them in my hand. But up in the airplane, looking down at the ground, I imagine seeing myself down there, amidst the giant hands of valleys and forests. My body would be a vanishingly minuscule speck-lump. Indeed, Husserl notes that to perceive block-earth as a whole, or enough of its parts to perceive it as an unfolding whole, I have to fly away in a plane or a spaceship. But then it is no longer perceptible as the vast thing it is—it vanishes into a little ball! The astronauts in Samantha Harvey's remarkable novel *Orbital* (2023) keep seeing how *hard* it is to perceive cities and regions of Earth as *themselves*. They flash by in tiny glints and glimpses, versus being revealed their own bigness—in the way that people seen from skyscrapers look like ants that can't be people. (Husserl doesn't get how *fast* spaceships must move to orbit Earth and how that alters perception of Earth).

For Husserl it appears that we must either leave material Earth behind to glimpse it or subsume it into our human world, as the constituting ground of our motion-sense. Or so argue critics such as Andrew Tyler Johnson (2014) who also contends that Husserl's transcendental philosophy simply cannot open to natural-Earth. In his lectures on *Husserl at the Limits of Phenomenology* (*HLP*[7]) Merleau-Ponty seems to concur. He notes that *Erdboden* as ultra-thing that grounds *Erdraum* (the earthspace in which we move) is such that "the earth as a whole has no place in this *Allraum*" (universal space), not because the Earth "is riveted, nailed down, but precisely because it is not a *Körper*"—Earth as *Erdboden* is not *a body at all*. For Merleau-Ponty, then, *Erdboden* is "un englobant," it engulfs or englobes the natural Earth, swallows it up (*HLP* 82/68).

Merleau-Ponty, though, opens a different sense of Earth in Husserl, in his characteristically rich and unusual reading of Husserl. Indeed, the *Umsturtz* text is profoundly important in his encounter with Husserl, which is shaped by the historical contingency of being able to study Husserl only by going to the archive that Von Breda smuggled from Germany to Belgium, to prevent the Nazis burning it. Most people begin with Husserl's relatively small canon of officially published books, then discover these are the tip of a vast hidden iceberg of complicating texts. Merleau-Ponty inverts this: he began deep in the archival iceberg, there found the *Umsturz* text, written along with *The Crisis of the European Sciences*[8]—and found a Husserl seeking a radically new conception of the relation between the lifeworld, science, and transcendental philosophy.

Merleau-Ponty's "inverted Husserl" is friendly to Anthony Steinbock's (2017) Husserl, who verges over the limits of phenomenology into what is prior or beyond. The abstractly transcendental Husserl can take seriously the thought experiment of the destruction of the world and contend that transcendental structures would nonetheless remain. Steinbock reveals another Husserl, of the *Crisis*, who, facing the actual destruction of the world, realizes that the unsurpassable limits and structures of the lifeworld must be approached as arising from a generativity beyond our experiential limits, that is indexed within our limits, by phenomena such as natality and death. Phenomenology does not live indoors. It limits mark us as belonging to what constitutes *us*—without us constituting *it*. For Steinbock, the ultimate point of the *Umsturz* is not lumping Earth into a *großer Klotz* that can only be seen from orbit, or engulfing Earth in *Erdboden*. Instead, Husserl reveals, within our experiential limits, a *Stammboden*, a stem- or root-abode that abides within us as our shared home-ground, even as we move over or away from it.[9] This *Stammboden* would, however, still conceal what is at stake in what I call Geosphere.

I'll approach this through Merleau-Ponty's reading of Husserl's *Umsturz* text as crucial to the juncture between phenomenology and ontology. This issue was broached in earlier points about *sur-réflexion* flaring over/on/against the ontological rootedness of reflection in a concrete planet, a *Stammboden* in *nature* that remains silent in Husserl. Here, I take advantage of Frank Chouraqui's (2016) astute analysis of Merleau-Ponty's reading of Husserl's *Umsturz* (in *HLP*).

Chouraqui crucially notes how Husserl's text hinges on a phenomenological analogy between the human body and the Earth. In this analogy, *Leib* (the lived body) stands to *Körper* (the human body as material thing) in the way that *Erdboden* (grounding-earth) stands to Earth as *großer Klotz* (block-earth). In the schematic notation of ratios between ratios:

Leib : *Körper* :: *Erdboden* : *großer Klotz*

Let me explicate. We do not experience our lived-body as a thing. In the lived-body, material parts envelope one another in a topological body-schema that grounds our oriented sense of the world and ourselves.[10] For Husserl, the lived-body is thus the zero-point of all orientation, and it engulfs material body-parts within it. (It's not as if my left hand is a part beside my right hand in my lived body, as the integral here from which I orient). Similarly, *Erdboden* as *Erdraum* is not located in geometrical space, it is the zero-point of that. It would beg the question if we claimed *Erdboden* functioned as a zero-point due to some geometrical arrangement of its parts.

Instead, we should think of *Erdboden*'s "ontological topology."[11] By this I mean the ontological relation of the being of its parts, as a structural whole that connects with the topology of the body. *Erdboden* would thus be an endogenous elaboration of the reversible, chiasmic topology of what Merleau-Ponty calls flesh, in which our bodily activity inherent crosses over into domains of passivity, and vice versa. But with *Erdboden*, our body is open to the generative passivity of Geosphere, as an abiding basis of our pushing off and around, in a rhythmic network of locomotory activity. A concrete, natural *Erdboden*, we could say, is the outer envelope of generative passivity, within and by which the envelopment of our body parts within our body schema is able to achieve active motion. This is because Husserl's *Erdboden* reveals that the lived-body in isolation, on its own, *cannot move*. The "I can" of an astronaut in outer space has no grounds for moving about. To move, the astronaut literally has to throw something away, a wrench, propellant.

On the other hand, we do not normally experience our own body as *Körper* from within ourselves, precisely because our *Körper* is enveloped in our living. For example, it is typically only if my arm goes numb that I experience it as a mass with weight: I feel the weight of things through my arm, not *its* weight. And, of course, I cannot get certain parts of my body in perceptual view because I can get no distance from my own body. I can never see my head as a *kleiner Klotz*, although I can see your head that way. This is analogous to not being able to experience Earth as *großer Klotz* from within.

Chouraqui (2016, 55–56) notes how this *Leib-Erd* analogy leads to questions that help me reframe the discussion. These begin with two alternatives: "[S]hould we say that *Leib* is different from the [transcendental] Ego that precedes it, and therefore regard the *Erde* as only a mode of the cogito (Husserl)"—Johnson's view of Husserl's error. Or "should we reduce the Ego to the *Leib*, and therefore to the *Erde* (Merleau-Ponty)?"—Merleau-Ponty's concept of the cogito and the lived-body in his earlier *Phenomenology of Perception*, at least as he looks back on it later.

In fact, due to his attention to nature, science, passivity, time and depth as preceding us, and his methodological commitment to radical reflection, I argue (contra Chouraqui and Merleau-Ponty's own self-criticism), that even in the *Phenomenology* Merleau-Ponty grasps that neither alternative is satisfactory, because the lived body opens to and reveals something far older, deep within it. Consider his remarks on time in the final sentences of each chapter of the second part of the *Phenomenology*. These begin with the famous conclusion of the sensing chapter, that "reflection only fully grasps itself if it refers to the prereflective fund it presupposes, upon which it draws, and that constitutes for it, like an original past, a past that has never been present" (*PhP* 289). The series ends with the conclusion of "Others and the Human World," that

> On the level of being, we will never understand that the subject is simultaneously creating [*naturant*] and created [*naturé*], and simultaneously infinite and finite. But if we uncover time beneath the subject, and if we reconnect the paradox of time to the paradoxes of the body, the world, the thing, and others, then we will understand that there is nothing more to understand.
>
> (*PhP* 424, right after the point that phenomenology turns into phenomenology of phenomenology)

Almost word for word, we could make the same points about the Geosphere we've been trying to un-earth within everyday *Erdboden*. Geosphere points to an original past (indexed by geological deep time) from which Ge itself is born; to a toposphere–Earthscape that is simultaneously creating a sphere of movement, yet is also created; that is simultaneously infinite, boundless as immovable grounding-earth, and finite, moving as a block-earth. But noticing this requires that our bodily, historical openness to the paradox of time, crosses over with the topology of our openness to Geosphere, as abiding abode of our bodies. This requires "transcendental geology," beyond the mere paradox of time.

But let me put this point of contention aside, to return to Chouraqui, who approaches this point through Merleau-Ponty's later ontology. Chouraqui notes that "in the very months of his lectures on Husserl's concept of the earth, Merleau-Ponty exhorts himself to 'say that I must show that what could be regarded as 'psychology' (*Ph. of Perception*) is really ontology.'" This move from psychology to ontology requires breaking the phenomenological cogito from the abstract transcendental Ego. This is why he seeks to "explore the extent to which Husserl's criticism of the Copernican earth can be taken up in a non-subjectivist perspective." All of this requires, Chouraqui argues, "a softening of the distinction Leib/Körper."

> This is not to say that they should not be seen as distinct—indeed, our experience offers us the *Leib*, but it is constitution that makes it a *Körper*—but rather that this movement of constitution must now be seen as a *structure of being itself*, and not as a simple psychological fact… We shall see that Merleau-Ponty will be led to regard the Earth as a pure principle of generation without origins, a pure relation of productivity.
>
> (All quotes from 2016, 56)

Put schematically, the point is this. It looks like Husserl's *Umsturz* boils down to the analogy $\textit{Leib} : \textit{Körper} :: \textit{Erdboden} : \textit{großer Klotz}$, and the terms on each side are forever cleft and closed from one another. But, for Merleau-Ponty, the very way to his later ontology hinges on what Husserl is starting to broach in the *Umsturz*: that we are open to Geosphere in the way we are open to time and history. The schema must be rewritten. I use χ to indicate chiasmatic, topological crossovers of active and passive being:

$$(\textit{Leib}\ \chi\ \textit{Körper})\ \chi\ (\textit{Erdboden}\ \chi\ \textit{großer Klotz})$$

But this also means that on a methodological level we need to understand that philosophy hinges on the following schema:

$$(\text{time}\ \chi\ \text{earth})\ \chi\ (\text{transcendental}\ \chi\ \text{nature})\ \chi\ (\text{activity}\ \chi\ \text{passivity})\ \chi\ (\text{phenomenology}\ \chi\ \text{ontology})$$

In this notation, however, each $\text{p}\ \chi\ \text{q}$ chiasmically couples with its own reverse, for example: $\text{activity}\ \chi\ \text{passivity}\ \chi\ \text{passivity}\ \chi\ \text{activity}$.[12] By crossing $\text{time}\ \chi\ \text{earth}$ with $\text{phenomenology}\ \chi\ \text{ontology}$ the schema above thus (I suggest) opens into the project of transcendental geology (see the final section below). The schema and project also require *sur-réflexion*: if your phenomenology crosses with ontology in this way, then you cannot detach the activity and passivity of your reflection from reflecting on/over $\text{time}\ \chi\ \text{earth}\ \chi\ \text{earth}\ \chi\ \text{time}$.

What Chouraqui detects in Merleau-Ponty's engagement with the *Umsturz* on his way to the *Visible and the Invisible* thus opens to bigger philosophical stakes. As Chouraqui puts it,

> To Merleau-Ponty, what this Husserlian text achieves is more than Husserl himself intended. It is not only a reflection on the relations between experimental science and phenomenology, one in which the former would appear as a moment of the latter. This text also illustrates the movement that arrives at the realization that the reduction can only ever reveal an always-already present that precedes the cogito itself.
>
> (2016, 63)

Here Chouraqui is commenting on Merleau-Ponty's concluding paragraphs of *HLP*:

> Because he leaves next to one another (correlatively), the realist-causal order and the idealist-constitutive order, Husserl is obligated to sustain this nearly "crazy paradox." One would have to take up the concrete relation of these two orders by turning both of them not into a physical world relative to the idealistic *Sinngebung*, but into two correlative aspects of Being.
>
> The Earth which is first is not the physical earth (by definition it is homogenized); it is the source Being, the *Stamm und Klotz* being, in pre- restfulness; the mind which is first is not the absolute Ego of *Sinngebung*. It is the *Denkmöglichkeit* and they are *Ineinander*, entangled [*enchêvtrés*].
>
> (*HLP* 92/76)

Crucially, the physical world and the idealistic *Sinngebung* are two correlative aspects of being. And it is this chiasmatic relation, I think, that Merleau-Ponty is indicating in the final remark that "they" are *entangled*. But in fact, I think *all* of the terms mentioned above are entangled. (The plural form of *enchêvtrés* seems to support this, there are multiple entanglements).

My key contention is that this "always already present that precedes the cogito itself" is inclusive of and unsurpassably rooted in Geosphere as a twofold being: physically visible Earth, with its scientifically studied geometry; and that Earth understood as manifesting a principle of being that needs to be understood in terms of its ontological topology, its network of being as generative of this sort of visible Earth.

Put otherwise, for Merleau-Ponty flesh names an ontological topology of elemental being wherein: visible beings manifest an inborn, invisible, generative being; and active, visible, delimited things and activities are shadowed by generative passivity. What we are encountering with Geosphere is a specifically geological node of flesh.

A way to understand this point is to grasp that the ontological topology of flesh unfolds not just as a reversible chiasm of body and world; all of that is topologically open to Geosphere, which is, in turn, spatially open to nature and cosmos (corresponding to our temporal openness to history and others). Schematically:

$$(\text{body } \chi \text{ world}) \, \chi \, (\text{Geosphere } \chi \text{ cosmos} - \text{nature}) \, \chi \, (\text{others } \chi \text{ history / deep time})$$

6.7 Revealing Geosphere in Total Solar Eclipses and Orbital and Navigational Experience

Beyond Merleau-Ponty's ontological complications of Husserl there are several phenomenological inadequacies in the descriptions underlying Husserl's *Umsturz* text.

object. The interweaving of your time, the time of your object, your space, and the space of your object, are the unsurpassable condition of geology. Descartes can pretend his philosophy must begin in his own time of reflection, by ejecting himself from the world, as if he casts no shadow beyond himself. The geologist must begin by trudging their own shadow *sur* (over/on) the greater shadow of Ge.

Merleau-Ponty's note urges that radical *sur-réflexion* must eventually trace its beginnings not just to a past that has never been a present, but to an *Urstiftung* complex of space and time. By calling this *sur-réflexion* "transcendental geology," Merleau-Ponty is, I think, arguing that the transcendental field (this-here experiential givenness, taken as the unsurpassable beginning of all science) must deepen not just into pre-scientific and pre-philosophical history but into its concrete roots in this cosmos—including our Geosphere as a liminally felt toposphere on which we are born/borne.

Transcendental geology would reveal that Geosphere is not simply subsumed to the human world or contained within its limits. Our experience and reflection flare out into spheres and times well beyond and before human life. It is not phenomenology that is anthropocentric. What is anthropocentric is the contention that human life is so much born within itself that it cannot get outside itself. That is Merleau-Ponty's problem with Sartre, Meillasoux and anybody who takes as obvious what and who we are, or takes it as obvious that we can sort that out by reflecting on ourselves from *within* ourselves. As if that had ever been possible! It is this hyper-anthropecentrism that eco-phenomenology must reject.

That is what total solar eclipses can briefly bring to light in collective experiences, if noticed in a phenomenological way: we are motes in the light of the cosmos. But that does not get in the way of knowing, it is the way to knowing. Ultimately, the project of transcendental geology involves science and philosophy deepening into a *sur-réflexion* that embraces and reveals the contingency of "this very time that is space, this very space that is time," as the ultimate field and borne of philosophy and science and all sense.

Acknowledgments

I thank my dad, L.R. Morris, follower and scholar of eclipses, who has published several articles on the history and significance of specific solar eclipses, took me to my first partial eclipse and sent me eclipses glasses for the 2024 eclipse. Without this and his infectious passion for eclipses, I would not noticed what I glimpsed in those few minutes. That philosophy hinges on thanks of this sort is another side of transcendental geology, as discussed by Amoroso.

Notes

1 Here cited by the pagination in the translation by Fred Kersten, revised by Leonard Lawlor, included in Merleau-Ponty (2002).
2 *VI* is cited by the pagination in the original French, Merleau-Ponty (1964a), then the English, Merleau-Ponty (1968).
3 Cf. Casey (2007), Casey (2017) .
4 *PhP* is cited by the French pagination that is included in the margins of the English translation cited her, Merleau-Ponty (2012).
5 See Oliver (2015) for a discussion of the significance of this photo to philosophy.
6 For a crucial and exemplary study of this text, see Steinbock (1995), 109–122. See Steinbock (2017), ch 3, for an immaculate discussion of the context of this text in relation to the *Crisis*.
7 *HLP* is cited by the pagination in the original French, Merleau-Ponty (1998), then the English, Merleau-Ponty (2002).

8 See Steinbock Limit, ch 3 for an immaculate situation of the writings around the crisis.
9 See especially Steinbock (2017), 120–21, where Steinbock connects *Stammboden* to a "geo-historical delimited sphere"—but as taken up in the lifeworld, whereas here we are tracing an internal opening of that to what is temporally and spacially prior to it.
10 Here I am explicating Chouraqui's points drawing on my own results about the body and our sense of space in Morris (2004).
11 See the October 1959 note on Ontology, that begins "Take topological space as a model of being," *VI* 264/210. A few pages later is the note on "The chiasm."
12 For the significance of passivity in this context, see Morris (2010a).
13 See Everett (2023), 49–50, in a chapter rich with significant points about variations in language and cognition regarding space.
14 Gladwin (1970). See Morris (2010b) for more on this in relation to Merleau-Ponty, animality, and place.
15 Aveni (2017) and Russo (2012) give helpful introductions to the science, experience, history, and psychology of eclipses.
16 Calculations based on NASA's *Five Millennium Canon of Solar Eclipses: −1999 to +3000*.
17 Some tips for future phenomenologies of eclipses: rehearse donning and doffing eclipse glasses. If you wear glasses, try to find eclipses glasses that flip up and down. The problem is the eclipse glasses are needed to see the sun without going blind, but with them on, you can see nothing except the sun. So you are blind to the world. Of course, don't bother taking photographs. Photos of the sun are for professionals. No machine can document what can be seen around you during this event. The only record is your lived body and memory. Pay attention.
18 See Paumgarten (2024).
19 See Aveni (2017) ch 16 for a discussion of zoological study of animal behaviour in eclipses.
20 See Russo (2012) ch 6 for an all too psychologistic account.
21 Exceptions are Amoroso (2021), whose assessment of the philosophical significance of transcendental geology complements mine—but who elaborates through Merleau-Ponty's engagements with childhood in and around his discussions of institution and personal history, which is an important dimension of this, but not the only one. De Fazio (2022) connects this theme with issues of the Earth and the Anthropocene, but the focus is more on linking Earth science to history, somewhat the opposite direction I pursue here. Thanks to Amoroso for sharing her article and exchanges on the topic.

References

Abram, David. 2011. *Becoming Animal: An Earthly Cosmology*. New York: Vintage Books.

Amoroso, Prisca. 2021. La geologia trascendentale come ecologia del pensiero. *Chiasmi International* 23: 223–235.

Aveni, Anthony F. 2017. *In The Shadow of the Moon: The Science, Magic, and Mystery of Solar Eclipses*. New Haven: Yale University Press.

Bergson, Henri. 2023. *Creative Evolution*. Translated by D. A. Landes. London: Routledge.

Bjornerud, Marcia. 2018. *Timefulness: How Thinking Life a Geologist Can Help Save the World*. Princeton: Princeton University Press.

Casey, Edward. 2017. *The World on Edge*. Bloomington, IN: Indiana University Press.

Casey, Edward S. 2007. *The World at Glance*. Bloomington: Indiana University Press.

Chouraqui, Frank. 2016. Merleau-Ponty and the Order of the Earth. *Research in Phenomenology* 46 (1): 54–69.

De Fazio, Gianluca. 2022. Abitare il deserto. La geologia trascendentale di Merleau-Ponty alla prova dell'Antropocene. *Chiasmi International* 24: 125–138.

Everett, Caleb. 2023. *A Myriad of Tongues: How Languages Reveal Differences in how we Think*. Cambridge, Massachusetts: Harvard University Press.

Frank, Adam, Marcelo Gleiser, and Evan Thompson. 2024. *The Blind Spot: Why Science Cannot Ignore Human Experience*. Cambridge, Massachusetts: The MIT Press.

Gallagher, Shaun, Bruce Janz, Lauren Reinerman, Jörg Templer, and Patricia Bockelman. 2015. *A Neurophenomenology of Awe and Wonder: Towards a Non-Reductionist Cognitive Science, New Directions in Philosophy and Cognitive Science*. Basingstoke, Hampshire: Palgrave Macmillan.

Gladwin, Thomas. 1970. *East is a Big Bird: Navigation and Logic on Puluwat Atoll*. Cambridge, Mass.: Harvard University Press.
Harvey, Samantha. 2023. *Orbital: A Novel*. New York: Grove Press.
Johnson, Andrew Tyler. 2014. A Critique of the Husserlian and Heideggerian Concepts of Earth: Toward a Transcendental Earth that Accords with the Experience of Life. *Journal of the British Society for Phenomenology* 45 (3): 220–238.
Meillasoux, Quentin. 2010. *After Finitude: An Essay on the Necessity of Contigency*. London: Continuum.
Merleau-Ponty, Maurice. 1964a. *Le visible et l'invisible*. Paris: Gallimard.
Merleau-Ponty, Maurice. 1964b. *Signs*. Translated by R. C. McCleary. Evanston, IL: Northwestern University Press.
Merleau-Ponty, Maurice. 1968. *The Visible and the Invisible*. Translated by A. Lingis. Evanston, IL: Northwestern University Press. Original edition, 1964.
Merleau-Ponty, Maurice. 1993. Eye and Mind. In *The Merleau-Ponty Aesthetics Reader: Philosophy and Painting*, edited by G. A. Johnson. Evanston, IL: Northwestern University Press.
Merleau-Ponty, Maurice. 1998. *Notes de cours sur L'origine de la géomterie de Husserl*. Paris: Presses Universitaires France.
Merleau-Ponty, Maurice. 2002. *Husserl at the Limits of Phenomenology*. Translated by L. Lawlor. Evanston, IL: Northwestern University Press.
Merleau-Ponty, Maurice. 2011. *Child Psychology and Pedagogy: The Sorbonne Lectures 1949-1952*. Translated by T. Welsh. Evanston, IL: Northwestern University Press.
Merleau-Ponty, Maurice. 2012. *Phenomenology of Perception*. Translated by D. A. Landes. Oxon, UK: Routledge.
Morris, David. 2004. *The Sense of Space*. Albany, NY: State University of New York Press.
Morris, David. 2010a. The Enigma of Reversibility and the Genesis of Sense in Merleau-Ponty. *Continental Philosophy Review* 43 (2): 141–165.
Morris, David. 2010b. The Place of Animal Being: Following Animal Embryogenesis and Navigation to the Hollow of Being in Merleau-Ponty. *Research in Phenomenology* 40: 188–218.
Oliver, Kelly. 2015. *Earth and World: Philosophy after the Apollo Missions*. New York: Columbia University Press.
Paumgarten, Nick. 2024. Watching the Eclipse from the Highest Mountain in Vermont. *New Yorker*, April 9, online.
Russo, Kate. 2012. *Total Addiction: The Life of an Eclipse Chaser*. New York: Springer.
Steinbock, Anthony J. 1995. *Home and Beyond: Generative Phenomenology after Husserl*. Evanston, Ill.: Northwestern University Press.
Steinbock, Anthony J. 2017. *Limit Phenomena and Phenomenology in Husserl*. London: Rowman & Littlefield.
Toadvine, Ted. 2024. *The Memory of the World: Deep Time, Animality, and Eschatology, Posthumanities*. Minneapolis: University of Minnesota Press.

7

AESTHETIC PRACTICES AND CULTURAL CHANGE

Challenging Environmental Neuroticism

Bryan E. Bannon

The idea that the ecological crisis humanity now collectively faces is at its root a cultural crisis is not a new one. Writers across time and traditions have in various ways tried to understand how aspects of specific cultures have contributed to the domination or despoliation of their ecological environments. In the Western philosophical tradition, the understandable tendency has been to criticize conceptual commitments of the culture, rationalism, dualism, patriarchy, etc.—all of which are real problems—and recommend those be reformed or abandoned.[1] Despite the validity of these analyses, the trouble with advocating a transformation of the culture by means of thinking differently about these problems neglects the fact that cultures have developed for historically and environmentally specific reasons, and those reasons make such simple conceptual displacement highly improbable. In many cases, the very things that make a culture successful—that is, enabling its members to navigate the difficulties of living in an environment and to establish meaningful lives there—are the very sources of the culture's internal conflicts and ecological woes. Hence, though I agree strongly with the need to develop counter narratives to those of dominant Western cultures, if a culture is going to change, it isn't simply a matter of thinking differently or even of altering policy; to change a culture is to alter members' *perceptions* of what pathways lead to a life of meaning. The aforementioned litany of conceptual traps Western culture lays for its members with respect to their relationship to their environment and one another must be disarmed by reforming the perceptual norms of the culture—that is, altering the affordances within the lived environment—with respect to what a good relationship to the rest of the natural world is like.

Eco-phenomenology presents itself as a challenge to the everyday practices and understandings that have shaped European culture and its colonial offshoots' relationship to the rest of the natural world in an effort to improve that relationship. In this project, it has many qualities that recommend it: a lack of ontological assumptions to counter the culture's pervasive dualism, an ecological understanding of both individuals and systems, and an approach to the relationship of individuals and systems that does not reduce the action of one to the other. These qualities move counter to the individualizing, dualizing culture eco-phenomenology seeks to analyze and change. If cultural change requires a shift in perceptual norms, phenomenologists' analyses of how aesthetic experiences create and reformulate such perceptual

DOI: 10.4324/9781003538097-9

norms will enable developing pathways of action oriented toward inhabiting the landscape in a manner more befitting the contemporary ecological situation.

I approach this project by taking up the dilemma with which the paper began: the tools that a given culture possesses to make sense of the world and members' place within it engender pathological results when those tools are used in less apt contexts. Here I am expanding on John Russon's (2014) concept of neurosis: a neurotic situation is one in which an individual's adapted behavior for navigating their developmental situation becomes incapacitating either when that situation changes or when it is applied to a different, less appropriate situation. In this sense, North American culture displays pathological neuroses with respect to many environmental situations. I will begin by arguing this point so as to better understand the culturally specific perceptual norms that require reformation. I then proceed to develop a model for transforming perceptual norms by drawing from Merleau-Ponty's views on culture and history. These views direct us to examine how aesthetic experiences can transform perception and, through it, alter habitual actions. I close with an analysis of two sets of cultural practices—landscaping and environmental art—that are currently contributing to such transformations.

7.1 Neurosis and Nature

So what is a cultural neurosis, and how does one address it? Neurosis, to summarize briefly Russon's (2014) argument, is a characteristic of *situations* in which an individual's interpretation of reality is leading them either to experience some form of psychic distress or to be thwarted in their projects. One's interpretation of a situation is the product of developing perceptual, affective, and behavioral habits that allow one to navigate situations in a way that meets one's need for human recognition. One thereby interprets and acts within specific situations through this developed habitual repertoire, a repertoire that developed in the context of familial and social life. A situation becomes neurotic when one's interpretation of the situation is out of joint with how the situation demands to be interpreted in order to engage in one's projects successfully.[2]

If situations can be neurotic because of the way in which a developmentally adapted interpretation becomes maladaptive when employed in specific milieu, so too can a culture become neurotic when the various strategies for action it comprises come to work against their reasons for existing. To see how this works, it will be helpful to examine both the sociological and phenomenological dimensions of culture because culture is not primarily an ideological guide to behavior. From a sociological perspective, culture is a set of norms, practices, institutions, symbols, and artifacts that aid in the interpretation of the environment, which in turn orient individuals' agency within their environments. As Pierre Bourdieu (1977) would have it, we develop "habitus," a historically grounded set of interlocking strategies for dealing with the demands of specific domains within which we operate. Inhabiting our cultural habitus, we perceive the things and people we encounter as having particular meaning and value and understand ourselves in terms of our relationships to these. Culture thus serves as both a facilitator and a constraint upon one's activity: it enables specific pathways of action within domains, but conceals from us alternatives as a result of their incongruity with the operant interpretations of one's situation. As Ann Swidler argues, "culture appears to shape action only in that the cultural repertoire limits the available range of strategies for action," and, hence, "a culture has enduring effects on those who hold it, not by shaping the ends they pursue, but by providing the characteristic repertoire from which they build lines of action" (1986, 284).

From a phenomenological perspective, we may add to this account how culture creates such lines of action by establishing *perceptual norms*. In this use, a perceptual norm refers to the way culture comes to inform our perception of how our world ought to be. As Swidler and Bourdieu indicate, culture structures how we interact with our environments by developing habitus and thereby having some courses of action appear possible while foreclosing other opportunities. But individuals do not translate a thought into action; one perceives cultural affordances and, in perceiving them, are drawn into action (Merleau-Ponty 1964a, 18). Being so drawn, perception is a form of "primordial expression:" even at the most basic level, perception is expressive of a way of individuating beings and seeing those beings as meaningful in terms of how we learn their relationships with one another (Merleau-Ponty 1960, 267). In short, we learn how to see within a culture, and seeing is always laden with meaning (250). Hence, what a culture provides is an architecture of sense that vision traces, but, like any structure we encounter and interact with, the meanings we perceive are never complete or cut off from further development (67–8). In this way, the history of culture is not only a catalogue of norms, beliefs, and institutions but also a history of how beings are individuated and interpreted over time. Culture here serves as an instituted order with our individual acts serving as the expressive gestures that confirm the existing architecture of sense or inaugurate an alternative order that would transform the institution (108).

To summarize, culture is an instituted order that educates our perception through habitual uses of the body to reckon with the environments that confront us. Cultural habitus functions through perception such that some affordances are apparent to us while others may be beyond imagination. Culture thereby creates perceptual norms that may be transformed or affirmed through members' actions. What makes a culture neurotic, environmentally speaking, is when perception is educated in such a manner that the developed habitus, the lines of action the perceptual norms establish to interact with the environment, no longer lead to mutually beneficial interactions with it—perhaps even creating the very problems that members then experience a need to address.

It should be clear that North American settler culture displays such neurotic tendencies with respect to the environment. The habitus of the dominant culture includes a variety of pathways for action that create, compound, or make intractable environmental problems even as they are enabling in other domains. Here, I will consider just a one of the more pronounced aspects of the cultural habitus that creates environmentally neurotic situations by way of example: moral judgment. Moral judgment is a cultural practice in which individual's actions are assessed according to more or less clearly articulated principles. It tends to be bi-modal insofar as judgment is rendered in a "right-wrong" framework. As a practice, it is effective in having individuals self-regulate their behavior in ways that preserve social cohesion. In this way, moral judgment may provide a heightened sense of autonomy, responsibility, and regard for others. Since moral judgment concerns itself primarily with individual behavior, it encourages us to think of agency and the value of our lives in individualistic terms. In an environmental context, this tendency promotes an individualistic approach to environmental problems that prevents the formation of the political community necessary to addressing them and obscures the contributions of social structures and institutions. Individualistic practices and ways of thinking are deeply entrenched in North American settler culture, even among some of its environmental heroes like Thoreau and Emerson. This individualism encourages members of the culture to impute the causes of environmental problems to individual behaviors and therefore seek out solutions to them by correcting those behaviors. Environmental problems are therefore traced back to moral failings of individuals who utilize pathways of action deemed immoral. But the issue here isn't a philosophical commitment to individualism. Rather, the trouble with moral judgment is how it conditions perception itself: one sees the person drinking water from a disposable plastic water bottle as morally defective rather than as, say, someone drawn into a choice based on what they perceive to be healthy in a world where municipal infrastructure is chronically underfunded, our social institutions fail to ensure the safety of common goods, and corporate advertising has successfully influenced people to *see* the bottled water as a safe, healthy, responsible choice. What's more, moral judgment takes a problem of collective action and transforms it into a problem of individual agency. By doing so, it promotes other individualistic cultural strategies for action as responses to environmental problems. For example, the person drinking bottled water thinks buying something is a viable solution to the lack of social protection of public goods. The consumeristic culture encourages them to believe that they can buy their way out of the problem.

The historically developed patterns of cultural habitus are neurotic in the environmental context due to how they shape perceptual norms and thereby the affordances for action in specific situations in a way that undermines our ability to inhabit the land or to address emergent environmental problems. The question is, how to reform the perceptual norms that give rise to these lines of action if the culture's perceptual norms preclude one from seeing how to act in ways that are not ecologically harmful?

7.2 Transforming Vision

To reform a neurotic culture, it is necessary to learn to see anew, but, if culture is more than a set of beliefs and values, re-educating vision is no simple matter. One approach is to think of culture as an "adventure," a notion Maurice Merleau-Ponty develops in the context of the philosophy of history but is appropriate insofar as perception is an inherently historical phenomenon. I take my cue for this reconstruction from Merleau-Ponty's observation in "Eye and Mind" that the essential characteristic of painting—that is, its stylized, expressive structure—is the essential characteristic of all culture (1964a, 15). In this way, cultural practices are analogous to traditions in painting insofar as they both represent institutions that serve as the interpretive context within which expressive acts stand out (or don't).

But what is an adventure, and how does it help us to understand cultural shift? Merleau-Ponty calls culture, "an original order of advent," meaning that culture provides a structure with respect to which historical transformations occur (Merleau-Ponty 1960, 267). The cultural order emerges through perception, how the relations in the world are perceived and unified into a meaningful whole (251, 255). But our perception is our own; our experiences lead us to have a specific relationship to that cultural endowment. In this way, our own lives become a "coherent deformation" of the cultural endowment. The culture thereby serves as "a promise of events"—that is, the individual lives it structures take it up and remake it through their expressive action, spurring it to *develop* in other directions. This redirection of the cultural order and subsequent development is the adventure of history (Merleau-Ponty 1964a, 74). Adventuring in a new direction also allows for viewing the old habitus in a new light, allows reinterpretation of one's history (Merleau-Ponty 1960, 95). This process is, of course, forever ongoing since the environment and the situations in which we find ourselves are always developing as well (Merleau-Ponty 1964a, 89–90).

However, just because a new cultural practice inaugurates a new order to which others will respond with their own expressive gestures does not ensure that such practices are addressing the culture's neurosis. For that to happen, the dialectical movement of history, its adventure, needs to be a "good" or "hyper" dialectic that opens a "landscape of praxis" (Merleau-Ponty 1955, 276). If one of the characteristics of neurosis is that it makes effective agency more difficult in specific situations, the kind of gestures sought out are ones that allow us to perceive a new horizon of possibility for living and acting with respect to those same situations. What the transformative gesture offers to open this landscape of praxis is a reconfiguration of the plurality of relations, which in turn invites a reflexive stance toward historical development insofar as it reveals that *things could be otherwise* (Merleau-Ponty 1964b, 127). Seeing the set of cultural relations in an unforeseen way is what allows us to see which cultural paths are errors and which lead to greater possibilities, itself an assessment that can change based upon the current cultural situation (Merleau-Ponty 1955, 36–9). Transformative practices will address the conditions it confronts as a result and thereby serve as the "noble form of memory" that gives new life to the old (276–7 and Merleau-Ponty 1960, 95).

Summarizing these ideas in a less technical way, we can say that the transformation of perceptual norms takes place through practices that simultaneously display the existing cultural habitus in a new light while also inaugurating a new horizon for action where the habitus is as yet indeterminate. In trying out new practices that reconfigure everyday relations, a model is instituted that others can take up and make their own. These new institutions will inevitably have their own problems, but only history will tell if the newly established habitus is an error or if it possesses the internal resources to address those problems. We can transform this account into a process of sorts: 1) experiencing the neurotic tendency of the culture leads to 2) engaging in a practice that displays the cultural habitus as neurotic thereby 3) instituting a new landscape of praxis for others to pursue in their own way or in community and 4) reforming perceptual norms with respect to the initial pathway of action.

To address humanity's ecological neurosis, it has become necessary to redirect cultural practices that have us persist down the path of mass extinction, of the degradation of ecological relationships that sustain life, and of deluding ourselves that happiness can only be achieved through acquisition. One might say we seek to break up the existing cultural habitus in the name of a "truer relation" to things and one another (Merleau-Ponty 1960, 257). By a truer relation, I am not referring to one that more closely corresponds to some inner organization of the world. After all, all appearances of truth conceal other appearances. Rather, a truer relation is one in which the perceptual norms bring to the fore concealed meanings that enable the current situation to be addressed in a manner that relieves neurotic tensions. Because every situation is by definition distinct, that is, it will be open to showing itself in myriad ways that admit to myriad interpretations, there is no one set of practices that one could identify to bring about this truer relation. Nevertheless, in the next section I will examine how two aesthetic practices reveal some contours of what the relationship to nature should be like to maintain the openness to interpreting the environment that makes possible addressing neurotic situations.

7.3 Seeing Nature Anew

The natural world surrounds us, envelops us, is us, but we rarely encounter it explicitly. Phenomenologists recognize that the backgrounding of things is constitutive of perception and what allows us to interpret the world and function within it. As Sue Rechter argues, one of the fundamental aspects of a culture is how "'Nature' is contained within each cultural horizon and at the same time exceeds that horizon as another 'total part' of the World" (2007, 41). In other words, the culture contains within itself an understanding of the nature of nature, but this understanding never exhausts the possible meanings of the environment. The cultural understandings of nature are expressed through the cultural habitus: different pathways of action embody an interpretation of the character of nature that inevitably conceals other possible interpretations. So, if we are to inaugurate, to borrow Val Plumwood's phrase, a "culture of nature" in which the ecological community and its agency are made visible along with a more egalitarian picture of human wellbeing, it is necessary to form reflective and reflexive communities around these practices (2001, 32). There is no totalizing practice that will accomplish all phases of the project of cultural transformation in one gesture, especially when one considers the complexity and diversity of environmental problems humanity collectively faces. But the project of transforming cultural perceptual norms is already underway in myriad contexts. Here, I will call attention to two sets of practices that highlight the contribution of aesthetics to the transformation of perceptual norms.

7.3.1 Earth Art

Following Amanda Boetzkes, I will refer to aesthetic practices that accentuate the human world's imbrication in the rest of the natural world as "earth art" (2010, 16). Boetzkes's study of earth art reveals the ethical encounter such works engender: by staging in various media the bodily confrontation with the elemental forces of the earth, earth art displays the limitations of the cultural representations of nature, opening a horizon in which the human relationship to the earth is laid bare. What is of interest here is how, in doing so, earth art challenges the perceptual norms of the culture in the service of reformulating the strategies of action that undermine the ecological systems on which the culture relies.

Earth art refers to a diverse set of aesthetic practices comprising artworks that minimize their environmental impact (e.g., Aurora Robson's (2008) *The Great Indoors*), publicize or otherwise make visible environmental harms (e.g., Maya Lin's (2021) *Ghost Forest*), integrate the environment of the work into the work itself (e.g., Christo and Jeanne Claude's (1983) *Surrounded Islands*), utilize ecological processes as part of the work (e.g., David Nash's (1977–ongoing) *Ash Dome)*, and display the potential of human collaborations with nature (e.g., Andy Goldsworthy's (1997–98) *Storm King Wall*). In most cases, these categorizations overlap. It is a broad category and, at times, one at odds with itself. Olafur Eliasson and Minik Rosing's (2014, 2015, 2018) *Ice Watch, 2014*, for example, is a dramatic presentation of the reality of climate disruption where twelve blocks of glacial ice calved from the Greenland ice sheet were transported to three symbolic locations (Copenhagen, Paris, and London). The creators acknowledged and tracked the carbon footprint of transporting hundreds of tons of ice from arctic waters to these European capitals, but justified the cost in terms of the increased attention, and concreteness, the works brought to the problem.

Whether deliberately or incidentally, earth art creates a space in which the culture's neurotic tendencies can be experienced. Chris Jordan's (2009–13) *Midway: Message from the Gyre* consists of images of albatross with their young alongside images of the carcasses of dead chicks. The chicks die from having been fed plastic that the parent has found floating in the ocean and has mistaken for food. The decaying corpses filled with bottle caps, cigarette lighters, and various other plastic detritus are disconcerting to say the least. But, beyond that, the works directly link the melancholia of the images to the world's dependence upon and carelessness with plastics. It is one thing to *know* that widespread use of disposable plastics is bad for the oceans, it's another entirely to witness its life-denying character (Foster 1992). So the work makes visible the cultural neurosis with respect to its subject: the material on which our lives, as currently organized, depend is, literally, antithetical to life.

But the value of earth art goes beyond creating this experiential space and engaging in critique. As David Wood describes, earth art is actively engaged in a project of *dehabituation*: in bringing to light the relations that the world comprises and displaying them in an unexpected manner, art presents an opportunity to break with established habitual patterns of thinking, feeling, and acting (2013, 128). Edward Casey argues that earth art accomplishes this dehabituation by providing "maps" of various forms of relationships, for example, accentuating the relations among the various elemental forces that comprise the work, or establishing a novel relationship between the work and the perceiving bodies with which it interacts (2004, 262, 269). In following these maps, we enter into the relational space of the work, which challenges habitual norms of perception, and come to see according to the norms it establishes. James Turrell's skyspaces (e.g., *Dividing the Light* (2007)) demonstrate this notion well: in looking up into the framed sky, one feels dis-oriented and drawn into another bodily relation to the environment where one sees the sky in a completely unaccustomed way.

Works of earth art not only open space for imagination but also provide guidance into the forms more ecologically sensitive inhabitation of the earth might take. The maps much earth art provides recommend what Emily Brady (2007) calls "an aesthetic regard for nature." For Brady, an aesthetic regard for nature is an interpretive stance of sorts; it is a way of attending to nature such that one displays an esteem for or deference to nature in one's interactions with it in the service of cultivating specific kinds of relationships between the artist (or viewer) and the environment (289). In calling for human *attentiveness* to our relationship with the forces of the earth, the perceptual norms embodied in earth art lead to alternative ethical ideals to guide humanity toward less exploitative ways of living that acknowledge our finite nature and interwovenness in ecological systems (Bannon 2011). Even though not *all* earth art presents a map that leads to a less neurotic relationship to the rest of the natural world, even these maps can prompt reflection on our neurotic failings to inhabit the earth well.[3]

But one of the most radical suggestions of earth art is its egalitarian promise. Although not everyone possesses the knowledge and skill to be a painter, thoughtfully placing leaves and flower petals, stacking stones, digging holes, arranging sticks, etc. are within most people's power. And even if the cairns one might find built by amateurs in the forest do not match the symmetry and grandeur of Goldsworthy's, by practicing their construction we cultivate an attentiveness to the specificity of the places we inhabit. These are perceptual skills that directly challenge the neurotic perceptual norms of the culture where every place is seen as an interchangeable space for economic exploitation and individuals are seen as the owners and masters of a separate natural world.

7.3.2 Rewilding Landscaping

Rewilding is the practice of taking landscapes humanity has altered and allowing them to resume a path of development outside of human dominion. At first, this sounds like a dualistic enterprise: removing the marks of human impact so that nature can pursue its course independent of humanity, keeping both in their respective places. But Andrea Gammon and Martin Drenthen have both done extensive work analyzing how rewilding can challenge dualism. Drenthen speaks of rewilding's "moral core:" "its ability to morally challenge us to reinterpret ourselves and our history of involvement in the landscape, by changing the interpretational horizon and situating ourselves within a wider ecological history" (2018, 418). Gammon (2018) likewise argues that rewilding is a self-reflexive act that compels us to acknowledge our membership in the larger ecological community while also calling for us to reform culture so as to be a better member of it. In both cases, rewilding invites a reinterpretation of ecological relations to situate humanity more squarely within the ecological community, a feature that fits well with undoing the cultural ecological neurosis.

But how is rewilding an aesthetic practice, and how does it challenge the everyday perceptual norms of culture? Consider the phenomenon of the lawn. As Joan Nassauer points out, the end of large-scale public investment in conservation lands relocates the nexus of struggle over maintaining landscape-scale ecological integrity and health from massive tracts of land to the local lot (1997, 71). The lawn therefore becomes a site of political contestation over community ideals. According to Ted Steinberg (2006) and studies like his, a well-manicured lawn is a symbol for individual integrity and community pride. However, lawns frequently require practices that serve as drivers of environmental destruction in order to preserve the specific manicured aesthetic (Saito 2020). In pursuing the aesthetic cultural ideal, members of the culture impoverish the local ecological relations: habitat is transformed; herbicides, pesticides, and fertilizers are applied, affecting the soils and waters; and significant time and resource are expended to maintain an appearance that does not have much ecological value. But since one establishes a sense of oneself in dialogue with those others with whom one has contact, the goodness of the practice is *apparent* in the ubiquity of the practice.

The first step in challenging this situation is to experience the practice as neurotic. To return to Nassauer, lawns embody an aesthetic of care; they display that someone has paid sustained attention to a landscape so as to have it display the values of that person (Nassauer 1997, 75). As such, practices that challenge the current perceptual norm may capitalize on this aesthetic by showing that *one can care otherwise*. That is, landscaping can become an adventure! But how to collaborate with the land one inhabits in order to create a coherent deformation of the lawn, thereby challenging the aesthetic regime of the suburb? Transforming lawn spaces into wild gardens creates such opportunities in a way that invites personal engagement and communal participation.

Rewilding can take many forms from large-scale rewilding projects like Adirondack Park in New York state or something as simple as planting a pollinator garden in lieu of grass. For current purposes, the latter type of project provides a clearer sense of the transformative power of the practice. Many American homes with lawns already landscape with various, usually decorative and frequently non-native, shrubs and bushes. So extending those planting areas and replacing them with more ecologically salubrious species that still offer the kinds of aesthetically pleasing effects is only a small shift. But in making this shift, the rewilding gardener displaces grass monoculture—with the accompanying maintenance regime it requires—in favor of a colorful and varied landscape. The rewilding gardener can take pride in their accomplishment, and their labor presents a model for neighbors to follow. This modeling function is also critical to the practice insofar as the success of rewilding requires a collective activity in order to operate at the scale necessary for the practice's ecological goals. Here is a hidden potential not only to challenge the perceptual norms around landscaping but also prompt reflection on other sources of cultural neurosis such as the institution of property, the primacy of the liberal individual, and the accompanying anemic sense of community.

At first, the term "rewilding gardener" sounds oxymoronic, but it calls attention to an important aspect of rewilding that Gammon highlights, the need to rewild human culture. She describes two ways this may occur: (1) through a "reflexive rewilding" in which members of the culture cultivate contact with conditions wilder than these with which most in North American settler culture are accustomed in the name of rediscovering capabilities for living within such environments and (2) through a "primitivist rewilding" in which people seek to remove themselves from the destructive cycles of activity established in the culture in the name of reestablishing a conscious connection with the rest of the natural world (2018, 336–38). Though hardly primitivist in orientation, the rewilding gardener strives for the same goals, at least with respect to landscaping. David Cooper (2004) describes some of the ways gardens can facilitate such rewilding: (1) as sites of meaning produced through the collaboration of human and other natural agencies, they call attention to human imbrication in larger ecological communities, (2) they are transient and ever evolving, which invites reflection on human finitude, and (3) since they require care to maintain themselves over time, they cultivate a greater sense of responsibility for preserving the natural world for preserving the natural world. By reintroducing wilder relations in an aesthetically appealing way into the local landscape, the rewilding gardener seeks to work with natural processes in a cooperative and non-dominating manner.

What the rewilding gardener accomplishes in concert with a variety of other natural agencies is altering the perceptual norms for what constitutes a well-cared for landscape in the culture. They have in common certain characteristics: (1) they all beautify the inhabited land, (2) they embody less control over the land, (3) they make space for a variety of other living things, (4) they do not require continual maintenance to remain what they are, and (5) they all operate according to a distinct temporality that transcends the imposition of a human timescale. Although the lawn thus rewilded presents an aesthetic affront to the lawn and the power structures and relationships it represents, it cannot be said that these landscapes are uncared for. They represent the possibility of developing norms for cohabitation with the rest of nature, and perhaps most importantly, the necessity of human cooperation to realize those goals.

7.4 Conclusion

Addressing a cultural neurosis is no easy task given how obvious the various habits of thought, perception, feeling, and acting that create neurotic situations are within the culture. There is no single practice that is going to resolve definitively the culture's environmental neuroses. After all, environments are always changing and so the cultural habitus needs to develop along with them. Since neurotic habits stem from the structure of experience itself, the best that can be done is to address them as they arise.

Throughout, I have emphasized the potential of aesthetic practices—engaging with or creating earth art and gardening—to transform culturally neurotic perceptual norms. In addition to what I describe above, there are several other virtues to this approach. Engaging in small-scale practices not only strengthens relationships between the various parties involved but can also cultivate capabilities among the parties in relation. For example, in caring for a pollinator garden, the gardener becomes invested in the wellbeing of the plants, insects, birds, etc. that make up the wildness of the place. Seeing how quickly wildness can be restored provides the gardener with a heightened sense of their environmental agency *through* the flourishing of the ecological system. Such practices are critical to addressing the environmental problems the culture's dominant pathways of action create when one considers how enervating it is to be faced with global environmental devastation and feel as if there is nothing that can be done. Sometimes the political will to address such large-scale problems involving structures and networks of relations can only be mustered by building a sense that doing something matters.

It may appear counterintuitive to recommend individual aesthetic practices to address the cultural neurosis at hand given that the above critique of how individualism contributes to situations becoming environmentally neurotic. But practices individual people undertake are not necessarily individualistic. The aesthetic practices described here are seldom individualistic in the relevant sense, even if individual people undertake them. They are inherently social in orientation given their public nature and ecological structure. Isis Brook, for example, has described how gardening creates social virtues through the sharing of a garden's excess, exchange of materials, passing on of knowledge, and camaraderie (2010, 22–23). So, even in situations where a person is gardening alone in front of their home or has an allotment, the individual garden invites conversation, serves as a model for others, and, perhaps most importantly, depends upon the collective agency of the natural world.

Additionally, since individual practices such as these are social in orientation, we can see how micro-political practices are inherently related to changes in macro-political contexts. Many environmental problems transcend personal choices and behaviors. For example, political choices such as housing and zoning policies shape the context of personal choices and will therefore have greater environmental impact. Some have consequently argued that alteration of personal habits are no substitute for political engagement (e.g., Maniates 2001). Although there is some truth to this point, it conflates personal choices with an individualistic orientation. After all, attending a protest or participating in local government are still personal choices. The current analysis highlights three ways in which micro-political practices are related to macro-political actions: (1) personal habits are formative of the perceptual norms that enable specific environmental situations to be perceived as problems, (2) the political will to address a problem is frequently dependent upon the existence of a public, which is in turn formed through personal experiences, aesthetic encounters and the like, and (3) personal aesthetic practices, insofar as they challenge aesthetic norms, are themselves modest political acts. Micro-political practices neither preclude, serve as a substitute for, nor take precedence over macro-political responses to environmental problems. Ideally, the two serve to reinforce and encourage one another.

As phenomenological philosophy has long recognized, we are all involved in the project of being a person. That is, we seek to live meaningful lives in the context of others who also seek the same. Our choices influence theirs, theirs our own, and they all take place within nested networks of social relations that constrain and enable specific choices. The dominant cultural interpretations used to determine the kind of person one will be, to commit oneself to the projects that will be the substance of that life, and to help to shape the collective life we all share are, with respect to our ecological existence, especially conducive to producing neurotic situations. The aesthetic practices I have described here are intended to serve as examples of the work necessary to transforming the received perceptual norms and thereby therapeutically address neuroses. What specifically will emerge is uncertain, of course, and will inevitably have its own problems. But, whatever form cultural change takes, we will at least be more mindful of our impact on the earth.

Notes

1 Contemporarily we may look to Val Plumwood's (2002) call for an environmental culture or Carolyn Merchant's work; in more recent history to Lynn White Jr.'s (1967) and Lewis W. Moncrief's (1970) exchange in *Science* concerning the cultural roots of the ecological crisis or Aldo Leopold's (1949) attempt to diagnose the malaise; or a lengthier look at history could cite Henry David Thoreau, William Blake, or Seneca.

2 It is important to note that Russon's conception of neurosis differs significantly from the traditional psychoanalytic conception of it in a number of ways. Most notably for current purposes, it is not a

quality of an individual psyche. Neuroses are a complex relationship between individual, world, and others within a situation, though they may still be characterized by anxiety and depression for the individual within that context.

3 Brady agrees with Allen Carlson's (1986) assessment of Michael Heizer's (1969) *Double Negative* as an example of a work of earth art that fails to display an aesthetic regard for nature. Though I agree with them that the work itself recommends the sort of masculinist mastery of the land that characterizes the dominant culture's relationship to the land, seeing that attitude embodied in the 1500′ × 50′ × 30′ trench may prompt reflection on whether such an attitude makes any sense. In presenting the ideal of mastery in this way, perhaps its absurdity is laid bare.

References

Bannon, Bryan E. 2011. "Re-Envisioning Nature: the Role of Aesthetics in Environmental Ethics." *Environmental Ethics* 33: 415–436.

Boetzkes, Amanda. 2010. *The Ethics of Earth Art*. University of Minnesota Press.

Bourdieu, Pierre. 1977. *Outline of a Theory of Practice*. Translated by Richard Nice. Cambridge University Press.

Brady, Emily. 2007. "Aesthetic Regard for Nature in Environmental and Land Art." *Ethics, Place & Environment* 10 (3): 287–300.

Brook, Isis. 2010. "The Virtues of Gardening." In *Gardening: Philosophy for Everyone*, edited by Fritz Allhoff and Dan O'Brien. Wiley-Blackwell.

Carlson, Allen. 1986. "Is Environmental Art an Aesthetic Affront to Nature?" *Canadian Journal of Philosophy* 16 (4): 635–650.

Casey, Edward S. 2004. "Mapping the Earth in Works of Art." In *Rethinking Nature: Essays in Environmental Philosophy*, edited by Bruce V. Foltz and Robert Frodeman. Indiana University Press.

Christo and Jeanne Claude. 1983. *Surrounded Islands*. Biscayne Bay, Greater Miami, FL, United States. https://christojeanneclaude.net/artworks/surrounded-islands/

Cooper, David E. 2004. *A Philosophy of Gardens*. Oxford University Press.

Drenthen, Martin. 2018. "Rewilding in Layered Landscapes as a Challenge to Place Identity." *Environmental Values* 27: 405–425.

Eliasson, Olafur and Minik Rosing. 2014. *Ice Watch, 2014*. Multiple Locations, Copenhagen, Denmark. https://olafureliasson.net/artwork/ice-watch-2014/

Eliasson, Olafur and Minik Rosing. 2015. *Ice Watch, 2014*. Multiple Locations, Paris, France. https://olafureliasson.net/artwork/ice-watch-2014/

Eliasson, Olafur and Minik Rosing. 2018. *Ice Watch, 2014*. Multiple Locations, London, England, United Kingdom. https://olafureliasson.net/artwork/ice-watch-2014/

Foster, Cheryl. 1992. "Aesthetic Disillusionment: Environment, Ethics, Art." *Environmental Values* 1: 205–215.

Gammon, Andrea. 2018. "The Many Meanings of Rewilding." *Environmental Values* 27: 331–350.

Goldsworthy, Andy. 1997–98. *Storm King Wall*. Storm King Art Center, New Windsor, NY, United States. https://collections.stormking.org/Detail/objects/401

Heizer, Michael. 1969. *Double Negative*. Overton, NV, United States. https://www.moca.org/exhibition/michael-heizer-double-negative-1969

Jordan, Chris. 2009–13. *Midway: Message from the Gyre*. Midway Islands. https://www.chrisjordan.com/Midway/

Leopold, Aldo. 1949. *A Sand County Almanac*. Oxford University Press.

Lin, Maya. 2021. *Ghost Forest*. Madison Square Park, New York, NY, United States. https://www.mayalinstudio.com/memory-works/ghost-forest-msp

Maniates, Michael. 2001. "Individualization: Plant a Tree, Buy a Bike, Save the World?" *Global Environmental Politics* 1: 31–52.

Merleau-Ponty, Maurice. 1955. *Les aventures des dialectique*. Gallimard.

Merleau-Ponty, Maurice. 1960. "Le langage indirect et les voix du silence." In *Signes*. Gallimard.

Merleau-Ponty, Maurice. 1964a. *L'oeil et l'esprit*. Gallimard.

Merleau-Ponty, Maurice. 1964b. *Le visible et l'invisible*. Gallimard.

Moncrief, Lewis W. 1970. "The Cultural Basis for Our Environmental Crisis." *Science* 170 (3957): 508–512.

Nash, David. 1977 (ongoing). *Ash Dome*. Cae'n-y-coed, Wales.
Nassauer, Joan. 1997. "Cultural Sustainability: Aligning Aesthetics and Ecology." In *Placing Nature: Culture and Landscape Ecology*, edited by Joan Nassauer. Island Press.
Plumwood, Val. 2001. "Nature as Agency and the Prospects for a Progressive Naturalism." *Capitalism Nature Socialism* 12: 3–32.
Plumwood, Val. 2002. *Environmental Culture: The Ecological Crisis or Reason*. Routledge.
Rechter, Sue. 2007. "The Originating Breaks Up: Merleau-Ponty, Ontology, and Culture." *Thesis Eleven* 90: 27–43.
Robson, Aurora. 2008. *The Great Indoors*. Rice Gallery, Houston, TX, United States. https://www.aurorarobson.com/the-great-indoors
Russon, John. 2014. *Human Experience: Philosophy, Neurosis, and the Elements of Everyday Life*. SUNY Press.
Saito, Yuriko. 2020. "Everyday Aesthetics and World-Making." *Contrastes: Revista Internacional de Filosophía* 25: 35–54.
Steinberg, Ted. 2006. *American Green: The Obsessive Quest for the Perfect Lawn*. W.W. Norton.
Swidler, Ann. 1986. "Culture in Action: Symbols and Strategies." *American Sociological Review* 51 (2): 273–286.
Turrell, James. 2007. *Dividing the Light*. Pomona College Museum of Art, Claremont, CA, United States. https://www.pomona.edu/museum/collections/james-turrell-skyspace
White Jr, Lynn. 1967. "The Historical Roots of Our Ecological Crisis." *Science* 155 (3767): 1203–1207.
Wood, David. 2013. "Can Only Art Save Us Now?" In *Interpreting Nature: The Emerging Field of Environmental Hermeneutics*, edited by Forrest Clingerman, Brian Treanor, Martin Drenthen, and David Utsler. Fordham University Press.

8
THE ACTIVITY OF HUMAN PASSIVITY AND THE PROCESS OF NATURE

Renaud Barbaras' "Eco-Phenomenology" vis-à-vis the Anthropocene Paradox

Petr Prášek

8.1 The Anthropocene Paradox and Eco-Phenomenology

Across both the natural and the social sciences as well as humanities, the term Anthropocene is now widely used to capture the transformation with regard to humans and their place in the world: the term refers to a revolution in the relationship between humans and nature, a new geological epoch or event in which humans have become a planetary force comparable to geological forces.[1] Despite this, Anthropocene discussions are still marked by remnants of the traditional dualism culture/nature, and this is probably necessary in that the Anthropocene is primarily a paradox pointing to the fact that the new epoch does not imply the emergence of some undifferentiated homogeneous entity inseparably comprising the human and the non-human, in which it would no longer make sense to distinguish the two: humanity's unprecedented power, its becoming-nature, is paradoxically associated with a huge increase in the unintended consequences of human action; human activity intending to overcome the passivity inherent in our thrownness into the world leads to an even more intense passivity, to a profound ecological crisis for which there is no solution. The dilemma usually imposed on us today – that between modern titanism manifested by an excessive faith in technology, that would eventually overcome this crisis too, and despair or the apocalyptic moods of especially younger generations suffering from environmental grief – must be exposed as a *false* one (Abram, Hine 2011, 7; Coeckelbergh 2021, 96). Apart from pretending as if there were no other alternatives, humanity today does have a certain privileged position vis-à-vis other beings of this planet, thanks to which it can at least moderate if not remedy the current situation. And the belief in technological solutionism should be ruled out because it has obviously been the desire for too much control and selfish management of the planet that landed us in the present situation of ecological devastation.

It seems that the only way to deal with such a state of affairs will be meticulously to untangle the puzzle of passivity and activity. In other words, there is certainly hope for dealing with the crisis, with that paradox of the Anthropocene but we should put aside all the

DOI: 10.4324/9781003538097-10

quickly offered solutions, including the very way of looking at the crisis as a problem with a (technological) solution. Technology in itself is certainly not inherently bad, it is certainly not *a priori* opposed to the possibility of life in accordance with the character of nature (Kohák 1987, x), but any dogmatic belief in it and its interventions is based on deep assumptions about what nature is or what we are, and should be accompanied by challenging philosophical questions (Evernden 1999, 35; Kohák 2000, xxiv–xxvi). In short, it seems that there is another way to tackle the ecological crisis, an eco-phenomenological way which consists in uncovering all the assumptions burdening the relationship between humans and nature, in its new ontological description and the resulting new attitude of humans towards themselves and nature. Using the methods of phenomenology, eco-phenomenology sets aside all assumptions and beliefs about what nature might be and, on the basis of a description of lived experience with it, it seeks a new conception of the relationship between humans and nature, as well as its implications for ethics (Brown, Toadvine 2003, xi–xiii; Toadvine 2017a, 175–176).

The main branch of eco-phenomenology today seems to be the Anglo-American one, and very influential within it are especially the texts based on the work of M. Merleau-Ponty. If we focus directly on the rebus of passivity and activity, we can refer to the work of probably the most prominent contemporary eco-phenomenologist David Abram. According to his famous book, *The Spell of the Sensuous: Perception and Language in a More-Than-Human World* (Abram 1997),[2] human activity has unplanned, overwhelming impacts, since it rests firmly on abstract thinking and its conception of reality and does not take into account the animated reality of the senses, where everything speaks or gesticulates to us (cf. Abram 1997, 9). More precisely, the key moment in the alienation of humans from sensuous nature came when humans began to separate their speech from its sensuous context by writing it (Abram 1997, 63–85) for this created an entirely new – purely human – space and time of abstract meanings that drowned out the mute call of sensible nature (Abram 1997, 86). This, according to Abram, is the beginning of the ecological crisis. In his critical response, Ted Toadvine, another Maurice Merleau-Ponty scholar, objects to Abram, among other things, that he cannot corrupt the unity of Merleau-Ponty's flesh of the world by introducing a new dualism of sensation versus writing and reflection (Toadvine 2005, 161). In other words, he claims that human alienation from nature has its origin not in phonetic writing, but first and foremost in the situation of the human body located in nature, whose correlate is the irrevocable transcendence of sensible nature (Toadvine 2005, 167–168). Thus, ecological consciousness arises not from sensible proximity to nature but rather from awareness of its transcendence (cf. Toadvine 2003, 139–153).

The main thesis of this chapter is that Renaud Barbaras, also a follower of Merleau-Ponty, who does not explicitly engage in eco-phenomenology or in the debate between Abram and Toadvine, has quite remarkably taken it up and synthesised all its important points in an original eco-phenomenological theory. Barbaras would probably formulate the problem of passivity and activity as follows: passivity towards the unintended consequences of human activity is linked to our finitude, our senses, and our body, and should one aim to understand it thoroughly, one must pose the usually unasked question of where this passivity comes from. This chapter will try to demonstrate that by virtue of his original answer to this question Barbaras has made a significant contribution to eco-phenomenology. A number of eco-phenomenological implications of his original theory will be listed, which offer new perspectives on the Anthropocene paradox of activity and passivity.

8.2 The Activity of Passivity: Desire as the Essence of Human Subjectivity

In searching for the essence of passivity, we often forget that it is the *capacity* of receptivity: the passive, from the Latin *passivus*, is one who is "capable of feeling or suffering".[3] According to Barbaras, who focuses on this active dimension of passivity as requiring something from the one who is passive, the passivity of sensory receptivity comes obviously from the body and the sense organs. But what does it mean to have a body and sense organs? (Barbaras 2019a, 103) According to Barbaras, it does not mean – as we usually believe – to let reality pass through subjective filters, but, on the contrary, to be able to receive a reality that is itself sensible. The sensible being is the world in the sense of the depth of sensory appearances, the third dimension of space and spatial things (in addition to width and height), their inexhaustible reservoir guaranteeing the continuation of our sensation (Barbaras 2019a, 89), which – although not actually there – is constantly present for us. In other words, sensation, that chaotic flow of sensations which we usually consider to be the most intimate part of the self, must be interpreted as an intrusion of something that is not the self, a receipt of the exteriority or transcendence of the world. Sensations are not the material of consciousness from which it constructs the meaning of the world, but various expressions of the human encounter with the world. Barbaras asks the old philosophical question of how such a capacity for receiving the world in the intimacy of consciousness (i.e., intentionality) is possible, and answers that consciousness must have a capacity of going beyond its limits – and he calls this capacity "desire" (Barbaras 2019a, 108–109). It is already implied in this that, following Jan Patočka, he rejects Edmund Husserl's solution: the core of intentionality cannot be subjective acts of consciousness, *noeses*, which are themselves immanent to consciousness (cf. Patočka 2015, 34). Barbaras sees Husserl's explanation as a fatal step that completely inverts the sense of subjectivity: instead of being a moment of the phenomenal field, the recipient of the appearances of the world is considered interiority in which the phenomenal field is located (Barbaras 2019b, 84–85). A subjectivity capable of receiving the sensible being of the world must, on the contrary, have at its base a stepping out of itself, a movement by which it transcends each of its actual appearances towards the world as the reservoir of all possible appearances. Desire is thus not a need that can be satisfied; on the contrary, every satisfaction, every phenomenalization of the world deepens and intensifies it, since its true object, what it desires, is the world as such. The activity behind the passivity of sensory receptivity is thus this movement of transcending, of aspiring to what is not yet given, but desire already communicates with it, in such a way that it wants more each time, that it is always disappointed, that it is always open to the new (Barbaras 2019a, 109–112). "As the intention of a presence that is in principle unpresentable, desire opens up the pure transcendence of the world; its inextinguishable thirst is pure reception." (Barbaras 2006, 113). But it is not necessary to stay on such an abstract level of description of desire as essence of subjectivity. Desire as an activity in the depth of passive receptivity is nothing but the movement of the body and its sense organs. The receptivity of perception, for example, of some actual seeing or some actual touching, presupposes an interdependence of sensation and movement that cannot be reduced to objective displacement: we see through a certain movement of the eyes that search, fix and accommodate, we feel the texture of a surface through the delicate work of the fingers running over the surface and pressing against it in just the right way (Barbaras 1994, 64–65). Every actual perception is immediately polarised by its horizon – which again triggers the movement of desire – whether it is in the form of what Husserl calls the inner horizon (other

8.4 Barbaras an Eco-Phenomenologist?

An objection to our interpretation of Barbaras as an eco-phenomenologist might be as follows: it is just another phenomenological theory of nature, which can be found in every major author, but not an eco-phenomenology, which, in addition to its explicit link to ecological issues, includes ethical considerations related to the human–nature relationship – and this is absent in Barbaras. This objection is correct, and yet formulated in this way not accurate because what Barbaras does in his cosmology comes close to other authors who are classified under eco-phenomenology. They might indeed state an explicit interest in ecological topics but then they carry out a phenomenological project by broadly building on potentially useful concepts and considerations of classical phenomenological authors – such as Husserl's "natural world" (Brown 2003), Heidegger's "being-in-the-world" or his interpretations of nature as *phusis* (Schalow 2022; Zimmerman 2003), Merleau-Ponty's "corporeality," (Abram 1997; Toadvine 2014) or Lévinas' notions of the "face" and "nutrients" (Benso 2000; Pelluchon 2015).

This is also true of the aforementioned Abram and Toadvine. Abram's thought is an "applied" phenomenology not in the sense that it works with empirical data from ecology or other sciences, but in the sense that it undoubtedly extends phenomenology to new topics by asking how Merleau-Ponty's philosophy might contribute to solving the ecological crisis. Barbaras implicitly builds on Abram's discussion with Toadvine, develops further Merleau-Ponty's ontology of flesh by answering a question that Merleau-Ponty himself did not answer – and this has many, if rather subtle, implications for eco-phenomenology as such. Certainly, it is not an attempt to use phenomenology explicitly to solve dilemmas in ecology or ecological philosophy, as is the case with eco-phenomenological authors who approach new original analyses and attempts at synthesis that could contribute to a new understanding of current social challenges, usually summarised under the term environmental crisis (e.g., Bannon 2014; Booth 2020; Hess 2023; Pelluchon 2020; Wood 2019). But implicitly Barbaras also contributes to contemporary debates: as this text has been at pains to show, he contributes to the clarification of the Anthropocene paradox of activity and passivity and, in general, he does exactly what philosophy first should do, that is, illuminate basic relevant concepts such as living being, body or nature.

Moreover, by addressing the question of how human alienation from nature phenomenologically appears, Barbaras completes what Toadvine finds missing in Abram's work.[4] In an ontology of nature founded on the monism of the flesh, human alienation cannot be explained by reference to written language alone. Such alienation must in some way already have been prefigured in the human body, in its difference from non-human bodies. Barbaras has long been trying to explain humans and animals anew, within the framework of a single process of life that produces both beings, without their difference being one of degree or quality. In what consists, then, the anthropological difference? He ponders his answer: this difference can be neither one of degree nor of quality: humans are not animals that are more adapted, more cunning or more efficient, because they obviously live and act in a different way from animals, but their difference is not qualitative either; humans are not animals that possess one more quality, for example reason or language, because that would be incompatible with the profound ontological continuity that unites them with animals (Barbaras 2008, 234–235, 2011, 168). Many years before his book on belonging, Barbaras concludes that difference must consist in the negation or deprivation of life in human beings. In other words, when Barbaras asks what must have happened to life for it to become human life, he answers that human life consists in a negation of the animal's original openness to the world: humans are able to repress the immediate call of the world and see it from a distance, that is, as an object situated in front of them (Barbaras 2008, 243–246), and even theoretically speculate about it and their own existence.

This is why human culture appears as a negation of *phusis* (Barbaras 2011, 174) and Barbaras' first conception of anthropological difference resembles the traditional conception of humans as excluded from nature. But there is a turning point in the book on belonging. Barbaras himself knew that thinking of human difference as a negation rather than a positive attribute was not enough, unless he demonstrated, against the background of his monistic conception of life, how human difference originated in life itself. And it was this problem that led him to change his conception. Now he argues that the human being does not differ by being more separated from the world, but on the contrary by belonging to nature the most, and therefore phenomenalizing it the most. This might seem as a return to traditional humanism or anthropocentrism but that is not the case. Indeed, Barbaras adds to this, in line with his previous position on anthropological difference, that "the more phenomenalization, the more occultation" – which is a purely human possibility; in an animal, the ontological ground (*sol*) can never be obscured by objectification (Barbaras 2019b, 107; cf. Deudon 2022, 72). We will return later to this human obfuscation of the world. Now it is necessary to emphasize that Barbaras' entire cosmological perspective, his prism of corporeality, does not lead to any form of anthropocentrism but, on the contrary, to the fact that all beings are equal in it but without humans losing their distinctiveness and responsibility. Different beings vary in the degree of belonging to the world (Barbaras 2019b, 76) (and, consequently, in the degree of its phenomenalization), but they all inhabit the world and all phenomenalize it in their own way: from the viewpoint of "cosmophany," differences do not exist, the only differences are topophanous, that is, they consist in the fact that each place necessarily phenomenalizes the world in a different way, with regard to its particular site. This is why one cannot reduce the one to the other and hold, for example, the traditional thesis of a hierarchy of being, according to which stones are inanimate and trees, although alive, are soulless, not sentient, which is in turn a quality that animals have, etc. – within Barbaras' conception, one cannot reduce, for example, an animal being to the simpler being of plant.

It can be said that Barbaras develops much more precisely, with more thorough phenomenological argumentation, the thesis found in Abram: his intuition of a single consciousness of the world (Abram 2011, chap. Mind). This implies, among other things, that the value of nature cannot be derived from its value to humans, from their consciousness. Nature is a process that can be expressed by different types of bodies.[5] According to Abram, through the fold of their bodies humans are part of the only consciousness of the Earth, into whose depths they penetrate, so that even what they consider to be their own thoughts does not really belong to them: thoughts do not come from some "inside" of thought, but arise through walking the open terrain of the landscape of thought, where human bodies meet the body of nature – just like other non-human bodies that gather the world in their own way (Abram 2011, 126). Barbaras supported this with a thoroughly developed theory of a single flesh of the world and its disintegration into multiple sites in the process of deflagration.

8.5 Conclusion: On the Ethical Dimension and Limits of (Barbaras') Eco-Phenomenology

Barbaras almost never writes about ethics, because he believes that phenomenology is a purely descriptive rather than a prescriptive discipline. At the same time, however, he is one of the authors who believes that it is possible to draw ethical implications from phenomenological descriptions, and he himself suggests how this might be done in the case of his own phenomenologico-cosmological theory. This is how even the last important dimension of eco-phenomenology appears in his work. In a 2016 interview, when Barbaras still thought that the essence of humanity lay in the highest degree of separation from nature, he claimed that evil is a consequence of our metaphysical situation, finitude, and separation from the world (Barbaras 2019a, 292). With regard to the latest development of his conception, this can be reformulated as follows: human action is so destructive in relation to nature because humans draw all their power from nature, and this power is enormous because of all beings they are closest to nature, they are characterised by the highest degree of belonging, and so they are able to phenomenalize the world as such; at the same time, however, human phenomenalization is afflicted by the separation, by the specific form of the human body as site, in which is embedded the danger that this power might be abused in favour of purely human objectifying phenomenalizations of nature. In other words, the human being, because of its specific site, occupies a cognitive relation to nature – nature as ontological ground and our belonging to it are obscured by the duality of sovereign consciousness/objective world (Barbaras 2019b, 107–108).

That is why technology cannot be the solution to the Anthropocene paradox. Barbaras addresses explicitly this question in an essay where he claims that technology is not an adaptive tool of life itself, but rather a purely human solution to overcoming the separation and surviving in spite of it (Barbaras 2011, 169–175). This is why the technological practices of our management of the planet have had so many unintended consequences – human technology harnesses the power of nature, but it is not a tool of nature itself in that it is based on its objectification. On the other hand, this does not mean that humans should succumb to despairing of being irreparably separated from nature – of the inevitability of ecological destruction. As has been already said, technology (or objective knowledge) is certainly not bad in itself, it is "only" its removal from its ontological context that is dangerous. Barbaras even suggests a way out of the hopelessness of the Anthropocene paradox when Mathias Goy explicitly asks him about the ecological theme in the above-mentioned interview:

> "Given your assertions about human beings being separated, exiled from life [...] how can we understand that humans destroy the environment in which they live, forgetting the life that they are?", to which Barbaras replies: "This is a complex set of questions, which I find difficult to answer in this context. Nevertheless, it seems to me that you yourselves are outlining the answer, or at least the principle behind it. If our situation is one of ontological exile, if we are separated from the arche-life [...] it is not surprising that life has become alien to us and that just about everything is permitted. It is therefore absolutely necessary and urgent to take the measure of our cosmic belonging [...] to understand that if we are exiled from life, the whole force of our desire derives from its power."
>
> (Barbaras 2019a, 293–294)

Quoting this interview, Grégori Jean interprets Barbaras' words so that the ethical treatment of nature (and thus the possibility of an eco-phenomenological ethics) is inscribed in our belonging to it (Jean 2026). Barbaras is indeed clear here: we are destroying the environment because we are separated from the life of nature. When we understand the sense in which we are part of it, this will lead to a correction of our actions – everything will no longer be permitted. It is not enough, then, to draw implications for our actions from phenomenological descriptions. Nature has different meanings in our experience, it is certainly not just the ontological ground to which we belong, but also a mere object that we exploit for our selfish purposes. The possibility of an eco-phenomenological ethics is thus bound up with the ontological priority of one phenomenological description or register over others. The notion of register is Richir's: registers are distinct systems of unifying phenomenological plurality – distinct ways of experiencing what appears. From one point of view, all registers, all the different ways in which nature appears, are equal, they are all part of human experience, but from another point of view they are not: one description has ontological priority, whereas others are merely obscuring nature as ontological ground. When describing such a cosmological ethics in more detail, Barbaras refers to Eugène Minkowski, in whose work ethics also converges with cosmology, involving an overcoming of human finitude, a kind of dispossession of the self, and opening to the flow of the anonymous life of nature – the most general ethical rule is to be worthy of the flow of life and its "humanity" (Barbaras, Ascárate 2022, 193–194). Erazim Kohák similarly speaks of the need for humanism, while abandoning its traditional, Enlightenment meaning. Both Minkowski's and Kohák's approach is, therefore, *posthumanist* rather than a humanist.

Of course, there might be one last objection: the whole eco-phenomenology is sufficient in theory, but does very little in terms of the urgency of today's ecological challenges. Given these, we need to know as soon as possible how to act in particular situations. This is a legitimate objection. The eco-phenomenology that we have been working on so far, that of Abram, Toadvine, or even that of Barbaras, is an important starting point, the ontological basis of all discussions. If only because it recognises that "problem-solving" itself belongs to the old thinking that brought us to the crisis: thinking that the crisis can be controlled and counted and solved, that we must reduce emissions by exactly so much and so many. Reducing emissions and other precisely calculated solutions are undoubtedly important, but if our primary concern is sustainable living and a relationship with nature, we must not succumb to the economic-technical paradigm.

But an eco-phenomenological stance seems to be not enough on its own; we need to find a way to apply it more widely. What we need is more specific design of eco-phenomenological practice. But will it then still be eco-phenomenology? According to what M. Burch suggests in his recent article on applied phenomenology, yes (Burch 2021, 275–294). In his opinion, in order for (eco-)phenomenology to be truly applied, it would be necessary to incorporate phenomenological findings into the descriptions of that problem by other sciences. Eco-phenomenology as a purely philosophical enterprise is not irrelevant, but it will use its potential to the maximum only if an eco-phenomenologist, as C. Pelluchon demands, goes beyond phenomenology (Pelluchon 2020, 55–59). For example, if she – as an economist (or in cooperation with him) – deals with the economic system and look for ways to incorporate the sense-experienced empathy towards the non-human world into the reforms of industrial mass production of meat or eggs. This seems to be the only way to definitely solve the conundrum of the Anthropocene paradox: to set limits to concrete forms of human activity by practising a new posthumanist, eco-phenomenological stance.[6]

Notes

1 See Crutzen, Stoermer 2000. The debate has not been stopped by the recent refusal of the Subcommission on Quaternary Stratigraphy (part of the Anthropocene Working Group) to use the word Anthropocene as a name for a geological epoch – for the Anthropocene is, according to many scientists, not an epoch but rather a so-called geological event (see e.g., Walker et al. 2024).

2 The following summary of the dialogue between Abram and Toadvine is a reformulation of passages in Prášek 2023, 370–371.
3 https://www.etymonline.com/word/passivity.
4 Abram responded to Toadvine's critique at length in a separate article, but its main impact seems to be his later greater emphasis on the role of the transcendence of nature. Without explicitly mentioning Toadvine, in his second major book, *Becoming Animal* – in contrast to what was said in *The Spell of the Sensuous*, in which the focus of his eco-phenomenology is the bodily entanglement with nature – Abram carries out an extensive phenomenological analysis of the transcendence of both living and non-living natural beings (cf. Abram 2005, 2011).
5 B. Bannon, who develops Merleau-Ponty's conception of nature from his lectures, comes to something similar: he feels the need to extend the ability to express the process of nature to non-living beings – the only difference with Abram is that Bannon does not attribute to them a conscious awareness (Bannon 2014, 133–134).
6 This work was supported by the European Regional Development Fund project "Beyond Security: Role of Conflict in Resilience-Building" (reg. no. CZ.02.01.01/00/22_008/0004595), which was implemented by the project partner, the Institute of Philosophy, Czech Academy of Sciences (IP CAS).

References

Abram, David. 1997. *The Spell of the Sensuous, Perception and Language in a More-Than-Human World.* New York: Vintage Books.
Abram, David. 2005. "Between the Body and the Breathing Earth, A Reply to Ted Toadvine," *Environmental Ethics* 27, 171–190.
Abram, David. 2011. *Becoming Animal.* New York: Vintage Books.
Abram, David, Hine, Dougald. 2011. "Coming to Our Animal Senses, A Dark Mountain Conversation Between David Abram and Dougald Hine," *Dark Mountain* 2, available at: https://www.academia.edu/38457103/Coming_To_Our_Animal_Senses_David_Abram_and_Dougald_Hine_2011_pdf
Bailly, Jean-Christophe. 2017. "Souveraineté des pierres," in J. C. Bailly, C. Benzaria, J. Castex (eds.), *Être pierre, cataloque de l'exposition au Musée Zadkine du 29 Septembre 2017 au 11 février 2018.* Paris: Musée Zadkine.
Bannon, Bryan E. 2014. *From Mastery to Mystery. A Phenomenological Foundation for an Environmental Ethics.* Athens (Ohio): Ohio University Press.
Barbaras, Renaud. 1994. *La perception. Essai sur le sensible.* Paris: Hatier.
Barbaras, Renaud. 2006. *Desire and Distance. Introduction to a Phenomenology of Perception.* Trans. Paul B. Milan. Stanford: Stanford University Press.
Barbaras, Renaud. 2008. *Introduction à une phénoménologie de la vie.* Paris: Vrin.
Barbaras, Renaud. 2011. *La vie lacunaire.* Paris: Vrin.
Barbaras, Renaud. 2013. *Dynamique de la manifestation.* Paris: Vrin.
Barbaras, Renaud. 2019a. *Recherches phénoménologiques.* Paris: Beauchesne.
Barbaras, Renaud. 2019b. *L'appartenance. Vers une cosmologie phénoménologique.* Louvain-la-neuve: Peeters.
Barbaras, Renaud. 2021. "Étude sur la cosmologie d'Eugène Minkowski," in E. Minkowski (ed.), *Vers une cosmologie.* Paris: Les éditions de compagnons d'humanité, pp. 267–377.
Barbaras, Renaud, Ascárate, Luz. 2022. "Discussion," in C. Riquier, C. Bobant (eds.), *Donner lieu. Conférences et débats sur la cosmologie phénoménologique de Renaud Barbaras.* Paris: Les éditions de compagnons d'humanité, pp. 191–195.
Benso, Silvia. 2000. *Face of Things. A Different Side of Ethics.* New York: SUNY Press.
Bergson, Henri. 1977. *The Two Sources of Morality and Religion.* Trans. R. Ashley Audra, Cloudesley Brereton. Notre Dame: University of Notre Dame Press.
Booth, Robert. 2020. *Becoming a Place of Unrest. Environmental Crisis and Eco-phenomenological Praxis.* Athens (Ohio): Ohio University Press.
Brown, Charles S. 2003. "The Real and the Good: Phenomenology and the Possibility of an Axiological Rationality," in C. Brown, T. Toadvine (eds.), *Eco-phenomenology. Back to the Earth Itself*, New York: SUNY Press, pp. 3–18.
Brown, Charles S., Toadvine, Ted. 2003. "Eco-Phenomenology: An Introduction," in C. Brown, T. Toadvine (eds.), *Eco-phenomenology. Back to the Earth Itself.* New York: SUNY Press, pp. ix–xxi.

Burch, Matthew. 2021. "Make Applied Phenomenology What It Needs to Be: An Interdisciplinary Research Program," *Continental Philosophy Review* 54, pp. 275–293.
Coeckelbergh, Mark. 2021. *Green Leviathan or the Poetics of Political Liberty, Navigating Freedom in the Age of Climate Change and Artificial Intelligence*. New York/London: Routledge.
Crutzen, Paul J., Stoermer, Eugene J. 2000. "The 'Anthropocene'," *IGBP Newsletter* 41, 17–18.
Deudon, A. 2022. "La vivante différence entre les vivants," in C. Riquier, C. Bobant (eds.), *Donner lieu. Conférences et débats sur la cosmologie phénoménologique de Renaud Barbaras*, Paris: Les éditions de compagnons d'humanité, pp. 55–78.
Evernden, Neil. 1999. *The Natural Alien, Humankind and Environment*. Toronto: Toronto Unniversity Press.
Heidegger, Martin. 1996. *Being and Time*. Trans. Joan Stambaugh. New York: SUNY Press.
Hess, Gérald. 2023. *Conscience cosmique. Pour une écologie en première personne*. Paris: Dehors.
Husserl, Edmund. 1989. *Ideas Pertaining to a Pure Phenomenology and to a Phenomenological Philosophy, II*. Trans. Richard Rojcewicz, André Schuwer. Dordrecht: Kluwer.
Jean, Grégori. 2026. "Counter-belonging. Environmental ethics and the phenomenology of the thwarted conatus," in P. Prášek, K. Novotný, B. E. Bannon (eds.), *The Faces of Nature: Eco-Phenomenology and the Philosophy of Nature*, to be published by Lanham, Maryland: Lexington Books.
Kohák, Erazim. 1987. *The Embers and the Stars, A Philosophical Inquiry into the Moral Sense of Nature*. Chicago: Chicago University Press.
Kohák, Erazim. 2000. *The Green Halo, A Bird's-Eye View of Ecological Ethics*. Chicago: Open Court.
Merleau-Ponty, Maurice. 1964. "Eye and Mind," in *The Primacy of Perception*. Trans. James. M. Edie. Evanston: Northwestern University Press.
Merleau-Ponty, Maurice. 1968. *The Visible and the Invisible*. Trans. Alphonso Lingis. Evanston: Northwestern University Press.
Patočka, Jan. 2015. "Husserl's Subjectivism and the Call for an Asubjective Phenomenology," in L. Učník, I. Chvatík, A. Williams (eds.), *Asubjective Phenomenology*. Trans. Ivan Chvatík, Matt Bower, Kenneth Maly. Nordhausen: Verlag Traugott Bautz GmbH, pp. 17–40.
Pelluchon, Corine. 2015. *Les Nourritures. Philosophie du Corps Politique*. Paris: Seuil.
Pelluchon, Corine. 2020. *Réparons le monde. Humains, animaux, nature*. Paris: Rivages.
Prášek, Petr. 2023. "A Theological Turn in Phenomenology? Janicaud and Contemporary French Phenomenology," *Studia Phaenomenologica* 23, pp. 351–375.
Schalow, Frank. 2022. *Heidegger's Ecological Turn. Community and Practice for Future Generations*. London: Routledge.
Toadvine, Ted. 2003. "The Primacy of Desire and Its Ecological Consequences," in C. Brown, T. Toadvine (eds.), *Eco-phenomenology. Back to the earth itself*, pp. 139–154.
Toadvine, Ted. 2005. "Limits of the Flesh: The Role of Reflection in David Abram's Eco-phenomenology," *Environmental Ethics* 27, pp. 155–170.
Toadvine, Ted. 2014. "The Elemental Past," *Research in Phenomenology* 44, pp. 262–279.
Toadvine, Ted. 2017a. "Phenomenology and Environmental Ethics," in S. M. Gardiner, A. Thompson (eds.), *The Oxford Handbook of Environmental Ethics*. Oxford: Oxford University Press, pp. 174–185.
Toadvine, Ted. 2017b. "Naturalism, Estrangement, and Resistance: On the Lived Sense of Nature," in G. Kuperus, M. Oele (eds.), *Ontologies of Nature*. Dordrecht: Springer, pp. 181–198.
Walker, Michael J. C. et al. 2024. "The Anthropocene is Best Understood as an Ongoing, Intensifying, Diachronous Event," *Boreas* 53, pp. 1–3.
Wood, David. 2019. *Reoccupy Earth. Notes toward an Other Beginning*. New York: Fordham University Press.
Zimmerman, Michael E. (2003) Heidegger's Phenomenology and Contemporary Environmentalism, in C. Brown, T. Toadvine (eds.), *Eco-phenomenology. Back to the Earth Itself*, pp. 73–102. https://www.etymonline.com/word/passivity

SECTION 3

Place and Cosmos

9
A "TASTE FOR THE INFINITE"
Passivity as a Lens to Interpret Place, Transcendence, and the Anthropocene

Forrest Clingerman

Introduction

This essay began in the middle of a meadow, with my eyes closed and the sun shining.

During an evening walk in summer, I suddenly decided to stop and close my eyes. Stopping and closing myself from sight meant I was disoriented. I was no longer fully able to seize the "where that I am" in an overtly active way. It meant opening myself to vulnerability, because I needed to feel how I was receiving the surroundings. It was an invitation to experience the place differently: birds that were unnoticed only moments before became present through song, the faint smell of wildflowers now pervaded the engagement with my surroundings, the warmth of the sun and slight breezes touched me. I was not *encountering* the landscape by focusing my attention, but rather I felt *encountered by place* by opening myself. Or perhaps there was an intertwining of encountering being encountered, especially when I spent a few more minutes closing and opening my eyes. The continual refocusing of the senses was a rhythm, an inhalation and exhalation of place.

The rhythm of the self in nature's opening is what I'd like to meditate on, by reflecting on passivity as a practice of interpreting environments in everyday experience. Such an experience seems to align with the definitions of passivity in phenomenological literature. Indeed, we might conceptualize passivity as a "mediating term," which (in the words of Fiona Hughes in her discussion of Merleau-Ponty) "...emerges at the interstices of the passive and the active, the conscious and the unconscious, when these are understood as opposition terms..."[1] Seeing active passivity, figures like Husserl and Merleau-Ponty developed a sense of passivity as more than a mere opposite of activity. I would like to be a bit less methodical and creatively misread some phenomenologists, in order to reflect on passivity as a practice of transcendence, place, and self.

Passivity, Reflection, and the World

Even though this reflection started in a material space of nature, its path winds through a set of thinkers, who reflect further on the practice of passivity as an openness to the transcendent dimension of the world. I wish to explore the passivity of three philosophers

 DOI: 10.4324/9781003538097-12

and theologians: Martin Heidegger, Martin Buber, and Friedrich Schleiermacher. These three offer descriptions of the space between passivity, receptivity, and the sense of the transcendent.

As a start, we can look to Martin Heidegger, and how he warns of the need for a meditative thinking of "releasement" in the face of technology and instrumentalist thinking of the world. In the "Memorial Address," Heidegger claims that humans are beings with the capacity to think, but in the modern age we have become "thought-less"—humanity is now "in *flight from thinking*."[2] Concepts of the world have changed due to modern science and technology, making the world now an object of calculative thinking.[3] Nature itself is an example of this: so he says, "Nature becomes a gigantic gasoline station, an energy source for modern technology and industry."[4] Thus modern technology takes on the power to define its relationship to humanity—it rules the whole material world. The one type of thinking we now do frequently is calculative, characterized as instrumental planning and organizing.

Heidegger encourages us toward another type of thinking: meditative. He writes that only unceasing meditative thinking can combat calculative thinking. Meditative thinking does not stick to a single side of an idea, but engages in a letting-go—a mediation. As an example, technology is at this point indispensable to our thought process (perhaps even more now than when Heidegger was writing). But while we use technology, we can act otherwise—"...yet with proper use also keep ourselves so free of them, that we may let go of them any time."[5] In other words, we learn to say yes and no to technology. "I would call this comportment toward technology which expresses 'yes' and at the same 'no,' by an old word, *releasement toward things*."

Heidegger's releasement is a somewhat subject-focused understanding of passivity as a sense of abiding in the space of thinking. Jewish philosopher Martin Buber's discussion of the I–Thou relationship is a more personal and ethical understanding of a practice of passivity. In Buber's influential work *I and Thou*, he meditates on the self-world relationship, wherein the I can only fully manifest itself in a radically open relationship to the other. That is to say, Buber explains the separation between an instrumental vision of the world and an open receptivity to the Other. Only in the latter relationship does the human become fully human, or as Buber explains, the human "...becomes an I through a You."[6]

Buber's philosophy is predicated on relationality. He compares what he calls two word-pairs: I-It (Ich-Es) and I-Thou (Ich-du).[7] The I-It is instrumental, reductive, and requires only part of the self. The I-It world is orderly, reliable, objective, and defines the world of general experience, according to Buber. However, unlike Heidegger's generally negative description of calculative thinking, Buber recognizes that the I-It is necessary, even though it has obvious limits: it is also not an uncovering of the truth of the self or the world, but rather is "... primally alien both outside and inside you."[8] This is an utterance that sees the world in terms of finite objects, but it is still built on relation.

The I-Thou word-pair, in contrast, is a different relation—and a profoundly ethical one. When saying I-Thou, there is no object, only the relation between self and other outside space, time, and causality. It is complete, absolute, and infinite. Thus when speaking to another human as I-You, Buber says, he or she is not an object among objects—"Not as if there were nothing but he; but everything else lives in *his* light" (59). Throughout most of our everyday experience, we need to live in the It world; we can't live solely in the pure present of I-You. On the other hand, Buber suggests that every It has the potential to be a You. We require both activity of shaping the world and its passive embrace, for "...with It a human being cannot live. But whoever lives only with that is not human." (85)

We see the intertwining with the Transcendent in the work of theologian Friedrich Schleiermacher.[9] In the second of the five speeches that constitute the influential book *On Religion*, Schleiermacher zeroes in on the question, "what is religion?" (p. 18) His answer is to describe religion as an openness to the infinite or the experience of oneself in the totality of the universe. In Schleiermacher's words, religion "...wishes to intuit the universe, wishes devoutly to overhear the universe's own manifestations and action, longs to be grasped and filled by the universe's immediate influences in childlike passivity." (p. 22). Thus, religion has a core object: "the universe and the relationship of humanity to it." (p. 19). Indeed, one cannot have philosophical speculation or moral reasoning without religion—that would be "rash arrogance" (p. 23)—because

> man [sic] has merely stolen the feeling of his infinity and godlikeness, and as an unjust possession it cannot thrive for him if he is not also conscious of his limitedness, the contingency of his whole form, the silent disappearance of his whole existence in the immeasurable.

Schleiermacher concludes that an intuition of the universe requires us to be open to the world: as he puts it, religion is "the sensibility and taste for the infinite." (p. 23) In other words, religion is the word we use to describe our passivity in the midst of the Transcendent, as experienced through our sense of being an individual in the universe.

In his later dogmatics, *The Christian Faith*, he offers a more explicit framework for seeing this passivity in the face of Transcendence. Evolving the language of his earlier work, religion is an "immediate self-consciousness" or feeling. Schleiermacher says consciousness consists of two activities of the self, which alternate: "an abiding-in-self (*Insichbleiben*) and a passing-beyond-self (*Aussichheraustreten*) on the part of the subject" (p. 8). Likewise, self-consciousness has a self-caused side and a non-self-caused side (p. 13). What Schleiermacher attempts to explain is how the self relates to the Other. Feeling or immediate self-consciousness is "entirely an abiding-in-self," in the realm of receptivity. For the subject, the sense of receptivity becomes a feeling of dependence (p. 13-14). Religion or piety is the experience when we are conscious of a "feeling of absolute dependence"—otherwise put, the experience of being in relation to God. This is the highest form of immediate self-consciousness, an essential part of human nature, which becomes the basis for communion with the other. Such a "feeling of absolute dependence" shows a connection to how passivity opens us to transcendence, especially when associated with his earlier description of religion as a sensibility and taste for the infinite. This feeling is connected with our experience of the world, not separated from it.

Passivity and the Hermeneutics of the Anthropocene

I began with an example of a prairie walk, which intimated the passivity of an everyday experience. Next, I quickly summarized how three thinkers can offer facets of how to understand this as *a practice of passivity*. This interpretive passivity is fragmentary: it takes as a fragment from Heidegger a need for releasement, from Buber a moral embrace of the I-Thou, and from Schleiermacher a feeling of absolute dependence in the face of the Absolute. These fragments are not exhaustive, but provides the contours for understanding passivity as a practice, not a thinking.

For these three thinkers, the need for this practice is made explicit by how frequently we seize our surroundings through calculative, technological mastery or placing the self beyond other (we might, of course, question this too easy reading of technology). From that, we see a possible response: Schleiermacher, Buber, and Heidegger all contrast a reductive, acquisitive, and dis-embodied thinking with one that taps into a depth of embodied existence. For instance, Schleiermacher and Heidegger show a means to understand the capacity to think and feel as a *ground* on which to encounter with the other. Heidegger, of course, is more interested than Schleiermacher in focusing on the self as thinking being, while Schleiermacher focuses our attention on how we intuit the infinite as finite creatures. Buber emphasizes what is somewhat missing: practicing the relational nature of passivity, which centers on a fundamentally moral relation wherein our whole being embraces the whole being of the other. Furthermore, in each, this practice is embodied or grounded. Heidegger uniquely disagrees with Buber and Schleiermacher in how passivity and instrumental thinking ought to be intertwined, yet even he avoids disembodied thinking. The most significant difference is the realm in which each thinker would place passivity: Buber's work seems to hold ethics and relationality as primary, Schleiermacher's work gives preference to the pre-conceptual intuition or feeling, and Heidegger defines the open-region through a passivity of thinking as being.

What is more, while not in complete agreement, together they offer a vision through which we return to inhaling and exhaling in a prairie clearing, slowly opening and closing our eyes, wherein passivity is finding ourselves in the *how* we might rest in a reflexive *transcendence in place*.

To conclude, I'd like to offer three points to ponder that emerge from these vignettes about passivity, transcendence, self, and place. First, while it is not immediately obvious in the brief example of the prairie, Heidegger, Buber, and Schleiermacher point to *passivity as a means to open ourselves to a placial multivalence of interpretation*.[10] Each one sees passive thinking as an interpretation of experience through metaphors of materiality and location—of the world, the other, and finally the universe. For the case of nature, these levels of materiality require an openness to competing visions of being encountered by place. Furthermore, it requires identifying how passive encounters are mediated by differing levels of the world, the material, or the landscape. In other words, we see passivity occurring on a diversity of scales: for Schleiermacher it is the large-scale confrontation with the universe, for Heidegger it is the opening or clearing, and for Buber the intimate scale of the other. The practice of passivity in nature includes receiving the gift of an encounter with individual birds through song, the collection of beings in the prairie landscape clearing, and ultimately through the reception of the infinite itself. What is unique about the experience of passivity in each of these levels is that the self is open to a profound Otherness of the environment, which in turn leads to an openness by the relationship of self and environment to the Transcendent Other. There is a reflexive attitude in passivity, wherein the Transcendent appears through the reception of the world.

Second, through the interpretive practice of passivity in place, we can uncover the depth dimension of nature. The metaphor of a "depth dimension" is sometimes associated with theologian Paul Tillich,[11] but in the present circumstance it is Schleiermacher and Buber that more appropriately identified with seeing nature in openness and passivity. For Schleiermacher, the active words of "thirst" and "sensibility" for the world are paradoxically connected with a passivity toward what is not caused by the self. Rather, the infinite—the Transcendent, the depth of existence—comes through the feeling of absolute dependence, as this is experienced in connection with the individual sensing the self in the universe. We are given an opportunity to either interpret the world as an I-It, or be embraced by our environments in the utterance of I-You.

Finally, the passivity discussed here offers a choice about our current world of the Anthropocene, but not the choice we might expect. The concept of the Anthropocene—the Age of the Human—has been a point of lively discussion for almost two decades.[12]

We are sometimes tempted to explain that the Anthropocene as a fork in the road—an either/or that is between active and passive modes of human existence. One path of human thinking would be to see the Anthropocene as the world of overwhelming human action—an I-It, a world of instrumental thinking, and a perspective based on bounded freedom in the face of finite existence.[13] Eyes wide open and filled with activity, humans see the humanly caused Anthropocene as a dystopia bereft of wildness and the Other, which Heidegger, Buber, and Schleiermacher would caution us against. The other path, then, would be one that we enter into with inert passivity as we close our eyes and receive the world. This is a passivity that attempts to simply turn away from instrumental and reductive thinking in order to define more humane—rather than merely human—world. In the works of the thinkers we've discussed, there are hints of this supposed alternative: Heidegger's releasement looks to the nature as something besides a gas can, for example, or Schleiermacher's taste for the infinite is a means to see what is beyond the self in itself. Passive thinking here is a roadmap toward the thinking of the idea of a benevolent, "good Anthropocene," wherein humanity is still the sole totalizing force, but one that stewards to the threshold of being open to the more-than-human.[14]

Certainly, seeing passivity as a bifurcation of the Anthropocene might appear advantageous, because it presents the dilemma in stark relief. Yet it also approaches passivity as an opposition to activity, rather than a mediating term. This accepts the presuppositions of the inevitability and human-centered Anthropos of the Anthropocene: for instance, an acceptance that humans are, in fact, an overwhelming geologic force, and that we are capable of redefining both our species and our planet. Perhaps more to the point, this either/or neglects the third relation to the Transcendent. And in the foregoing description of passivity implies something a bit more paradoxical, something more ultimate. It is an inhalation and exhalation, as birds and flowers envelop us. It is a "yes and a no" for Heidegger. A relationship defined entirely by the other for Buber. Or the taste—a very physical and personal ingestion—of the infinite—something beyond our very capacity to understand. If passivity is the relation of passive and active, would it not present a means to at least for an instant *overcome* some of the presuppositions of the Anthropocene? Or, more forcefully, should we see passivity as mediating and reframing the hermeneutics of the Anthropocene, by seeing it as a pursuit or attitude of how our embodied particularity overturns overhumanization? While the foregoing hasn't provided a map for this path, it does seem to suggest an invitation.

Notes

1 Fiona Hughes, "A Passivity Prior to Passive and Active," *Mind* 122 (2013): 421 [419-450].
2 Martin Heidegger, *Discourse on Thinking* (New York: Harper Torchbooks, 1966), 45.
3 See further, Forrest Clingerman, "Humanity, Relationality and Justice in the Anthropocene," in *Dimensions of Intra- and Intergenerational Justice in the Debate about Sustainability*, ed. by Silviya Serafimova (Sofia: Avangard Prima, 2020), 52–68; Forrest Clingerman, "Walking into the Frame: A Theological Exploration ofPilgrimage along Anton Mauve's *A Dutch Road.*" *Literature and Theology* 23 (2009): 18–32; Forrest Clingerman, "Wilderness as the Place between Philosophy and Theology: Questioning Martin Drenthen on the Otherness of Nature." *Environmental Values* 19 (2010): 211–232.
4 Forrest Clingerman, "Humanity, Relationality and Justice in the Anthropocene," 50.
5 Forrest Clingerman, "Humanity, Relationality and Justice in the Anthropocene," 54.
6 Martin Buber, *I and Thou* (New York: Scribners, 1970): 80.
7 See further, Clingerman, "Humanity, Relationality and Justice."

8 Clingerman, "Humanity, Relationality and Justice," 83.
9 See further, Forrest Clingerman, "Natural or Political Public Space? Spaces of Interpretation, Environmentalism, and the Sacred Depth," in *Environment, Space and the Public*, ed. by Mark Luccarelli and Sigurd Bergmann (Amsterdam: Brill/Rodopi, 2015), pp. 133–147; Clingerman, "Wilderness as the Place between Philosophy and Theology."
10 See further, Forrest Clingerman, "Memory, Imagination, and the Hermeneutics of Place," in *Interpreting Nature: The Emerging Field of Environmental Hermeneutics*, ed. by Forrest Clingerman, Brian Treanor, Martin Drenthen, and David Utsler (New York: Fordham University Press, 2014), pp. 245–263.
11 On Tillich and "depth dimension" see further, Forrest Clingerman, "The Intimate Distance of Herons: Theological Travels through Nature, Place, and Migration." *Ethics, Place & Environment* 11 (2008): 313–325; Clingerman, "Natural or Political Public Spaces?"; Clingerman, "Wilderness as the Place between Philosophy and Theology."
12 For discussion concerning the Anthropocene as a "hermeneutical concept," see Forrest Clingerman, "Place and the Hermeneutics of the Anthropocene." *Worldviews* 20 (2016): 225–237; Forrest Clingerman, "Sacrifice Zones and Interpreting the Anthropocene" in *Environmental Hermeneutics in the Anthropocene: Nature and the Conflict of Interpretations*, ed. by David Utsler, Forrest Clingerman, and Brian Treanor (London: Routledge, 2026), pp. 96–106.
13 On the Anthropocene as "overhumanization," see Clingerman, "Place and the Hermeneutics of the Anthropocene"; Clingerman, "Sacrifice Zones and the Anthropocene"; and Forrest Clingerman, "Theologies of Creation and the Anthropocene," in *T. & T. Clark Handbook on the Doctrine of Creation*, ed. Jason Goroncy (London: Bloomsbury, 2024), pp. 608–622.
14 See further, Clingerman, "Humanity, Relationality and Justice in the Anthropocene"; Clingerman, "Place and the Hermeneutics of the Anthropocene"; and Clingerman, "Theologies of Creation and the Anthropocene."

References

Buber, Martin. *I and Thou* (New York: Scribners, 1970).

Clingerman, Forrest. "The Intimate Distance of Herons: Theological Travels through Nature, Place, and Migration." *Ethics, Place & Environment* 11 (2008): 313–325.

Clingerman, Forrest. "Walking into the Frame: A Theological Exploration of Pilgrimage along Anton Mauve's *A Dutch Road*." *Literature and Theology* 23 (2009): 18–32.

Clingerman, Forrest. "Wilderness as the Place between Philosophy and Theology: Questioning Martin Drenthen on the Otherness of Nature." *Environmental Values* 19 (2010): 211–232.

Clingerman, Forrest. "Memory, Imagination, and the Hermeneutics of Place," in *Interpreting Nature: The Emerging Field of Environmental Hermeneutics*, ed. by Forrest Clingerman, Brian Treanor, Martin Drenthen, and David Utsler (New York: Fordham University Press, 2014), pp. 245–263.

Clingerman, Forrest. "Natural or Political Public Space? Spaces of Interpretation, Environmentalism, and the Sacred Depth," in *Environment, Space and the Public*, ed. by Mark Luccarelli and Sigurd Bergmann (Amsterdam: Brill/Rodopi, 2015), pp. 133–147.

Clingerman, Forrest. "Place and the Hermeneutics of the Anthropocene." *Worldviews* 20 (2016): 225–237.

Clingerman, Forrest. "Humanity, Relationality and Justice in the Anthropocene," in *Dimensions of Intra- and Intergenerational Justice in the Debate about Sustainability*, ed. by Silviya Serafimova (Sofia: Avangard Prima, 2020), pp. 52–68.

Clingerman, Forrest. "Theologies of Creation and the Anthropocene," in *T. & T. Clark Handbook on the Doctrine of Creation*, ed. Jason Goroncy (London: Bloomsbury, 2024), pp. 608–622.

Clingerman, Forrest. "Sacrifice Zones and Interpreting the Anthropocene," in *Environmental Hermeneutics in the Anthropocene: Nature and the Conflict of Interpretations*, ed. by David Utsler, Forrest Clingerman, and Brian Treanor (London: Routledge, 2026), pp. 96–106.

Heidegger, Martin. *Discourse on Thinking*. New York: Harper Torchbooks, 1966.

Heidegger, Martin. *Country Path Conversations*. Bloomington: Indiana University Press, 2010.

Heidegger, Martin. "The Question Concerning Technology," in Basic Writings, ed. by *David Farrell*, 2nd edition (San Francisco: Harper San Francisco, 1993), pp. 308–341.
Hughes, Fiona. "A Passivity Prior to Passive and Active." *Mind* 122 (2013): 419–450.
Merleau-Ponty, Maurice. *Phenomenology of Perception*. New York: Routledge, 2012.
Schleiermacher, Friedrich D.E. *The Christian Faith*. Translated by H.R. Mackintosh and J.S. Stewart. Edinburgh: T&T Clark, 1989.
Schleiermacher, Friedrich D.E. *On Religion: Speeches to Its Cultured Despisers*. Translated by Richard Crouter. New York: Cambridge University Press, 1997.

10

THE PASSIVITY OF AWE AS A WAY TO A YET-TO-BE-DISCLOSED-PLACE

Martin Mølholm

Introduction

This chapter will address the sensation of awe[1] as a fundamental attunement that *predates* the Sense of Wonder.[2] It is a state of Being which can be described as an 'awakened pre-awareness' that *may* strike us moments *before* we begin to contemplate and wonder, and in a way that forces us to hold back, give way and maintain a respectful distance. I will argue that awe is prerequisite for a particular kind of 'seeing' the *no*-thing, *prior to* it becoming some-thing and a-thing. It is the doorstep to a world that is not yet, and eventually to what I call – and later will unfold – the yet-to-be-disclosed-place (Mølholm 2019). It is, therefore, intrinsically relational: as a 'doorsteps' it *connects* that which is 'there' with the 'I-here', but *before* 'the there' approaches 'the here', or vice versa thus leading, Wood (2003) writes in *What is Eco-phenomenology*, 'to such issues as dependency, cooperation, symbiosis, and synergy' (p. 221). I will further also argue that the attunement of awe is fundamental for us to experience this kind of co-dependent, co-operating, symbiotic, and synergetic co-existence, but that this fundament is gradually and inevitably undermined by an accelerating society that impedes the establishing of deep, intimate, and respectful relationships between the human being and other living beings. It is a development that has driven the late-modern human being away from its 'essential possibility of Da-sein' as that which is '*not yet* real and *not always* necessary' (Heidegger 1927, pp. 134–135); that, according to the American eco-phenomenologist Tara Kennedy (2014), has led us astray, away from 'our essential nature' and who we are as Dasein, by the ontotheological view of ourselves and the world according to which we are nothing other than 'intrinsically-meaningless resources merely standing by to be optimized' (p. 110), and where nature, Martin Heidegger (1959) writes, is nothing but 'a gigantic gasoline station, an energy source for modern technology and industry' (p. 50).

An important and essential 'gateway' to this possibility and potentiality is *awe*. But to fully understand its significance it is important to make a clear distinction between awe and wonder, as awe is *not* (more or less) synonymous with wonder as might be suggested in sentences such as 'that which is regarded with the greatest awe and wonder' (Abram 1996, p. 9). And it is *not*, as Swanton (2010) claims, a phenomenon subordinate to wonder as the totalizing: 'The understanding of wonder as totalizing is secured by contrasting wonder with several other

DOI: 10.4324/9781003538097-13

states with which wonder might easily be confused: admiration, marveling, astonishment, awe, amazement, and curiosity' (p. 164). While Swanton rightfully argues that wonder is 'an attunement to the world as a whole', I will in this chapter show how awe is that which may enable us to get a glimpse of *the awe-some* of the wondrous, *before* we *do* wonder: contemplate, act, and engage *with* and *in* the world.[3]

The chapter will begin with a short story of an experience of awe, *followed by* deep contemplative wonder that I had in the summer of 2023. From there, I go on to describe how I, in the ensuing month and as the experience settled, came to realize the significance of the experience and how it came to signify something that we as human beings have lost, namely the ability to 'sense', as we have succumbed to the exploitation and optimization of every available resource that we can lay our hands on, including the human.

Coming Home as Being on the Way in Dwelling

I will begin this chapter with a short story about a lived experience I had when I, in 2023, walked the *Camino Via de la Plata* from Sevilla to Santiago de Compostela in Spain.[4]

> *It was a scorching Sunday morning in the beginning of July just outside Torremejia, a small village a couple of hundred kilometers north of Sevilla, Spain. Still early in the day and only a couple of hours after sunrise, the temperature had just passed 35 degrees Celsius. Above my head a burning sun; beneath my feet a parched and dusty road; before my eyes an infinite and naked horizon, flickering in the heat haze; behind me a feverish buzzing world fading further and further away. And while my body was an overabundance of sensations: vital strength and a dawning fatigue; sizzling muscles and a searing skin; the thrusting force of the ground as my feet touched down, and the compression from the backpack on my shoulders, hips and back, I myself was pervaded by a deep calm. From a permeating feeling of peace and tranquility, I heard an inner poet singing from a place within me, that I had not before had the fortune of receiving visits from, but who now sang:*
>
> *Today I have felt the dents, cracks and chinks*
> *beneath my feet;*
> *My soul has bathed in the sounds of the Symphony of Nature;*
> *The gaze of my eyes has swum to the horizons of the world,*
> *and beyond the firmaments cloister;*
> *My skin has been caressed by the hot panting of gentle winds;*
> *My tongue has tasted the dry warmth of the rays of the sun.*
>
> *But suddenly the poet-in-me stopped singing, the sensations of my body fell silent. The sun held back its heat, the dusty road came to a halt, and the world stopped turning. Time stood still, as my eyes caught glimpse of the tiniest of movements between pebble rocks on the right-hand side of me. Everything appeared to freeze: time, the surrounding space and my body, soul and spirit, as I was held back by the smallest of creatures scuttling about withered sprigs and miniature cactuses. Right here in this hot, dry, barren and hostile part of the world where any sign of life seemed almost impossible to find, a mouse went about its day and business as if completely comfortable with its situation and not a worry to bother it.*

As I stood there nailed to the ground, I sensed that both the poet-in-me and the mouse represented something unique and important; something of profound significance that I had yet to discover – something that I patiently had to await. In that moment life felt as vibrant, meaningful, and perfectly happy as it had ever done before. A sparkling sensation proliferated my body and filled it with an exuberant joy. The encounter with the mouse and the poet-in-me shook and moved me so profoundly that an irresistible smile from a deep and hidden place inside me erupted on my face with a chuckling sound, announcing a strike of wonder so forceful, that it felt as if I almost lost my foothold in the everyday world. Perhaps this was the kind of experience that the German philosopher Josef Pieper (1947b) had in mind when he wrote that he 'who undertake to live under the sign and constellation 'wonder' (why *is* there such a thing as being?) must certainly be prepared to find themselves lost, at times, in the ordinary workaday world' (p. 113). I was on my third Camino-walk and had been 'lost' before yet, this was 'more' than just being 'lost in wonder'- much more. What I had experienced was something that *predated* my experience of wonder, and which I have not found more adequately described to match my experience, than in Heidegger's *Remembrance* (1943a). It was an experience of the fundamental attunement of awe which for an infinitesimal moment that seemed to be frozen in time, 'nailed me' at an unsurpassable distance keeping that, which had appeared, 'ready for its arising from its source' (Heidegger 1943a, p. 153), but not yet 'there' as *some*-thing or *a* thing.

Month later I read the American environmental- and eco-phenomenologist David Abram's *The Spell of the Sensuous* (1996). And in it I suddenly found my encounter with the mouse and the poet-in-me reflected in his depiction of the wondrous sight he had late one evening, of the stars and the fireflies in a Bali rice paddy, and how he felt as if he was 'falling through space' (pp. 3–4). He describes it as a 'magical moment' in which he had the 'experience of existing in a world made up of multiple intelligences' (Abram 1996, p. 9). My encounter with the *mouse* (who) *went about its day and business as if completely comfortable with its situation and not a worry to bother it* was indeed a 'magical moment' not that dissimilar. Right there under the burning sun, I intuitively knew that not only the mouse and the poet-in-me, but also the cactuses, olives, and the withered grass (yes, perhaps even the rocks and the sand) were experiencing lifeforms, each with their 'predications and sensations, albeit sensations that are very different from our own' (Abram 1996, p. 10). In that moment I was simply *looking* at the mouse, in a sense much alike the one that Pieper calls the *contemplative act*. An act during which you are 'awake and active', but without your attention being strained; an act during which you 'open one's eyes receptively to whatever offers itself to one's vision, and the things seen enter into us, so to speak, without calling for any effort or strain on your part to possess them' (Pieper 1947a, p. 26). I too, was looking. But for the briefest of a moment, just before I started contemplating and wondering, there was *no*thing but a state of silent, non-moving, and non-acting awe. First, I was in awe, then in deep, contemplative wonder (Hansen 2025, Chapter 15); first I was passive, then active; first I was receptive, then perceptive, and with no desire to capture, control or keep neither the poet and/or the mouse nor the knowledge and/or the truth about them. It was a dwelling much like the one that Kennedy (2014) describes, in which we let 'things ripen of their own accord', and where we resist 'forcing a conceptual interpretation onto them' (p. 127). A dwelling that allows us to see the world as 'mysterious, radiant, awesome, and not something to be totally 'ordered about', calculated, manipulated for our own ends' (Swanton 2010 p. 160; quoted in Kennedy 2014, p. 127). But while Abram (1996) was 'perplexed and confused' about how 'everyone appeared [...] to be strangely unaware of happenings unfolding all around them *in the present*' (p. 201), what has puzzled me is that the fundamental attunement of awe seems to receive only scant attention in the eco-phenomenological literature. If we are to recognize 'ourselves as entities capable of appreciating things as more than instrumentally and calculatively valuable' (Kennedy 2014, p. 110), awe is – as I will elaborate on in the following – fundamental. We must first learn to resist the urge to *do* something, take *action, manipulate*, and *transform*, if we want to reach the point where we recognize that 'the well-being and flourishing of human and nonhuman Life on Earth have value in themselves', which are 'independent of the usefulness of the nonhuman world for human purposes' as Sessions and Naess formulates it in *The Basic Principles of Deep Ecology* (1986, p. 14).

Returning home, Abram (1996) appeared somewhat disillusioned:

> They seemed utterly oblivious to all those phenomena to which I had had to sensitize myself in order to communicate with the indigenous magicians in the course of my fieldwork: the lives of other animals, the minute gestures of insects and plants, the speech of birds, the tastes in the wind, the flux of sounds and smells....
>
> (p. 201)

For reasons that will later be unfolded, I agree that the late-modern human being is increasingly oblivious to the lives of every other living creature, including the human being itself. But I am hopeful, and place my hope not in *doing* something else, but in *Being* and letting *Be*. For even though I was walking across desert land, I felt more at home than ever before. Suddenly I understood what is meant by the words, that it is not the final destination of the Camino that is the goal and purpose of the walk, but The Camino – The Way – itself. For being on the way, Kennedy (2014) writes, *is* being at home:

> When Da-sein seeks to understand his ontological homeless, when he engages in genuine thinking, meditative thinking, when he *gives thought* by giving thanks, he is already underway toward his essence, and, being on the way home, a home that consists in being on the way, he is homeless no more.
>
> (p. 125)

Walking the Camino, I was in more than one way *on* the way and thus, in more than one way *at* home. I was experiencing homecoming as not only a spiritual, contemplative movement, but also a physically and bodily one. I was literally engaged in a process I find, is best described by the Danish philosopher Søren Kierkegaard, in a letter that he wrote to his sister-in-law in 1847: 'I have walked my way to the most excellent of my thoughts, and I know of no thought so dire that, after all, you can walk from it'.[5] And so, a few minutes after the encounter I sat down and slowly, patiently, and almost hesitantly started to write down the words of the poet-in-me and the brief description of the experience with the mouse that initiates this chapter, not yet realizing the significance of it, but nonetheless having the feeling that it might possibly and somehow have not only changed *me* but also my understanding of the problems facing the late-modern human being. Problems that might be narrowed down to one, single word: *loss*!

Holding Back the Breath and Dwelling in the Clearing

It was not until weeks later that I began to realize the significance of the encounter. It came to me in the middle of the night and may have dawned on me in a dream that I have now long forgotten, but which I nonetheless suddenly woke from, filled with the strangest, most wonderful and joyous experience of what can best be described with one of Heidegger's concepts, namely, as a *clearing*, where 'what is cleared is the free, the open' (Heidegger 1966–1967, p. 161). It is a state of awareness where 'things come to presence of their own' (Kennedy 2014, p. 126). Something had been puzzling in the undergrowth of my mind that I, without knowing so, must have been dwelling on but careful to safeguard and preserve; to spare so as to, with the words of Heidegger (1951) 'leave something beforehand in its own nature', as we 'return it specifically to its being' (pp. 146–147). For a moment it was as if, figuratively speaking, a whole lot of trunks, branches, twigs and leaves had been *cleared* and something presented itself to me whose presence I had only vaguely sensed, but not yet seen: *But suddenly the poet-in-me stopped singing, the sensations of my body fell silent. [...]. Time stood still, as my eyes caught glimpse of the tiniest of movements between pebble rocks on the right-hand side of me. Everything appeared to freeze...* That night, waking up from a dream, half awake and half asleep, I was back on the parched and dusty road, stupefied and holding back my breath, as if even the smallest of motions would break the spell and dissolve the magical clearing of the free and the open that had been concealed. Something of tremendous significance had happened that I, literally speaking, struggled to come to terms with – and still am. But I understood as much that something that had before been 'covered over' and concealed, was now being un-concealed as *aletheia*. Not in the meaning that everything stood, or stands, bright, and clear to me as 'truth' – Heidegger (1927) explicitly warns us to translate the term *aletheia* with truth (p. 202). But in the meaning that, for the briefest of moments a clearing had opened and allowed for something to emerge: 'Light and fire can first find their place only in the clearing', as Heidegger explained it in a conversation with contemporary German philosopher Eugen Fink (1966–1967, p. 161). A clearing has to be made, in order for that which hides in the undergrowth to come to light and to a presence of its own. That night I didn't yet know what that was, and probably never fully will, as clearing is *both* the coming to be of beings in their beingness, the 'un-/a-' in un-concealment/a-letheia, *and* at the same time the keeping out of view that which remains hidden – the '-concealed/-letheia'. Some-thing steps into the clearing that we then endeavor to apprehend (*Vernehmen*) in the double meaning of the word: to *passively* let that which appears and shows itself come to us, and to *actively* 'determine and set fast how things are going and how things stand (Heidegger 1935, p. 153).

In that way, there is in Heidegger's way of describing the human beings' being-in-the-world, almost a kind of 'gentleness' that I recognize from not only my dream, but indeed also from the meeting I had with the poet-in-me and the mouse. A gentleness that is further emphasized when he in *Introduction to Metaphysics* (1935) writes that the process of apprehension, among other things, comes in the shape of a *seeming* (Heidegger, 1935): a '*letting-be-*seen' (own italic) in the form of a manifest presence. A presence which, in my situation, took the shape of the poet-in-me and a mouse. Two creatures – one intrinsic, the other extrinsic – who have stayed with me ever since, so as if to 'tell me' something imperatively essential – about me, about the human existence and, perhaps more importantly, nature and life itself.

Only later it dawned on me, that my experience with the poet-in-me and the mouse occurred as a dwelling upon what Heidegger (1949) calls The Thing (*Das Ding*), in a world made up of The Fourfold (*Das Geviert*): the earth and the sky, the divinities and the mortals (p. 11). Swanton (2010) explains that what Heidegger means is, that we as human beings are 'open to the mystery of the world as a *dwelling* place, and as a dwelling place the world consists of the "four-fold"' (p. 150). The revealing and disclosedness, un-concealment, of the world has a four-part structure through which, Being as truth reveals itself, and where each part is necessary for the truth of Being to emerge. As an example, Heidegger (1949) mentions the jug: 'The jughood of the jug', he says, 'essences in the gift of the pour' (p. 10) – the grace of the jug is that it gives. As an object, the human being uses the jug to, for example, contain wine. As such it can serve both as a 'socializing lubricant', a cultural experience and a religious symbol of the blood of Christ, that is, divinities. Divinities 'need not be supernatural' (Swanton 2010, p. 150). They can be anything that holds a quality, a character, that enables us as human beings to transcend the ordinary, mundane, familiar, as the divinities serve as 'the hidden messengers of godhood' (Heidegger 1949, p. 16). The wine is made of grapes nurtured by the earth and the sky, and 'the gift of the pour is a libation for the mortals. It quenches their thirst' (Heidegger 1949, p. 11).

My 'jug' was the Camino, and what it endowed me with was the overwhelming sensation of awe leading me to dwell in deep, contemplative wonder at the sight of the mouse and the song of the poet-in-me. The 'Camino-hood' of the Camino essences therefore in the gift of awe. For me as *a mortal* to walk in praise[6] and contemplation, the walk symbolizing *the divinities* which brought out the mouse and poet-in-me that struck me with awe. I owed my gratitude to *the earth* that provided me with the materials, foods and drinks that sustained my life across these hard and inhospitable parts of Spain, and to the sun in *the sky*, the rain, the wind, and the air that nurtures the crops, and to every living creature, including *the mouse*, that upholds the balance which keeps it all together.

I cannot say that I fully understand the significance of that moment of awe and wonder, of that unconcealment, that the mouse and the poet-in-me evoked. After all, unconcealment is *a seeming*, Heidegger (1935) emphasizes, that only for a moment surfaces 'and just as fleetingly and unsteadily disappears again' (p. 108). What I *do* understand, though, is, that apparently I had to go to what at first glance appears to be the most godforsaken place on earth, for me to 'lose my foothold in the everyday world' (Pieper 1947b, p. 113) and fully experience and appreciate the miracles and mysteries of life, and that we exist 'in a world made up of multiple intelligence' (Abram 1996, p. 9). Why is that? Are such drastic measures necessary? The German philosopher Byung-Chul Han's answer might very well be yes. According to Han, we have now managed to create a high-speed society that is so saturated with information, communication, and innovation that need to be processed and materialized and then, just as quickly let go of again so that we can catch on to the next bits of information, communication, and innovation, that dwelling and contemplating is de facto impossible for they who want to maintain their relevance (Han 2012, 2019). Perhaps it is virtually *impossible* not to lose ourselves in what Heidegger (1927) describes as 'the publicness of the they and its idle talk', and thereby fail to hear our own self (Heidegger 1927, p. 250), unless we are able to break away from the buzzing world: *before my eyes an infinite and naked horizon, flickering in the heat haze; behind me a feverish buzzing world fading further and further away...* Perhaps we are gradually in the process of dismantling the ability to be in awe, to wonder and to appreciate the significance of *The Fourfold*. Hence the before mentioned loss.

In Awe as Astonished Remembrance of That Which Is Near...

Above I wrote, that being in awe comes prior to be in wonder. Let me explain... When reading through Heidegger's work it becomes clear that he had a reconning with the *metaphysical thinking*. He repeatedly, Capobianco (2010) elucidates, criticized the metaphysical thinking for not moving beyond the representation of '"being" as beings in terms of *a* being, namely as *eidos, actualitas*, ego, nature, Spirit, and the like' (p. 16), and think 'the ontological difference in its primordiality: Being (namely presencing itself) and beings (that which comes-to-*presence* in any form)' (p. 17). In *Postscript to 'What is Metaphysics?'* Heidegger (1943b) writes:

> Metaphysics is the history of this truth. It says what beings are in bringing to a concept the beingness of beings. In the beingness of beings, metaphysics thinks being, yet without being able to ponder the truth of being in the manner of its own thinking. [...] it is necessary to ask what metaphysics is in its ground. This questioning must think metaphysically and at the same time think out of the ground of metaphysics, i.e., in a manner that is no longer metaphysical.
>
> (p. 232)

To 'think' a thinking that thinks *beyond* and out of the ground of metaphysics, Heidegger introduces in *Remembrance* (1943a) the concept of *awe* (Die Scheu).[7] Awe is the sensation that we feel, when we are literally speechless and feel as if we are nailed to the ground; when we stand 'hesitant and does not rush toward the source', and by not doing so instead 'undertakes the authentic path to it' (Heidegger 1943a, p. 154). As such, when in awe we are passive and awaiting, keeping a respectful distance and holding back any thought and talk, meandering around the phenomenon in an apophatic approach so as not 'to capture it by appropriation' (Visse, Hansen & Leget 2020, p. 2) – not yet, that is. It is a sensation that, Capobianco (2010, p. 85) notices, shares similarities with Heidegger's description in *What is Metaphysics* (1929) of the existential anxiety as the 'shrinking back before … that is surely not any sort of flight but rather a kind of bewildered calm' that 'takes it departure from the nothing' and 'brings Da-sein for the first time before beings as such' (Heidegger 1929, pp. 102–103). When in awe[8] we are, Heidegger (1955) later writes, held fast by that from which we step back. Literally, no-thing moves. That day on the Camino, *everything appeared to freeze: time, the surrounding space and my body, soul and spirit, as I was held back by the smallest of creatures…* I was nailed to the ground and held speechless when overwhelmed by what in that moment was not (just) a mouse or a poet-in-me, but 'nothing', a no-thing: not a rodent, a pet, a Disney character or an enemy; not an insight, an understanding, a knowledge or an epiphany; not any-one-thing. It was as if I was being guided, or pulled, back to a 'first beginning'; back to 'die Anfängnis des Anfangs' from out of which 'das Seiende in das Da aufstehen läßt' (Heidegger 1941, p. 16);[9] back to an origin of Being. No thoughts or words tried to capture the moment, conceptualize it, grasp it, define it, and explain it. Something deep inside me understood that that, in lack of a better word, would be sacrilegious. Instead, I was in 'a bewildered calm', dumbfounded and right there and then simply in a relationship with the no-thing. No more, no less. It was, as described in the introduction, an 'awakened pre-awareness': all of my senses were fully awake, for a minuscule fragment of a moment it was an awakenedness that was *not yet* aware in the form of a knowledge or perception of a *particular* (kind of) situation or fact. I was in awe. A fundamental attunement that Heidegger (1943a) describes as

> that reserved, patient, astonished remembrance of that which remains near in a nearness which consists solely in keeping something distant in its fullness, and thereby keeping it ready for its arising from its source. [...]. Awe is the knowledge that the origin cannot be directly experienced.
>
> (p. 153)

The no-thing is there as 'ready for its arising'. Still kept at a distance in its fullness, it has not risen yet and when in awe we know, that the origin here glimpsed will not let itself experience directly. Awe is the doorstep, and as soon as we *start* to move or *are* moved, it is all over: the no-thing becomes some-thing and, eventually, a-thing. Becoming *is* 'the essence of appearing', Heidegger (1935) wrote, and it 'involves stepping-forth and stepping-away, this hither and hence in the genuine demonstrative, indicative sense. Being is thus dispersed into the manifold beings' (p. 112). It is no wonder, therefore, that Heidegger (1943a) further describes awe as a homecoming that celebrates and remembers the origin (p. 153). So even though awe is not the direct experience of the origin, it is nonetheless a homecoming that has affinities to what he a couple of years before called 'the Event' (*Das Ereignis*). One that brings us toe to toe with the first beginning, allowing us a glimpse of where it all starts and comes from, and which he describes as 'Das Seyn als der Abgrund der Verschenkung'[10] (Heidegger 1941, p. 15). In that situation, starring down into the dark and seemingly bottomless deep of nothingness that, nonetheless, appears to be pregnant with promises of secrets not yet revealed – 'a prize so rich an frail', as Heidegger quotes the German poet Stefan George in *The Nature of Language* (1957–1958, p. 60ff) – but ready to be given away to the daring, awe might very well be the word that best describes the feeling that overwhelms us. When in awe we are literally, at least for a while, left speechless and empty of words as we are 'held fast by that which we have stepped back from', gasp and hold our breath. 'Das Ereignis', Heidegger concludes, 'ist das Seyn' (Heidegger 1941, p. 16). In that sense, awe as a fundamental attunement, is the *pre-metaphysical, pre-ontological* 'state of Being' which takes Being into the clearing of the unthought, and 'get into view what metaphysical thinking simply could not: the pure appropriating, giving, letting of what appears' (Capobianco 2010, p. 50). It is a leaving be in a non-calculative thinking that does not try to capture, contain, own, utilize, and consume; that is gentle and holds a reverent attitude toward the world and everything in it. It is a leaving be that has no agency or agenda, not even to contemplate. *Then* we may come to wonder *in* astonishment, where we step back and restrain ourselves, but nonetheless are attuned (Heidegger 1955, pp. 84–85); where a clearing clears and a dawning apprehension comes to light. Thus, and once again: first in awe, *then* in wonder.

The Late-Modern Society: Out of Awe and Wonder, into the Calculative Thinking

If it was only that simple… As early as in the 1920s when writing *Being and Time*, Heidegger (1927) had a reconning with a world gone astray. A world in which *The They*, in its 'inconspicuousness and unascertainability' unfolded 'its true dictatorship' through the utilization of public transportation and the use of information services that 'dissolves one's own Da-sein' (p. 118). After the Second World War he sharpens the tone when writing, that the world has now been reduced to 'an object, open to the attacks of calculative thought', and where nature has become 'a gigantic gasoline station, an energy source for modern technology and industry' (Heidegger 1959, p. 50). A world in which we do not even 'stop to consider that an attack with technological means is being prepared upon the life of nature and man' (Heidegger 1959, p. 52).

Are we, then, doomed and inevitably headed toward a technologically driven apocalypse? As is the case with almost anything else, that depends… But if we want to stop the late-modern human being from becoming still more alienated and disenfranchised, perhaps we should listen more carefully to what Heidegger (1959) more than 75 years ago encouraged us to do, namely to 'pit contemplative thinking[11] decisively against merely calculative thinking' if we are to avoid to 'fall into bondage to them (the technical devises)' (Heidegger 1959, pp. 53–54). To do that we must once again learn to appreciate the richness of patience and reservedness, of the courage to go slowly, and of hesitation and being speechless in awe. Awe is the 'spark' that lights up the fire illuminating the clearing where 'things come to presence of their own' (Kennedy 2014, p. 126) – things that are now 'some thing' and not 'no-thing', yet still not *a* thing, and upon which we now dwell in contemplative wonder as a passive-receptive being-with-the-world (Mølholm & Hansen 2024b); it is the 'gateway' that connects the above mentioned abyss of Beyng with the 'beingness of being' of the *Human* being (lower case b) as 'a *thinking*, that is a *contemplative* being' (Heidegger 1959, p. 47). Thus, awe – and astonishment – is *nothing like* a faculty of the human being but a fundamental *attunement* (Heidegger 1927, § 29);[12] one that, like an invisible hand forces us to stop, *before* we start to contemplate that which brought us to a halt, and *then* begin to apprehend and once again sets the world in motion. When contemplating, says Pieper (1947a)

> our soul is passive and receptive. We are, to be sure, awake and active, but our attention is not strained; we simply 'look' – in so far, that is, as we contemplate it and are not already 'observing it'.
>
> (p. 26)

Contrary to the calculative thinking, which has affinity to what Pieper calls *observing*, when contemplating, the human being is careful *not* to begin figuring out, counting, measuring, and to weighing up. As soon as we go beyond contemplation and begin the process of apprehension (which we, according to Heidegger (1935) inevitably do), we move from the purely passive letting that which appears appear, to the active *acquisition* of what *ontically* is familiar to us as knowledge, insight and understanding and thus something that we have now taken possession of so that we can now begin to figure out, measure, etc. Our relation to the world become *instrumental.* Or, with a term by Heidegger (1936–1938), *machinated* (p. 88ff) meaning the 'abandonment of beings by being' where 'man is so fully blinded by what is objective and machinational that beings already withdraw from him' (Heidegger 1936–1938, pp. 77 & 78). The beings of the world: the living creatures, the soil and minerals, as well as the human being itself, becomes objects of his desire to understand, manipulate, control, utilize, grow, or wipe out.

The act of contemplation *has* no agenda. It is an *ontological* 'state of Being' in which we simply 'look'; where we open our 'eyes receptively to whatever offers itself to one's vision, and the things seen enter into us, so to speak, without calling for any effort or strain on our part to possess them' (Pieper 1947a, p. 26). I imagine that my experience on the Camino of being *pervaded by a deep calm*, and with *a permeating feeling of peace and tranquility* could very well be one that 'paves the way' to the contemplative seeing. It was as if I was literally 'learning to see', as Nietzsche (1889) expresses it: 'accustom(ing) the eye to quietness, to patience, to reserve; to postpone judgement, to survey and comprehend each case from all sides; [...] to defer decision', instead of following every impulse (p. 36). That day in the beginning of July, under the burning Spanish sun, I patiently listened to the silent voice of the poet-in-me, postponed any judgment and comprehension of what appeared in front and inside of me. It was not, and by no means need not be, 'high-flown. It is enough if we dwell at what lies close, [...] upon that which concerns us, each one of us, here and now' (Nietzsche 1889, p. 47).

Back Home, Through Awe and Wonder, to the Yet-To-Be-Disclosed-Place

For an eternal moment of awe that kept the time from going, put the hands on the clock to rest, laid quiet the wind and brought the rotation of the earth to a halt, I stood there under the sizzling sun. Gradually and seamlessly, without noticing the transition, awe was succeeded by a dwelling in contemplation at the wonder that caused me to hesitate and sojourn. I stood in marveled correspondence with a poet-in-me and a mouse that appeared and shined forth in the clearing of the Event. I gazed in a soundless, bewildered admiration, and calm, wondering: *What is that!* Not as a question *about*, but as a wondering *at* something (Visse, Hansen & Leget 2020, p. 7), hence the exclamation mark. Or rather: at some-thing that was still not *a* (specific) thing, yet no longer a no-thing. It was a wondering at 'a living phenomenon seen as a wonder' (Visse, Hansen & Leget 2020, p. 7). It was a wondering at the wonder of life. It was an overwhelming experience, permeating me with a feeling of deep gratitude and peace.

In that eternal moment I was 'at home' in what I call the *yet-to-be-disclosed-place* (Mølholm 2019). It is a concept that was originally developed to signify the vaguely designated state of being-in-the-world that the Discourses[13] on stress, depression, and anxiety suggests, but does not indicate, and to where we are 'brought back' by, what Heidegger (1927) describes as the 'call from afar, to afar' of the existential conscience, back to our 'ownmost potentiality-of-being-a-self' (p. 251 & 249). The stress reactions of the late-modern human being are, seen from this existential perspective,

> a pro-active 'summoning' (*Anruf*): an inner and silent request to answer the existential yearning to un-weave from the fabric of the world, when 'loosing itself in the publicness of the They and its idle talk it *fails to hear* its own self' .
>
> (Heidegger 1927, p. 250; Mølholm & Hansen 2024b, p. 124)

At the yet-to-be-disclosed-place we are brought back to our potentiality of *being* in awe and wonder; of *being* in contemplation and dwelling in or at whatever lies closest, instead of losing ourselves in the averageness of the They, where 'every other is like the next' (Heidegger 1927, p. 119). In the yet-to-be-disclosed-place we are once more brought to see, not observe; to postpone judgment, not (yet) apprehend. Out there in complete solitude and the total silence of the wasteland, where the only living organisms besides the mouse (and, of course, the poet-in-me) were a million olives, and cut off from the buzzing, hectic, noisy, flickering, and constantly changing work-life I was, if only for a fleeting moment, able to not only 'hear' but also answer, the call of my existential conscience, and to allow that, as Heidegger (1929) calls it, 'peculiar calm' (p. 100) that comes from it, to pervade me. For a moment I had managed to un-weave from the fabric of the world. I was 'at home', liberated from the stress of the everyday work–life.

The Late-Modern Life: A Suffering of the Loss of Sense-Ability and Attain-Ability

In a late-modern society where the human being has succumbed to the imperative of performance and exploitation, where matters have to be smoothly integrated into 'the stream of capital, information and communication', and actions have to be 'made operational – subordinate to a calculable, steerable and controllable process' (Han 2012, p. 1), everything that calls us to take a deep breath, slow down, be patient and reserved, hesitate, contemplate, and let our hearts fill with wonder and awe is perceived as a poisoning of the well. Exploitation and contemplation are each other's negation. In a late-modern society where the ontic curiosity demands: *Tell me! Reveal your secrets and make yourself transparent! Let me know so I can control!*, everything that causes the late-modern human being to be dedicated to that which gives itself to it is deemed a dangerous friction to unstrained development, growth, innovation, creation, and consumption. It is a society of disclosedness… a phenomenon where, Heidegger (1923) wrote more than a hundred years ago, 'what is being encountered is there as "a means to", "useful for", "of importance for". Its being-*there* is based on its being-there-in-order-to-do-this and being-there-for-this' (p. 75). In the society of disclosedness there is no room for *kairological moments* of simply being-there. Moments where 'what we are concerned about and attend to is there as not yet, as to be… for the first time, as already, as approaching, as until now, as for the time being, as finally' (Heidegger 1923, p. 78).[14] Moments where *un*-concealment and disclosing walk hand in hand with *con*cealment and hiding away.

In the late-modern performance- and exploitation-society, Mølholm & Hansen (2024b) writes, 'the scientific and technological pursuit of the unattainable have brought the late-modern human being in a situation and relation with the world where they somehow have lost their connection to it, and to the rhythm of life itself' (p. 127). The world and everything in it: the flora and fauna, materials and minerals have been turned into a storage room with no closing hours, ceilings, or walls; with no-one to guard and run it, and with no-one currently taking responsibility of filling it up again.[15] Or, as Heidegger was quoted earlier, as *a gigantic gasoline station* treated with no consideration for the life of nature and men.

The late-modern human being is no exception. It, too, is a resource to be exploited and utilized to its fullest (Han 2010; Mølholm & Hansen 2024a). Tapping into what is today known as the *cognitive capital*, for example, cogito, emotions, dedication, commitment, curiosity, attention, fantasy, innovation, creativity, and empathy (Andersen 2013, p. 27ff), the human being is now living in what Han (2010) calls the *Achievement Society* with its credo 'yes, we can' (p. 8). Guided by the ethical imperative that 'if better is possible, then good is not good enough' (Hjortkjær 2020, p. 15), it has become 'an *animal laborans* that exploits itself – and does so voluntarily, without external constraints […] it is predator and prey at once' (Han 2010, p. 10); a performance subject that is both holding and using the whip that whips it (Han 2021, pp. 27–28). To the individual human being the result is an alarming increase in the rate of stress, depression, burnout, and anxiety. And not only psychologically but to an extended degree also – and maybe first and foremost – existentially. A deep sense of emptiness and alienation, and a feeling of increased isolation, pervades the late-modern human being, so that it has become still more difficult to *actively seek out a relationship* with that which rushes us by in still larger numbers and in a still more accelerated pace (Rosa 2013; Rosa 2017). For it cannot establish a deep and intimate relation to that which it only has a shallow connection and a fleeting, superficial contact. With no time, no calm and no patience to dwell and mesh with that which appears, it is impossible to be touched and affected – to be, with the words of Levinas and Levinas (1984), ordained to serve the Other (p. 195). Consequently, the late-modern human being is losing its ability to be 'tuned through-and-through in addressedness' and attain to 'an immediate and spontaneous, intensive relation to the world and nature'[16] (Rosa 2016, cp. V.3). Instead, the relation between the late-modern human being and the world (including the self) has been alienated so that the world has now become 'silent, cold, indifferent or even repulsive' (Rosa 2013, p. 100).

The mounting crisis of the late-modern world, and not only the human societies, is a crisis of both detachment and dis-attachment stemming from the cutting off from authentic Being: from patience and reserve; from hesitation and leaving be; from the joyous clearing and the bewildered calm; from the marveling and wondering of the marvels and wonders of life. From awe. From understanding the insignificance of human life and 'avoid emphasizing any particular, individual being, not even focusing on the human being', as Heidegger writes in *Introduction to Metaphysics* (1935, p. 4). He continues:

> For what is this being, after all! Let us consider the earth within the dark immensity of space in the universe. We can compare it to a tiny grain of sand; more than a kilometer of emptiness extends between it and the next grain of its size; on the surface of this tiny grain of sand lives a stupefied swarm of supposedly clever animals, crawling all over each other, who for a brief moment have invented knowledge. [...]. And what is a human lifespan amid millions of years? Barely a move of the second hand, a breath. Within beings as a whole there is no justification to be found for emphasizing precisely *this* being that is called the human being and among which we ourselves happen to belong.
> (Heidegger 1935, pp. 4–5)

But more than anything else, the crisis of the late-modern society is a crisis of *sense-ability*. It is a crisis stemming from a late-modern life in a high-speed society that has become so intense, that we have no time for rest, yet nonetheless crave for more because intensity 'can be indistinguishable from presence' and 'speed is an intoxicant' (Skårderud 1998, p. 18); where the human being has no defense- and regulatory system, no 'mental immune system' to ward off or limit the afflux of *the positive* (Han 2010, p. 1), for example, personal and professional growth and development, influence, flexibility, dedication, involvement, enthusiasm, which ultimately leads to the 'constipation' of an intoxicated, un-sense-able mental system. As a result, it is a crisis where a growing number of people do not realize how serious a mental condition they are in, before they are literally burned out, unable to perform even the simplest trifles of their everyday life; where they on Tuesday are speeding in the outer fast lane, but on Wednesday lying help- and hopelessly wrecked on the hard shoulder. Furthermore, it is therefore also a crisis of *attainability*. For without a 'mental immune system' regulating the intake, and a system that is mistaking speed and intensity with presence, and hyper-attention with intimacy, love and care, the patience and gentleness, reservedness and nearness that it takes to become present, attentive, intimate, loving and caring is becoming still more unattainable. Not to mention being able to actually take the time that it requires to reach the patience, gentleness, reservedness and nearness that it takes to get there at all. Therefore, this chapter is a story of loss: the loss of senseability and attainability; the loss of the ability to be in awe and wonder; the loss of the time and circumstances to see and look in contemplation; the loss of the possibility to dwell and find serenity; the loss of tranquility and peace of heart; the loss of life in the broadest sense of the word.

Notes

1 Awe is translated from the German word *Die Scheu* (Heidegger 1943b), as it is applied in Heidegger's 'Remembrance' (1943a). Directly translated it means *shy*. When I, nevertheless, translate *Die Scheu* with *awe* it is because I agree with Capobianco (2010) that this 'better captures the resonance of the word' (p. 159, endnote 25). Capobianco does not further elaborate why he believes so. But in *Andenken* (1943c) Heidegger writes that 'Die Scheu ist wohl ein anderes als de Schüchternheit', the latter meaning *shyness, uneasiness, coyness or modesty*, 'die im allem, was ihr begegnet, nur zaghaft und unsicher bleibt' (p. 131). I will later argue that *awe* is neither a faculty, nor a skill or competence that we can willfully activate or strengthen through hard training, but rather a *fundamental attunement* or sensation, an inherently human capacity that, if the circumstances are right and allow it, may 'come over us'.

2 See Chapter 15 by Finn Thorbjørn Hansen, for an elaborate and thorough examination and account of Wonder/Thaumazein.

3 Hansen (chapter 15) describes awe as one of four dimensions of deep, contemplative wonder; as 'a direct seeing (Scheu) of the phenomenon in and by itself', where it 'becomes a Without-Why-Wonder'. I maintain that awe as a fundamental attunement does not *do* any-thing, not even deep contemplation, as *the act* of contemplation itself involves *an acting* which we can have some kind of awareness of. Awe is rather 'the spark' which lights up the clearing that allows for the deep, contemplative wonder of, as Hansen (chapter 15) formulates it, wisdom-, meaning- and wonder-seeking wonderment to occur.

4 By drawing from my own experience, I base this article on inspirations from the auto-ethnographic methods, 'auto' referring to the introspection of the researchers' own observations, reflections and examinations, 'ethno' to the extrospection of culture and, finally, 'graphic' to the scientific process (Baarts 2015, p. 171). In this article I am pulling from a concrete, personal experience in order to exemplify an existential phenomenon and metaphysical problem related to the late-modern (western) society, and thus to the topic of my research. The story is a translation from Danish of a part of the journal/diary that I wrote while walking the Camino Via de la Plata.

5 Translated by the author from Danish: *Jeg har gaaet mig mine bedste Tanker til, og jeg kender ingen Tanke saa tung, at man jo kan gaae fra den.*

6 Praise here refers to both the religious and the secular meaning of the word: as an expression of gratitude, tribute to, and appreciation of, being bestowed with the privileges of the few, allowing me to devote six full weeks with no other agenda than walking the Camino.

7 As I have previously argue (see above, footnote 1) *awe* is not a skill that can be trained, but something that 'happens to us', and often when we least expect it. Thus, 'to "think" a thinking that thinks…' might just be another example of the human entrapment into rational thinking and reveal my limited ability to 'think' out of The Cartesian Box.

8 *In What is Philosophy* from 1955, Heidegger talks about *astonishment* instead of awe. I agree, though, with Capobianco (2010) when he says that Heidegger's 'phenomenological description of the mood of astonishment is virtually *identical* to his description of awe *(die Scheu)*' (p. 84).

9 English: 'the Originating of the beginning […] beings are let to arise in the there'. I am using the translation by Capobianco (2010, p. 43) instead of that of Peter Hanly in the 2023 publication of *On Inception* from Indiana University Press, as I believe that 'Originating' best captures the neutrality and impersonality of the German word 'Anfängnis' that Capobianco (2010, p. 43) emphasizes to be essential to understanding Heidegger's concept of 'Das Ereignis'.

10 In English: Beyng as the abyss of giving away.

11 In the English translation here quoted, the German term 'Besinnliches Denken' is translated to 'meditative thinking'. I believe that to be a distortion of the meaning in German as 'Besinnliches' is related to the word 'Besinnung' which means to 'regain one's composure' and 'peaceful reflection', and 'Besinnliches' thus can only rightfully be translated with *contemplative*.

12 It is important to emphasize that Heidegger (1927) in *Being and Time* uses the term *attunement* as the ontological term for what is 'ontically most familiar and an everyday kind of thing: mood, being in a mood' (p. 126), Thus, when I argue awe to be a pre-metaphysical and pre-ontological state of Being then it even predates Heidegger's (1927) notion that 'in attunement, Da-sein is always brought before itself, it has always already found itself […]' (p. 128). It is more of a *fundamental* attunement in the sense brought forward by Swanton (2010) through which we are allowed access to Being and without which 'whole aspects of the world would remain hidden from us' (p. 159).

13 Discourses with a capital D indicate the reference to primary discursive formations of social meaning across space and time.

14 Heidegger calls these moments *kairologial moments* (Heidegger 1923, p. 78).

15 Each year, the *Global Footprint Network*, an international research organization, provides a report documenting 'the date when humanity's demand for ecological resources and services in a given year exceeds what Earth can regenerate in that year', called *Earth Overshoot Day*. 'In 2023, Earth Overshoot Day lands' on August 2nd· See: https://www.overshootday.org/about/. Richardson et al. (2023) conclude in a research article published on September 13th 2023 that

> the scientific updates and analyses presented here confirms that humanity is today placing unprecedented pressure on Earth systems. Perhaps most worrying in terms of maintaining Earth system in a Holocene-like interglacial state is that all biosphere-related planetary boundary processes

providing the resilience (capacity to dampen disturbance) of Earth system are at or close to a high-risk level of transgression. [...]. This update of the planetary boundaries' framework may serve as a renewed wake-up call to humankind that Earth is in danger of leaving its Holocene-like state. (p. 11)

16 From German (own translation): *durchstimmende Angesprochenheit*; *einen unmittelbaren und intensiven Welt-beziehungsweise Sachbezug*.

References

Abram, D. (1996). *The Spell of the Sensuous, Perception and Lanuage in a More-Than-Human-World.* New York: Vintage Books/Random House Inc. 1997.

Andersen, M.F. (2013). Når omfavnelse bliver til kvælertag: hvordan kan arbejdet blive et spørgsmål om liv eller død (Eng: When embracing becomes a stranglehold: how can work become a question of life and death). In Andersen, M.F. & Brinkmann, S. (eds). *Nye perspektiver på stress* (Eng: *New perspectives on stress*). Aarhus: Forlaget Klim.

Baarts, C. (2015). Autoetnografi (Eng: Autoethnography). In Brinkmann, S. & Tanggaard, L. (eds.). *Kvalitative Metoder* (Eng: *Qualitative Methods*). Copenhagen: Hans Reitzels Forlag.

Capobianco, R. (2010). *Engaging Heidegger*. Toronto: University of Toronto Press Incorporated.

Han, B.C. (2010). *The Burnout Society*. Stanford: Stanford University Press 2015.

Han, B.C. (2012). *The Transparency Society*. Stanford: Stanford University Press 2015.

Han, B.C. (2019). *The Disappearance of Rituals*. Cambridge: Polity Press 2020.

Han, B.C. (2021). *The Palliative Society*. Cambridge: Polity Press 2022.

Hansen, F.T. (2025). The Sense of Wonder as Key to Eco-Buildung. In Heneise, Michael T., Cassandra Falke, Espen Dahl, Alice Sundman, and Edvard Lia, *The Routledge Handbook of Eco-phenomenology*. London: Routledge.

Heidegger, M. (1923). *Ontology – the Hermeneutics of Facticity*. Bloomington: Indiana University Press 2008.

Heidegger, M. (1927). *Being and Time*. New York: State University of New York Press 1996.

Heidegger, M. (1929). What is Metaphysics? In Heidegger, M. (ed.) (1927–1964). *Basic Writings*, pp. 93–110. New York: HarperCollins Publishers 1993.

Heidegger, M. (1935). *Introduction to Metaphysics*. New Haven: Yale University Press 2000a.

Heidegger, M. (1936–1938). *Contributions to philosophy (from enowning)*. Bloomington: Indiana University Press 1999.

Heidegger, M. (1941). *Über den Anfang (GA 70)*. Frankfurt am Main: Vittorio Klostermann 2005.

Heidegger, M. (1943a). Postscript to 'What is Metaphysics?'. In Heidegger, M. (ed.) (1919–1961). *Pathmarks*, pp. 231–237. Cambridge: Cambridge University Press 1998.

Heidegger, M. (1943b). Remembrance. In Heidegger, M. (ed.) (1936–1968). *Elucidations of Hölderlin's Poetry*, pp. 101–173. New York: Humanity Books 2000.

Heidegger, M. (1943c). Andenken. In Heidegger, M. (ed.) (1936–1968). *Erläuterungen zu Hölderlins Dichtung*, pp. 75–151. Frankfurt am Main: Vittorio Klostermann 1981.

Heidegger, M: (1949). Insight Into That Which Is, Bremen Lectures 1949. In Heidegger, M (ed.) (1949/1957). *Bremen and Freiburg Lectures*. Bloomington: Indiana University Press 2012.

Heidegger, M. (1951). Building, Dwelling, Thinking. In Heidegger, M. (ed.) (1971). *Poetry, Language, Thought*. New York: HarperCollins Publishers Inc. 2001.

Heidegger, M. (1955). *What is Philosophy*. Albany, New York: NCUP Inc.

Heidegger, M. (1957–1958). The Nature of Language. In Heidegger, M. (ed.) (1959). *On the Way to Language*, pp. 57–108. New York: HarperCollins Publishers 1971.

Heidegger, M. (1959). *Discourse on Thinking*. New York: Harper & Row Publishers 1966.

Heidegger, M. & Fink, E. (1966–1967). *Heraclitus Seminar 1966/67*. Tuscaloosa: University of Alabama Press 1979.

Hjortkjær, C. (2020). *Utilstrækkelig - hvorfor den nye moral gør de unge psykisk syge* (Eng: *Inadequate - why the new moral makes the young mentally ill*). Aarhus: Forlaget Klim.

Kennedy, T. (2014). *Heidegger and the Ethics of the Earth: Eco-Phenomenology in the Age of Technology*. https://digitalrepository.unm.edu/phil_etds/7

Levinas, E. & Levinas, E. (1984). Ethics and Infinity. *CrossCurrents*, Vol. *34*, No. 2 (summer 1984), pp. 191–203.

Mølholm, M. (2019). The Existential 'Anruf' as the Agency of the Anti-Objectives of the Discourse on Stress. *Zeitschrift für Diskursforschung, Heft* Vol. *2*, pp. 190–214.

Mølholm, M. & Hansen, F.T. (2024a). The Compulsion to Achieve as a Way to Immortality, The Discourse on Stress and Burnout seen in the Perspective of an 'excess of positivity'. *Zeitschrift für Diskursforschung, Heft*, Vol. *1*, pp. 97–120.

Mølholm, M. & Hansen, F.T. (2024b). The Sound of Wonder as a Way of Ontological Homecoming: The Discourse on Stress and Burnout seen from the perspective of an 'excess of negativity'. *Zeitschrift für Diskursforschung, Heft*, Vol. *1*, pp. 121–141.

Nietzsche, F. (1889). *Twilight of the Idols*. Mineola, NY: Dover Publications Inc. 2019.

Pieper, J. (1947a). Leisure – The Basis of Culture. In Pieper, J. (ed.) (1947). *Leisure – The Basis of Culture*, pp. 15–74. San Francisco: Ignatius Press 2009.

Pieper, J. (1947b). The Philosophical Act. In Pieper, J (ed.) (1947a). *Leisure - The Basis of Culture*, pp. 75–143. San Francisco: Ignatius Press 2009.

Richardson, K., Steffen, W., Lucht, W., Bendtsen, J., Cornell, S.E., Donges, J.F., Drüke, M., Fetzer, I., Bala, G., Bloh, W.V., Feulner, G., Fiedler, S., Gerten, D., Gleeson, T., Hofmann, M., Huiskamp, W., Kummu, M., Mohan, C., Nogués-Bravo, D., Petri, S., Porkka, M., Rahmstorf, S., Schaphoff, S., Thonicke, K., Tobian, A., Virkki, V., Wang-Erlandsson, W. L., & Rockström, J. (2023). Earth Beyond Six of Nine Planetary Boundaries. In: *Science Advances*, *9*, 13th of September 2023. Accessed September 18th 2023 at: https://www.science.org/doi/10.1126/sciadv.adh2458

Rosa, H. (2013). *Alienation and Acceleration – Towards a Critical Theory of Late-Modern Temporality*. Copenhagen: NSU Press & Nordiskt Sommaruniversitet.

Rosa, H. (2016). *Resonanz – Eine Soziologie der Weltbeziehung (e-book)*. Berlin: Suhrkamp Verlag.

Rosa, H. (2017). Dynamic Stabilization, the Triple A. Approach to the Good Life, and the Resonance Conception. *Questions de Communication*, Vol. *31*, pp. 437–456.

Sessions, G. & Naess, A. (1986). The Basic Principles of Deep Ecology. *The Trumpeter*, Vol. *3*, no. 4 fall, p. 86. ISSN: 0832-6193.

Skårderud, F. (1998). *Uro, en rejse i det moderne selv* (Eng: *Restlessness, a journey in the modern self*). Copenhagen: Tiderne Skifter 1999.

Swanton, C. (2010). Heideggerian Environmental Virtue Ethics. *Journal of Agricultural and Environmental Ethics 23*, pp. 145–166. DOI: 10.1007/s10806-009-9186-1

Visse, M., Hansen, F.T. & Leget, C. (2020). Apophatic Inquiry, Living the Questions Themselves. *International Journal of Qualitative Methods 19*, pp. 1–11.

Wood, D. (2003). What is Eco-phenomenology. In Brown, C.S. & Toadvine, T. (eds.). *Eco-phenomenology, Back to the Earth Itself*. Albany: State University of New York Press.

11

THE ECO-PHENOMENOLOGY OF DREAMING

Flesh, Relation, and Perception beyond the Human

Michael T. Heneise

Introduction

At 69°N, where the polar night stretches on for months, I find myself struggling to fall asleep. My body, long entrained to the Sun's rhythms, drifts unmoored, its internal clock faltering in the absence of its celestial anchor. Even with fish oil supplements and the artificial glow of an office "light therapy" lamp, rest remains fragmented. It feels imposed, an unnatural rhythm pressing itself upon a body once attuned to a more reciprocal order. Yet insomnia is not merely deprivation; it is a perceptual rupture, a limit-experience that casts into relief the otherwise continuous interplay between dreaming and wakefulness.

For while dreaming and waking differ in their qualities of attention and coherence, they are deeply intertwined modes of perception. Dreaming is not a departure from perception but a revelatory intensification of it. It lays bare the structures of embodiment and world-relation that also sustain waking life. Insomnia, by contrast, interrupts this rhythm. It neither reveals nor conceals but withholds. It marks the disarticulation of perception's usual flow, exposing its fragility.

I read Walter Benjamin's words as an injunction: "The realization (Verwertung) of dream elements in the course of waking up is the paradigm of dialectical thinking" (1999, 13). Benjamin is not merely describing a process of thought but an embodied mode of perception, in which elements of the dream reassert themselves in wakefulness. This dialectical slippage aligns with Maurice Merleau-Ponty's insistence that perception is never wholly transparent but always emerges through a chiasmic play of presence and absence, wakefulness and reverie.

In the Arctic winter, where wakefulness and inattention blur - indeed, do we not retreat to the daydream even amid our most engaging staff meetings? - this dialectic is not metaphorical but phenomenological: a corporeal experience of perspectival disjunction. My mind loosens its grip not by volition, but by necessity. Perception slips, and the darkened landscape - muted, recursive, and unending - claims the body not through visibility, but through a kind of possession. As Benjamin saw in history, so too in perception: recognition arises not through clarity alone, but through movement - between presence and its withdrawal, between self-possession and dispossession. The polar night does not merely disturb sleep; it forces a

 DOI: 10.4324/9781003538097-14

confrontation with the ecology of perception, where my embeddedness in the world - its intertwining - discloses other modes of seeing and knowing.

Indeed, sleep and dreaming lend themselves to philosophical inquiry - not as abstract puzzles, nor as retreats from a coherent real, but as modes of ontological disclosure. Dreaming reveals the underlying structures of perception and embodiment, exposing the interwoven character of self and world. When sleep finally pulls me into its bosom, whether during my office nap or at home, it is not merely the function of an exhausted mind or body. It is a pivot into another deep, authentic mode of perception. Insomnia clarifies how perception always exceeds the individual subject; a wakefulness that, as Levinas (2001) puts it, refuses retreat and reveals the uncontainable fact of being-there. This vigilance corresponds to what Levinas elsewhere describes as *il y a* - the impersonal presence of being that overwhelms without form or exit. In insomnia, the self is exposed not to an encounter with the Other, but to the sheer, uncontainable fact of being-there. In this exposure, the limits of subjectivity begin to tremble. Perception no longer appears as a sovereign act, but as a form of dispossession, revealing our embeddedness in a shared field of being.

Western philosophy has typically dismissed dreams as illusion. Plato, in the *Republic*, suggests that dreams represent an ungoverned state in which unruly desires surface, disconnected from the rational soul's dominion (Republic, 571c). René Descartes, in his *Meditations*, famously invokes the dream argument to cast doubt on perception altogether, presenting dreams as the very paradigm of deception (Descartes [1641] 1996, 19). Even Sigmund Freud, while reclaiming dreams as worthy of interpretation, ultimately reduces them to latent content rather than recognizing them as genuine modes of perception (Freud [1900] 1997, 311). Each of these perspectives, in its own way, reinforces a strict boundary between mind and world, appearance and reality.

Merleau-Ponty offers a radically different account. He situates dreaming within a perceptual ontology grounded in the lived body, open to the world even in sleep. His late work, especially *The Visible and the Invisible*, dissolves the boundary between self and world through his concept of "flesh" (la chair), the medium of intertwining between perception and being. "Where are we to put the limit between the body and the world, since the world is flesh?" he asks (Merleau-Ponty 1968, 138). In this view, perception is not the act of a sovereign subject, but a chiasmic process that precedes and exceeds the ego. Dreaming, then, is not reducible to private imagery or neural residue; it is a site of intensified perception in which the entanglement of self and world becomes legible.

While *Phenomenology of Perception* (2012) grounds perception in the body's intentional arc, it is in *The Visible and the Invisible* that Merleau-Ponty develops the deeper implications of perception as reversible, passive, and porous. As James Morley observes in his analysis of *The Visible and the Invisible*, "dreaming is crucial to the central argument" of this later text (Morley 1999, 97),[1] because it reveals the ontological continuity between sleeping and waking, and initiates Merleau-Ponty's most radical insights into perception. Dreaming, like perception more broadly, unfolds through shared flesh, an elemental reversibility in which self and world fold into each other.

In this chapter, I argue that Merleau-Ponty's carnal ontology, and particularly his concepts of reversibility, encroachment, and promiscuity, offers a framework for understanding dreaming not as a neurocognitive anomaly but as a mode of intersubjective, more-than-human entanglement. When read alongside psychoanalytic theories of multiplicity in dream-subjectivity and indigenous accounts of dreaming as ecological encounter, dreams emerge not as escapes from the world, but as intensified participations in it. What arises is an eco-phenomenology of dreaming; one that moves beyond the sovereign human subject and toward a participatory ontology grounded in the permeability of body, world, and cosmos.

Merleau-Ponty's Flesh and Écart

Merleau-Ponty's untimely death at 53 left his most radical ontological insights tantalizingly incomplete. Yet in the fragments - his working notes, lectures, and *The Visible and the Invisible* - we witness a deepening preoccupation with perception as fundamentally structured by *écart*: a gap, divergence, or spacing that both enables and complicates experience. *Écart* is not simply separation, but a generative spacing, which Morris (2010, 142)[2] describes as "a gap or excess in being". It is a divergence through which perception and sense emerge. It opens perception to a world that is never fully given, never entirely present. I see, but never all at once. My movements anticipate what lies just beyond view, and in that delay, perception unfolds - not as completion, but as a continual becoming.

Jennifer McWeeny (2016, 215)[3] defines *écart* as "that divergence that founds the mutual dependence between the ambiguities of experience, including unity and difference, visibility and invisibility, vision and touch, transparency and opacity." To this I would add temporality. For if perception is structured by *écart*, then time itself must be understood not as a linear flow, but as a fissured, multilayered field. Dreaming, in this sense, stages écart in its most unbound form, where temporal coherence dissolves and experiential layers interweave. Past folds into present; the future brushes against memory. The dreamer is both inside the scene and floating above it. This is no mere trick of narrative logic, but a structural feature of perception rendered visible in the dream (cf. Merleau-Ponty 2012, 306).

This helps explain why Merleau-Ponty turns increasingly to *écart* in his late work. This is not as a detour but as a necessary term to account for how experience occurs in the flesh. *Écart* marks the delay between stimulus and response, intention and realization. In wakefulness, it allows perceptual life to stabilize itself, however momentary. But in dreaming, this stabilization falters. This is not because perception fails, but because it becomes more fluid, more openly entangled. We enter what Gail Weiss (1999, 170) calls a "perforated zone" of perception, where self and other, subject and object, flicker and shift. Dreaming thus radicalizes *écart*, not as pathology or regression, but as ontological disclosure - *as the visibility of an otherwise concealed genesis of sense* (Morris 2010, 141).[4]

This brings us to reversibility, which must be understood not merely as a reciprocal exchange, as in touching and being touched, seeing and being seen, but as a deeper condition of being structured by non-coincidence. "There is between the seer and the visible a sort of reversibility," writes Merleau-Ponty, "which is the secret of perception" (1968, 263). But crucially, this reversibility is never symmetrical. I touch with one hand and feel it with the other, but the two never fully coincide. The self is always entangled with the world, yes; but also always at a slight remove from it. As Morris explains, this divergence is not an external relation between two entities but an internal inflection of being. In dreams, this asymmetry becomes profound. The reversibility of the dream extends beyond bodily sensation, into the very fabric of time and selfhood. One is both actor and observer, subject and scene, caught in a chiasm where distinctions stretch and warp, but never quite collapse.

Despite its radical break with classical phenomenology, Merleau-Ponty's later ontology never fully abandons incarnate subjectivity. "I am always on the same side of my body," he writes, "it presents itself to me in one invariable perspective" (1968, 148). At first glance, this sounds like a regression; a return to the subject of perception, fixed in its own point of view.

But I would argue it's something else: a methodological move, a deliberate anchoring from which to explore the immense variability of perceptual life. The body is not the end of the story, but the place where the story begins. From this "same side," we are thrown outward into a dynamic, plural, and participatory world.

Understanding this requires us to grasp Merleau-Ponty's notion of flesh not as a substance, but as a field; a chiasmic fabric in which self and world interpenetrate without reducing one to the other. It is *écart*, the generative divergence within being, that opens this field - not as lack, but as the very spacing through which perception unfolds. As Morris (2010, 141) puts it, Merleau-Ponty's late ontology marks "a radically new philosophy," in which being is no longer framed by identity or presence, but by divergence; an internal differentiation that conditions perception and the genesis of sense. Reversibility, in this view, is not a mere symmetry but a fundamental chirality within being; a reversible fold through which sense comes into being. Dreaming, I propose, enacts this divergence in its most intensified form: not by bracketing *écart*, but by dilating it; exposing the interval where sense emerges and identity flickers into multiplicity. Morris insists that reversibility is not a perceptual condition imposed by consciousness, but an ontological movement internal to being itself. Reversibility enables, rather than results from, perception.

Écart, then, is not an aberration or supplement. It is the very logic of perception *rendered visible in the dream*. And when placed in conversation with reversibility, encroachment, and indeed promiscuity, it reveals dreaming not as some neural anomaly, but as a revelatory modality of eco-phenomenological insight. The dream is not merely about us; it is an event of the flesh, a site where the world touches us otherwise.

Dreaming, Passivity, and the Entrelacs of Perception

Merleau-Ponty encourages us to rethink the relation between waking and dreaming not as binary opposites, but as structurally interwoven modes of perception. While dreaming may lack the practical engagement of waking life, it remains, for Merleau-Ponty, a genuine modality of perceptual experience. In *Phenomenology of Perception*, he insists that "the difference between perception and dream [is] not absolute," and that both belong to "our experiences" (2012, 38). Dreaming, far from being a distortion of perception, renders visible its more elusive dynamics, particularly the complex entanglement of activity and passivity, self and world.

Dreaming is marked by what we might call 'operative passivity'. That is, perception in sleep continues to function even when conscious intentionality recedes. Merleau-Ponty refers to this as operative intentionality; a pre-reflective mode of embodied awareness that structures our engagement with the world without becoming its object (2012, 415). In dreaming, this operative field is not suspended but reoriented. The dream body senses, moves, responds; though its movements are not constrained by external resistance, and its environment, while real in experience, lacks objective fixity.

This situates dreaming within Merleau-Ponty's later ontology of the flesh. Rather than viewing perception as a one-way relation from subject to object, Merleau-Ponty introduces the notion of intertwining (*entrelacs*) - the mutual implication of self and world in a shared ontological fabric. In *The Visible and the Invisible*, he writes: "the body is made of the same flesh as the world," and perception occurs in this zone of reciprocal emergence (1968, 248). Dreaming, I suggest, draws this *entrelacs* into view. The dreamer is not merely the recipient of images but participates in their constitution. In dreams, the relation between seer and seen, agent and patient, becomes more fluid, shifting, and oscillating in ways not possible in waking life.

As Morley (1999, 97–99) argues, the dream self-moves between positions: now actor, now observer; revealing the permeability of perceptual structures, a movement I take to be indicative of what Merleau-Ponty later formulates as the chiasm of being. This echoes Merleau-Ponty's insistence that the subject is never autonomous, but always already entangled in the world it perceives. The self in dreams may be dispersed, fragmented, or multiplied, not in the sense of loss, but as an intensification of perception's relational structure.

Psychoanalytic literature has long observed this multiplicity. Freud ([1900] 1953, 312–313, 507, 561–562) describes the dream not as pure fantasy, but as the symbolic transformation of internal and external stimuli, mediated by what he calls the *Nachtwächter*, the night watchman. This figure regulates the thresholds of perception during sleep, allowing disruptive stimuli, such as a loud noise, to be translated symbolically rather than waking the dreamer (1953, 565). In this way, the dream does not block perception but modulates it, maintaining a structured, if altered, relation to the world.

This dynamic finds resonance in Merleau-Ponty's notion of the body as both perceiving and perceived; a structure he explores most clearly in the reversibility of touch (1968, 133–143). But rather than focusing solely on reversibility's symmetry, dreaming foregrounds the asymmetry and latency inherent in perceptual experience. The dreamer, like the touching hand that cannot touch itself directly, perceives from a displaced position. One sees oneself from the outside or inhabits multiple roles within the dream. Morris (2010, 144) likens this structure to a glove turned inside out; a metaphor for reversibility as an internal operation within being, not merely a spatial inversion but a chiral inflection that differentiates without separating. Dreaming brings this structure into relief, making perceptible the divergence that silently underlies all perception. This echoes Morley's interpretation of dreams as "revelatory of the structure of perception" (1999, 104), though here I develop this claim further through the lens of reversibility and eco-phenomenology.

Merleau-Ponty considers this kind of perceptual decentering in *Phenomenology of Perception*, particularly in his discussion of "the synthesis of one's own body" (2012, 149–155), where one becomes visible to oneself in ways ordinarily obscured in waking life. This decentering is conceptually close to Husserl's idea of appresentation, the indirect givenness of the other's subjectivity, which implies that I too am visible from without (*Cartesian Meditations*, 1960, §§42–55). But dreaming radicalizes this dynamic. The dream does not merely simulate being seen; it stages it. The self is not only a perceiver but becomes one image among others, caught in a shifting gaze that resists singular ownership.

Through *Verdichtung* (condensation) and *Verschiebung* (displacement), the dream synthesizes fragments of meaning, organizing them into new perceptual forms. In this light, Freud's *Nachtwächter* can be understood not as an obstacle to perception but as a figure of modulation, managing the porous boundary between dream and world; a task not unlike the perceptual latency Morris sees in the genesis of sense (2010, 141).

Subsequent psychoanalysts such as Sullivan (1953) and Willock (2007, 88) have extended this view, emphasizing that the dream self is neither entirely passive nor fully autonomous. It engages with an environment that remains relational, mutable, and at times resistant. This aligns with Merleau-Ponty's broader assertion that perception, even in sleep, is a structure of lived reciprocity. It is an interlaced exchange shaped not by mastery, but by vulnerability.

Passivity, in this context, does not imply inertness but radical receptivity. Bergo (2009, 31–54), drawing on both Levinas and Merleau-Ponty, describes this as a mode of subjectivity that is seized by the world, permeated by its forces; what she, following Levinas, terms "a passivity more passive than all passivity" (2009, 32). The dreamer does not simply choose to perceive but is drawn into a perceptual field where agency emerges only through relational exposure. In dreams, as Bergo notes, perception operates in an altered modality, still active, but without mastery; still embodied, but without ground. Although Bergo carefully distinguishes Merleau-Ponty's ontological exploration of passivity from Levinas's ethical framing, I propose that dreaming exemplifies a convergence of these dimensions. Dreaming is a lived structure that is simultaneously perceptual and ethical. What Morris calls a "gap or excess in being that allows for a genesis of sense" (2010, 141) becomes, in dreaming, vividly manifest—disorienting yet generative.

Morley (1999, 104) similarly emphasizes that dreaming makes the structure of perception visible. Whereas waking life often conceals the reversibility and relationality of perception beneath pragmatic action, dreaming foregrounds these dynamics. One's perspective shifts, one's body is felt from without, and perceptual boundaries blur, revealing the *entrelacs* of being in ways waking consciousness rarely permits.

This has implications for how we understand nature and perception more broadly. Merleau-Ponty (1968, 128) asserts that the body is not an instrument of perception, but the very medium through which the world appears. The dream thus does not isolate us from the world but exposes our embeddedness in it. As Morley (1999, 109) notes, the dream world is not simply a projection of the inner self, but an expression of the body's ongoing relationality with its environment.

From this perspective, dreaming contributes to a radically ecological account of perception; one in which subject and world, self and nature, are co-constituted in a shared ontological field. Dreaming does not negate perception but intensifies its basic logic. The dream is not an escape from reality but an opening onto its deeper structures. As Merleau-Ponty writes, "the flesh is not matter, is not mind, is not substance. To designate it, we should need the old term 'element' in the sense it was used to speak of water, air, earth, and fire" (1968, 139). In this sense, dreaming discloses the elemental texture of perception; one that binds us to a world that is always already inside us, even in sleep.

From Reversibility to Promiscuity

Reversibility challenges conventional notions of subjectivity by showing that perception is not a private, volitional act, but a dynamic exchange between self and world. Yet Merleau-Ponty does not stop there. Reversibility is only a point of departure. In *The Visible and the Invisible*, it functions as a threshold concept, introducing a broader ontological shift. Rather than an end in itself, it is a methodological framing that prepares the ground for more radical reflections on embodiment. Merleau-Ponty begins with reciprocity but soon complicates it through notions such as *empiétement* (encroachment) and possession – concepts that move us beyond the reversible interplay of perceiver and perceived.

These ideas suggest that subjectivity is not merely relational but unstable; its contours are blurred, its sovereignty compromised. In this model, the self is not just a locus of perception, but a porous site of interpenetration, where world and other impress themselves from within. In *Phenomenology of Perception*, Merleau-Ponty already hints at this when he writes that "I am not in front of my body, I am in it, or rather I am it" (2012, 200). We do not simply inhabit the world; we are inhabited by it. The flesh, understood as the element common to seer and seen, thinker and world, generates a mode of being in which agency and alterity are not separate poles but co-constitutive forces.

If reversibility marks this mutual implication, possession intensifies it by introducing asymmetry. Here, the self is not only open to the world but is overtaken by it. Jennifer McWeeny (2016, 221) characterizes possession as an "excess of subjectivity," a condition in which the first-person perspective is always already permeated by alterity, destabilizing the assumption of sovereign selfhood. Crucially, she situates this excess in the body schema itself, which she reads not as stable ground but as a site of ontological vulnerability; a medium through which possession becomes possible (McWeeny, 215–216). In dreaming, this dynamic becomes especially apparent. The dreamer is not simply perceiving, but is also, in a sense, possessed; the dream unfolds with its own volition, often beyond the dreamer's control. Possession does not operate within mutual exchange; it unsettles it. Unlike reversibility, which maintains some degree of perspectival differentiation, possession disorients. One speaks in the dream and hears words that are not one's own. One acts, but without authoring the action. McWeeny finds an analogue to this in Freud's case of Dora, where the speaking subject is displaced, as if taken over by another's voice (McWeeny, 223).

Such moments challenge the phenomenological emphasis on self-affection and first-person perspective. But do they have ontological weight, or are they confined to the realm of ephemeral subjective states? One could argue that dream experience lacks philosophical traction. Yet Merleau-Ponty insists that dreaming and waking both belong to perception's domain: "The difference between perception and dream not being absolute, one is justified in counting them both among our experiences" (2012, 38). The dream may not replicate the world's resistance, but it nonetheless enacts the structural dynamics of perception; estranged and intensified.

In this light, possession reveals a subjective field always in flux. A dream in which your hands trace symbols you don't recognize, or where you speak with a voice not your own, demonstrates the way the self is already entangled with alterity. The dream-self is not a diminished version of the waking self; it is a disclosure of its instability. This kind of experience, where one oscillates between agency and subjection, between self and other, calls into question the assumption that perception is always anchored in the "same side" of the body (Merleau-Ponty 1968, 148).

Promiscuity marks the furthest point in this ontological progression. Merleau-Ponty writes: "The ontological relation of promiscuity goes further than possession" (1968, 234). Here, the distinction between self and world does not merely blur but dissolves. Whereas possession implies the incursion of otherness, promiscuity describes a fusion so complete that the terms lose their separation. Promiscuity, McWeeny writes, is like the blending of pigments: "as the combination of yellow and blue paint makes green," the contributing elements become ontologically indistinct (2016, 226). In this state, the body no longer acts as boundary or barrier but is refigured, in McWeeny's terms, as "a point of passage" (2016, 225). A body is no longer bounded by sides but open to incorporation and interpenetration.

Dreaming, I suggest, vividly enacts this mode of experience, extending McWeeny's notion of promiscuity into the fluid thresholds of sleep and oneiric subjectivity. In certain dreams, one does not merely move through a landscape but becomes it. One does not see an animal but lives its hunger, feels its limbs, breathes with its lungs. Perceptual position multiplies and migrates—bird, stone, sky—without rupture or interval. The dream no longer contains the self; it absorbs it. Boundaries cease to function. There is no interior and exterior, only the flux of sensation and sense.

McWeeny (2016, 227) writes that promiscuity puts in question "the first-person nature of experience—its mineness, its auto-affection." Without this guarantee, the continuity of personal identity can no longer be assumed. And if that continuity falters, then so too does the methodological foundation of traditional phenomenology. As Merleau-Ponty's own later work suggests, the body is never wholly perspectival, never fully one's own. The attempt to maintain univocal subjectivity becomes less a phenomenological given than a strategic anchoring point; one that is continually undone by the very ontology it seeks to clarify.

If the self is porous, if its boundaries are always under negotiation, then responsibility cannot be reduced to volition. Levinas posits that responsibility for the Other is not a matter of voluntary commitment but an inherent ethical demand that arises independently of one's intentions (1969, 48).[5] While Levinas maintains a strict distinction between ontology and ethics, I take his insight as compatible with a more ontologically porous framework, where subjectivity is already constituted through relational exposure, as also suggested by Merleau-Ponty's notion of flesh. We are already answerable before we choose to be. In dreams, this ethical exposure is radicalized. The self is addressed, summoned, and interpellated by forces it cannot master. A voice calls out; a presence intrudes. One does not respond as much as is response. Dreaming becomes not just a perceptual but an ethical event.

This insight resonates with many indigenous cosmologies, including those found in the eastern Himalayas, where the dream-self is never solely one's own. Among several Highland Asian societies, the dream is a relational domain populated by ancestors, animals, and spirit-persons, where agency circulates and subjectivity is shared (Heneise 2018, 56). Here, too, the self is not bounded but enacted through contact. Dreams do not occur within the self; they compose it.

If subjectivity is already open, then perception is never ethically neutral. It is exposure, vulnerability, and invitation. The dream reveals this with particular clarity - not as a metaphor but as a mode of being. To dream is not merely to imagine or to retreat; it is to participate in the world's own unfolding, to become permeable to others, human and nonhuman alike. It is to discover, again and again, that perception is not possession but relation. And that, perhaps, is where phenomenology must begin again.

Dream Cosmologies and Porous Selves

In the previous section, we observed how subjectivity can overflow its own bounds, becoming inhabited by the world in ways that trouble any neat account of a discrete, individuated self. Here, we briefly return to the notion of possession, not to reiterate what has already been said, but to offer firmer conceptual grounding before we move more fully into the territory of promiscuity. Possession, unlike reversibility, is marked by asymmetry. While reversibility retains a structural tension, where the perceiver and perceived remain distinct even as they fold into each other; possession introduces a more unilateral encroachment, wherein alterity does not merely touch the self but permeates it.

McWeeny (2016, 221) writes that "possession signals the inherent openness of subjectivity to the world, a condition in which the first-person perspective is always already inhabited by an Other." This resonates with anthropological accounts in which possession is not merely a disruption of subjectivity, but an expansion; an event in which the self is stretched to accommodate external agencies. The subject may be overtaken, yes, but not erased. Possession thus remains ontologically useful: it stages a transformation, not a dissolution. The self remains a site, or envelope, within which something else arrives.

Marc Augé's (1999) study of African dream cosmologies offers a particularly illuminating point of comparison. In many such traditions, dreams are not confined to the interiority of the individual but are relational events; encounters with spirits, ancestors, and other dreamers, often structured through communal interpretation. Augé distinguishes between two enigmatic configurations of the self: the *enigma of presence*, in which the dreamer retains a dim awareness of their dreaming body, and the *enigma of absence*, in which another force occupies the self in states of possession (Augé 1999, 30). Crucially, the latter does not signal the annihilation of the subject but its productive displacement. "The possessed is not enigmatic but absent," Augé writes, and it is precisely in this absence that a different configuration of agency emerges; one in which the body remains active while interiority is suspended (Augé 1999, 30).

This resonates strongly with Merleau-Ponty's late ontology. Just as Augé's possessed subject is "an actor" whose behavior becomes overdetermined by forces beyond the ego (Augé 1999, 30), Merleau-Ponty's concept of *empiétement* names the condition in which perception is not a balanced reciprocity, but an uneven interpenetration. In such moments, the flesh becomes a conduit, a site of transit rather than containment (Merleau-Ponty 1968, 234).

Furthermore, Augé observes that dreams and possession imply different logics of the self: dreams require memory and narration, while possession is marked by forgetting. "The dream exists only through the memory that was kept of it," he writes (Augé 1999, 36), distinguishing it from possession, which draws its force from erasure. Dreams, in this sense, are both events and acts of interpretation. They remake the self not through multiplicity, but through what Augé calls an "ephemeral collection of elements of diverse origins"; a form of recomposed singularity (Augé 1999, 36). This conception, too, dovetails with Merleau-Ponty's flesh, where identity is not substance but spacing; a site of divergence and folding.

While possession maintains a trace of the self, promiscuity enacts a deeper ontological breach. In possession, one may feel something else acting through them. But in promiscuity, one does not merely receive the other - one becomes it. As Merleau-Ponty writes, "The ontological relation of promiscuity goes further than possession" (Merleau-Ponty 1968, 234). The distinction here is subtle but vital. Promiscuity does not just indicate closeness, or even mutual interpenetration. It denotes a condition in which the difference between self and other, perceiver and perceived, is no longer intelligible. McWeeny (2016, 226) explains that "In promiscuity, bodily sidedness gives way to incorporation and interpenetration. The body's erotic openness to the world is fully realized, for better or for worse."

Dreaming, I suggest, offers a phenomenological inflection point of this state. Merleau-Ponty affirms that dreaming is not a deviation from perception but part of it: "The difference between perception and dream not being absolute, one is justified in counting them both among our experiences" (Merleau-Ponty 2012, 38; cf. Morley 1999, 97). The dream-self does not merely shift roles; it dissolves. You may become another person, merge with objects, inhabit multiple perspectives at once without rupture or contradiction. This is not merely symbolic: it reveals the structure of subjectivity as inherently porous. As McWeeny notes, "To put the first-personal perspective in question is also to put the methods of traditional phenomenology in question" (2016, 227); a not-so-subtle suggestion that Merleau-Ponty's ontology of flesh may exceed even the methodological bounds of Husserlian and Sartrean frameworks.

Among Himalayan cosmologies, the porousness of the self is neither anomalous nor restricted to sleep. Drawing on my own fieldwork among the Angami Nagas, the *ruopfü*—the life force or animating principle—is not confined to the human body but moves fluidly through dreams, landscapes, and other beings. This is most vividly illustrated in the *tekhumiavi* phenomenon (from the Tenyidie: "man in tiger form"), a form of therianthropy wherein the *ruopfü* temporarily assumes the body and perceptual apparatus of a predator (Heneise

2016). These are not dreams in which one imagines being a tiger; they are events in which the dream-self is the tiger.[6]

The *tekhumiavi* experience unfolds as a kind of distributed perception. The *ruopfü* leaves the human body, assumes the tiger's sensory-motor capacities, and returns, often with embodied memories of the hunt: exhaustion, thrill, even the taste of fresh blood. Colonial ethnographers such as Hutton (1921, 208–216) recorded multiple accounts in which injuries to the animal body were experienced simultaneously in the human form. In one 2017 case I encountered near the Indo-Myanmar border, a village youth unknowingly wounded a leopard believed to be linked to a known *tekhumiavi*. At the same moment, the man's body was seized by sudden pain. No wound was found. It was understood: had the leopard died, so too would the man (Heneise 2016, 2018, 76; Hutton 1921).

Such events invite us not to judge their truth status, but to consider what kinds of being they disclose. Do they reflect a form of possession, where the human self remains, but is overtaken? Or are they closer to promiscuity, where boundaries dissolve entirely? The *tekhumiavi* may best be seen as a hybrid: a movement between these states. The self does not vanish, but nor does it remain intact. It enters the world of the tiger, acts, feels, and remembers as the tiger, and returns with the trace of that life in its body.

This ontological movement collapses the distinction between dream and waking, between human and animal, between interior and exterior. In the *tekhumiavi*, we glimpse an alternative model of perception; one that is not grounded in an isolated subject but in a relational field. It exemplifies what Merleau-Ponty might call a chiasm—a folding and crossing of self and world. The dream-self here does not reflect on the world; it is of it. This, as Morris (2010, 144) reminds us, is the deeper structure of reversibility: not reciprocal contact, but an internal divergence of being; a turning-inside-out that reveals the chiral complexity of perception. Promiscuity, then, can be seen as the fullest expression of this divergence, where no fold remains exterior to another.

If perception is not self-contained but always already entangled, then it is also vulnerable. To perceive is to be exposed, to be altered, and to be encroached upon. This resonates with Levinas's critique of self-possession, where ethical responsibility emerges not from autonomy but from the self's exposure to the Other: "responsibility is not a matter of my commitment, but concerns me despite myself" (Levinas 1969, 48). In Merleau-Ponty's ontology, this vulnerability is not a failure but a condition of the flesh. Perception is promiscuous: it absorbs, transforms, and exceeds the subject.

If promiscuity is the condition of perception, then responsibility cannot be confined to moral deliberation; it becomes ontological. The self, already open to the other, cannot choose whether to be responsible; it already is. In the *tekhumiavi*, this is vividly rendered. The hunter's act reverberates across forms, registering not only in the wounded animal but in the doubled human body. Here, perception is not mere observation but participation. It does not end at the limits of the skin.

This is where the eco-phenomenological implications become clear. If perception is not simply human, but unfolds in a broader field of relational entanglement, then dreaming—as understood in these indigenous contexts—offers an epistemological and ontological resource. Dreams are not distortions of reality; they are disclosures of being. They reveal a world in which selves are multiple, subjectivity is open, and the borders between human and nonhuman are porous, reversible, and shared.

Merleau-Ponty's late ontology, in this light, is not merely an extension of phenomenology but a reconfiguration of it. He invites us to think not of perception as the act of a stable ego, but as a mutual encounter: chiasmic, promiscuous, and exposed. If perception is already a site of entanglement and transformation, then responsibility becomes not a task but a condition; one that binds us not only to other humans but to the entire field of being through which perception is always unfolding.

Dreaming, Insomnia, and the Eco-Phenomenology of Perception: Ethical Entanglements

In my darkened office, where I take my early afternoon nap after a lecture, where the rhythms of wakefulness and dreaming are often uneven and unsure, I return to an earlier question: what does it mean to perceive when perception itself is destabilized? My body, once bound to the steady cadence of the Sun, lingers in a state of unrest—my own heterotopia. And yet, in the oscillation between sleep and wakefulness, I glimpse another ontology, one where perception is neither the exclusive domain of a self-enclosed body nor a purely external affair but emerges from a shared field of intertwining. It is precisely this intertwining that insomnia, as a limit-experience, exposes in its breakdown. Insomnia, then, is not merely deprivation but an epistemic disturbance, a site where the boundaries of subjectivity and world begin to tremble. If perception is our most foundational comportment toward the world, then dreaming—and its refusal, insomnia—brings into relief the ecological embeddedness of that comportment, and the precarity that sustains it.

It is precisely here that Merleau-Ponty's notion of flesh finds its most radical articulation: not as a possession of a Cartesian subject, but as a shared element - "not matter, is not mind, is not substance" (Merleau-Ponty 1968, 139) - but rather the condition of perception itself. Flesh is the chiasmic milieu in which the perceiver and the perceived cross, encroach, and interpenetrate. In restless nights, I do not merely lie awake but am drawn out of myself into the textures and silences of the Arctic night. In dreams, this entanglement deepens: boundaries between self and world dissolve, and what appears is not fantasy, but the very interweaving of being.

And yet, these dream worlds are not merely aesthetic: they are sites of ethical implication. Promiscuity, in Merleau-Ponty's terms, means that perception always carries with it the presence of the other. "The ontological relation of promiscuity goes further than possession," he writes (Merleau-Ponty 1968, 234), describing a situation in which the boundary between perceiver and perceived is no longer discernible. In dreams, I do not merely observe a world; I am claimed by it. I do not construct the dream; I am drawn into its logic, inhabited by its voices. In this way, dreams do not obscure the ethical demand but radicalize it. I am not sovereign in the dream; I am entangled.

Insomnia, however, offers another mode of exposure. If dreaming enacts the permeability of perception, its relational openness, insomnia renders its opposite: a rigid, unrelenting consciousness that refuses to dissolve. Where the dream loosens temporal and bodily cohesion, insomnia imposes duration, stillness, and solitude. In insomnia, I am not claimed by the world but abandoned to its distance. The radiator clicks, not rhythmically but mechanically. The polar night does not envelop me; it watches me from without. Levinas evokes this condition with precision:

> It is not that there is my vigilance in the night; in insomnia it is the night itself that watches. It watches. In this anonymous night watch… I am, one might say, the object rather than the subject of an anonymous thought.
>
> (Levinas 2001 [1947], 66)

Levinas here is not describing a psychological state but a metaphysical structure of exposed existence, where subjectivity is no longer a refuge but the object of an impersonal vigilance. Insomnia thus inverts the dynamic of possession. It is not simply another modality of perception but a disruption; a state of vigilance that arrests rather than opens. The self is hyper-attuned, not relationally open but locked in a condition of suspension, where the intersubjective rhythm of perception no longer holds.

This contrast is illuminating. If dreaming offers a model of perception as relational and open—where boundaries fold, where subject and world blend—insomnia underscores the labor required to sustain perception as a coherent, bounded act. In Merleau-Ponty's *Phenomenology of Perception*, he writes, "I lie down in my bed, on my left side… I close my eyes, breathe slowly… I too call forth the visitation of sleep by imitating the breathing and posture of the sleeper" (2012, 189). Yet this imitation does not guarantee entry into sleep. Like the initiate awaiting the god, one may remain suspended: embodied, alert, but untouched by transformation. The *tekhumiavi*, I would suggest, occupies the inverse of this: a dreaming body that does not imitate, but becomes other; not mimetic but ontological, not summoning the spirit but crossing into it. The threshold of dreaming, then, is not merely a passive state but a portal of becoming; something insomnia, for all its intensity, withholds.

This, then, raises the ethical question: what kind of responsibility emerges in these divergent conditions? Levinas argues that ethical relation arises not from action but from exposure: "Responsibility is not a matter of my commitment… it concerns me despite myself" (Levinas 1969, 48). In the dream, I am overtaken, addressed, disturbed by alterity. But in insomnia, I am returned to a hyper-presence in which I cannot be otherwise. Where dreaming enacts an ethics of openness, insomnia renders its counter-image: a wakefulness in which no encounter comes, and no dissolution of self occurs. If Levinas imagines insomnia as an ethical vigilance, my experience suggests something more tragic: a state in which the demand of the Other is never heard, where my very awareness forecloses relation.

Still, this is not to suggest that dreaming simply displaces Levinas's ethical vision. His emphasis on the human face—its radical alterity, its resistance to comprehension—has no analogue in dreaming, where the other may be nonhuman or more-than-human. But perhaps this is precisely where dreaming complements Levinas. If the face is the index of ethical relation, then the dream extends that relation beyond the human, into a broader ecology of being. In dreaming, I do not only encounter other persons; I encounter landscapes, animals, spirits, and other selves that do not reduce to moral abstractions. The world, in its multiplicity, addresses me.

In this way, Merleau-Ponty's ontology of flesh reaches its most expansive scope. The dream becomes an ontological disclosure: it reveals not merely a symbolic world, but a world shared; where perception is distributed, porous, and promiscuous. The *tekhumiavi* does not simply dream of being-animal; he becomes animal. The *ruopfü* crosses species, time, and terrain, revealing perception not as interiority but as interbeing. Such states expose the limits of phenomenology, where the self is never only subject but a site of encounter, of encroachment, of ethical exigency.

In returning to the Arctic night, I realize that perception is never merely a faculty of the individual. In dreaming, I am drawn outward. In insomnia, I am fixed, withheld, still exposed. But in both cases, perception refuses to be confined to the individual body. It spills across the threshold of skin and memory, into a world that addresses, alters, and implicates me.

Dreaming lays bare an entangled world in which humans, animals, and the elements of wakefulness and sleep interweave. Insomnia, in turn, exposes the fragility of that openness.

Both states remind us that perception is never neutral: it is structured by vulnerability, folded into the flesh of the world. Where dreaming discloses the promiscuous excess of this chiasmic flesh, insomnia reveals its precarity: its capacity to fray, to disarticulate. Ethical responsibility, then, is not the result of relation. It is its condition. And the dream may be its first unfolding.

Notes

1 Morley's (1999) article draws on *Phenomenology of Perception*, *The Problem of Passivity*, and *The Visible and the Invisible*, and remains the most systematic treatment of dreaming in Merleau-Ponty's corpus. My reading follows and at times extends his insights toward an eco-phenomenological interpretation of dreaming and perception.

2 David Morris (2010) offers the most sustained account of reversibility in Merleau-Ponty's late ontology, linking it to the genesis of sense and the internal divergence (*écart*) of being. His reading emphasizes reversibility as a chiral structure within being—not merely perceptual, but generative of perception itself. I extend this ontology into the temporally and affectively intensified domain of dreaming. See also his monograph Ref. "Merleau-Ponty's Developmental Ontology (2018)" is cited in text but not provided in the reference list. Please provide details in the list or delete the citation from the text. *Merleau-Ponty's Developmental Ontology* (2018) for a broader account of embodiment, temporality, and sense formation.

3 McWeeny's (2016) article is one of the most sustained engagements with the ontological implications of promiscuity in Merleau-Ponty's later work. Her tripartite framework—reversibility, possession, and promiscuity—guides my own analysis, which extends her insights into the domain of dream perception and its implications for ecological and decolonial phenomenology.

4 James Morley's (1999) reading of *The Visible and the Invisible* foregrounds the "oneiric" as a category that resists the binary of real and imaginary, showing how dreaming discloses the structure of perception through reversibility, relationality, and passivity. While I follow Morley in affirming the philosophical significance of dreaming, I extend his framing into an ontological register shaped by Merleau-Ponty's concept of flesh and situated within an eco-phenomenological framework attentive to more-than-human entanglements.

5 While Levinas's ethical philosophy resists ontological framing—particularly in *Totality and Infinity* and *Otherwise than Being*—his accounts of exposure, insomnia, and responsibility "despite myself" offer critical resources for rethinking subjectivity as fundamentally relational and non-sovereign. My use of these concepts, especially in relation to dreaming and the porous structures of perception, draws them into ontological terrain Levinas may have deliberately avoided. This extension remains aligned with his ethical orientation, while traversing a phenomenological register shaped by Merleau-Ponty's ontology of flesh.

6 While some readers may interpret these events metaphorically or as cultural belief, I approach them as ontological disclosures aligned with relational and perspectival ontologies in contemporary anthropology. My reading is informed by the work of Viveiros de Castro (2004), Kohn (2013), and others who take seriously the ways non-Western worlds articulate alternative configurations of selfhood, perception, and interspecies being. These insights resonate with Merleau-Ponty's emphasis on reversibility and promiscuity, and underscore dreaming's value as both an epistemological and ontological resource.

References

Abram, David. 1997. *The Spell of the Sensuous: Perception and Language in a More-Than-Human World*. New York: Vintage Books.

Augé, Marc. 1999. *The War of Dreams: Exercises in Ethno-Fiction*. Translated by Liz Heron. London: Pluto Press.

Barbaras, Renaud. 2004. *The Being of the Phenomenon: Merleau-Ponty's Ontology*. Translated by Ted Toadvine and Leonard Lawlor. Bloomington: Indiana University Press.

Benjamin, Walter. 1999. *The Arcades Project*. Translated by Howard Eiland and Kevin McLaughlin. Cambridge: Harvard University Press.

Bergo, Bettina. 2009. "Radical Passivity in Levinas and Merleau-Ponty (Lectures of 1954)." In *Radical Passivity: Rethinking Ethical Agency in Levinas*, edited by Benda Hofmeyr, 31–54. Dordrecht: Springer.

Descartes, René. (1641) 1996. *Meditations on First Philosophy*. Translated by John Cottingham. Cambridge: Cambridge University Press.

Freud, Sigmund. (1900) 1997. *The Interpretation of Dreams*. Translated by James Strachey. New York: Avon.

Freud, Sigmund. (1900) 1953. *The Interpretation of Dreams. In The Standard Edition of the Complete Psychological Works of Sigmund Freud*, Vols. 4–5. Translated and edited by James Strachey. London: Hogarth Press.

Grotstein, James S. 2009. *"Orphans of the Real": The Work of Melanie Klein and Donald Winnicott with Lost and Unclaimed Children*. London: Karnac Books.

Heidegger, Martin. (1927) 1962. *Being and Time*. Translated by John Macquarrie and Edward Robinson. New York: Harper & Row.

Heneise, Michael T. 2016. "The Naga Tigerman and the Modern Assemblage of a Myth." In *Anthropology and Cryptozoology: Exploring Encounters with Mysterious Creatures (Multispecies Encounters)*, edited by Samantha Hurn, 91–106. London: Routledge.

Heneise, Michael T. 2018. *Agency and Knowledge in Northeast India: The Life and Landscapes of Dreams*. London: Routledge.

Hofmeyr, Benda, ed. 2009. *Radical Passivity: Rethinking Ethical Agency in Levinas*. Dordrecht: Springer.

Husserl, Edmund. (1960) 1977. *Cartesian Meditations: An Introduction to Phenomenology*. Translated by Dorion Cairns. The Hague: Martinus Nijhoff.

Husserl, Edmund. (1936) 1970. *The Crisis of European Sciences and Transcendental Phenomenology*. Translated by David Carr. Evanston, IL: Northwestern University Press.

Husserl, Edmund. (1913) 1982. *Ideas Pertaining to a Pure Phenomenology and to a Phenomenological Philosophy*. Translated by F. Kersten. The Hague: Martinus Nijhoff.

Husserl, Edmund. 2001. *Analyses Concerning Passive and Active Synthesis: Lectures on Transcendental Logic*. Translated by Anthony J. Steinbock. Dordrecht: Springer.

Hutton, J. H. 1921. *The Angami Nagas*. London: Macmillan & Co.

Kant, Immanuel. (1781) 1998. *Critique of Pure Reason*. Translated by Paul Guyer and Allen Wood. Cambridge: Cambridge University Press.

Kohn, Eduardo. 2013. *How Forests Think: Toward an Anthropology Beyond the Human*. Berkeley: University of California Press.

Levinas, Emmanuel. (1969) 1969. *Totality and Infinity: An Essay on Exteriority*. Translated by Alphonso Lingis. Pittsburgh: Duquesne University Press.

Levinas, Emmanuel. (1947) 2001. *Existence and Existents*. Translated by Alphonso Lingis. Pittsburgh: Duquesne University Press.

McWeeny, Jennifer. 2016. "Flesh Possessed: On the Promiscuity of Subjectivity in Merleau-Ponty's Ontology." *Chiasmi International: Trilingual Studies Concerning the Thought of Merleau-Ponty* 18: 215–231.

Merleau-Ponty, Maurice. (1968) 2002. *The Visible and the Invisible*, edited by Claude Lefort. Translated by Alphonso Lingis. Evanston, IL: Northwestern University Press.

Merleau-Ponty, Maurice. (1945) 2012. *Phenomenology of Perception*. Translated by Donald A. Landes. London: Routledge.

Merleau-Ponty, Maurice. 2020. *The Sensible World and the World of Expression: Course Notes from the Collège de France, 1953*. Translated by Bryan Smyth. Evanston, IL: Northwestern University Press.

Morley, James. 1999. "The Sleeping Subject: Merleau-Ponty on Dreaming." *Journal of Consciousness Studies* 6 (11–12): 87–107.

Morris, David. 2010. "The Enigma of Reversibility and the Genesis of Sense in Merleau-Ponty." *Continental Philosophy Review* 43: 141–165.

Morris, David. 2018. *Merleau-Ponty's Developmental Ontology*. Evanston, IL: Northwestern University Press.

Plato. 1992. *Republic*. Translated by G. M. A. Grube, revised by C. D. C. Reeve. Indianapolis: Hackett Publishing.

Sullivan, Harry Stack. 1953. *The Interpersonal Theory of Psychiatry*. New York: Norton.

Toadvine, Ted. 2009. *Merleau-Ponty's Philosophy of Nature*. Evanston, IL: Northwestern University Press.

Toadvine, Ted. 2021. *Nature's Edge: Merleau-Ponty and Environmental Philosophy*. Albany, NY: State University of New York Press.
Viveiros de Castro, Eduardo. 2004. "Perspectival Anthropology and the Method of Controlled Equivocation." *Tipití: Journal of the Society for the Anthropology of Lowland South America* 2 (1): 3–22.
Weiss, Gail. 1999. *Body Images: Embodiment as Intercorporeality*. London: Routledge.
Willock, Brent. 2007. *Comparing Psychoanalytic Psychotherapies: Developmental, Self and Object Relations Perspectives*. London: Karnac Books.
Zeillinger, Peter. 2009. "Radical Passivity as the (Only) Basis for Effective Ethical Action: Reading the 'Passage to the Third' in *Otherwise than Being*." In *Radical Passivity*, edited by Benda Hofmeyr, 135–158. Dordrecht: Springer.

12
THE EARTH IS THE ESSENCE OF ALL BEINGS
Eco-Phenomenology in the Upaniṣads

Meera Baindur

Introduction: Thinking and Writing through Eco-Phenomenology, Premodern Texts, and Experience

Eco-phenomenology addresses the experience of a human being's encounter with nature. As I analyze and articulate the subject matter of the *Upaniṣads* through an eco-phenomenological framework for this chapter, it is commonly understood that my writing will describe concepts, analyze texts, and draw conclusions related to the topic. However, I believe that Environmental Humanities may benefit better from creating transformative texts akin to poetry and literature rather than texts that are only limited to conceptual ideas and disengaged with experience. Therefore, I aim to present a form of writing that combines a narrative and expository genre in its composition. The expository insights discuss the ecophenomenological perspectives found in the Upaniṣads and some epistemic outcomes. At the same time, the intervening narrative focuses on my edited and rewritten autobiographical notes originally recorded between 1992 and 1997 while living in a forested area of the Himalayas called Govind Pasuvihar. This form of writing is not new; earlier ecological writers such as Henry David Thoreau and Aldo Leopold have already created such texts, and many others follow this tradition. I am grateful for their inspiration, particularly Leopold's Sand County Almanac. In the context of eco-phenomenology, these autobiographical notes illustrate how personal experiences influence environmental awareness. This approach uncovers the intricate connections between the self, the world, and meaning, providing a deeper and more nuanced understanding of phenomenological concepts. Phenomenologists have often pointed out that the subjective experience is inherently connected to the broader context of the lifeworld (*Lebenswelt*). For me such reflections (of my pre-academic experience) highlight my own situatedness of knowledge and make eco-phenomenological insights concrete and accessible. In some sense these narratives are ways in which my own consciousness "structured" experience revealing a pre-reflective and intentional nature of embodied perception. The narratives also deliberately displace abstract ideas and reinforce philosophy as a lived experience.

In this essay, I also draw insights from Wood's (2001) seminal paper on eco-phenomenology for some conceptualizations that can be applied to these pre-modern texts. I would attempt to extend new ecophenomenological perspectives that may arise from this intercultural

DOI: 10.4324/9781003538097-15

philosophy project of phenomenology, ecology, and Indian traditions of thought. Wood (2001, 78) proposes that eco-phenomenology provides a "middle ground"—one that opposes both scientific reductionism and ecological fundamentalism while upholding a profound commitment to comprehending nature from a phenomenological standpoint. He (Wood, 2001:92) critiques Deep Ecology for its inclination towards inflexible holism and closure, contending that eco-phenomenology presents a more flexible, relational perspective on nature. He (Wood, 2001:93–94) recognizes that Deep Ecology highlights interconnectedness but cautions that it may foster totalitarian-like ideologies—particularly when promoting extreme actions such as population control to safeguard the environment.

The Upaniṣads are often viewed through metaphysical absolutism, treating nature merely as **a** symbol of Brahman. However, due to their association with Brahman and Atman, core concepts in Advaita philosophy, they lend themselves to a deep ecology framework.

Reconciling spiritual teachings with the notion of nature as sacred often leads to a detached view, overlooking landscapes, people, animals, and other entities described in these texts as unimportant metaphors for discovering the supreme self or *Brahman*. This is a form of mystification. By exploring differently, we can demystify the Upaniṣads' connection with nature, as the self's mystification—often overshadowing other scholarly descriptions—can be reconsidered. Uncovering alternative themes allows for a focus on genuine human encounters with nature. This results in interpretations engaging with the natural world, extending beyond naïve realism and mysticism.

By moving beyond esoteric readings of Upaniṣads that are always seen as directed towards the goal of liberation, demystification also makes the texts more relevant to contemporary environmental philosophy, showing how concepts like *prāṇa* and *rasa* reflect deep ecological awareness. Additionally, it enables a bridge between ancient Indian thought and modern phenomenological discourse, fostering an ethical and embodied relationship with the natural world. Demystification allows for us to interpret the text from a sense of wonder that is created through perceptive experience. Mystification on the other hand reduces nature to an instrument that supports liberation. This causes nature to be interpreted as a metaphysical symbol, making it passive. On the other hand recognizing the textual descriptions of nature as a lived-reality encountered phenomenologically—through its textures, rhythms, and interactions—allows for an alignment with eco-phenomenology's call to engage with the world as it is. In this way, demystification transforms the Upaniṣads from purely spiritual treatises into texts that offer profound ecological and philosophical insights.

The interconnectedness of humans and nature is a significant theme in the Upaniṣadic texts, particularly those associated with the forest, known as *Āraṇyaka*. By shifting focus from the mystified notion of a cosmic self to actual encounters between humans and nature, other themes emerge, such as the sacredness of nature, the experiential aspects of existence, and the psychological and ecological immersion these encounters evoke. This approach fosters a transformative understanding of the texts, moving beyond purely epistemic interpretations to embrace a more poetic and experiential narrative that reflects the profound relationship between humanity and the natural world. Finally, the concept of *demystification* in the *Upaniṣads* aligns with phenomenology's critical method of bracketing taken-for-granted assumptions (*epoché*), allowing for a direct encounter with experience as it is given.

We particularly discuss two of the earlier Upaniṣads, called *Āranyakas*—meaning "of/from the forest" the Chāndogya and the Bṛhadāraṇyaka are part of the same lore and contain similar themes (Olivelle, 1998, 12). These earlier texts engage with natural wilderness landscapes, portraying experiences that can be phenomenologically described. The *Chāndogya*

Upaniṣad depicts encounters with nature that are not merely incidental but serve as contexts for reflecting on the relationship between humans and the natural world. While these encounters are not *intrinsically* phenomenological, they can be understood through a phenomenological lens—examining how they reveal structures of experience and meaning. In Husserlian terms, phenomenology elucidates how phenomena appear to consciousness, while Merleau-Ponty emphasizes the embodied nature of perception. The *Upaniṣad* captures this engagement through symbolic myths and narratives that reflect on the interactions between nature and humanity. Additionally, much of the text explores a distinct form of pedagogical experience where nature becomes a site for learning, contemplation, and transformation. The reflections on cosmic oneness for the students examined in the texts include numerous descriptions of nature, such as rain, earth, birds, animals, and landscapes. Philosophical truths appear to be intertwined within both the landscape and the bodyscape. This chapter explores the eco-phenomenological hermeneutics of these texts, unravelling a profound conceptualization of nature that extends beyond merely serving as a habitat for humans.

Texts of Deities, Nature, and the Forest

Within the Vedic texts, philosophers distinguished the subject matter of the verses from their purpose. They classified the corpus thematically as *karma-kāṇḍa*, those sections dealing with action, and *jñāna-kāṇḍa*, those dealing with knowledge. The action section encompasses the prescriptive and descriptive parts of the Vedic literature that address rituals, including "fire rituals" or *yajña*, as well as daily rituals and ceremonies essential for a person's Vedic life. The transition from Vedic ritualism to Upaniṣadic introspection signifies a shift from a mythic-external engagement with the world to an inward, phenomenological reflection, where meaning arises through direct lived experience and contemplation.

MacCulloch (1994, 201–202) suggests that even in the ritualistic parts of Vedas, many verses eulogize parts of nature as deities. According to him, a sense of wonder about the natural elements made the Vedic people ascribe divinity and power to various natural elements, including the Earth, the Heavens, the Clouds, the Sun, and the Moon. Some kinds of these deities were powers that controlled nature, such as the ruler of clouds and rain called "Indra," the master of all waters called "Varuna," and the power of death called "Yama." These deities are embedded in the cultural memory of Indian philosophy for centuries of practice and transmissions.

> We trekked across a high altitude "Darva top" pass above Donditaal in the Himalayas. The porters had gone on ahead to set camp, and we were slowly dragging ourselves along. Suddenly, we chanced up a cloud perched on our footpath. It was a small mass of sparkling white about the size of an elephant. We had never seen such a solid-like cloud up so close before. Even though we knew logically that we could walk through the vapor, it exuded an otherworldly presence that made us hesitate. One of my companions let out a nervous laugh, attempting to encourage the rest of us to move forward; he said, "Hey, Indra, thanks for the welcome." We saw the sky light up with a crack of lightning and heard a loud rumble behind us as if on cue. We fearfully turned back to see if Indra had indeed turned up; this was a wilderness at a high altitude, but we only saw large cumulus clouds darkening up the sky ominously around snow-clad peaks. We turned around, and the cloud had vanished, and we walked on in silence, humbled by this experience.

In the Upanishads, early notations of nature ontology, indicate that knowledge is not abstract but emerges as relational, entangled with a more-than-human world. This narrative records the layers of meaning that arise from encountering nature as the other, marked by a hesitation stemming from the interplay between perception, mythic imagination, and the lived immediacy of wilderness. These encounters occur at the intersection of embodied cognition and cultural memory, transforming nature into a field where past, present, body, and myth are interwoven to create a certain kind of givenness or disclosure. In this context, cultural memory becomes a lived structure that informs perception, which is seldom direct and always subject to interpretation.

In Sanskrit, *araṇya* means forest, and many Upaniṣads are known as *āraṇyaka*-s, or "forest texts." The contrast between *araṇya* (forest) and *grāmya* (settlement) during the Vedic period underscores the Upaniṣads' association with wisdom cultivated in nature. For the Indian traditions, the forest was not only a "wild" space but also a place not bound by the rules of society. Thapar (2012, 113) suggests that the forest was not just a jungle beyond the settlement but also a place "where the hierarchies and regulations of the *grāmya* were not observed." The *Aranya* was an effortless place that inspired reflection and was free of the restrictions of society, including duties of caste and community. The original residents of the forest were often non-Aryan people who were designated with mystical and dangerous characteristics. One of the first references to the deity of the forest is found in the *Ṛg Veda* in a hymn to *Aranyani*, or the goddess of the forest. She is described as someone who does not till the land to produce food; she is unafraid and not violent to those who are not her enemies (Ṛg Veda Trans. Griffiths, 1973, Hymn CXLVI, 640). In India, designated areas called sacred groves are sections of forest regarded as sacred or as belonging to a God, sometimes still preserved as a part of religious and cultural traditions. Shaman rituals associated with these deities are found in many parts of rural India.

It has always been a custom for the seekers of truth to venture into the forest for contemplation or meditation. This can be a temporary journey, such as a pilgrimage or a retreat. Sometimes, this can be a more profound quest of a lifetime to seek the goal of liberation. Whatever the duration of this journey, the quest is always traditionally undertaken in a forest, which is a sacred landscape. This *aranya* becomes a territory where one is willing to let go of the familiar ways of separating oneself from nature, defending oneself in many ways against the natural.

One can say that the kind of sacred that one experiences in nature is the experience of being nature, being *prakrti*. This experience of nature is lost to us in the *grāmya* [settlement] and its extension the *nagara* [the city]. This keeping out seems to establish order and maintain human culture apart from nature and is the first political move that allows a relationship of exploitation with nature. To return to the sacred, in the Upaniṣad is first to reject the polis, the political act of separation from nature. This is where going to the forest becomes a symbol of accepting nature within us and outside of us. In this sense, the spiritual becomes the opposite of the political. If political is to separate from nature, the spiritual or the sacred is to unite ourselves with it, in a bodily entanglement. And because nature allows this unity, it is sacred. Such a nature was not separate from the human being.

> Walking in a forest differs significantly from walking in a city, village, or fields. Although there were footpaths, one had to venture off the path to gather firewood or fodder. Gradually, I learned to walk "with" the forest. A crucial aspect of developing this understanding was to drop all preconceived notions of order and chaos. Each excursion

> into the jungle revealed new perspectives; no tree resembled another, no path mirrored the last, and even the sky transformed daily. Conversely, when I lost my way, everything looked the same. Every tree looked similar, and the rocks resembled each other. Nature changed, yet it also held a sense of constancy. Grasping this concept was the first step to experiencing the sacred essence of nature. In my previous actions—marking trees, clearing pathways, and cutting fodder grass in straight lines—I felt a sense of comfort, but I had yet to cultivate a genuine relationship with nature; it was always a struggle—me against nature. I perceived nature as the "other," leading my mind towards separation and creating a dynamic driven by power, where I was the active agent and nature merely a passive recipient of my actions, sometimes even resisting my attempts. Even when I sought to "see the sacred" in nature, it perpetuated an unequal relationship—an imbalanced interaction that lacked true sanctity. My pursuit of peace, beauty, or truth in nature still revolved around utilizing it for my personal fulfillment.

This relationship is political because the human being sees, gets, or uses nature separated from her. Bruce V. Foltz (1995, 331) critiques modernity's conception of nature as a mere external entity, devoid of depth or interiority. He argues that modern thought reduces nature to a "second-order exteriority"—a surface without inherent meaning or inner life. This view does not fragment nature into discrete faceless parts but renders nature as a whole into an external object, standing apart from human cognition. In this framework, nature becomes an absolute externality, with "nothing behind it and nothing within it."

We never know the true nature, the other side that nature chooses to reveal to us, without a mythopoetic sensibility:

> …That is, such an awareness apprehends nature as a whole as being wondrously given and bestowed in a marvelous way. For this sensibility, nature possesses a kind of causality so extraordinary that it is perhaps best not to call it causality at all but rather a radical givenness, a bestowal or—to employ a mytho-poetic word, an aesthetic-theological word that onto-theology has tried to appropriate—creation.
>
> (Foltz, 1995, 335)

Through an eco-phenomenological lens, my personal encounters in the forest reveal that the impulse to impose order is symptomatic of this separation. Instead of experiencing nature as a dynamic, lived reality—full of variations and interconnections—our conventional approach reduces it to a static, external object. By engaging directly with the forest, with its simultaneous variety and sameness, we glimpse a way of relating that transcends this externality. In eco-phenomenology, such encounters are not merely empirical events but are understood as openings into a more profound relational mode. According to Wood (2001) eco-phenomenology invites us to reconsider our relationship with the natural world by engaging with it through embodiment, perception, and ethical responsiveness, rather than through domination or detachment:

> Ecophenomenology asks us to move beyond the disembodied gaze, to reinhabit our sensing, feeling bodies, and to experience the Earth not as object but as presence, as something with which we are already entangled.
>
> (Wood 2001, 211)

In the Upaniṣads, seekers journey into the forest, waiting in conversation and teachings as nature reveals the sacred that is a form of this entanglement. In some passages, nature speaks directly to humans; in others, it illustrates creation's reality. The seekers do not pry; they wait for nature to reveal. Knowing is a transformative experience; they emphasize at the end of each chapter, "he who knows… will gain."

The Nāsadīya Sūkta stands apart in the Vedic corpus as a profound meditation on existence, epistemology, and cosmic origins. It invites contemplation rather than offering a definitive answer, aligning with later Indian philosophical traditions that emphasize experiential insight (*anubhava*) over dogma. The text acknowledges the limits of both human and divine knowledge, demonstrating a rare philosophical humility in premodern thought by concluding that perhaps even the supreme overseer (adhyakṣa), if such an entity exists, may not know.

> In my early days in the Himalayas, I found myself relating to nature as if it were something to be managed. My initial experiences centered around learning to live as a **part** of nature rather than just alongside it. Over time, my perspective transformed. One had to engage **with** nature, not merely exist within it (I certainly didn't want to become the bear's dinner). Living with nature entails an attitude of relationality. For example, my hands do not exist separately from my legs or my head; together, they form my being. I don't approach my arms and feet as distinct entities. As I connected with nature, I stopped stumbling over rocks that felt "in **my** way." Instead, I walked harmoniously with the landscape, fostering a relationship rather than wishing obstacles away or struggling to fit in; I learned to brush off insects that landed on my face peacefully instead of swatting them away frantically. I welcomed the rain, gathering water from the roof for storage alongside my companions. I settled comfortably on the muddy forest floor; mud was no longer "dirty" to me. The most significant shift was discovering food in the forest. There was an abundance of edible treasures that sustained us through snow and rain: mushrooms, wild berries, a grove of Dalchini trees, wild apricots, fern fronds reminiscent of okra, and even the stinging nettle of the Himalayas known as Kandayali, which could be turned into a delicious curry. We simply needed was to live as nature, shedding our pretenses and the impulse to modify everything in sight. To enter the forest and meditate is to step into the realm of undivided nature, a sacred landscape that exists both within and around us. I now dwelt in nature in the Himalayas, experiencing a profound connection rather than merely visiting or observing it. Every tree, rock, and bamboo patch felt familiar, as they simply existed in my world. I didn't need to know their scientific names, label birds, document wildlife, or name everything I encountered. It was sufficient that they were present.

The mythopoetic elements of the Upaniṣads reverse traditional hierarchies, positioning humans outside the typical dominance. This is achieved by reversing roles in pedagogy, where the teacher is the figure of epistemic authority who imparts knowledge to those with less understanding. In the Upaniṣads, teacher and learner roles often reverse; nature becomes the teacher, and humans the taught.

In the Chāndogya Upaniṣad (trans. Olivelle 1998, 219), a bull teaches a spiritual seeker in the story of Satyakama Jabala. Satyakama approaches Gautama Haridrumata to become his student despite not knowing his father's identity and is accepted despite lineage rules. The second part is particularly intriguing: the teacher, concerned about Satyakama's vulnerability to bullying, sends him to the forest with cattle, instructing him to return only when their number reaches a thousand.

Years pass as Satyakama grazes the cattle in the forest, waiting for their numbers to increase. When they finally reach a thousand, the head bull speaks to Satyakama, becoming his first teacher. The bull imparts knowledge about nature and space's vastness and encourages Satyakama to seek further wisdom from other teachers. Satyakama continues his journey, receiving insights from fire, a swan, and a waterfowl, each revealing deeper secrets of the universe and cosmic interrelatedness.

His body becomes luminous and transformed through the knowledge he receives from his first teachers. Nature nurtures Satyakama's understanding and facilitates his growth. This type of knowledge, known as *jnāna* in Indian philosophy, is transformative, unlike *Vidya*, another word for knowledge which focuses on the craft of learning. In this narrative, the bull symbolizes wisdom embedded in nature, imparted to humans through contemplation, reflection, and engagement with the environment. Notably, Satyakama does not retreat into solitude to contemplate his inner nature; instead, he lives in harmony with the forest, caring for the cattle and embracing a pastoral life while engaging with his surroundings. His teacher, recognizing his transformation, acknowledges his wisdom, and the humble Satyakama requests further instruction. The Upaniṣads demonstrate that knowledge arises not solely from human instruction but from communion and entanglement with the environment (C.U. 4.4-9, trans. Olivelle 1998, 219–223).

Essence Moving through Everything

The phenomenology of the more-than-human world seeks to uncover the fundamental characteristics of our experiences. It aims to reveal a universal essence by stripping away the contingent and subjective elements that shape our perception of reality. In a similar vein, the Upaniṣadic texts endeavor to accomplish this goal. They attempt to distill the essence of the inhabited universe, referring to it as the "life of the life," the "sun of the sun," or the "water of the water."

In some places, especially in the chapter that I am dealing with, in the verses that I am dealing with, it is called *rasa* or savor. *Rasa* in Sanskrit refers to the juicy flavor, sap, or extract of something, the vital essence of plants, fruits, or herbs. The rasa signifies the quintessence, of something, what gives it being, vitality, or purpose. The quest for a universal or cosmic essence forms the basis of most of the dialogues of the Upaniṣad.

> After the rains, the forest flourished with vibrant green undergrowth. Local children from the nearby village explored this terrain with small water bottles. They wet their hands, rubbed the stem of a plant, and scraped the black-gray sap clinging to their palms into a small bottle or box. After sunset, they rolled these into little balls and sold them to a local trader by weight for a few rupees each. Some even mixed cow dung with the dried resin to increase the weight but were quickly caught, as the trader often crushed the balls to check whether they contained plant sap or something else. "What is this?" I asked the curious children. "Bhang-ki-goli," they replied in Hindi, referring to the raw resin ball from the cannabis plant. This was my first encounter with the cannabis plant, the source of bhāṅg, an intoxicant, which was sometimes used to flavour food and drinks during spring festivals. I had consumed bhāṅg in a drink during Holi, the festival of colors, but I had never encountered the source, the plant from which this

> bhāṅg originated. I examined the plant—just a bushy specimen with fronds boasting eight to ten long, serrated leaflets. "The sap is the essence of its function," I thought. Although I couldn't bring myself to smoke or consume this sap as cannabis, I regularly used it as a painkiller and skin ointment for burns and insect bites after crushing the leaves and stems. My encounter with the rasa of the cannabis plant was truly an encounter with the sapient.

Just as phenomenology seeks to articulate the structures of consciousness and being through lived experience, the Upaniṣads offer a reflective, experiential insight into the nature of reality through symbolic narratives and meditative inquiry. Both traditions resist reduction of meaning to external objectivity and instead privilege direct, first-person encounter with the real—whether through silent meditation or descriptive reflection on phenomena. Thus, the Upaniṣadic method parallels phenomenology in seeking not just to know *about* the world, but to dwell *within* its depths. The Chāndogya Upaniṣad (henceforth C.U.) begins with a verse that explains the syllable *Aum* that is called the *Udgītha*, or the High chant. The following section introduces the "*Rasa*" text of the Upaniṣad. The most generalized essence of creation, according to the Upaniṣad, is *Aum*. The seeker of this essence is within a forest, surrounded by herbs, beings, the sun, the moon, the stars, the rivers, the mountains, and the sky—a being essentially surrounded by nature.

> The essence of these beings here is the earth; the essence of the earth is the waters; the essence of the waters is plants; the essence of plants is man; the essence of man is speech; the essence of speech is the Rg verse; the essence of the Rg verse is the Saman chant; the essence of the Saman chant is the High Chant. This High Chant is the quintessence of all essences; it is the highest, the ultimate, the eighth.
>
> (C.U. 1.2, trans. Olivelle 1998, 171)

The vision of nature in this Upaniṣad is not that of other than humans but a vision that views the universe as an interconnected series of journeys of the self or life force that moves through beings and nature. Unlike the poetry of romanticism, which saw nature as sacred as a reflection of the heaven's own beauty or a beauty created by God, the poetry of the Upaniṣad engages with the environment in a bodily way, looking beyond a static and theological inspiration. It is to be stressed here that nature in the Upaniṣads is not the abstract conceptualization of a non-human nature or wilderness that is so commonly associated with the term. Instead, nature is the realm of the experienced, the earthly world where we cohabit with the more-than-human. These verses seem to pre-empt the process of the **eidetic shift** in Edmund Husserl's phenomenology. First, they describe the move from empirical instances to phenomena's essential structures (or "essences"). Second, the finding of the essence moves from one layer of nature to another, ending in language and life, both of which are embodied in the human being endowed with speech.

Husserl describes this process in *Ideas Pertaining to a Pure Phenomenology and to a Phenomenological Philosophy* (1913):

> We effect a peculiar "change of attitude" in which we disregard the particular facts of experience and ascend to the universal essences that give them meaning. This is the essence-intuition (Wesensschau), by which the eidos of a phenomenon becomes accessible.
>
> (Husserl, 1913/1982, §3)

The Eidetic shift is not possible as long as we are "in" nature. Thus, the Upaniṣads ensure we are "with" nature when articulating this verse. While being "in nature" emphasizes the placement of the human in the natural world, the Upaniṣads are situated from the perspective of a being with nature. The human being is not a passive recipient inside nature or contained within it; instead, he or she is engaged "with" nature in an entangled way. This entanglement is possible due to the movement of *Rasa*, or the essence, from one layer of this universe to another—from the plants to the humans, from the earth to the plants, and so on. Being with nature implies a relationship-oriented perspective. The process of this relationship is articulated through the sharing of essence, which is ultimately equated to prāṇa, or the vital force.

> As one wakes up to a morning in nature, the predawn silence seems to give in to the sounds of the stirring forest, birds, insects, people, and even the cracking fires, all of which together create a symphony of sounds of life. Individually, they are recognizable as individual sounds. But together, they form inspiring evidence of the life of the vital force that moves them. These natural sounds are heard together like a hum, the sound of AUM. "The unifying darkness of the night is drawn aside by the sun, which shines through like the truth, flowing downwards on the snow-clad peaks like a golden curtain. The light moves down into the valley, chasing out all the darkness. But the shadows of the mountain valleys only get darker as the sun climbs overhead. Everything reflects the interplay of unity and diversity at the same time."

The essence of all beings is articulated as the final couple of the organ of speech, vǎk, and of prāṇa, the vital force. Engaging with the world of essences requires us to articulate, name, or understand the particularities in language. Speech becomes central to meaning making and is deeply connected to the life force. Om, the sacred syllable in the Upaniṣads, is directly formed out of this coupling of speech, work, and prāṇa life. The Upaniṣad recognizes that perception and experience are deeply tied to the idea of language. Distinguishing things from each other makes relatedness possible. There would be no need to relate to anything if everything were one homogeneous self; that is the state of final liberation, not that of an enworlded being.

Water, Food, and Fire: Those That We Live with and Consume

Humans inhabiting the world, while recognizing the oneness of the cosmos related to the essence or the inner self, are also aware of and should contemplate what makes life possible and desire possible. The Upaniṣads do not stop with this coupling of life force and speech; they also create conditionalities. In another passage, while defining each of the letters, it is declared:

> Now, then, one should venerate the syllables of the word *udgitha*—High Chant—namely *ud*, *gi*, and *tha*. The syllable ud is simply breath, for people rise up (*ud-stha-*) by means of breath; the syllable *gi* is speech, for speech utterances are called words (*gir*); and the syllable *tha* is food, for this whole world rests (*sthita*) on food.
>
> (C.U. 1.3.4, trans. Olivelle 1998, 175)

Elsewhere in the later part of the Upaniṣad, in the sixth chapter, a father teaches his son and remarks that he should drink water, even if he fasts: "Do not eat for fifteen days, but drink water at will. Breath is made of water; so it will not be cut off if one drinks." (C.U.6.7.1, trans. Olivelle 1998, 251).

The father instructs the son that food is connected to the mind, so after not eating food for fifteen days and drinking only water, his memory fails him, and he is unable to recite verses that he has learned. (C.U. trans. Olivelle 1998, 251). The Chāndogya Upaniṣad also describes that fire is connected to speech. Here it means that energy burns to create thought and language. Fire, as an experience, transforms what it touches. Fire is analogous to consumption, and therefore, speech consumes the world. Our senses consume the world through speech because we sense what we articulate and distinguish. We sense the particularities, the thingness of the world. From a phenomenological point of view, we recognize the world we inhabit through speech and the act of meaning-making and then articulate our relationship to it. This is captured beautifully in the verses of the Upaniṣad.

The most important part of learning to be with nature was learning the importance of these three things: food, water, and fire. Nature provided all three resources in a forest. The relationality with the landscape was through how and where we gathered them.

> The ways of the water were difficult to understand. We lived near a perennial river, but water availability is never certain in a forest. As we walked up the slopes, the river was left behind in the deep valley, away from easy reach. Rainy season would make the river turbulent with mudslides and rolling rocks, and generally dangerous to approach. On the other hand, it was little springs that quenched our thirst. We would find a broad leaf and fix it to the bare surface of a wet rock or mossy slope at an angle, and the water would collect on the leaf and flow off its tip. During rains, the turbid river was brown with silt, and we drank rainwater collected in buckets under the open sky. In winter, when the banks were slippery with ice, we melted snow on the fire to make chai and watched the embers die out.

Water and food are the pathways through which life sustains itself. The essence flows through the cosmos. The Chāndogya Upaniṣad makes many references to water and food, and the Upaniṣad clearly links water and the life force.

The civilization that treats water with disrespect and wastes it certainly loses touch with the importance and value of life. We ignore its significance beyond our ***use*** of it. Water connects us not only across all beings but also to the evolution of life itself on this planet. It is water that sustains life. Like the old king Midas, we exchange the freshness of water for the market of petrol and power, losing our connection with our earliest history and origin as life forms. We are destroying systematically the very womb that bore us, the water-places. Nature is a term that captures a "whole." The idea of wholeness is not about a sum of its parts alone, and it is an emergency that transcends it.

In eco-phenomenology, nature is not an inert background or a resource to be used, but a **living context**—a meaningful field of relations within which the self is constituted. Water, as described in the *Chāndogya Upaniṣad*, becomes not just a substance but a carrier of life, a sacred medium of connection. This corresponds with David Abram's phenomenological insight that elements like air, water, and earth are not passive matter but active participants in our perceptual and embodied experience:

> The life-world is the world of our immediately lived experience, as we live it, prior to all our thoughts about it. It is that which is present to us in our everyday tasks and enjoyments—reality as it engages us before being analyzed by our theories and our science. (Abram 1996, 33)

The passage evokes a loss—civilizations "lose touch with the value of life" when they commodify water. Trading the "freshness of water for petrol and power" signals a phenomenological rupture—a shift from being *in* and *with* the world, to standing *over and against* it.

In the Chāndogya Upaniṣad passage 5.11–16 (trans. Olivelle 1998, 239–245), the universe is perceived as a ***being*** whose feet are the earth, sustained by food and water. Proper contemplation of nature requires viewing the whole being of *Vaiśvānara*, Ātman, or Brahman, rather than focusing on individual components.

A story in this text recounts how five householder Vedic scholars approach a king to understand this being. Each contemplates different aspects of nature but cannot find the truth. King Ashvapati asks them to describe the aspect of the cosmic being they contemplate. They reply with various aspects: heaven, sun, air, ether-space (*akasha*), water, and earth. After listening, the King advises them to consider the whole being—a being whose feet are the earth and whose head is the heaven. This Being consumes all food in the universe, to which all food is offered. The deepest understanding of this life force principle is that it connects not to rationality, consciousness, or sentience, but to interconnectedness.

What is missing in the modern idea of nature is the perception of this link between "us" and nature as the "other." We in the cities do not perceive the connection between our electricity consumption and the tiger or the honey bee. The separation is so deep that this lack of connection leads to a non-sacred connection with nature. In eco-phenomenology, nature is not conceived as an inert background or a mere resource to be exploited, but rather as a living context—a meaningful field of relations within which the self is fundamentally constituted. As Wood observes, *"The world is not a neutral container, but a dynamic field of significance in which we find ourselves always already involved"* (Wood 2001, 212). This reconceptualization challenges anthropocentric and instrumentalist paradigms, calling instead for an ontologically grounded, embodied engagement with the more-than-human world.

> Lifebreath, undoubtedly, is greater than hope, for all this is fixed to life breath, as spokes are fixed to the hub. Lifebreath proceeds by means of lifebreath; lifebreath gives lifebreath and gives to lifebreath. The father is lifebreath; the mother is lifebreath; a brother is lifebreath; a sister is lifebreath; a teacher is lifebreath; and a Brahmin is lifebreath.
>
> (C.U. 7.15.1, trans. Olivelle 1998, 269)

Relating to nature involves two processes of contemplation. First, we must recognize that we are part of this cosmic life and our interrelatedness with it. To discover this essence externally, one must look within for the source of one's being—the inner self. These ideas likely contributed to the development of *Advaita* philosophy. The state of being united with the absolute essence of life is considered sacred. The initial step is to acknowledge that we embody the life force within. Once we feel at one with and recognize the cycles of life and death within ourselves, we can begin to look outward and perceive various names and forms. However, the Upaniṣads do not limit themselves to describing the inner self. After each process of contemplation, they emphasize the wholeness of the cosmic being and the interconnection of all entities within the cycle of life. Thus, the second process is to perceive unity in diversity and the interconnectedness of life. By recognizing this unity within ourselves, we can identify the same life force present in all beings and creations, deepening our understanding of the interconnectedness of everything.

> One year, the people around the forest where I lived decided to pursue apple orchard farming. The large-scale removal of local trees and shrubs left the slopes bare, exposing mud and loose soil. That year, the rains triggered unprecedented landslides, and a massive rock struck the suspension bridge that connected us to the outside world. The bridge twisted like a Möbius strip, rendering it unusable. As I slowly walked through the ravaged forests, observing the eroded earth and the potential orchards, I discovered a small green grove, an oasis amidst the destruction. It was the Devtha Pani place, the local spring beneath a tree. I pointed it out to the villagers. "See how the trees hold the land and the water," I said. Then, an elderly villager, who embodied the voice of nature, echoed sentiments reminiscent of Upaniṣadic lore:
>
> "The tree holds the water, the earth holds the tree, and the water holds the earth. This is Pratyaksh Bhagvan(Seen-God)."

The Upaniṣads called this a special kind of "knowing" (*pratyakṣa*) and this is not mere learning but is experiential. Similarly, Wood argues that "Knowing is not primarily a matter of conceptual representation, but of responsive engagement, of being touched and moved by what we encounter" (Wood 2001, 214). Use the verb "to see" to emphasize the empirical nature of this experience. This state of being in nature is to dissolve the boundaries and confines of the narrow mind to see the enworldedness and the embodiment of life interconnected with all beings. The sacred nature is thus not a disembodied abstract event of consciousness. Still, for the Upaniṣads, it is the experience of nature as a totality through its body of which we, too, are an integral part.

Temporalities and Order

> In the city, weather served merely as an inconvenience or convenience for other activities. The clock time was calibrated, and the days of the week was segmented with appointments, and life went on according to work, college, and movie show timings. In the forest, however, existence was governed by nature's timetable. Even the simplest broom adhered to the rhythm of the seasons. The larger grass broom required harvesting in the fall, just as it began to yellow. Conversely, pine needles were gathered in the summer, when they reached their peak length and strength, useful for crafting finer brooms for the kitchen. Life followed an annual cycle; other dates and days held little meaning. Seasonal markers for us included abrupt cloudbursts, gusty winds, snowfall, river flooding, blooming apricot flowers, and the monkeys' migration. Our perception of time was in sync with nature's rhythm; winter days were shorter, often allowing us only seven to eight hours of wakefulness. Among the mountains, I became acutely aware of the precariousness of existence within nature. Some creatures would not survive until the next snowfall, harvest, or seasonal shift. Cows frequently tumbled off cliffs into the river below. Adventure climbers exploring higher peaks often met similar fates, their bodies discovered downstream. Others succumbed to the cold, suffered accidents as buses plunged into valleys, or died from undiagnosed ailments in the absence of medical care. Regularly, dogs were taken by leopards that roamed the region. People hunted, killing goats for festive celebrations. Death intertwined with life, an inextricable aspect of existence. This acceptance of death fostered a contemplative relationship with

life's fragility—not merely viewed through the lens of risk and danger, but recognized as an inherent part of living. In every moment, our being could cease. Yet, as I delved into the Upaniṣads from the Advaitic perspective, I began to entertain the possibility of existing as a self beyond the death of the body.

Wood (2001, 94) suggests that:

> …our grasp of Nature is significantly altered by thinking through four strands of time: the invisibility of time's plexity—The invisibility of time, the celebration of finitude, the coordination of rhythms, and the interruptions and breakdown of temporal horizons.

Similar to this, we can see that the Upaniṣad also recognizes time's plexity, albeit in a pre-modern mythopoetic way. The experience of time is articulated through repetitions of cycles, such as seasons and the repetition of day and night. Time is also ordered because it is rendered visible through experiences of nature cyclically and endlessly in a year (*saṁvatsara*) repeats itself. The word saṁvatsara signifies a complete cyclical period of time, particularly a year, and is rooted in the idea of dwelling or abiding through a full cycle.

In this sense, Indian philosophical time is not progressive but cyclical; the seasons come one after another, creating a rhythm of human activity. Sometimes, this rhythm is broken and collapses, as in the case of the description of a famine in one of the chapters. At the same time, narratives and stories break the idea of a lifetime through myths. Some beings meditate for thousands of years, some for a year, and some for a day.

In the Upaniṣad, celebrating finitude—death—is common. Beings, especially embodied ones, die based on the philosophy of pre-time. Time begins after pre-time, post-creation, making death possible as things exist and end within lifetimes. Time creates diversity and distinction; *kāla*, or time, also means death. It transforms and disengages the vital force, separating essence from form, while the vital force remains eternal. Embodied beings pass through time, and time flows through the vital force, which is the essence of all. Time flows, poetically referred to as *Ābda*, derived from *abda*, meaning water. These notions of eternality lay the groundwork for the philosophical idea of an imperishable self, the essence that persists beyond the body's death.

Many passages in the Upaniṣads discuss death and its different modalities. Sound arises, exists, and dies. A tree can be cut down and lie dead. As humans die, they lose intelligence and consciousness. The Upaniṣad describes death as a natural process.

This is summarized in the act of chopping a tree, described as a process rather than a before-and-after. The lifeforce departs branch by branch; death ebbs away:

> Now, take this huge tree here, son. If someone were to hack it at the bottom, its living sap would flow. Likewise, if someone were to hack it in the middle, its living sap would flow, and if someone were to hack it at the top, its living sap would flow. Pervaded by the living (jīva) essence (Ātman), this tree stands here ceaselessly drinking water and flourishing. When, however, life (jīva) leaves one of its branches, that branch withers away. When it leaves a second branch, that likewise withers away, and when it leaves a third branch, that also withers away. When it leaves the entire tree, the whole tree withers away. "In exactly the same way, 'he continued,' know that this, of course, dies when it is bereft of life (jīva); but life itself does not die."
>
> (C.U.6.11.1–3, trans. Olivelle 1998, 253, 255)

13

THE PASSIVITY OF TIME OPENS THE SPACE FOR ECO-POLITICAL ACTIVITY

Edvard Lia

Introduction

It has become a generic feature of academic texts in environmental philosophy to begin by enumerating a list of usually depressing statistics and natural disasters that have occurred recently or might soon be on the horizon.[1] The intention behind this gesture of enumerating aspects of the wayward way of the world is usually to enact an affective and effective response on behalf of the readers or listeners. Yet this gesture can just as easily inhibit as invigorate the response it is supposed to provoke. The paradoxical problem of environmental action stems partly from an overwhelming abundance of information concerning the present crises. We suffer not from a lack of information but from an excess of it. This issue is inextricably linked to temporality. Many of us are all consciously aware that action is urgently needed yet radical change seems, broadly speaking, to be postponed indefinitely into an ever-receding future. All our actions necessarily rely on imperfect information, but it is wrong to think that perfect information is what we need or want. If we could predict the future with complete accuracy, there would be no freedom to act. The alternative is that we must experience a necessary, minimal moment of what Jacques Derrida calls undecidability for there to be a question of ethical or political decisions in the first place and from the first moment. If we could simply follow a programmatic rule of action, there would be no actual decision to be made, and we would not be free in any meaningful sense of the term. This is why the experience of undecidability inherent in the passivity of time is "the nonethical opening of ethics" (Hägglund 2008, 97). In what follows, I seek to argue against the position proclaiming that all we need is to wait—either for the apocalypse, a God to save us, or until we have gathered all the potentially infinite information to be used as firm foundation for acting.

To that end, this chapter offers a theorization of the link between time's passivity and environmental activity. If it is true that time heals all wounds, it is because time also opens those wounds. This is why the passing of time is the ultra-transcendental condition of possibility for everything good and bad alike. Time limits our avenues for action but equally gives rise to all possibilities in general. That we are exposed to the unpredictable coming of time, which we can neither arrest nor reverse, does not prevent acts of decision-making but instead

 DOI: 10.4324/9781003538097-17

opens the possibility of discriminating between good and bad decisions, while simultaneously making those decisions constitutively urgent.

It is important to note that this chapter does not aim to intervene in debates concerning the accuracy of Martin Hägglund's reading of Derrida.[2] The main reason why I rely on Hägglund's reading is that it offers a coherent and systematic account of Derrida's works and that this particular reading offers useful resources for thinking through central issues animating this handbook. Accordingly, my more determinate aim with this chapter is to indicate the relevance of Hägglund's reading of Derrida for conceptualizing environmental activity within the context of eco-phenomenology, though the argument's import is not necessarily restricted to eco-phenomenology. For this precise reason, the contested nature of Hägglund's reading of Derrida is largely beside the point. Meanwhile, the eco-phenomenological context in which my argument is fleshed out is important insofar as the conception of time articulated in this chapter contrasts starkly with the Heideggerian vein of eco-phenomenology. The final section of the chapter briefly considers this difference and how my discussion indirectly illuminates the problems inherent in Heideggerian eco-phenomenology. The question motivating my inquiry is: If we follow the account of time that Hägglund develops in his reading of Derrida, what practical consequences follow for how we conceptualize environmental activity in relation to contemporary debates in eco-phenomenology over the relation between activity and passivity?

Tracing Time

In his early critique of the metaphysics of presence, Derrida teases out time's internal division from itself, which he designates as its trace structure. In *Margins of Philosophy*, he explains that "[a]n interval must separate the present from what it is not in order for the present to be itself, but this interval that constitutes it as present must, by the same token, divide the present in and of itself" (Derrida 1982, 13). Hägglund elaborates on this point by clarifying that "[t]he succession of time requires not only that each now is superseded by another now, but also that this alteration is at work from the beginning. The purportedly single now is always already divided by the movement of temporalization" (Hägglund 2008, 16). The name that Derrida gives to this internal division is spacing—a shorthand which encapsulates both the becoming-space of time and the becoming-time of space, aspects on which I elaborate below. Derrida insists that there is no such thing as an undivided, full presence. At the very moment I type these words, they have already been inscribed as traces in the past. Due to time's constitutive division, we cannot master the flow of time to arrest or reverse its deterioration of our planet.

These are implications of being committed to the idea that time consists in every now being succeeded by another now. Hägglund clarifies that Derrida does not offer an ontological or transcendental theory of temporality: "The trace is not an ontological stipulation but a *logical structure* that makes explicit what is implicit in the concept of succession" (Hägglund 2011a, 265).[3] Derrida's account of time based on temporal succession is, according to Hägglund's rendition, not an ontological theory of time. Moreover, succession should not be conflated with a linear conceptualization of time. Succession "rather accounts for the constitutive delay and deferral of any event" (Hägglund 2011a, 265). The concept of succession—and Derrida's analysis of it—is not limited to a linear conception of time because even, for instance, a circular conception of time presupposes that one moment comes before or after another. A circular conception of time in which everything rhythmically unfolds and folds up again in a repetitive pattern nonetheless presupposes that something is repeated. The first occurrence of the rhythm comes before the second occurrence. The first is succeeded by the second. The only way to conceptualize time without relying on the concept of succession is to abolish time in an eternal presence, whereby the radius of the circle in the circular notion of time shrinks to a single point, completely immune to alteration. This eternal point remains forever unalterable, resistant to all change and thus encapsulates the absence of time, whereas time itself—by contrast—remains constitutively rooted in succession. Succession is therefore not necessarily bound to chronological linearity but is implicit in every intelligible form of time at all.

What Derrida names the spacing of time remains opaque thus far in my discussion. What does it mean that spacing is a shorthand for the becoming-space of time and the becoming-time of space? The co-implication of time and space means that neither can exist without the other, but also that each comes to be through the other. The traditional distinction between the spatial and the temporal comes from the distinction between simultaneity and succession. Hägglund glosses this point as follows:

> The spatial can remain the same, since the simultaneity of space allows one point to co-exist with another. In contrast, the temporal can never remain the same, since the succession of time entails that every moment ceases to be as soon as it comes to be and thus negates itself.
>
> (Hägglund 2011a, 269)

However, to speak of time's passing or flowing already indicates the co-implication of time and space via spatial metaphors. As Hägglund neatly puts it: "Time is nothing but negation, so in order to be anything it has to be spatialized" (Hägglund 2011a, 270). Time cannot be anything in itself because it is simply the irreducible process of alteration by negation, whereby any given thing is subjected to the relentless process of becoming other than itself, of being altered. That we deploy spatial metaphors whenever we speak of time is, however, "not a failure of language to capture pure time but follows from an originary *becoming-space of time*. The very concept of duration presupposes that something remains across an interval of time and only that which is spatial can remain" (Hägglund 2011a, 270). In fact, the word "metaphor" is, unsurprisingly, itself a spatial metaphor—etymologically meaning "a carrying over" of something from one domain to another (*Online Etymological Dictionary*, n.d.).

Meanwhile, the becoming-time of space, the other half co-constituting the spacing of time, is evident in the necessity of temporalization for a "[spatial] point to *remain* the same as itself or to exist *at the same time* as another point. The simultaneity of space is itself a temporal notion" (Hägglund 2011a, 270). Two points in space are not simultaneous if space is not always already undergoing the process of temporalization. A spatial point consequently requires a minimal form of temporalization to be intelligible *as* spatial, which indicates how the becoming-space of time and the becoming-time of space are co-originary. Whereas the succession of time is always already a spatial notion, the "simultaneity of space is itself a temporal notion" (Hägglund 2011a, 270). This unity constitutes what Derrida calls the spacing of time.

In *Radical Atheism*, Hägglund confines his analysis of Derrida's account of time to the time of life specifically, but in subsequent publications he has distinguished explicitly between the time of life and the time of non-living entities. The spacing of time pertains to anything and everything; it is "the condition for anything that is subject to succession, whether animate or inanimate, ideal or material" (Hägglund 2011a, 265). Even a lifeless rock is subjected to the process of alteration, through forces of erosion, as is a historical event like the Haitian revolution, whose historical significance is always exposed to the threat of being forgotten. In this context, Hägglund deploys the concept of survival in a sense logically prior to its biological determination. Survival here means the general co-implication of persistence and destruction of something in its mode of being. To survive thus means to be "already marked by the destruction of a past that is no longer while persisting for a future that is not yet" (Hägglund 2011a, 274). In this precise sense, the logic of survival does not depend upon a biological conception of life. Hägglund's example of something which is not living but nonetheless surviving, that is, which remains and disintegrates over time, is an "isotope that has a rate of radioactive decay across billions of years," which is "*surviving* [even though] it is not alive" (Hägglund 2011a, 274). The term half-life is often used to describe the rate of radioactive decay, and this is the sense in which survival should be understood in this context. To survive here means to be temporally split, exposed to the relentless passing and coming of time. Even the rock is marked by its past erosion and points toward its own future persistence. Likewise with the Haitian revolution: it was once nothing but a future promise that may never have happened but now lies in the past being perennially threatened by the risk of being neglected or forgotten. Promises and memories alike are accordingly exposed to the passing and coming of time.

However, this determination of survival as common to both what is living and what is not living does not imply that we need to reject all discrimination between what is living and what is not. What distinguishes the time of life from the time of non-living entities is that living beings are fundamentally characterized by caring for their own survival. This is essential for comprehending the relevance of the previous discussion for pressing environmental issues. Whereas the radioactive isotope is indifferent toward its own survival, which is an empirical question, even the most elementary forms of life exhibit a care for their own survival through their metabolic interactions with their environment, with which a living being exchanges material to sustain itself as a living form. Hägglund insists that this conceptual distinction should not be conflated with an empirical determination of which entities are alive and which are not (Hägglund 2011a, 275). Care for one's own survival is not a biological determination of life but a logical concept, based on which an empirical investigation can be conducted.

Living Autoimmunity

The structure of care exhibited by living beings through our very organic corporeality is only possible because we are passively exposed to the spacing of time. For Derrida, being exposed to time in this manner is the ultra-transcendental condition of life in general, which means that life is constitutively autoimmune. This requires some terminological clarification. For Immanuel Kant, time is a transcendental condition of possibility for any finite experience, meaning that it is inconceivable for us, by virtue of being finite creatures, to experience or cognize something outside of time. Kant insists that time does not apply to things in themselves but only to how things appear for a finite consciousness. By contrast, calling the spacing of time an ultra-transcendental condition of possibility means, in Hägglund's gloss of Derrida, that "[t]he spacing of time is the condition not only for everything that can be cognized and experienced, but also for everything that can be thought and desired" (Hägglund 2008, 19).

To clarify what this ultra-transcendental status encompasses, it is instructive to look at what Derrida calls autoimmunity, which arises from the passivity of time. He does not use autoimmunity as a technical biological term, like with autoimmune diseases whereby an organism attacks itself from within because it mistakenly "thinks" it is being threatened from the outside. There is nonetheless a logical connection between the biological determination of autoimmune diseases and the ultra-transcendental sense in which Derrida deploys the term. Hägglund glosses the relation between the tracing of time and autoimmunity as follows:

> To think the trace as an ultratranscendental condition is [...] to think a constitutive finitude that is absolutely without exception. From within its very constitution life is threatened by death, memory is threatened by forgetting, identity is threatened by alterity.
>
> (Hägglund 2008, 19)

Concerning the exemplary case of life, death is an internal and constitutive possibility that always threatens life from within at every moment of time. To be alive is to be mortal, which means that I *must* have been born and that I *can* die at any given moment. This is not a negative predicament preventing me from living life to its fullest, but an ultra-transcendental condition that opens the space for every moment of joy and mourning alike. Hence, life's constitutive possibility of destruction inherent in mortality is a condition for a living being to care about anything. Life's constitutive mortality is what makes it autoimmune in Derrida's sense. Put differently, life is immune to absolute autonomy. Being fully self-sufficient is a kind of immunity to the world, but by virtue of being finite, life is constitutively autoimmune. In fact, Derrida clarifies that "*pharmakon* is another name, an old name, for this autoimmunitary logic" whereby something is simultaneously "self-protecting and self-destroying, at once remedy and poison" (Derrida 2003, 124), such as other-dependent, mortal life. *A life which is absolutely safeguarded from death is no life at all.* It lacks the very animation in the form of temporal urgency which makes it possible to care for others and oneself. Deepening even further our understanding of autoimmunity requires a detour via the concept of the unconditional.

Scholarship on Derrida has traditionally drawn attention to an ethical or religious turn, which is supposed to indicate a fundamental shift of Derrida's attention. His later texts appear far more explicitly concerned with ethical, political, and religious issues than his early critique of the metaphysics of presence. However, as Hägglund insists, there is in fact a strong continuity between the early and late Derrida. To understand what Derrida means by "the unconditional" in his later texts, it is necessary to clarify the connection between unconditionality and his early analysis of time. I follow Hägglund's assertion that it would be a mistake to assimilate Derrida's unconditional to Kant's regulative ideas, which several commentators on Derrida have done.[4] The regulative ideas of reason arise, according to Kant, because of reason's failure to think the unconditioned, outlined in the antinomies of pure reason in his first *Critique*. Kant resolves the antinomies by disproving the two antithetical propositions which each antinomy consists of. For instance, the third antinomy concerns spontaneity and causal determinism: either the universe can be explained in fully deterministic terms *or* there exists spontaneous freedom that can initiate new causal chains (Kant 1998, 484–9). In the present context, Kant's actual exposition for disproving these two antithetical claims is not directly relevant, but the conclusion Kant draws from the antinomies in general is still important. Concerning the third antinomy, Kant argues that transcendental freedom cannot be a *constitutive* principle of reason but must be regarded as a *regulative* principle of reason, which means that we necessarily act *as if* we are free. We cannot theoretically know for certain that we *are* free, but we must act practically *as if* we are. His moral philosophy in general is intended to flesh out the consequences of the human being's practical (in contrast to transcendental) freedom.

When we speak of ideals like justice in a colloquial sense, they often sound like regulative ideas which means that they are thinkable in theory but impossible to achieve in practice due to our limitations as finite creatures. In his discussion of Kant and Derrida, Hägglund goes through examples like justice, hospitality, and the gift to show how our conception of these radically changes if we no longer see them as regulative ideas. The issue revolves around the notion of purity. If we consider what real justice would be, we might be tempted to imagine a utopian society in which everything is perfectly harmonious. Critics would point out that such an ideal is impossible due to human nature, implying that humans are intrinsically greedy, selfish, etc. The idea is that a perfect society is desirable, but because we are finite creatures—limited, for instance, by our supposed natural vices like greed or selfishness—it is impossible for us to achieve such a state.

To show why this understanding of the issue is mistaken, the example of the gift is particularly instructive. A pure gift is, for Derrida, unthinkable even though it must be thought. For a gift to be pure it would exceed time's passive alteration. In a neat phrase, Hägglund remarks that the gift is always a "gift of time." He explains further:

> The purportedly good gift must run the risk of being a poison [...], and the purportedly generous gift must run the risk of dealing with counterfeit money. The point is not only that the gift *cannot* be secured against these risks but also that it *ought not* to be secured against them, since such security would cancel out the gift. If the gift were secured against being corrupted it could never be given, since nothing could happen to it.
>
> (Hägglund 2008, 39)

The "ought" that Hägglund uses here is not a moral ought. He is not claiming that a gift ought not to be secured against risks for the reason that it would be morally wrong. On the contrary, this ought follows from the logical structure of the gift, which is that it is a gift of time. A gift must be uncontaminated by the logic of reciprocal economic exchange to be a gift. Even the idea that one should be grateful for a gift received would count as an act of reciprocation. Yet by being exposed to time, the gift must always run this risk. Otherwise, nothing is given and everything remains the same: a pure eternity where nothing ever happens or is given. That a gift is always a gift of time is accordingly what opens the space for both gratitude and its opposite.

The difference between Kant's and Derrida's respective conceptions of the unconditional has consequences for how we conceive of the relation between activity and passivity. Hägglund makes this point remarkably clear:

> For Kant, the unconditional is the Idea of a sovereign instance that is not subjected to time and space (e.g., God). For Derrida, on the contrary, the unconditional is the spacing of time that undermines the very Idea of a sovereign instance.
>
> (Hägglund 2008, 19)

Undermining the idea of a sovereign instance is precisely what is meant by Derrida's notion of autoimmunity; it is an immunity to complete autonomy, the absolutely sovereign. Sovereignty is here understood as utterly unrestricted control. This rejection of absolute sovereignty as a desirable ideal is completely in line with Derrida's analysis of the gift. If we could decide ahead of time whether a gift will be well received, the gift's status as a gift would be annulled. It would no longer be a gift. But we are then left with an aporetical conundrum: if we cannot be regarded as fully sovereign agents mastering our decisions, why should we do anything or conduct ourselves responsibly?

I draw attention to this reading because I claim that it opens a productive way to understand the relationship between time and activity/passivity in an eco-phenomenological context. The core issue concerns the relationship between the passivity or givenness of time and environmental, practical activity. More precisely, it concerns what the gift of time opens or discloses for us as living beings exposed to the passing and coming of time. A common reading of Derrida's conception of time highlights his remarks on the messianic without messianism, which he discusses, for instance, in *Specters of Marx* (Derrida 2006). In writings like these, Derrida emphasizes the distinction between the future and what is to-come (*à-venir*), whereby the latter never arrives in its full consummation but constitutes a kind of virtual futurity which is never actualized. Accordingly, when Derrida speaks of the messianic without messianism, he does not propose that the Messiah one day will arrive, but that this to-come of arrival as a structural promise is a necessary component of ideas like justice, hospitality, democracy, etc. According to the standard reading of Derrida, this requires us to always wait in anticipation without any expectations that our hopes will be fulfilled. It is in this seemingly passive waiting which I think one can discern a striking but not surprising resemblance with the late Heidegger's influence on a strand of eco-phenomenology.

The analysis I have provided above of how the passivity of time opens the space for environmental activity is, I claim, a much more fruitful point of departure for contemporary eco-phenomenology than the late-Heideggerian strain. In general, eco-phenomenologists are critical of the early Heidegger's insistence on willful decisionism and instead tend to joyously celebrate the late Heidegger's notion of *Gelassenheit* (releasement) as the fundamental concept of eco-phenomenological practice.[6] This Heideggerian strand in eco-phenomenology is particularly striking in its sometimes uncritical, celebratory mood concerning Heidegger's critique of modern technology's instrumentalization of everything between heaven and earth, including heaven and earth themselves. According to this line of thinking, the instrumentalized form of rationality informing modern technoscience has, by following its own internal logic to its end, given rise to our current era's environmental predicament. Modern technology is thereby rejected out of hand. The baby is thrown out with the bathwater by this eco-phenomenological current.

There are two major interlinked issues which Heideggerian eco-phenomenologists must address. First, as living beings, they cannot but care about the fate of more-than-human life on planet Earth.[7] If we grant this point, how exactly does *Gelassenheit* do anything at all about the current predicament? There is much debate surrounding the implications of *Gelassenheit* and I cannot give a full overview of the breadth and depth in these debates, but the core issue revolves around the ambiguity in the meaning of the term concerning its supposed passivity. Second, if *Gelassenheit* is understood along Heideggerian lines, it is supposed to be an openness toward a new epoch of Being, an other beginning, for which *Gelassenheit* is a preparation and accordingly *a means toward an end*, which is the new beginning. In that case, is *Gelassenheit* not thereby always already entangled in the very instrumentalization it was supposed to abolish?

One could perhaps charitably read the concept of *Gelassenheit* as precisely what Heidegger seems to suggest: that it lies beyond the distinction between activity and passivity. In his main text on *Gelassenheit*, Heidegger stages a conversation between a scientist, a scholar, and a guide, who are discussing the nature of thinking and the possibility of conceiving of thinking as non-willing, but not as a simple negation of willing. Without rehearsing the whole conversation, it is worth noting that the scientist denies the identification of releasement with passivity in the sense of inactivity. *Gelassenheit* is "in no way a matter of impotently letting things slide and drift along" and the scientist follows up by proclaiming that "releasement lies—if we may still speak of a lying here—outside the distinction between activity and passivity" (Heidegger 2010a, 70).

In my view, the problem lies not in the concept of *Gelassenheit* in general but in its deployment by Heidegger and his followers, whereby only the passive dimension of the concept is usually appreciated. This appreciation of the concept's passive dimension has two interlinked justifications: (1) to avoid the problematic connotation of willing with Nazi ideology in the early Heidegger and (2) to avoid the will's tight entanglement with technological mastery over nature. In both cases, the will is the big bad wolf. It is precisely the notion of the will which *Gelassenheit* is intended to dissolve, even though "[t]he transition out of willing into releasement is what seems to me [the scientist] the genuine difficulty" (Heidegger 2010a, 70). The late Heideggerians will normally proclaim that *Gelassenheit* is beyond the distinction between activity and passivity, but their eco-phenomenological practice does not always seem to be aligned with this theoretical position. Rather than transcending the distinction between activity and passivity, *Gelassenheit* sometimes seems to end up as a contemplative, passive waiting for the only god that can save us. This is a reference to what Heidegger infamously proclaims about the practical efficacy of philosophy, or lack thereof, in an interview with the German newspaper *Der Spiegel* conducted in 1966, but only published, at his own request, after Heidegger's death in 1976:

> If I may answer briefly, and perhaps clumsily, but after long reflection: philosophy will be unable to effect any immediate change in the current state of the world. This is true not only of philosophy but of all purely human reflection and endeavor. Only a god can save us. The only possibility available to us is that by thinking and poetizing we prepare a readiness for the appearance of a god, or for the absence of a god in [our] decline, insofar as in view of the absent god we are in a state of decline.
>
> (Heidegger 2010b, 57)

According to the late Heidegger, ontic phenomena like the climate crisis are only superficial phenomena, not worthy of much attention in comparison with the deeper ontological problematic concerning the question of the meaning of Being and the possibility of a new beginning. Thus, the climate crisis is allegedly only the most recent manifestation of modern technology's inherent drive to enframe our earthly home, exemplified by proposed pseudo-solutions like geoengineering. For Heidegger, this world is already doomed by the advent of planetary nihilism, which was initiated in ancient Greece but only consummated in the past couple centuries. Thus, we can do nothing but wait for the saving god to liberate us and begin anew.

The source of this position is the distinction Heidegger makes between calculative and meditative thinking, modes of thinking which are mutually exclusive and fundamentally at odds with each other. Calculative thinking "always reckon[s] with conditions that are given. [It] take[s] them into account with the calculated intention of their serving specific purposes" (Heidegger 1966, 46). Meditative thinking, by contrast, is "thinking which contemplates the meaning which reigns in everything that is. [...] It is worthless for dealing with current business. It profits nothing in carrying out practical affairs" (Heidegger 1966, 46). Although he mentions the latter part of this quotation as an accusation from hypothetical opponents, Heidegger regards this non-instrumental aspect of meditative thinking as a strength rather than a weakness. That this kind of thinking is not concerned with immediate results based on pre-determined ends is a sign that meditative thinking is more essential than the calculative kind, in Heidegger's view.

The problem is not this distinction, or description, of these modes of thinking but what tends to be the outright rejection of calculative thinking in practice. Admittedly, Heidegger seems to concede that "each [is] justified and needed in its own way: calculative thinking and meditative thinking" (Heidegger 1966, 46), but the source of the justification for calculative thinking is what Heidegger finds troubling. Calculative thinking, machination, and modern technology have ultimately consummated a state of planetary nihilism. Heidegger-inspired eco-phenomenologists likewise often neglect the ineradicable aspect of calculation for living beings that care about leading good lives. This can most readily be discerned in the widespread interest in contemplative practices among eco-phenomenologists, such as mindfulness meditation, which are supposed to serve as antidotes to the hurry-scurry form of mindless activity characteristic of our historical moment. Heideggerians will understandably hurry to dissociate mindless machination/calculation with Jewishness, a connection which their master emphasized in his notebooks. To be clear, it is certainly possible to re-frame acts of meditation as means toward leading good lives, but this requires disconnecting the activity from its metaphysical soil, whether that may be found in Buddhism or elsewhere. In any case, Derrida's nuanced discussion of calculation far supersedes Heidegger's, in my view. Derrida fully acknowledges that reducing all thinking to calculation erases everything worth caring about and that it would be an unintelligible form of thinking at all. Yet he nonetheless insists that we cannot finally do away with calculation.

What about the following suggestion regarding eco-phenomenological practice? *Gelassenheit* understood in a concrete eco-phenomenological context is a way of letting nature be what it is. To achieve that end—*the instrumentalization is always already in play!*—we may certainly need more thoughtful engagement with, and appreciation of, the more-than-human natural world. One suggestion would be to let living nature be by disassembling fossil capital's predatory energetic extraction of natural resources. Instead, humans may make use of forms of energy which are more sustainable, not only for the human species but also more sustainable for other forms of life. Similarly, although human beings need to consume food to live, this fact does not by itself necessitate the mass slaughter of animals on an industrialized scale.

These are examples which indicate how decision-making is both necessary and urgent, but also that no decision, for instance concerning our socio-economic forms of life, is ever final or sheltered from critique. Nor do these suggestions imply that we need to erase all distinctions by making abstract comparisons based on our contempt for modern technology, as Heidegger sometimes seems to do. One egregious example can be found in his infamous homogenizing comparison of modern agriculture with the Nazi gas chambers because both operate according to the parameters of calculative thinking:

> Agriculture is now a mechanized food industry, in essence the same as the production of corpses in the gas chambers and extermination camps, the same as the blockading and starving of countries, the same as the production of hydrogen bombs.
>
> (Heidegger 2012, 27)

For Heidegger, the horrifying aspect of these phenomena is that they are organized mechanically and industrially according to the logic of calculative thinking. To claim that all acts of calculation are equally malevolent is simply baseless. What is even worse is that Heidegger does not make this comparison to demonstrate the case for veganism by comparing the mass slaughter of animals with the Holocaust. Rather, he bulldozes the field to illustrate contemporary nihilism's own erasure of differences, instead of arguing why we need to maintain qualitative differences between phenomena. Contra Heidegger, we might claim that to discriminate is a fundamental act of responsibility, in which I take myself to prioritize one action, one person, or one form of life over another, thereby exposing myself to potential critique, to which I must respond by articulating the reasons for acting or revising my actions if I cannot justify myself.

So long as the human species survives, we will impact nature in some way or another. To think otherwise would be to fundamentally misunderstand humanity's embodied embeddedness in nature. Yet this constitutive embeddedness within, and impact on, nature does not preclude the possibility of altering *where* and *how* this impact becomes sedimented in nature. On the contrary, the passing gift of time opens the space for environmental activity and likewise constitutes "the nonethical opening of ethics." The problem arises when one thinks that *Gelassenheit* becomes exclusively about wandering in silence in the forest or on the mountain, listening to the many beautiful and sublime sounds or sermons of nature, like birds chirping, wind blowing, or thunder clapping. Such an appreciation of, and wondrous attitude toward, nature certainly pertains to eco-phenomenological practice, but it should not be thought of as exhaustive of such practice. If eco-phenomenological practice is *Gelassenheit*—understood minimally as letting nature be—then it must by necessity involve more than recreational activities. Facilitating nature's being through letting it be involves so much more. It is clearly true that we have some sense of negative certainty about the consequences of our actions, and this is precisely why we should act with a great degree of caution. But eco-phenomenological responsibility must include both passivity and activity. We are passive vis-à-vis time and nature, insofar as they are gifts bestowed upon us beyond our own choosing. We cannot avoid being temporal creatures embedded in a natural world, but we *can* act more responsibly on that basis than modern technology has tended to suggest. This is a logical consequence of Derrida's analysis of time and the autoimmunity of life.

To conclude, why does my title mention eco-political, and not simply environmental, activity? What about ethics? Are concrete policies supposed to heal the metabolic rift or our disturbed relation with more-than-human nature? I have intentionally neglected to use the term environmental ethics and instead opted for a more politically oriented term. I regard it as symptomatic that many environmental issues are framed as individual choices concerning lifestyles. Take an ever-increasing phenomenon in my hometown, Tromsø, like tourism, whereby relatively affluent people fly or cruise across the globe to experience the Aurora Borealis in the winter, midnight sun during summer, as well as Sámi culture all year round, among a host of other touristified attractions, like taking the cable car or climbing the Sherpa steps to the top of the mountain, or going for a sled ride with husky dogs. Moralistically condemning an individual's choice to spend some of their finite time, energy, and money to travel the world gets us nowhere. We could instead ask if it is reasonable that our socio-economic mode of production necessarily leads to people and companies on this planet making enormous profits from destroying the only planet that we know is hospitable to life—at least for the time being, in the double sense of that phrase. An unconditional hospitality toward terrestrial life requires more than merely passive letting go. Yet what we ought to do on a practical level is a question which neither a strictly eco-phenomenological description nor an ultra-transcendental account of time can settle. If eco-phenomenologists genuinely do care about life on this planet by virtue of themselves being alive, these questions must be discussed. There lies no ultimate salvation in waiting for an apocalyptic bang or whimper.

Notes

1 In addition to the editorial team, thank you to Sivert Vorren for feedback on an earlier draft of this chapter.
2 For the most extensive debate concerning this point, see John D. Caputo's lengthy critique (Caputo 2011) of Hägglund's *Radical Atheism* and the latter's subsequent response to that critique (Hägglund 2011b).
3 Unless otherwise stated, all emphases in quoted passages appear in the original.
4 For a related critique of this widespread conflation of Kant and Derrida concerning the unconditional, see La Caze 2007.
5 Cf. Caroline Levine's remark: "As delays and open-endedness are put to use by fossil fuel companies and climate denialists, it seems clear that there is no intrinsically resistant tempo, no way of approaching time that will itself get us out of climate trouble" (Levine 2023, 11).
6 Readers familiar with the eco-phenomenological tradition will readily recognize this influence, but newer readers need only take a glance at the number of chapters in this handbook that evoke Heideggerian *Gelassenheit* either explicitly or implicitly. Along with Merleau-Ponty, Heidegger is arguably the most influential philosopher informing contemporary eco-phenomenology. In the eco-phenomenological context, it is noteworthy that *Gelassenheit* or letting-be is discussed briefly in only two of the contributions to *Eco-Phenomenology: Back to the Earth Itself*, namely, the chapters by Zimmerman (2003) and Langer (2003). An exhaustive analysis of how *Gelassenheit* has been appropriated within eco-phenomenology unfortunately lies beyond the scope of this chapter. For a coherent account of Heidegger's transition in his works from willing to *Gelassenheit* in general, but not the latter term's eco-phenomenological appropriation, see Davis (2007).
7 Even such an affirmation seems to run counter to Heidegger's own thinking, as explained by Monika Langer, who discusses intrinsic values in nature:

> Heidegger's point here is that valuing something—no matter how positively—subjectivizes it, instead of letting it be. By valuing it, we reduce it to an object of our estimation. Heidegger's critique suggests that despite all claims to the contrary, "environmental ethics" is unavoidably anthropocentric.
>
> (Langer 2003, 112–113)

References

Caputo, John D. 2011. "The Return of Anti-Religion: From Radical Atheism to Radical Theology." *Journal for Cultural and Religious Theory* 11 (2): 32–124.

Davis, Bret W. 2007. *Heidegger and the Will: On the Way to Gelassenheit*. Evanston: Northwestern University Press.

Derrida, Jacques. 1982. *Margins of Philosophy*. Translated by Alan Bass. Chicago: The University of Chicago Press.

Derrida, Jacques. 1992. "Force of Law: The 'Mystical Foundation of Authority.'" Translated by Mary Quaintance. In *Deconstruction and the Possibility of Justice*, edited by Michel Rosenfeld Drucilla Cornell, and David Gray Carlson, 3–67. New York: Routledge.

Derrida, Jacques. 1998. "Hospitality, Justice and Responsibility: A Dialogue with Jacques Derrida." In *Questioning Ethics: Contemporary Debates in Continental Ethics*, edited by Mark Dooley and Richard Kearney, 65–83. New York: Routledge.

Derrida, Jacques. 2003. "Autoimmunity: Real and Symbolic Suicides. A Dialogue with Jacques Derrida." Translated by Pascale-Anne Brault and Michael Naas. In *Philosophy in a Time of Terror: Dialogues with Jürgen Habermas and Jacques Derrida*, edited by Giovanna Borradori, 85–136. Chicago: The University of Chicago Press.

Derrida, Jacques. 2006. *Specters of Marx: The State of the Debt, the Work of Mourning and the New International*. Translated by Peggy Kamuf. New York: Routledge Classics.

Hägglund, Martin. 2008. *Radical Atheism: Derrida and the Time of Life*. Stanford: Stanford University Press.

Hägglund, Martin. 2011a. "The Arche-Materiality of Time: Deconstruction, Evolution and Speculative Materialism." In *Theory after 'Theory'*, edited by Jane Elliott and Derek Attridge, 265–277. New York: Routledge.

Hägglund, Martin. 2011b. "The Radical Evil of Deconstruction: A Reply to John Caputo," *Journal for Cultural and Religious Theory* 11 (2): 126–150.
Heidegger, Martin. 1966. "Memorial Address." Translated by John M. Anderson and E. Hans Freund. In *Discourse on Thinking*, 43–57. New York: Harper & Row.
Heidegger, Martin. 2010a. "Ἀγχιβασίη: A Triadic Conversation on a Country Path between a Scientist, a Scholar, and a Guide." Translated by Bret W. Davis. In *Country Path Conversations*, 1–104. Bloomington and Indianapolis: Indiana University Press.
Heidegger, Martin. 2010b. "'Only a God Can Save Us': The *Spiegel* Interview (1966)." Translated by William J. Richardson. In *Heidegger: The Man and Thinker*, edited by Thomas Sheehan, 45–67. New Brunswick: Transaction Publishers.
Heidegger, Martin. 2012. *Bremen and Freiburg Lectures: Insight Into That Which Is and Basic Principles of Thinking*. Translated by Andrew J. Mitchell. Bloomington and Indianapolis: Indiana University Press.
Kant, Immanuel. 1998. *Critique of Pure Reason*. Translated by Paul Guyer and Allen W. Wood. Cambridge: Cambridge University Press.
La Caze, Marguerite. 2007. "At the Intersection: Kant, Derrida, and the Relation between Ethics and Politics." *Political Theory* 35 (6): 781–805.
Langer, Monika. 2003. "Nietzsche, Heidegger, and Merleau-Ponty: Some of Their Contributions and Limitations for 'Environmentalism.'" In *Eco-Phenomenology: Back to the Earth Itself*, edited by Charles S. Brown and Ted Toadvine, 103–120. Albany: SUNY Press.
Levine, Caroline. 2023. *The Activist Humanist: Form and Method in the Climate Crisis*. Princeton: Princeton University Press.
"metaphor (n.)" (n.d.) in Online Etymological Dictionary, https://www.etymonline.com/word/metaphor. (Accessed 10 June 2024).
Oliver, Kelly. 2015. *Earth and World: Philosophy After the Apollo Missions*. New York: Columbia University Press.
Zimmerman, Michael E. 2003. "Heidegger's Phenomenology and Contemporary Environmentalism." In *Eco-Phenomenology: Back to the Earth Itself*, edited by Charles S. Brown and Ted Toadvine, 73–101. Albany: SUNY Press.

14
HOW MUSIC LISTENING MAY NURTURE OUR RECEPTIVITY AND RESONANCE WITH NATURE

Charlotte Lindvang

Introduction

The German sociologist Hartmut Rosa points out that in late modern society, there is a constant demand for acceleration, expansion, and growth, and consequently a growing number of people experience stress and loneliness (Rosa 2017). If the human mind is too occupied with a rational kind of everyday life, too busy, too focused on being effective and productive – we might be cutoff from the world, unable to see or hear. According to Rosa, resonance might be an important part of the solution, understood as a state of being in touch with something in the world that is important, valuable, and meaningful.

In this chapter, I present four qualities that seem appropriate for opening ourselves to a more resonant and caring coexistence with all living beings – people, animals, plants, and landscapes: sensuousness and aliveness, contemplation, connectedness, and devotion to something greater. These four dimensions are not discrete, but intimately connected and interwoven in our lives. Together they form a kind of quadrant of receptivity and resonance – a mode of attunement that draws us into deeper relational presence with the more-than-human world. Listening to music is one human activity that may call us, touch us, and facilitate these phenomena. During the chapter, vignettes will illustrate how such experiences may emerge.

Care or Despair

According to Rosa, resonance does not necessarily mean harmony and pleasantness. To reach a mode of receptivity and resonance requires that we dare to be vulnerable; allowing ourselves to be touched and transformed in non-predictable and non-controllable ways (Rosa 2019). Hence, the opposite of resonance is not dissonance, but "alienation" – when we are alienated from the world, it appears mute and colorless, it does not touch or move us; there is no resonance (Rosa 2024, 24).

How can we change our mindset and reduce alienation? How can we move into a resonant mode of being and develop care and commitment to live in more earth-friendly ways that lead to sustainable futures on Earth? We are many who ask these questions and we are obliged to think about this inside every faculty and every profession and to look for cross-disciplinary

 DOI: 10.4324/9781003538097-18

collaborations to achieve change. In the field of music therapy, the noble focus is to care for people who may be affected by illness, loss, tragedies, people who may not have a voice, or whose voice is weakened and easily overheard in our society. Music therapy practice, apply music to decrease isolation and alienation to each other, and to the world. One of the burning questions is how empathy, inclusion, and resonance, carried forward in music therapy, can reach further into the future lives of our children and grandchildren, and further to embrace "more-than-human" life.

It happens in music therapy that people share their suffering from feeling alienated from the world, the world inside as much as the world around. One example is a woman who came to a music therapy session incredibly stressed and feeling "beside herself," and not present. In her stressful state she was not able to be in touch with her body sensations, her inner thoughts, and feelings. Furthermore, she felt alone and disconnected to other living beings. During the session we listened to music[1] and during the listening she imagined to be in a summer house in the countryside with sensations of light and colors, and of connectedness with the animals, birds in the sky, and a horse who just came to her, with no effort on her side. Through the music and her nature imagery, she moved from a kind of numbness to a sense of anchoring, homecoming, and resonance. She expressed the following words during the music listening:

> There is a curtain quietly flapping in the window – the light is shining through it – it is in the early morning, the bog wife brews, it is late in the summer… I see a golden color; the grass and the foliage are golden. Everything is quiet. Above me a big bird is hovering, circling.
>
> …
>
> Now a horse is looking at me, the eyes are marvelous, they are so gentle. We approach each other – the horse and I – we stand close – we look into each other's eyes… now it put its head down to me – I feel it – I just stand there, and I hold its head. We stand in silence, connected. There is nothing in the entire world that I am heading for except from being here.

Art Matters

Different genres of art offer opportunities to perceive, immerse in, and reflect on the changing world. However, art is not merely a mirror of our lives or a space for reflection – it can complement traditional forms of science as it represents a unique field of knowledge with the potential to generate insights, create change, and foster coexistence. Art engages with possibilities, uncertainties, lived relationships, and it explores what we can imagine differently. Art can vitalize our senses and help us to see and hear and to imagine a sustainable future (Christensen 2021).

We need to step out of our familiar and rational everyday consciousness to generate and share knowledge and wisdom in new ways. Instead of looking away from the catastrophe or even stick to denial, music, poetry, and other art forms can touch and move us and allow us to transcend conventional thinking. Art is a way to explore deeper, more intuitive forms of understanding and communication and thereby enhance intimacy with the living. Furthermore, engaging with art keep us connected to our energy and vitality, and provide a medium for expressing emotions and experiences, helping us to process and communicate complex feelings that emerge in these challenging times.

Music has a profound emotional impact on human beings, and it can provide intense, powerful, and memorable experiences (Gabrielsson 2011). Rosa points out that powerful concert experiences sometimes correspond to an experience of epiphany, a revelation where listeners undergo an immediate and fundamental transformation of their being-in-the-world (Rosa 2024). Lipe (2002) found in a review of 52 articles dealing with music, spirituality, and health that "engagement with music provides a way to access and energize the imagination, leading to new, creative ways of listening, thinking, and being" (Lipe 2002, 233). Furthermore, music is a common elicitor of awe: When listening to a moving piece of music, people may be transported to a different state of mind (Monroy and Keltner 2023).

The Climate Music Project[2] is an example of collaborative art projects where scientists and musicians collaborate to create musical and visual experiences guided by the scientific data on climate change. Music is a universal language that defies communication barriers, and music is accessible, and easier, for most people, to relate to, than articles about the climate crisis. Therefore, music provides a unique platform to reach people and connect people to climate science and make climate issues resonate on a personal and a community level. As statements from the audience illustrate, people embody the music through rhythm, tempo, dynamics, and pitch, instead of just looking at and away from the data.

Listening to the Sounding World

We live in a sounding world. Everything has a voice. In our ordinary mode of consciousness, we might not hear, but if we listen carefully, the voice of the wind, the insects, the leaves, might reach us. Often nature offers a multiplicity of voices at the same time and what we hear is duets or complex polyphonies. In a fundamental sense the sounds of the world are the first music and, in that sense, what we usually call music is a specific human configuration within the world of sounds (Ihde 2007).

An auditory experience is a pervasive bodily experience. We do not merely hear with our ears – the whole body is involved. This is very well demonstrated by deaf musicians like percussionist Evelyn Glennie, who has been profoundly deaf since childhood. Glennie has widened our understanding of what music is and shown us that listening is only partly to do with our ears since the body is a resonance chamber and we live in a universe of sound.

According to Ihde (2007) sight has been the preferred source of knowledge in the western society and with this focus on what we can see, we risk underestimating the richness of our listening capacity.

> Sounds permeates and penetrates my bodily being. It is implicated from the highest reaches of my intelligence that embodies itself in language to the most primitive needs of standing upright through the sense of balance that I indirectly know lies in the inner ear.
>
> (Ihde 2007, 45)

In the process of listening, we let the world come to us, and we let it in, as the soundwaves reach us and in a delicate touch call us into a relationship. The sense of hearing is the first to fully develop in a human being already during prenatal growth, allowing fetuses to listen to the sounds of the world and respond to sounds even before birth. According to Rosa and the phenomenon of resonance, listening is crucial; we can only experience resonance and a deeper sense of belonging by listening to people, nature, things, or the context we are in (Rosa 2019).

Rachel Carson, American pioneer and environmental advocate, writes in her book "The sense of wonder": "For most of us, knowledge of our world comes largely through sight, yet we look about with such unseeing eyes that we are partially blind" (Carson 1965, 59). It is almost 60 years since she claimed that. It is still relevant to ask: How can we get eyes that are able to really see and ears able to really listen? According to Carson the world needs humans to make a radical reshaping of our customary way of hearing: to listen more deeply with our whole body and mind.

Sensuousness and Aliveness

David Abram's contribution "The spell of the sensuous" is concerned with our pre-reflective, sensuous experience of the world. Abram (1996) points out that we need to return to the cultivation of our senses to restore our reciprocity with the planet. Referring to Merleau-Ponty, Abram focuses on the breathing-sensing body and calls us to remember and rejuvenate the organic basis of our thinking. He points to the importance of "giving voice to the world from our experienced situation *within* it, recalling us to our participation in the here-and-now, rejuvenating our sense of wonder…" (Abram 1996, 47).

Music can awaken our senses and influence our breath. It can also call us to a bodily presence and open to vitality (Kenny 2006). Carolyn Kenny confirms our embeddedness beneath our thinking by saying that: "Music represents a pre-verbal or primordial level of communication and therefore broadens the possibilities for communication and eliminates boundaries" (Kenny 2006, 12). To illustrate, I will share a wonderful and surprising experience I had in one of my private music therapy sessions.

> I listened to a piece of Bach,[3] a piano concert, while I was in a deeply relaxed state. As I was listening, I indulged in the music and gradually felt gravity and grounding in my body. The sensation developed to an experience of being a tree. I felt, I was a tree with an inner aliveness and with branches reaching out. I sensed my roots, and I felt like I had a thick and solid trunk. It was an amazing feeling of being alive, being calm and solemn, and of being rooted and deeply connected.

Hartmut Rosa has in his recent book *When Monsters Roar and Angels Sing* (Rosa 2024) combined his sociological theory about resonance with his personal intense experience of heavy metal music. Rosa reflects on the phenomenon of listening to music in general; it is neither active nor passive but something in between. The music comes to the listener and at the same time the listener receives the music, and a subtle and dynamic interaction transform the listener in that moment. However, it is not always like that to listen to music. We can listen to the same music on two separate occasions and experience it completely different. One day the music leaves us untouched and another day the music finds us open and vulnerable, and the experience of resonance emerges (Rosa 2019, 2024). Hence it is something inside of us that is susceptible and prepared to encounter the uncharted landscapes. According to Rosa presence and receptivity are prerequisites for resonance to occur. He describes that in a resonance experience, the relationship becomes fluid: If we listen intently to a piece of music and experience to be open to the unexpected, we are in that moment both inwardly focused and simultaneously directed toward the music and the surroundings, and it does not make sense to distinguish what is internal and what is external. Rosa writes:

> The music seems to come from within and from outside at the same time; it is inside us, and we are inside the music. It is closer to us than our own breath. It touches our inner core and simultaneously fills the space around us.
>
> (Rosa 2024, 95)

It often happens that people during music listening in music therapy sessions, have sensuous experiences with nature. They are imaging, for example, clouds in the sky, they sense the sunbeams, the warmth on their skin, they touch the green leaves and smell flowers, they meet animals and experience the power, sometimes horror of nature, or they dive in the ocean or even fly and looking down to the ground through a bird's eyes. In a study by music therapists Lars Ole Bonde and Bolette Beck (2019) they explored how spontaneous nature imagery is evoked and processed during music listening in the specific music therapy method Guided Imagery and Music (GIM: developed by Helen Bonny in the 1970s). Bonde and Beck found that nature imagery was a common feature and included minerals, plants, animals, landscapes and ecosystems, planetary and cosmic experiences evoked by music of any style and period (in GIM we use primarily classical music). The authors conclude that beyond serving as support and resource in a personal process of for example grieving loss or overcoming stress or burnout, experiences with nature imagery may provide an expanded dimension, assisting the music listener in contemplating, and seeing him/herself as part of a greater ecological system (Bonde and Beck 2019).

The Intrinsic Connection between Nature and Music

Music allows us to be immersed in it, to be one with it, to be enveloped by it, as well as it is possible to listen to the music from a more distanced position. "Take time to listen and talk about the voices of the earth and what they mean – the majestic voice of thunder, the winds, the sound of surf or flowing streams" (Carson 1965, 75). Nature has its own beats and melodies – it has its music in the sounds of animals, the rustling and whispering of the wind, the trickling, gurgling, and roaring of water. Carson writes: "No child should grow up unaware of the dawn chorus of the birds in spring…The first voices are heard before daybreak… In that dawn chorus one hears the throb of life itself" (Carson 1965, 77).

Throughout history, nature's auditory universe has inspired artists and composers. Nature has been portrayed in music as a kind of resonance with its forms and expressions over time. For instance, composers have tried to capture nature's beauty or translate the mathematics found in natural phenomena into musical expressions.

The art of incorporating birdsong into music finds its master in the French composer Olivier Messiaen. In addition to being a composer, Messiaen was an ornithologist, deeply fascinated by birds and their songs. Messiaen ventured into nature to study, observe, and listen closely to birds. He meticulously notated bird songs from around the world, integrating these translations into his musical language. This is audible for example in his 1958-piece, *Catalogue d'oiseaux* (Catalogue of Birds). In the score, Messiaen included his impressions and descriptions of nature just below the notes, guiding the musician (pianist) to immerse themselves in nature during the performance.

Due to his religious faith, Messiaen viewed birdsongs as more than mere sounds useful to a composer; he saw them as profound musical expressions. When Messiaen visited Copenhagen in 1977 to receive the Léonie Sonning Music Award, he participated in an interview where he described his passion for birds (Christensen 1992).

> I always have the birds with me. It has made many people smile – they do not understand why I note down bird songs. I do it because I believe that birds are great musicians. I even think they are more musical than we are – after all, they invented all kinds of music long before humans did. (…) Birds are great masters – and I love them for another reason: they sing spontaneously, naturally, without having learned it, without holding concerts, not for the sake of honor, without getting paid for it, without artistic rights. They sing for the sake of the song itself or to show they are present or to woo their beloved.
>
> (Christensen 1992, 17) (Author translation)

As Messiaen emphasized, the bird is not an actor; the bird is fully immersed in its existence. When we, as humans, are drawn to the bird's beautiful sounds and songs, it may be because they remind us of the essence of being alive – living life without performing, consuming, or striving for wealth. We enter a profound sense of presence, and a form of self-forgetfulness.

Another example of the encounter and connection between nature and music is a composition inspired from seismology created by the contemporary Danish composer Kim Helweg:

> During the Danish festival "Bloom," where nature, art, and science enlighten people about the universe, the world and us, attendees in 2020 experienced the interdisciplinary work "Future – The Earth Trembles"; People explored the phenomenon of earthquakes through lectures, percussion, and opera. The music composition was created in collaboration with a geophysicist, who explained how an earthquake begins with a rupture in the Earth's subsoil, followed by various tremors and types of waves. Sound, like earthquakes, consists of waves, making it possible to translate seismic waves into an audible range. Musically, the frequencies and tonal material generated during an earthquake are unpredictable. Thus, the composition invited the listeners into an unknown realm and facilitated reflections on how the Earth is moving, and its internal dynamics are fundamental to the foundation of all life on Earth.

Contemplation

The dynamic nature of life within us means that there is an ever-evolving environment inside us – a vivid life of feelings, memories, and imagination. As well as interpreting nature in creative ways, as exemplified above, composers have used music as medium for conveying the inner nature of being a human. When music is moving dynamically, with varying forms, rhythms and timbre, surprises and harmonies, it resembles nature including human nature. Therefore, listening deeply to music invites us to become aware of our body, our breathing, and our state of mind, potentially deepening our connection with nature.

According to Rosa's analysis our biggest challenge today is our western way of living where we feel trapped in a hamster wheel while our human activities have led to widespread ecological breakdown. Rosa points out that to be in a living exchange with others and with nature we need to take the risk of making ourselves vulnerable and let ourselves be touched in non-predictable and non-controllable ways (Rosa 2017, 50).

In a contemplative or meditative state, it is possible to reach a mode of existential balance where our fears are calmed down, our desires and feelings of alienation are silenced, and it is possible to experience connectedness and resonance with life, with the Earth, with the universe (Rosa 2019, 119–120).

> The ideal resonant experience is itself free from desire, insofar as desire always means striving to change, to transform and overcome one's present state or situation, while a resonant experience carries its fulfillment in itself.
>
> (Rosa 2019, 119)

Sometimes music can create a safe space of listening in which we can be still, be silent and open our hearts to a mode of just being and receiving what is there. For example, one person came to a series of music therapy sessions to find a way to reduce the level of stress in his everyday life and regain the ability to be present.

> Living with the challenge of being stressed at work and suffering from swarms of thoughts all the time hindered this person in keeping his attention and being present. For example, he did not manage to listen to the needs of his dog and go for a walk with his dog… We collaborated on a music playlist for him, so with carefully selected music in his ears, he gradually became able to walk in the forest, in a contemplative manner. He told me, that one day he had a deeply touching experience of equanimity, as he was walking and listening to a piece of music, he felt the gradual opening of sensing nature again – he was reconnecting with all the colors of green. He even found that he and the dog had the same pace as they walked.

Coherence/Connectedness

Our close relationship with music originates from the moment we arrive as pure nature in the world: through sounds of heartbeats and voices we are connected to the world and through our innate communicative musicality we build relationships. Through our life music can pave the way for our sense of connectedness, care, and hearty cohesion; that we are a part of each other and a part of nature (Kenny 2006). Music is something that most people *feel* connected to. In music, we often find a gateway to what truly matters, a doorway to understanding and experiencing the essence of life. That could be due to the ability of music to encompass both darkness and light and thereby create a space for the contrasting nature of life (Bonde and Vist 2013). Since the dawn of time music has gathered people and has also been a way to expand consciousness and a catalyst to unite with forces of nature. Music and nature are deeply intertwined in many cultures, reflecting the values and beliefs about the natural world. Carolyn Kenny who was a music therapist, phenomenologist, and a member of the indigenous eagle clan of Haida nation said that "music is not so much a connector as a vehicle of inspiration that reminds us that we are already connected." According to Kenny our sensibility can help us to realize that we belong (Kenny 2006, 22).

Following Hartmut Rosa resonance is to feel and listen to your own voice and to the voice of your neighbor, and to listen with curiosity to the kind of consonance, dissonance, or harmony, which will emerge in the moment. As the following narrative illustrate, the "neighbor" can be any living thing.

In a music therapy session, a client prepared for her music listening by telling me about her process of moving from a house where she had lived for more than 30 years. She expressed the following as the music accompanied her:

> I am sitting in the garden…, I love to sit here, it is sad that I will be leaving, but I am also thankful, that there is something to mourn about. Now I realize that it is the garden and all the plants here and not the house that I feel most sad to leave. I am looking at the apple tree that my son got for his Baptism. And then the small Japanese part of the garden – oh, the green ferns, the soil… I remember I once got the soil from my neighbor, and I made this plateau, then later green ferns came out, completely unexpected, the seeds must have waited in the ground for more than a hundred years, since the soil was from beneath the neighbor's house which was reconstructed.

D is crying while these deep memories emerge – she continues to walk around in the garden, she touches the plants, and contemplates that she has been with the flowers and trees all these years and followed their growth… and she said:

> Now I see, this is not my work…, every plant, every bush here has its own life, and I just got the gift of taking part in it all, … I have enjoyed it, all this creative power, I am just a part of it all …

Rosa talks about the fleeting nature of bodily and passionate experiences. However, our contemplation about the experience, may deepen the resonance. For example, witnessing a sunset can be a strong and breathtaking experience not least because we are *aware* of the essential transience of it. Likewise, a music listening experience is temporalized, a passing moment of vitality and maybe of beauty (Rosa 2019, 120). In the narrative about the connection between D and the plants, the possibility of remembering and re-connecting through the music provided her with expanded reflections on what it means to live together.

Devotion to Something Greater

Various meditation- and mindfulness practices deepens our connection with our inner selves. It can also heighten our awareness and spiritual presence in the world we are living in.

According to a music therapy researcher Giorgos Tsiris spirituality is neither a purely subjective inner experience, nor an objective reality "out there." Spirituality exists in the "in-betweenness" – it is an open space (Tsiris 2017, 297). Neither the music nor the listener alone creates a spiritual experience; it emerges from their interaction. In accounts with strong experiences with music, people use words like "magical," "mysterious," "eternity," or "cosmic" to describe it. For example, the listener can get a feeling of timelessness or a dissolvement of the ego, a merging into something greater or feeling at one with the universe (Gabrielsson 2011, 162).

Spiritual qualities have been attributed to music and music therapy for example by the violinist and music therapist Helen Bonny, who suggested that listening to carefully selected music may help us to let go of control, get in touch with the ineffable beauty of life and lead us to forgiveness (Bonny 2002; Bonny and Savary 1973). According to Bonny spirituality means thinking of others, an acceptance of each day of one's life, and a habit of gratitude for everything. Spirituality involves a process of looking deep inside and discovering our inner unity with all life (Bonny 2002, 178). Nature imagery was also central in some of Bonny's own experiences. For example, Bonny reports that she once during music listening had a deep transpersonal experience; a tree began to grow in front of her and flowers of all varieties sprang out of its branches. She felt an intuitive communion with the tree, and she was filled with a wordless knowledge. The imagery ended in a transpersonal experience of standing in a shimmering shower of golden light and a voice surrounded her with the words "I am in all, we are all one" (Bonny 2002, 90).

Sometimes music uncovers the depth in us, in our being, to an extent that we cannot express in words. Bonny said: "Music is a language of immediacy" (Bonny 2002, 105).

> Music is a combination of sound and silence... it is those moments of silence after the perfect blending of sound in music when all things are not only possible but are co-existent. Breaking through the barriers of limitation into "All-that-is." It is then that music becomes the language of immediacy.
>
> (Bonny 2002, 114)

The Norwegian author Karl Ove Knausgaard sometimes manages to write in "a language of immediacy." He often describes nature in such a vivid way that it removes barriers of limitations, and it feels like being there, and your inner is moving like nature as you read. Knausgaard describes music as a direct link to our sensitivity, our memory and emotions, our deep connection to nature. In the following excerpt from the novel "The Morningstar" (Knausgård 2020) we follow the character named Solveig, who lives in the Northern Norway. She is working as a nurse, and we follow her experience as she drives home from work, in the early morning after she has had a night shift at the hospital:

> I found in the car one of my old cd's and I put it on without looking at which one I took. I liked to let chance rule – at least the little coincidences, – it often gave me joy. ... Woohh, this one was so familiar. But what was it? I increased the volume as I reached the end of the valley and the climb up the mountainside began. Albinoni. Adagio.[4] Down on the other side, passing the water with the waterfall in white in all the green, between the rock walls and then another steep, narrow climb.
>
> I felt so happy that I did not know what to do. I could as well cry as laugh. I was singing along the music that rose and fell like the landscape. But the music did not come from the landscape, it came from the heaven. Or from the inside. Someone's inside. Everyone's inside. My inside.
>
> (Knausgård 2020, 165)

This is an everyday experience that unexpectedly elevates Solveig to an almost incomprehensible sense of connection and deep bond with the landscape, her life, and strong emotions, where all distinctions disappear. The moment of resonance arises as Solveig is receptive to the music and simultaneously the landscape. The process is uncontrollable; not only was her music choice random, but it was also a coincidence that she drove through a landscape that matched the music, with the same twists, contrasts, and interplay. Right there, it happened!

Once I experienced a kind of spiritual breakthrough during listening to music in my own therapy.[5] The images lead me to a moment of unity and enlightenment and a sense of hope for the future, that I desperately needed and still need. The following words describe my experience:

> I found myself in a landscape of curved dunes. The scent of sand, salt, and seaweed filled the air, and I could hear the waves rolling towards the shore just before I reached a top of a dune and the overwhelming beauty of the ocean in its deep blue colors met my eyes... Then I saw a large crowd of people gathering on the beach. They began holding hands and gradually forming a half-circle along the coastline. They stood there, everyone gazing towards the horizon, as if in a shared prayer or thanksgiving for life on Earth. My heart was completely open, and a powerful light from the sky reached me. The light sank into me, into my heart, into everybody's heart.

Discussion

Humanity is responsible for the current state of the planet, where there are likely no areas of nature left untouched by human activity. It is easy to feel despair and even feel hatred toward humanity or profound sorrow over how far things have come, with life on Earth being stifled by human influence. However, the nature crisis calls us to reorient, rethink, and reexperience the world. In these challenging times, listening and being receptive, offering dialogues, mutual support, and therapeutic spaces are essential for expressing our emotions and heeding this call. We must look within ourselves, taking responsibility for our inner balance and for nurturing positive aspects of our human nature. But how can listening to music help us, when the world is burning, and call us for action now? It is a paradox that we need to linger, hold back, make space, and let go, to find our way to resonance, while at the same time, the clock is ticking, and we witness humanity continuing its usual ways, living as heavy consumers without regard for others or the rest of the living beings on the planet. Rosa's concept of resonance suggests that true, meaningful change requires a slower, more contemplative, and receptive approach. However, there exists a profound tension between the urgency of immediate changes and the necessity of taking time to listen, reflect, and connect deeply with the world around us.

Consumerism and the growth paradigm result in a depletion of the Earth's resources and the production of unimaginable amounts of waste. Music is a part of a consumer culture used to manipulate humans to buy things. Music is powerful, and sadly; music can be weaponized. We rarely discuss that music can be used to such dark purposes like manipulating our deeper needs and values or to conduct terror in prisons. At Guantanamo Bay detention camp they played music extremely loud and in unpredictable time periods, so that prisoners were unable to find an aural space not invaded by loud music (The Guardian) (Anonymous 2015). Methods of "noise torture" or "sound torture" are well-documented forms of terror, highlighting that music is not inherently good and can be used for harmful purposes. Moreover, the same piece of music can evoke vastly different responses in different individuals, and even the same person may experience it differently depending on factors such as the intention behind the music listening, the context in which it is heard, and the listener's mood (Gabrielsson 2011).

Many people in our modern civilization experience a disconnection from nature. They do not feel connected or interested in it; plants and animals seem foreign, the forest echoes with loneliness, and the beach is only pleasant in summer when machines have cleared the seaweed, allowing for sunbathing on warm sand. Some people feel more attached to local shops and commercial spaces. When humans fail to see themselves as embedded within the ecosystem, they establish a sense of separation from the ecosystem (Schuurman-Olson 2023). How can we rekindle a deeper connection, curiosity, and care for nature and the planet's health? In our searching for resonance and reciprocity with the living planet, it might seem more authentic to experience nature directly; sit by the sea or in the mountains to immerse oneself in nature, rather than sitting in our homes or in therapeutic spaces listening to music created by humans to evoke inner images of nature. However, the reality of urban living means that people need to find ways to integrate nature into their daily lives and to maintain their inner connection to nature. Furthermore, music teaches us to let go of control, to stay in a receptive and listening mode offering the possibility of profound experiences of resonance and connection to ourselves, each other and the web of life and sonic environments of which we are already a part of and exist within (Schuurman-Olson 2023).

It is possible to criticize the romantic relationship with nature, where we humans see nature in our own image, idealizing, and humanizing it. Even imaging nature during music listening may be a harmonic romantic veil that masks the Earth in its extreme pain of mass extinction. On the other hand, first music has the ability of representing nature in all its aspects, embracing all dynamics from quiet streams to furious waves. Second, appreciating the beauty of nature may also foster a deeper emotional connection that encourages us to protect and cherish the natural world. This balance between idealization and genuine care is crucial in our relationship with nature. Far from the romanticizing and glamourizing nature we can discuss the importance of experiencing beauty. The poet and author David Whyte has written about beauty in his essays "Consolations", that "beauty is the harvest of presence" – "Beauty is an achieved state of both deep attention and self-forgetting; the self-forgetting of seeing, hearing, smelling or touching that erases our separation, our distance, our fear of the other..." (Whyte 2014). In this understanding beauty has nothing to do with something being pretty, harmonic, or nice-looking. Beauty is much closer to the phenomenon of resonance and our possibility of enhancing our presence, connectedness, and love for the awesome nature of the planet.

Conclusion

Our existence is deeply tied to the health and sustainability of the planet. If we become alienated to this fact, or become overwhelmed by despair, we risk turning our backs on both the urgency to act *and* the need to listen. On all levels it seems that *listening is the act* that precedes every other step that we will have to take to provide resonance, and ensure futures for the coming generations of humans and other living species on Earth.

Notes

1 Music: Pärt, "Spiegel im Spiegel".
2 https://climatemusic.org/.
3 Bach: Piano concert, F major. Largo.
4 Music: Albinoni, adagio in G minor.
5 Music: Rachmaninoff, "Vespers, Op. 37: VI. AveMaria".

References

Abram, D. (1996). *The spell of the sensuous: Perception and language in a more-than-human world.* Vintage.

Anonymous. (2015, 09-30). Music as a torture weapon: Exploring the dark side. *Guardian*. https://www.theguardian.com/music/tomserviceblog/2015/sep/30/bleak-history-music-as-torture-suzanne-cusick-morag-grant

Bonde, L. O., & Beck, B. D. (2019). Imagining nature during music listening. An exploration of the meaning, sharing and therapeutic potential of nature imagery in guided imagery and music. In E. Pfeifer (Ed.), *Natur in Psychotherapie und Künstlerischer Therapie Theoretische, methodische und praktische Grundlagen* (pp. 145–168 (Vol. II)). Psychosozial-Verlag.

Bonde, L. O., & Vist, T. (2013). "Then certain songs came": Music listening in the grieving process after losing a child. In L. O. Bonde, E. Ruud, M. S. Skånland & G. Trondalen (Eds.), *Musical life stories: Health musicking in everyday life* (pp. 139–164). Norges musikkhøgskole.

Bonny, H. L. (2002). *Music & consciousness: The evolution of guided imagery and music* (Summer L., Ed.). Barcelona.

Bonny, H. L., & Savary, L. M. (1973). *Music and. your mind: Listening with a new consciousness*. Harper & Row.

Carson, R. (1965). *The sense of wonder*. Harper & Row.

Christensen, E. (1992). Fuglene synger videre. *Dansk Musikterapi*, *67*(1), 16–19.

Christensen, S. M. (2021). Kunst er et særligt erkendelsesfelt, der har potentialet til at spille en afgørende rolle i verden. Essay*Art Matter*. https://artmatter.dk/journal/kunst-er-et-saerligt-erkendelsesfelt-der-har-potentialet-til-at-spille-en-afgoerende-rolle-i-verden/

Gabrielsson, A. (2011). *Strong experiences with music: Music is much more than just music*. Oxford University Press.

Ihde, D. (2007). *Listening and voice: Phenomenologies of sound* (2nd ed.). State University of New York Press.

Kenny, C. (2006). *Music & life in the field of play: An anthology*. Barcelona Publishers.

Knausgård, K. O. (2020). *Morgenstjernen*. Lindhardt og Ringhof.

Lipe, A. W. (2002). Beyond therapy: Music, spirituality, and health in human experience: A review of literature. *Journal of Music Therapy*, *39*(3), 209–240. doi: 10.1093/jmt/39.3.209

Monroy, M., & Keltner, D. (2023). Awe as a pathway to mental and physical health. *Perspectives on Psychological Science*, *18*(2), 309–320. doi: 10.1177/17456916221094856

Rosa, H. (2017). Available, accessible, attainable: The mindset of growth and the resonance conception of the good life. In *The good life beyond growth: new perspectives* (pp. 39–53). Taylor and Francis. doi: 10.4324/9781315542126

Rosa, H. (2019). *Resonance: a sociology of our relationship to the world*. Wiley.

Rosa, H. (2024). *Når monstre brøler og engle synger: heavy metal og resonans* [When monsters roar and angels sing] (G. Damgaard Trans.). Eksistensen.

Schuurman-Olson, S. (2023). Singing in the dark: Seeking connection in a time of climate crisis. *The Canadian Music Educator*, *64*(3), 34–41.

Tsiris, G. (2017). Music therapy and spirituality: An international survey of music therapists' perceptions. *Nordic Journal of Music Therapy*, *26*(4), 293–319. doi: 10.1080/08098131.2016.1239647

Whyte, D. (2014). *Consolations: The solace, nourishment and underlying meaning of everyday words* (Revised ed.). Many Rivers Press.

15
THE SENSE OF WONDER AS KEY TO ECO-BILDUNG

Finn Thorbjørn Hansen

Introduction

This chapter will focus on the sense of wonder as a key to understanding what *Eco-Bildung*[1] is and how it can be nurtured in practice. When talking about "Eco-Bildung" it lines up with former and contemporary attempts to describe what "Ecological Literacy" is (Orr, 1992, Orr & Stone, 2005; Paulsen et al., 2022), and why "Education for Wonder" and "Biocentric Wonder" are essential to education for a sustainable future (Washington 2018, 2019; Schinkel 2021). However, I will argue that these attempts can be qualified further, philosophically and phenomenologically, by putting a stronger emphasis on the importance of *passive-receptiveness* and a sort of *non-willing* and *non-doing* akin to what the German Christian Mystic Meister Eckhart called *Gelassenheit*, and what the Daoist calls *Wu Wei* (effortless action). What I want to develop is a new view of Eco-Bildung, seen from an eco-phenomenological perspective, that draws on the concept of *Bildung* by Meister Eckhart, the reception of *Gelassenheit* in late Heidegger, and the ways new Daoist eco-phenomenologists (Girardot, Miller & Xiaogan, 2001; Chai, 2020, 2022; Nelson, 2021, 2024) describe the special kind of nature connectedness that follows the practice of *Gelassenheit* and *Wu Wei*.

What holds these three approaches together is an intersection of wonder, love, and a strange sense of being at home in the world and befriending the world. The track I want to follow is to inquire into how the sense of wonder is a portal or way into a loving, caring, and befriending relationship with the world. Not the world of humans especially. On the contrary, humans are only a tiny (but damn disturbing and noisy) part of the living world and life on this planet. When I say "world", I mean the "the more-than-human-world" (Abram, 2017) – that is earth and the species and lifeforms (plants, animals, and nature elements) that modern man seems to have gotten "out of tune" with.

Eco-Bildung is a sort of "hope-concept" for a "Post-Anthropocene culture" where humans in the damaged Anthropocene Age (which may follow for the next two centuries or more because of long-term pollution and damages to the Earth's eco-system and biodiversity) can begin to live a more non-anthropocentric, humble, reverent, friendly, and respectful way of life with the other species on this planet. You could call this Post-Anthropocene culture and ideal of Eco-Bildung a beginning attempt to create more life-wise, life-friendly, and existential sustainable life forms between humans and other "earthlings".

 DOI: 10.4324/9781003538097-19

To have a sense of this kind of contemplative aesthetical, philosophical, and apophatic wonder (Schinkel, 2021; Hansen, 2023, Han, 2024) will be seen as key to this form of preparation for a Post-Anthropocene Culture and for nurturing Eco-Bildung.

In what follows, I will start with outlining the broader picture found in the theories of Environmental Humanities and Environmental Ethics where they point to the need to understand the radical change from living in a Holocene worldview (sustainability in the 1st Degree), to an Anthropocene worldview (sustainability in the 2nd Degree), which for some eco-philosophers (like myself) and environmental researcher also calls for a Post-Anthropocene worldview and, finally, a new and reverent culture (sustainability in the 3rd Degree). From there, I briefly describe how different environmental researchers and educators have talked about "eco-literacy" and "eco-bildung" with a special eye for those who place the sense of wonder in the center of eco-literacy or eco-bildung. This leads to the third part where I will dwell on the phenomenology of contemplative wonder and why this world-relation seems to open up for an "other-acknowledging" and befriending of the world and a love for and strange feeling of being at home in the world. In the fourth part, I show how some Heideggerian and Daoist eco-phenomenologists have been inspired by the concept of *Gelassenheit* and *Wu Wei*, and I relate the sense of wonder to their thinking. In the last and concluding part of the chapter, I point briefly to some possible ways of practicing wonder-evoking eco-Bildung and why this should, in my mind, be closely related to the concept of "existential sustainability" and "education for planetary well-being" (Aaltonen et al., 2023).

From a Holocene to an Anthropocene and Post-Anthropocene Worldview and Culture

Let me start with a concrete case. At a hospital in Denmark, the management wanted a special task force to develop a new sustainable climate strategy for the hospital. The members of this task force had a framework for understanding climate change and sustainability that was closely related to the United Nation's 17 Sustainable Goals. They encourage us to act in sustainable ways, which should ideally and in practice cover and contain three dimensions. (1) *Climate sustainability*: safeguarding the precondition for life: for example, do not use plastic gloves, driving in electric cars, etc.; (2) *economic sustainability*: circular economy, recycling, etc. and (3) *social sustainability*: human rights and being aware of social welfare, work–life balance, and equality among gender and ethnicity.

However, some in the task group were worried that they would not be able to "move" the hospital's more than 6.000 leaders and employees to work wholeheartedly for a sustainable hospital, if these people were not motivated to do so on a deeper, more emotional level. How, they asked, do we create such a radical mind shift? How do we enhance deep motivation to fundamentally change their daily work life habits and maybe also their living habits at home to become more sustainable?

I was invited to give some suggestions, and after listening to the task force's way of understanding and reflecting upon what to be done to create a more regenerative and sustainable mindset, I discovered that they primarily worked on three levels of reflection. These three levels of reflection can be described as:

First-order reflection. **Concrete problem-solving focus** here and now: How can we now and very concretely in our department do waste sorting more effectively?

Second-order reflection. **Problem-reflectivity**: How can we educate employees to become better at, and more innovative in, reflecting upon different ways of solving concrete problems of sustainability at the hospital?

Third-order reflection. **Systemic reflections**: Which systemic, political, structural, and management conditions, possibilities, and challenges are there, so that we can better solve first- and second-order problems?

However, when listening to regenerative and environmental researchers, designers, and eco-philosophers with a background in eco-feminism, deep ecology, or regenerative thinking, we learn that we must also be working with other levels of reflections than these three. Steven Harding, for example, talks about three fundamental dimensions when wanting to create an ecological change in human mindset and life forms. These are: (a) deep impression, (b) deep questioning, and (c) deep commitment (Harding, 1997, 2020). He claims that if you want people to radically change their lives and work style in a more ecological, regenerative, and sustainable way, you need to first let them experience a deep impression. That is, something in their own life, a life experience, that went straight to their heart, and connected them to or pushed their core values and life view. Next, you need to facilitate or give space and time and acceptance, and encouragement for deep questioning. This means that it is expected of the participant to be curious, enquiring, pondering, and dwelling in wonder about or at the deep impression they had. Finally, Harding suggests that this will lead these persons into a deep motivation for acting and making changes in their life, so they can be truer to the core values and new insights to which the deep impression and deep questioning have connected them to.

Now, how can we develop a sort of thinking and being in the extension of the former three levels of reflections mentioned before, that cultivate such deep impression, questioning, and motivation? My suggestion was to the task group at the Danish hospital that they also work with two more levels of reflection. These are:

Fourth-order reflection: **Existential reflectivity**: When reflecting from within a lived experience that made a special impression on you, what kind of core values are tacitly embedded in this lived experience? Which "lived values" guide your life and work? When solving concrete problems at your work, what is the deeper ethical or existential meaning of doing this? In which ways does this problem-solving give you and your colleagues a deeper sense of meaning? Why should you be proud when telling your grandchildren what you did at work?

Fifth-order reflection: **Ontological Awareness** (or Eco-Tonality): When listening to your heart, do these problem-solving activities help you in your everyday life to get into resonance with what you existentially long for and are deeply connected to? How do we learn to listen to the call of being, of what in this moment and this situation *calls* us to do with our whole person? How do we train this "musicality"? Is it not strange that sensing the presence of nature in these inspiring moments often makes us feel more attuned to life and to what matters in life?

Fourth- and fifth-order reflections have to do with, what I describe as "*existential sustainability*". By existential sustainability, I mean two things: (1) the ability to create a type of reflection and awareness that focuses on the existential (*meaning-making*) and spiritual (*meaning-receiving*) dimensions in life and work and human relationships with each other and with the more-than-human world; (2) the ability and care to "let be", to let be what is, and to have a sense of and musicality to hear the voices of the more-than-human species and elements on the earth.

"Eco-tonality"[2] is, I think, I nice word for describing the meaning–receiving–dimension and the ontological resonance and musicality you need to cultivate to become aware of, what I elsewhere have termed the "apophatic callings" in life (Hansen, 2022, 2024c). By that, I mean callings in life that evoke deep wonder and a strange feeling of connectedness with the world or the phenomenon that you wonder at.

"Apophatic" means the acknowledgment of the ineffable nature of these callings; they cannot be scientifically explained or philosophically defined in clear logical and epistemological languages. Apophatic callings are connected to the unsayable (Rhodes, 2014; Franke, 2014, 2018) and yet, they say so much to us, like speaking straight into our hearts. Typically, the language of fifth-order reflection will resort to poetic, symbolic, metaphorical, or other forms of aesthetical, religious or spiritual expressions and practices. Whereas the language of fourth-order reflection will be more philosophical in its expressions, not in an analytical academic sense, but in an existential and contemplative way.

Eco-tonality in its ontological interpretation, as I present it here, seems to be in kinship to what the British eco-philosopher Jonathan Bate term as "ecopoetic" when he describes what it means to dwell on earth or hear the "Song of the Earth" (Bate, 2000), suggesting a deeper ontological engagement with nature beyond more utilitarian, scientific and political interests, and concerns.

The dynamic or dialectics between fourth- and fifth-order reflection is that in fifth-order reflections, you are becoming so quiet and open and listening that you can hear the "Call of Being" (Heidegger, Marcel). In fourth-order reflections, you then try to give a personal answer or response (thoughts, feelings, or actions) to this calling. You could say that in fourth-order reflections, you prepare yourself (through deep questioning) to become open and receptive (through fifth-order reflections) to the calling (deep impressions), which then, in consonance, foster a deep motivation to act in resonance with this calling.

My point to the task force at the Danish hospital was that if they wanted to raise a deep ecological and sustainable mindset in the leaders and employees of the hospital, these two further levels should be added. Doing so might create a regenerative mindset and motivation on a deeply personal level, but it may also create a new "eco-professionalism" (Hutchins, 2022; Barnett, 2023) where their professional reflections, methods, and actions become more attuned with a connectedness to nature, or Being, as such, as the Late Heidegger would add, as we will see later in this chapter.

Connecting eco-tonality (fifth-order reflection) to eco-professionalism creates an interesting analogy. Normally, professionalism is understood as something skillful and competent, which technicians, craftsmen, clerks, consultants, or industrial innovators do – led by first-, second-, and third-order reflections, whereas the artists, when creating an art piece, go beyond the merely technical and craftsmanship approaches. An artist needs to have the necessary "artistic inspiration and talent" to make a product into an art piece. Likewise, one could say that to get into a deep motivation for working and living more regenerative and sustainable one needs to leave the instrumental problem-solving attitude and utilitarian production attitude (first- to third-order reflections) to become more like an artist with both an existential-philosophical and aesthetical-contemplative attitude (fourth- to fifth-order reflections).

What I suggested to the task force was, therefore, workshops ("Wonder Labs", Hansen, 2015, 2024b) that may foster this kind of fourth- and fifth-order reflections.

To sum up, the three forms of sustainability: climate, economic, and social sustainability need a fourth one – *Existential Sustainability* – that has to do with the *existential motivation* and *ontological relation* (or eco-tonality) with nature (or the apophatic callings of Being, cf. Heidegger).

Now, how do we create or facilitate spaces for fourth- and fifth-order reflections in practice? What is a Wonder Lab? What kind of "eco-literacy" or eco-Bildung can foster this kind of existential reflexivity and ontological awareness (or eco-tonality)?

The Sense of Wonder as Central in Eco-Literacy and Eco-Bildung

When discussing eco-literacy and eco-Bildung, we need first to reflect upon the difference between *literacy* and *Bildung*. Briefly, literacy refers to a positive ability to read, write, and communicate effectively and in according to the rules of grammar and spelling within a given language. It is a set of skills and competencies needed to function in society. Whereas Bildung, on the other hand, is a broader German concept that encompasses personal growth, self-cultivation, and not mainly a harmonization of the individual within communities (socialization) but rather with the world or life as such (Grinberg et al., 2022).

There are existential and spiritual dimensions connected to Bildung that are not contained in the concept of literacy. Literacy is more skills-oriented, while Bildung is a holistic personal and cultural maturation process. Literacy can be seen as a component of Bildung, whereas Bildung goes beyond just acquiring skills and capacities to include character formation, development of one's humanity, and a sense of transcendence. Literacy is often assessed through measurable outcomes, while Bildung is more of a philosophical concept focused on the journey of transcendence-inspired self-cultivation. In short, literacy connects to mainly positivistic quantitative knowledge, while Bildung connects to phenomenological qualitative knowing, or rather, to be more precise, to a certain "negative wisdom", that is, to a source for seeing and listening to what cannot be directly seen, heard, measured or "used", but sensed through art, philosophy, theology, and other wisdom traditions and spiritual practices.

In the first book made on *Ecological Literacy* (Orr, 1992) David Orr describes ecological literacy as an ability to ask, "What then?" and not just to read, count, and compute, but more importantly, ask why and to what greater purpose should we do this reading, counting, and computing. Thus, in the concept of eco-literacy, Orr has a philosophical bildung-dimension of eco-wisdom built into his concept. This becomes even more visible when he continues to write that some attitudes are especially central to eco-literacy. One of these is the sense of wonder, he claims: "If literacy is driven by the search for knowledge, ecological literacy is driven by the sense of wonder, the sheer delight in being alive in a beautiful, mysterious, bountiful world" (Orr, 1992, p. 22).

Orr points out that the sense of wonder is rooted in emotions or what E.O. Wilson (1984) has called "biophilia", that is, the love for life as such. In this sense, the phenomena of wonder and love are closely related.

Orr also refers to the American and iconic environmental researcher and activist Rachel Carson, author of *Silent Spring* (1962), who, in her very last book published one year after her death (as a kind of eco-pedagogical testament to coming generations) exactly points to the sense of wonder as essential for ecological education. The title of the book is *The Sense of Wonder: A Celebration of Nature for Parents and Children* (1998/1965). In this small and beautifully illustrated book, she writes:

> If a child is to keep alive his inborn sense of wonder without any such gift from fairies, he needs the companionship of at least one adult who can share it, rediscovering with him the joy, excitement, and mystery of the world we live in.
>
> (Carson, 1998, p. 45, 49)

She saw, similar to Wilson, a very close connection between the sense of wonder and the love of, and care for nature. In her book *The Sense of Wonder*, she also points to the healing dimension of the sense of wonder:

> Those who contemplate the beauty of the earth find reserves of strength that will endure as long as life lasts. There is symbolic as well as actual beauty in the migration of the birds, the ebb and flow of the tides, the folded bud ready for spring. There is something infinitely healing in the repeated refrains of nature – the assurance that dawn comes after night, and spring after the winter.
>
> (Carson, 1998, pp. 99–100)

Linda Lear writes in her introduction to the book, that Carson here shares her prescription on how to nurture a lifelong sense of wonder, how to maintain the freshness with which we see the natural world for the first time, and how to preserve awe and wonder in lives lived too often in opposition to nature. "Help Your Child to Wonder" was the title of the first draft of this book where she, as Lear writes: "…in these brief pages captures the essence of the wonder-filled world of children and stirs in us that ancient longing for unity with the living world" (Carson, 1998, p. 9).

Carson's courageous fight for higher ecological consciousness and for practical and political actions to protect nature and nonhuman life has since been of huge inspiration for generations of environmental researchers, activists, environmental ethicians, and eco-philosophers. Her emphasis on the centrality of the sense of wonder to create deep commitment has been taken up by modern and contemporary environmental researchers, activists, philosophers, and educationalists like, for example, Kathleen D. Moore and Haydn Washington.

Moore (2008) points to the moral significance of wonder and refers to a speech made by Carson in 1954, where Carson says: "I believe that the more clearly we can focus our attention on the wonders and realities of the universe about us, the less taste we will have for destruction" (Carson, 1954 – quoted in Moore, 2008, p. 28) Moore argues that the sense of wonder should be a central virtue and cornerstone in environmental ethics: "I think that a sense of wonder is a virtue in at least this sense, finding what it means to be fully human in a celebration of our place in the more-than-human world" (Moore, 2008, p. 273). With Carson she also points to the capacity and sense of wonder as the "… fulfillment of our potential as human beings in our awareness of the astonishing world, our care, and our thanksgiving" (Moore, 2008, p. 42). We shall later see how she phenomenologically makes a distinction between being curious, astonished and amazed and why she also connects wonder to friendship or a befriending of the world.

In 1974, an Australian young man went bivouacking with three friends in the Australian bushes. One early morning, this young man woke up in his sleeping back and stared right into the deep black eyes of an Australian Lyrebird. He felt that, at that moment, there was no fear in him or the bird, there was contact, a deep connection between them. It was a wonderful moment, he remembered. It was today's internationally known environmental researcher and activist, Haydn Washington. "We [the Lyrebird and him] existed at this moment and time, and the gulf between our histories and separate evolution was gone" (Washington, 2019, p. 2). He writes that at that moment, the sense of belonging and harmony was overwhelming. In and just after the moment, he was not only in a deep caring mode for the Lyrebird but also for the whole landscape and place where this bird belonged.

> I didn't know it then – but my life had just changed forevermore. Whatever communication passed between the Lyrebird and me, it made me feel I would have to speak for a place that had no human voice. At that moment and in the following sun-drenched days spent in that sandstone gorge, I fell in love with the land. Absolutely and indelibly. My sense of awe and wonder at the beauty of this place stunned me. I kept smiling at the sheer wonder of its existence. That moment launched me on five hectic years fighting to save the Colo Wilderness from dams, power-lines, possible power stations, and coal mines.
>
> (Washington, 2019, p. 2)

This event became life-changing for him, and several other deep impressions continued in his life as a mature man and environmental researcher activist. In the article "Education for Wonder" (Washington, 2018) and book *The Sense of Wonder Towards Nature: Healing the Planet through Belonging* (2019), he emphasizes the importance of enhancing a sense of wonder in education from kindergartens to universities. As he writes:

> ...We need to educate effectively to retain and encourage a sense of (deep) wonder. Given that people of all religions (and philosophies) can experience a sense of wonder at nature (and its often life-changing effects), it would be perfectly appropriate if all religions, secular beliefs, and education systems sought to rejuvenate society's sense of wonder. Abram observes that a "felt mystery" (Abram, 2010, p. 278) underlines all society's doctrines, and this should unify many social groups and beliefs.
>
> (Washington, 2018, p. 2)

What Is a Deep Contemplative Wonder (*Thaumazein*) and Its Relation to Love (*Philia*)?

Washington, in the quotation above, writes in front of the word "wonder" in brackets "(deep)". This is important, I think because the kind of wonder that has the healing effect of which Carson speaks and its ability to make us feel at home in the world need to be distinguished from other forms of wonder and wonderments.

When talking about deep wonder we could at first say, that this kind of wonder is connected to "contemplative wonder" as opposed to "inquisitive wonder" (Schinkel, 2021) and "Enchantment" (Curry, 2020, 2023).

Inquisitive wonder is knowledge-, cause-, and explaining-seeking wonderment, which we can find both in the more pragmatic and utilitarian problem-solving reflections as well as in the higher scientific forms of wonderment in natural and social science. You could say that it is basically the inquisitive wonder that you will find in first- to third-order reflections mentioned earlier in this chapter. Of course, you will also, in some parts of human science, find this kind of inquisitive wonder, but what differentiates the research in the Humanities (in the discipline of liberal arts, philosophy, theology, etc.) from natural and social science, as Gadamer makes clear in *Truth and Method* (Gadamer, 2006/1960), is the living *experiences* of philosophy, art, and of history itself (cf. human stories through millenniums of lived experiences and of what matters and what it is to be a human and in life). These special experiences (to be caught in an artistic moment of creation, to be immersed in a philosophical wonder, to be grasped by the spoken word in the life story) have a wondrous ability to touch upon a "truth experience" of a more existential, phenomenological-hermeneutic and apophatic nature. "These [the experiences from philosophy, art, history] are all modes of experience in which a truth is communicated that cannot be verified by the methodological means proper to science" (Gadamer, 2006, p. xxi).

When I say that contemplative deep wonder cannot be sufficiently described as "enchantment" it is because the emotion of enchantment most often is limited to reaching the level of a Wow-experience. Art, when it is not deep – but only cleverly tailoring your emotions and feelings with "artistic tools" – can easily create this enchanting wow effect in us. However, this is not the ontological and eco-spiritual (meta-physical and meta-psychic) event that deep contemplative wonder can bring us into. Heidegger writes in his book *The Origin of the Work of Art*:

> Experience is the source that is standard not only for art appreciation and enjoyment, but also for artistic creation. Everything is an experience. Yet perhaps experience is the element in which art dies. The dying occurs so slowly that it takes a few centuries.
>
> (Heidegger, 1975, p. 79)

Deep contemplative wonder is, you could say, in contact with an archaic (mystical) impression of the phenomenon or impulse (*die Sache Selbst*) that the artist is *trying* to express in an aesthetic form and art piece. That is *before* this impression becomes a sensuous human perception and living experience for the person who senses it. Art and the artistic experience die, as Heidegger indicates, if it is not in resonance with this mystical impression before it coagulate into a "lived experience" for the perceiver. The deep contemplative wonder, on the other hand, is connected to this mystical impression, which is not perceived yet on the horizon of the perceiver, but is in a "coming forth movement". Deep contemplative wonder lives by and through the resonance with this mysteriously "pre-experience", or "pre-sensing". Or, as Paul Cezanne formulates this enigmatic artistic impulse or calling: "What I am trying to convey to you is more mysterious, it is entwined in the very roots of being, in the impalpable source of sensations" (Cezanne – quoted in Merleau-Ponty, 1993, p. 121).

The deep contemplative wonder is a metaphysical experience. It is an experience of transcendence entangled *in* the immanence. This sort of wonder is precisely connected to the apophatic truth experience, which Gadamer and Late Heidegger talked about. The apophatic callings (of that, which is on the way to language) cannot be grasped epistemologically, in representative language, or verified methodologically through natural and social science, nor through what we psychologically understand as our subjective, lived experiences and psychic-based feelings.

You could also say that contemplative wonder is wisdom-, meaning- and wonder-seeking wonderment, which emerges through a fundamental mood or attunement (*Grundstimmung*) of leisure (Pieper, 1963) or as a deep sense of presence – as Being-there (Heidegger, 1962/1927). Contemplative wonder displaces the cognitive and affective everyday ego as well as the professional and scientific ego. It breaks with what the phenomenologists in general call "the natural attitude" (Husserl, 1982), which covers both the tacit taken-for-granted everyday understandings of and perceptions in our lifeworld, as well as when we scientifically still maintain a subject-object distinction, positioning ourselves as detached observers over and against a world of objects.

To be in a contemplative wonder is, as the great phenomenologist Eugen Fink and later Merleau-Ponty reminds us, intrinsically connected to the "phenomenological attitude", or to be more precise, to the so-called "phenomenological reduction" (Merleau-Ponty, 2007). The phenomenological reduction helps the phenomenologist to get in tune or resonance with the phenomenon in and by itself before it coagulates and becomes a perception or concept, as it were, in front of the intentional inquisitive and explanation-seeking eye (Marion, 1998).

In *The Poetics of Space*, the French phenomenologist Gaston Bachelard refers to the Dutch phenomenologist J.H. Van den Berg, who said: "Poets and painters are born phenomenologists" (van den Berg – in Bachelard (2014, p. 13). Bachelard wants to turn our attention towards the event that things and living phenomena "speak" to us. As a result, if we give this language (the languages of things and phenomena) its full value, "...we have a contact with things" (2014). What we need to learn, he adds, is to get into such contact and dialogue with the phenomenon as a "speaking being".

The problem then is to approach the otherness of the living meaning and language of the phenomenon, we want to get in resonance and dialogue with so that the meaning and language we hear are not already "dead meaning" killed by too anthropocentric perceptions, expressions, and languages. How do we get from the poor or derivative phenomenon, which only appears in front of the eye of the "natural attitude" – to the rich (saturated) or pure phenomenon (Marion, 2002a), where the phenomenon is released to speak with its own voice and radical otherness and we only become a witness or receiver?

We can, Marion (2002b) suggests, through deep (and not shallow) art and poetry get a stronger musicality for and receptiveness to hear the calling or voice of the phenomenon in and by itself. Hence, the sensuous and aesthetical wonder that the artists are captured by and create from, might give us one of the openings to an encounter and resonance with the phenomenon in itself. Philosophy can likewise – when a deep philosophical wonder drives this kind of philosophizing – get us in resonance with the phenomenon. Spiritual or religious experiences of awe and wonder can also pave the way to contact and dialogue with the phenomenon that ignited the deep and religious or spiritually sensed wonder. A kind of awe and wonder that is by no means only for a privileged few artists, philosophers, or spiritual practitioners, but can be experienced unexpectedly and suddenly and often on a volatile and prereflexive moment by children, young people, adults, and the elderly in everyday life, as the American psychologist Dacher Keltner (2023) has observed.

Therefore, it seems to me, that when we now reflect on how to understand different forms of deep contemplative wonder, we need a distinction between an aesthetical, a philosophical, and a spiritual/religious kind of contemplative wonder. Who and where are we, when we are "taken" by either an aesthetical, philosophical, or spiritual/religious wonder? In which way or ways are these different contemplative wonders and wonderments to be understood in relationship to each other and indeed to nature or the sense of belonging to nature?

If an essential dimension of deep contemplative wonder, as, for example, Haydn Washington experienced it, is a kind of ontological homecoming saturated with love and care for the place and phenomenon he experienced in the Australian bush, then how could an aesthetical, philosophical, and spiritual/religious approach to deep wonder be put into dynamic play in an eco-educational setting? How can these three forms of deep wonder interact and supply each other in an "Education for Wonder" or a wonder-based "education for planetary well-being"? How are we to understand the sense of wonder as key to Eco-Bildung if we want to think with an aesthetical, philosophical, and spiritual/religious wonder dimension?

Dubious Eco-Phenomenological Readings of Heidegger's Philosophy

In the pioneering 2003 book *Eco-Phenomenology: Back to Earth Itself* (Brown & Toadvine, 2003) some authors (including Michael Zimmerman) reflect on Heidegger's understanding of the human relationship to the world and the earth as an event of *Dasein*. *Dasein*, as defined by the early Heidegger in *Sein und Zeit*, is constitutive of how phenomena and the world as such appear to us. In Heidegger's early thinking, the fundamental mood of anxiety constitutes *Dasein*. Here, Heidegger posits that it is only through confronting anxiety, the felt sense of meaninglessness, and embracing the finitude of all values in life that we can achieve a true, authentic, and genuine personal existence and engagement in the world. This entails a recognition of anxiety as the most fundamental world-relation a human being is grounded upon and a willingness to look this nihilistic experience of nothingness straight into its eyes as inherent to our basic human condition.

In other words, in *Sein und Zeit*, the early and "existentialistic" Heidegger posited that the feeling of homelessness and uncanniness (or *Unheimlichkeit*) and lack of connectedness to the world and nature are constitutive of being human. To exist as a human being is to live in a kind of exile; to exist and live authentically is *not* to feel at home in the world but in this terrifying and anxiety-provoking exile. Consequently, many modern eco-phenomenologists choose to distance themselves from Heidegger's philosophy of existence, as it is articulated in *Sein und Zeit*, due to its lack of provision for a fundamental connection to nature and a basic sense of belonging in nature and life as such.

Another crucial aspect of Heidegger's philosophy of existence, as highlighted by the Spanish eco-phenomenologist Michael Marder (2013), is the absence of a conceptual framework for understanding the lives of entities that exist beyond the human realm, including plants, animals, minerals, and the elements of the earth. Heidegger, Marder asserts, renders the world and its myriad of living other-than-human existences "worldless". Only humans can "exist" and "have a world". However, this assertion is, Marder claims, open to question. For example, plants "have" their own world. There is, Marder notices, no evidence that Heidegger was aware of this. Even when we move from Heidegger's philosophy of existence in *Sein und Zeit* to the late Heidegger's philosophy of Being, the question remains unanswered if we are to follow Marder's reading of Heidegger.

In my view, the first eco-phenomenological critique of Heidegger is unduly narrow in its interpretation of the philosopher as an existentialistic thinker and lacks sufficient insight into Heidegger's later philosophy of Being. It is important to note the distinction between Heidegger's philosophy of existence and his philosophy of Being. As Richard Capobianco, an expert in the field of Heidegger studies, has persuasively argued (Capobianco, 2011, 2015, 2020, 2022), an examination of the late Heidegger's work on Being and care of the self reveals a fundamental shift in his approach.

While earlier works on the existential philosophy of Heidegger emphasized the primacy of *Dasein* and anxiety and "meaning-making", the later works demonstrate a shift towards a more comprehensive understanding of Being and care of the self as a fundamental form of being-*with*-life in a *wonder*-based, life witnessing and meaning-receiving way (Capobianco,

2011). From the mid-1930s onwards, Heidegger began to describe *Beyng* (spelled *Seyn* in German) as the deepest level of nature (in Greek: *Physis*). "That [which is marked by the word *Physis*] is the overabundance, the excess of what presences. Being lets beings come to presence" (Heidegger, 2012/1969, p. 59). In this thinking, the fundamental mood (*Grundstimmung*) is no longer anxiety but wonder (in Greek: *Thaumazein*), as Heidegger says "The Greeks name the relation to this thrust of presence *thaumazein*" (Heidegger, 2012/1969, p. 38). Through the fundamental contemplative wonder, when human beings experience openness and proximity to the world as a felt mystery, they can undergo an "ontological homecoming." This is characterized by a sense of connectedness to nature and even moments of encounter with the sacred (Capobianco, 2015, 2022). This understanding of the sacred is not aligned with traditional religious interpretations. Instead, it is an agnostic and apophatic experience of the sacred as something ineffable and also theologically inexplicable. It is, therefore, incumbent upon contemporary eco-phenomenologists (Harvey, 2009; Kennedy, 2014) to consider the late Heideggerian concept of nature (*physis*) as an apophatic concept and phenomenon that is, in fact, far less anthropocentric than the human–nature relationship described in *Sein und Zeit*.

The second criticism, as represented by Michael Marder (2013), is, in my view, more pertinent. It seems that the late Heidegger is aware of the non-anthropocentric perspective yet does not explicitly link it to the concept of "the more than human life." Instead, he appears to associate it with an abstract notion of existence in and of itself. To counter this limitation in Heidegger's linguistic approach to the more-than-human life, some eco-phenomenologists (Bannon, 2014) prefer to consider thinking with the French phenomenologist Merleau-Ponty and his conception of nature, although this is not the focus of the present discussion.

Eco-Bildung in a Late Heideggerian and New Daoist Eco-Phenomenological Perspective

Late Heideggerian and new Daoist eco-phenomenologists are especially concerned with the concepts *Gelassenheit* and *Wu Wei*, respectively.

Gelassenheit can be translated as "releasement" or "letting go" and "letting be". The Daoist concept of *Wu Wei* can be translated as "effort-less action". Both the Heideggerian and New Daoist eco-phenomenologists claim that in order to hear the voices of nature and "the more than human life", we need as humans to radically get out of our anthropocentric knowing, willing, and doing. We need to get into a Bildung process of becoming empty of our human-constructed knowledge (through an attitude of deep wonder), and let go of our self-centered mind and project-creating attitude and will to do and perform, to just "be", do nothing (with utilitarian or scientific orientations), let be and let come what may, and to stop being always a pro-active "doer".

The German philosopher Joseph Pieper describes this as to be in a state of contemplative leisure – as the basis of culture, which has a sense for wonders and experiences of transcendence. Leisure (in contrast to utilitarian and scientific work) is a spiritual attitude that implies "…an attitude of non-activity, of inward calm, of silence; it means not being 'busy', but letting things happen. Leisure is a form of silence which is the prerequisite of the apprehension of reality" (Pieper, 1963, p. 46).

This kind of contemplative leisure understood as non-utilitarian and non-explaining-seeking activity inspired by what calls the soul (not the ego) to act (or think or sing), is very analog to what the New Daoist eco-phenomenologists describe as *Wu Wei*, which translates to "effort-less action" led by Dao. Dao is here understood as the way to get into resonance with the ineffable existence of nature experienced as a mysterious life-nurturing and always life-renewing force in the very center of earthly life. New Daoist eco-phenomenologists like Chai (2020, 2022) and Nelson (2021, 2024) see clear connections between, what Heidegger describes as *Gelassenheit* and *Physis* as a hidden nature (Heraklit: "Nature loves to hide") on the one hand, and on the other hand, what Daoism describes as *Wu Wei* and *Dao* as the way to a hidden nature (*Ziran*), behind or beyond the version of nature that can be seen by the pragmatic and scientific mind. It is a kind of seeing with the heart, that both the Late Heideggerian and New Daoist eco-phenomenologists are pointing to as pivotal for a new and non-anthropological understanding of, and living relation to, nature, and with the living species on this earth.

The eco-theologian Sam Mickey (2019) also points to this connection between Daoism and the mystical element in Late Heidegger's thought. The root of this thinking in Western history is, as he shows, to be found in the German Christian medieval mystic Meister Eckhart. Eckhart was also the first theologian and philosopher, who terms *Bildung* as a spiritual self-transformation. Let us, therefore, dwell briefly on the thoughts that Eckhart has on the concept of *Gelassenheit*, *Bildung*, and what Eckhart describes as a "Wonder Without a Why", which will bring us back to look at the sense of wonder in Eco-Bildung in a more eco-spiritual way.

Meister Eckhart on Gelassenheit, Bildung, and Without-Why-Wonder

Gelassenheit for Eckhart encapsulates a profound spiritual attitude characterized by detachment and openness to God's will. Eckhart emphasizes that *Gelassenheit* involves releasing one's desires and ambitions, allowing individuals to become receptive to divine influence. It signifies a state of inner peace, free from personal will and ego. This attitude is – as we shall see developed through Eckhart's concept of *Bildung* – not merely passive-receptive; it actively fosters a sense of freedom and forgiveness, enabling individuals to let things be as they are without unnecessary interference or judgment. But the kind of "activity" or agency is not, Eckhart underlines, ignited through the human ego and will, but through a special "giving-in" to what is calling at the moment.

Eckhart's understanding of *Gelassenheit* also aligns with the idea of "poverty of spirit", where true fulfillment comes from aligning one's will with God's, thus achieving a harmonious existence rooted in humbleness, not-knowing ("*docta ignoratia*", learned unknowing, as it is later described by one of Eckhart's theological successors – Cusano), mercy or grace. In this way, *Gelassenheit* was, for Eckhart, essential for experiencing the mystical union with God that he advocates. In short, what Eckhart is calling God, or better the Hidden Godhead behind human's idols of God (Carabine, 1995), can be read as what Late Heidegger, in a philosophical-spiritual version, names *Beyng* (Heidegger, 1977, 1999) and the "openness to the mystery"(Heidegger, 1969, p. 55), and what Daoism calls *The Way* and the *Nameless* (Heidegger, 1971, 2010; Nelson, 2024).

Eckhart's understanding of *Bildung* is closely related to the concept of *Gelassenheit* and the idea that "the poor in spirit" (the person saturated in not-knowing, or deep wonder) is closer to hearing the divine callings than the person "rich in spirit", that is, the ones who *know* or want to explain and understand the universe through human rationality, methods, and language. One of Eckhart's artistic successors was a mystical Christian the poet Angelus Silesius (1624–1677) who wrote a short poem about a blooming rose:

The rose is without 'why'
It blooms simply because it blooms.
It thinks not of itself,
And no display it shows.

This poem must be seen in line with the Christian apophatic tradition, which has roots in the Christian and neo-platonic mystic Pseudo-Dyonysius (Rhodes, 2014). The apophatic thinking is connected to "negative theology", which briefly posits that God as a prime divine force in the universe and time cannot be pictured, said, and described directedly in spoken or written words or any images and idols (or as the Daoist world say: "Do not give name to the nameless"). The divine is ineffable and deeply mysterious. However, we can, the apophatic thinkers claim, in indirect, paradoxical, negative (dialectical), baroque, humorous, ironic, and poetic ways, and by ways of *living* in love, care, leisure, humbleness, silence, awe, and deep wonder get in resonance with the divine force in life.

Eckhart is the first in German history to use the word *Bildung* as an expression for inner development in humans toward this divinity (Frimberger, 2024). We can – in very special moments – get a glimpse of this divinity in our souls, Eckhart claims, but it is only through our hearts (when seeing, listening, and thinking with the heart), and when the mind of the soul has become empty, like a blank mirror or quiet forest lake that mirrors the sky above it. Thus, *Bildung*, for Eckhart, is a praxis and process of spiritual self-transformation, which consists of two dimensions: 1) when ignited by the grace of God (when passively receptive to letting go and letting come, hearing the mysterious calling) and 2) when the person then tries (actively and in a humble, loving and wonder-evoked way) to form his/her life so that he/she can get more in tune in his/her everyday life with the divine. Thus, with Eckhart, we see an integration of *Vita Contemplativa* (grace) with *Vita Activa* (effort) (cf. Kotva, 2022), because Eckhart wanted too to live a spiritual life *in* this earthly life now (Fox, 2000).

Importantly for the eco-phenomenologists that are drawn to this kind of Praxis of *Gelassenheit* (Harvey, 2009), is that Eckhart saw the divine in every creation, in every bird, lilies, rivers, sky, and all element and living species on this earth. It was, for Eckhart, through a close interconnectedness with nature and all living species on this earth, that the human had an opportunity to learn the "language" of God. Nature serves as a divine book, with each creature representing a letter or word within this text. "Every single creature is full of God and is a book about God. Every creature is a word of God" (Eckhart, quoted in Fox, 2000, pp. 57–59). Thus, the book of nature is the book we need to learn to read with a wondrous gaze, if we want to experience the divinity in nature and to feel at home in this world.

The Four Dimensions of Contemplative Deep Wonder

Late Heidegger dwells upon Silesius' poem *The Rose Without a Why* (Heidegger, 1996) to illustrate what he means by contemplative wonder at the deepest level. It happens when we experience *a* wonder and the attunement of *Scheu* in German, which can be translated as awe or rather an attunement of existential withdrawal or quiet shyness, diffidence, or bashfulness (Heidegger, 2012) that reveals a deeper encounter and engagement with *physis* itself. The attunement of *Scheu* happens to us in that moment of openness and receptivity, which emerges *before* we *relate* to the wonder we sense, and before we see ourselves in the light of this wonder (or act afterward by virtue of this wonder).

The kind of wonder we may feel on a psychological and "ontic" level as a feeling and lived experience, is, as I have already indicated in the previous part of this chapter, what I term a "Wow-Wonder". The kind of wonder that follows when we wondrously ask Socratic and existential-philosophical questions, we could then call "Negative Wisdom-Wonder". This kind of wonderment is the place where we try to relate to the calling or truth speaking from the wonder experience, and yet we, again and again, experience the finitude and limitation of "positive assertions" and human language. You could say – like Kierkegaard does (1989) – that this kind of Socratic Negative Wisdom-Wonder is a preparation and spiritual cleansing for the soul to be able to see the wonder pure and in itself (by emptying one's own self for human knowledge and certainties through negative dialectics). The kind of wonder, which is captured in the attunement named *Scheu*, may then be called "Without-Why-Wonder" to follow the mystical thinkers and poets of the apophatic tradition. And lastly, the kind of wonder-driven action, what the Daoist calls *Wu Wei* or effortless action, would then be a fourth kind of deep contemplative wonder, which could be named "*Wu Wei*-Wonder".

These are, I will suggest, the four dimensions of deep contemplative wonder. The Wow-Wonder will typically be the starting point based on a shallow level of lived experience, which can be deepened further by a Socratic Negative Wisdom-Wonder. This kind of Socratic wonder asks for tacit assumptions taken for granted in the lived experience. After the effort of the Socratic negative dialectics and deconstruction (preparing emptying process), a more spiritual horizon and direct contact with the wonder (a phenomenon) emanates. This is where the gaze of wonder as a direct seeing (*Scheu*) of the phenomenon in and by itself becomes a Without-Why-Wonder. Lastly, after being immersed in this kind of saturated silence and deep meaning-giving wonder, deep motivation may arise and call the person in deep wonder to act and engage in life.

How Can We Practice an Eco-Bildung That Is Existentially Sustainable? Ending Remarks

This leads me to the ending remarks. The point, that the American eco-theologian Sam Mickey (2019) wants to pass on is that the contemporary debate in Environmental Humanities and Ethics misses an eco-spiritual dimension that is grounded on a fundamental wonder-based approach.

In the Late Holocene approach (Sustainability of 1. Degree), the approach to sustainability and regenerative thinking is governed by basically a human-centered and *utilitarian* worldview that puts anthropocentric needs and values at the center asking for example: "What is the use of a rose"? In the Anthropocene and Trans-Humanistic approach to sustainability (Sustainability of 2. Degree) we are asked either to a) follow a bio-scientistic (bio-centric) worldview and knowledge of the *intrinsic* needs and values of life itself (or a rose), or b) we are asked to see the value of the rose in a *systemic* perspective to see how the rose contributes to the whole ecological earth system. What all approaches seem to miss is the rose seen as a present and living felt mystery (and grace).

> Instrumental, intrinsic, and systemic values are all missing something about plants, something that is astoundingly simple and obvious, yet impossible to grasp. It is like an open secret, hidden in plain sight. There is really nothing to it, like the empty hub of a wheel. What do I love when I love a plant? That is what I love: a plant as an open question, a mystery.
>
> (Mickey, 2019, p. 2)

I like to call this fourth approach to understanding sustainability and regeneration in a Post-Anthropocentric culture (sustainability of 3. Degree) the "Thauma-Centric worldview" (Hansen, 2023), referring to the Greek antique word *Thaumazein* for deep philosophical wonder. Here, the sense of wonder is put at the center, and so too is the awe, reverence, love, care, and befriending that may follow such an eco-spiritual sense of not-knowing and receptivity.

In the introduction of this essay, I called the concept of Post-Anthropocentric culture a "hope-concept". I did that because the kind of ineffable (unexplainable) hope that follows a worldview immersed in deep wonder allows us to hope for a balanced global health in the future that we – through our contemporary science and bio- and eco-centric worldviews – have difficulties believing in now. I follow Carson, Orr, Moore, and Washington when they point to the sense of wonder as being of pivotal significance for eco-education. I have, in this essay, tried to build further on this by developing a concept of *Existential Sustainability* and *Eco-Bildung* that is philosophically based on deep wonder described now by four dimensions: Wow-, Wisdom-, Without-Why- and *Wu Wei*-Wonder.

The existential fourth-order reflection and eco-spiritual fifth-order reflection, which I suggest the Danish hospital at the beginning of the essay to be guided by, can now be described through these four forms of deep wonder. Existential fourth-order reflection has to do with Wow- and Wisdom-Wonder. Eco-spiritual fifth-order reflection has to do with Without-Why-Wonder and *Wu Wei*-Wonder. Without-Why-Wonder helps us to hear the call and "eco-tonality" we need to tune in on. *Wu Wei*-Wonder helps us to respond and act in everyday life in attunement to this mysterious eco-tonality. This kind of receptive-passive listening to the apophatic callings in life – which is found in the eco-spiritual fifth-order reflection – is based on the thinking of late Heideggerian and New Daoist eco-phenomenologists.

These four forms of deep wonder, I want to conclude, are also forms of deep wonder that can ignite and qualify the kind of "Deep Impression, Deep Questioning and Deep Commitment" that the British eco-philosopher Steven Harding sees as the three most important supporting beams in building education for planetary well-being. Now, I would like to add a fourth supporting beam, which is "Deep Hope".

Notes

1 The concept "Eco-Bildung" is first reflected upon in the concept "Eco-Reflexive Bildung" (Sjöström, Eilks and Zuin, 2016) and Sjöström (2018, p. 154). Here these educational researchers interestingly want to make a post-positivistic, humanistic and ecological (biocentric as opposed to an anthropocentric) approach to science-education. The German word and the tradition of Bildung are difficult to translate into Anglo-American philosophy of education, but it is translated as "liberal education", "self-edification", "self-cultivation", "self-formation" or as a process of intertwined "personal and cultural maturation" where art, philosophy and humanities are seen as central to the learning process and self-transformation of the learner. By putting an "Eco-" before Bildung as "Eco-Bildung", the authors expand the traditional humanistic-philosophical understanding of Bildung to incorporate an ecological and biocentric dimension. To that they also add a critical reflexive dimension, so that this kind of "eco-reflexive Bildung' aims at developing a critical and reconstructive view on how modernistic, rationalistic and utilitarian Western educations, institutions, cultures and life forms should be changed, be reformed, in order to live in a society and have science educations that are more directed towards a biocentric and sustainable worldview.

2 Eco-tonality can also be interpreted, as do Hughes (2006), in a more landscape-ecological manner pointing to the way this landscape tune the captivated into a special mood exploring the emotional and sensory dimensions of the ecological experience. Others (Li, Steffensen & Huang, 2020) use the word to focus on eco-linguistics exploring how linguistic choices (the "tone of language") reflect and

shape our understanding of the relationship between language and ecological the environment. Some (Forrest, 2021) work with the concept of eco-tonality as a way to listen in new ways to the non-human's sounds and how to use those eco-tonalities as inspiration for artistic thinking and creation.

References

Aaltonen, V., Hiljanen, M., Layne, H., Lehtonen, A., Löyttyniemi, M., Mykrä, N., Virtanen, A., & Heikkinen, H. (2023). Education for planetary well-being. In: M. Elo, J. Hytönen, S. Karkulehto, T. Kortetmäki, J. Kotiaho, M. Puurtinen, & M. Salo (Eds.), *Interdisciplinary Perspectives on Planetary Well-being*. Routledge.

Abram, D. (2010). *Becoming Animal: An Earthly Cosmology*. Vintage Books.

Abram, D. (2017). *The Spell of the Sensuous: Perception and Language in a More-Than-Human world.* Vintage Books.

Bachelard, G. (2014). *The Poetics of Space*. Penguin Books.

Bannon, B. (2014). *From Mastery to Mystery: A Phenomenological Foundation for an Environmental Ethic*. Ohio University Press.

Barnett, R. (2023). The Very Idea of Academic Professionalism: At an End or Beginning Anew with an Eco-professionalism? In: Lybeck, E. & O'Connell, O. (Eds.), *Universities in Crisis: Academic Professionalism in Uncertain Times*, 23–40. Bloomsbury.

Bate, J. (2000). *The Song of the Earth*. Picador.

Brown, C. & Toadvine, T. (Eds.) (2003). *Eco-Phenomenology: Back to the Earth Itself.* SUNY Press.

Capobianco, R. (2011). *Engaging Heidegger*. University of Toronto Press.

Capobianco, R. (2015). *Heidegger's Way of Being*. University of Toronto Press.

Capobianco, R. (Ed.) (2020). *Heidegger and the Holy*. Rowman & Littlefield.

Capobianco, R. (2022). *Heidegger's Being: The Shimmering Unfolding*. University of Toronto Press.

Carabine, D. (1995). *The Undknown God: Negative Theology in the Platonic Tradition: Plato to Eriugena.* Wipf & Stock.

Carson, R. (1954). Speech to Theta Stigma Phi. April 21.

Carson, R. (1962). *Silent Spring*. Houghton Mifflin Company.

Carson, R. (1998/1956). *The Sense of Wonder: A Celebration of Nature for Parents and Children*. Harper Perennial.

Chai, D. (Ed.) (2020). *Daoist Encounters with Phenomenology*. Bloomsbury Academic.

Chai, D. (Ed.) (2022). *Daoist Resonances in Heidegger*. Bloomsbury Academic.

Curry, P. (2020). *Enchantment: Wonder in modern life*. Floris Books.

Curry, P. (2023). *Art and Enchantment: How Wonder Works*. Routledge.

Forrest, N. (2021). Ecotonalities: Sound Practices for Listening with the Inhuman. In: Lee, F., Campbell-Jones, C. & Paxman, K. (Eds.), *More Just, More Sustainable Futures: Artistic Research Symposium for PhD Students*, Vol. 2021, 4–14.

Fox, M. (2000). *Passion for Creation: The Earth-Honoring Spirituality of Meister Eckhart*. Inner Tradition.

Franke, W. (2014). *A Philosophy of the Unsayable*. University of Notre Dame Press.

Franke, W. (2018). *Apophatic Paths from Europe to China*. SUNY Press.

Frimberger, K. (2024). Education as the Art of Making Oneself at Home in the World with and Through Others. *Studies in Philosophy and Education*, 43, 5, 515–535.

Gadamer, H.-G. (2006/1960). *Truth and Method.* Continuum.

Girardot, N., Miller, J. & Xiaogan, L. (Eds.) (2001). *Daoism and Ecology*. Harvard University Press.

Grinberg, J., Rogers, D. & Gregory, M. (Eds.) (2022). *Philosophy of Education: Modern and Contemporary Ideas at Play*. Kendall Hunt Publishing Company

Han, B.-C. (2024). *Vita Contemplativa*. Political Press.

Hansen, F.T. (2015). The Call and Practice of Wonder: How to Evoke a Socratic Community of Wonder in Professional Settings, s. In: Noah Weiss, M. (Ed.), *Socratic Handbook: Dialogue Methods for Philosophical Practice*, 217–244. LIT Verlag.

Hansen, F.T. (2022). What would an Apophatic Action Research look like? *International Journal of Action Research*, Eikeland (Ed.), special issue on «Conceptualizing AR». 18(2/2022), 100–115.

Hansen, F.T. (2023). Apophatic and Existential Wonder as a Humanizing Force. In: Hansen, F.T., Eide, S.B., & Leget, C. (Eds.), *Wonder, Silence, and Human Flourishing: Towards a Rehumanization of Health, Education & Welfare*, 21–46. Lexington Books.
Hansen, F.T. (2024a). Existential Health and Spiritual Recovery: Two possible new and important concepts in Health Humanities. *The Nordic Journal of Arts, Culture and Health*, 6(1), 1–12. https://doi.org/10.18261/njach.6.1.3
Hansen, F.T. (2024b). Wonder and Philosophy as Grounding Sources in Health Humanities. In: Crawford, P., Kadetz, P. (Eds.), *Palgrave Encyclopedia of the Health Humanities*, 1–15. Palgrave Macmillan. https://doi.org/10.1007/978-3-030-26825-1_221-2
Hansen, F.T. (2024c). The Sense of Wonder as a Necessary 'Philosophical Literacy' in Healthcare. In: Lisen Dellenborg & Daniel Enstedt (Eds.), *Culture, Spirituality and Religious Literacy in Healthcare*, 217–231. Routledge.
Harding, S. (1997). What is Deep Ecology? *Resurgence* 185 (1997): 14–17. file:///Users/finnth/Downloads/Deep_Ecology_Stephan_Harding_WHAT_IS_DEE%20(2).pdf
Harding, S. (2020). *Animate Earth: Science, Intuition and Gaia*. Green Books.
Harvey, S. (2009). *Heidegger and Eco-Phenomenology: Gelassenheit as Practice*. VDM Verlag.
Heidegger, M. (1962/1927). *Being and Time*. Harper.
Heidegger, M. (1969). *Discourse on Thinking*. Harper.
Heidegger, M. (1971). *On the Way to Language*. Translated by Peter D. Hertz. Harper.
Heidegger, M. (1975). *Poetry, Language, Thought*. Harper & Row.
Heidegger, M. (1977). *The Question Concerning Technology & Other Essays*. Harper Perennial.
Heidegger, M. (1996). *The Principle of Reason*. Indiana University Press.
Heidegger, M. (1999). Letter on "Humanism". In: *Pathmarks*, 239–276. Cambridge University Press.
Heidegger, M. (2010). *Country Path Conversations*. Translated by Bret W. Davis. Indiana University Press.
Heidegger, M. (2012). *Contributions to Philosophy of the Event* (trans., Richard Rojcemicz & Daniela Vallega-Neu). Indiana University Press.
Hughes, J. D. (2006). The Mosaic of Culture and Nature: Organization of Space in an Inhabited Cosmos. *Nature and Culture*, 1(1), Spring: 1–9.
Husserl, E. (1982). *Ideas Pertaining to a Pure Phenomenology and to a Phenomenological Philosophy*, First Book (F. Kersten, Trans.). Martinus Nijhoff Publishers.
Hutchins, G. (2022). *Leading by Nature: The Process of Becoming a Regenerative Leader*. Wordzworth Publishing.
Keltner, D. (2023). *Awe: The Transformative Power of Everyday Wonder*. Penguin Books.
Kennedy, T. (2014). *Heidegger and the Ethics of the Earth: Eco-Phenomenology in the Age of Technology*. Dissertation: https://digitalrepository.unm.edu/cgi/viewcontent.cgi?article=1006&context=phil_etds
Kierkegaard, S. (1989/1841). *The Concept of Irony with Continual Reference to Socrates*. Princeton University Press.
Kotva, S. (2022). *Effort and Grace: On the Spiritual Exercise of Philosophy*. Bloomsbury Academic.
Li, J., Steffensen, S. V., & Huang, G. (2020). Rethinking ecolinguistics from a distributed language perspective. *Language Sciences*, 80, 10127713. https://doi.org/10.1016/j.langsci.2020.101277
Marder, P. (2013). *Plant-Thinking: A Philosophy of Vegetal Life*. Columbia University Press.
Marion, J.-L. (1998). *Reduction and Givenness: Investigations of Husserl, Heidegger, and Phenomenology*. Northwestern University Press.
Marion, J.-L. (2002a). *Being Given: Toward a Phenomenology of Givenness*. Stanford University Press.
Marion, J.-L. (2002b). *In Excess: Studies of Saturated Phenomena*. Fordham University Press.
Merleau-Ponty, M. (1993/1964). Eye and Mind. In: Johnson, G. (Ed.), *The Merleau-Ponty Aesthetics Reader: Philosophy and Painting*, 121–149. Northwestern University Press.
Merleau-Ponty, M. (2007/1945). *Phenomenology of Perception*. Routledge.
Mickey, S. (2019). Without Why: Useless Plants in Daoism and Christianity. *Religions*, 10(65), 1–14. https://doi.org/10.3390/rel10010065
Moore, K. (2008). The Truth of the Barnacle: Rachel Carson and the Moral Significance of Wonder. In: Moore, K. & Sideris, L. (Eds.), *Rachel Carson: Legacy and Challenge*. SUNY Press.
Nelson, E. (2021). *Daoism and Environmental Philosophy: Nourishing Life*. Routledge.
Nelson, E. (2024). *Heidegger and Dao*. Bloomsbury Academic.
Orr, (1992). *Ecological Literacy: Education and the Transition to a Postmodern World*. SUNY Press.

Orr, D. & Stone, M. (Eds.) (2005). *Ecological Literacy: Educating Our Children for a Sustainable World*. Sierra Club Books.

Paulsen, M., Jagodzinski, J. & Hawke, S. (Eds.) (2022). *Pedagogy in the Anthropocene: Re-Wilding Education for a New Earth*. Palgrave.

Pieper, J. (1963). *Leisure: The Basic of Culture*. Ignatius Press.

Rhodes, M.C. (2014). *Mystery in Philosophy: An Invocation of Pseudo-Dionysius*. Lexington Books.

Schinkel, A. (2021). *Wonder and Education: On the Educational Importance of Contemplative Wonder*. Bloomsbury Academic.

Sjöström, J. (2018). Science teacher identity and eco-transformation of science education: comparing Western modernism with Confucianism and reflexive Bildung. *Cultural Studies of Science Education*, 13(1), 147161. https://doi.org/10.1007/s11422-017-9807-3

Sjöström, J., Eilks, I., & Zuin, V. G. (2016). Towards eco-reflexive science education: A critical reflection about educational implications of green chemistry. *Science & Education*, 25(3–4), 321–341. https://doi.org/10.1007/s11191-016-9818-6

Washington, H. (2018). Education for Wonder. *Education Science*, 8(125), 1–18. https://doi.org/10.3390/educsci8030125

Washington, H. (2019). *A Sense of Wonder Towards Nature: Healing the Planet through Belonging*. Routledge.

Wilson, Edward O. (1984). *Biophilia: The Human Bond with Other Species*. Harvard University Press.

16

PHENOMENOLOGY OF FRIENDSHIP IN A COMMON WORLD

Marina Garcés

Friendship is a social and affective bond that is part of the world in a strange way: on the one hand, it is a type of relationship that is present and positively valued in all cultures and throughout history. On the other hand, it is the only stable social relationship for which no specific institutions or laws have been invented. This does not mean that it is on the fringes of public life, on the contrary: friendship relationships are a specific layer of bonds, different from those articulated through institutions such as the family, religion and the state, but also different from those of material and economic exchange related to needs and productive life. This has meant that friendship has been understood as a form of private bond between individuals, especially within the framework of modernity, which is based on the separation of private and public life.

The aim of this chapter is different: to understand friendship as constitutive of the "common world", beyond the duality between the private and the public, between the intimate and the institutional. That means, also, to approach friendship as an openness to the other that precedes intentionality and agency and to propose a reconceptualisation of symmetry and reciprocity not only among humans but also in a more-than-human world.

To this end, we propose first to analyse the concept of friendship from the philosophical notion of "form of life", and then to develop its consequences through a conception of friendship that is more than human. This will ultimately open up the possibility of understanding friendship as that set of practices in which the strangeness that constitutes us is elaborated as a bond rather than a threat.[1]

Friendship as a Form of Life

The last interview Foucault gave before his death was entitled "Friendship as a Way of Life" and was published in 1981 in the magazine *Gai Pied* (Foucault, 1981). It is short and focuses on the debate about homosexuality. The editors and readers of the magazine are young and ask a Foucault in his fifties about his homosexual experience. Foucault rejects an identity linked to both age and sexual practice, stating that "the problem of homosexuality is directed towards the problem of friendship". For Foucault, the question is not how to integrate this sexual difference into existing structures, but to what extent homosexuality, like other forms of affective bonding that occur at the limits of what is recognised, can be a gateway to the

DOI: 10.4324/9781003538097-20

creation and multiplication of relationships that bear no analogy to institutionalised ones, and thus become the starting point for a new culture and an ethics different from the one that has been socially perpetuated.

From this perspective, Foucault argues that friendship is not a type of relationship but a form of life. He uses the term "form of life" to refer to something that is already happening but not yet fully constituted. It does not refer to a style or a fashion, but to a sensibility that develops from the capacity to relate to others and their worlds. In this sense, the problem of friendship is posed as follows: it is not a question of "friendship", but of a "friendship" that is not yet fully constituted.

Foucault uses the concept of "form of life" in passing in some of his later writings. This last interview is too brief and colloquial to be able to specify and deepen its use. Nevertheless, it is a concept with a long history in 20th century philosophy, especially thanks to Giorgio Agamben's work on *homo sacer* and *nuda vita* (Agamben, 1998). The concept of form of life (*Lebensform*) comes from the philosophy of the second Wittgenstein, specifically from the *Philosophical Investigations* (Wittgenstein, 2009). In this labyrinthine and enigmatic book, composed like a bazaar of language games, this term appears recurrently. "Imagining a language means imagining a form of life" (§19). At other moments, he specifies that in order to be able to speak, one must participate in a form of life. Any use of language that can be shared with others rests on a basis which, for Wittgenstein, is not a foundation or essence, but a way of acting. There is a "framework on which our language functions" (§240) which is composed of the concrete relations between linguistic behaviour, non-linguistic behaviour, and the situations in which they are produced and acquire meaning. These are social and natural practices, objective and subjective, active and passive, which form the basis of our capacity for agreement. All understanding is not reduced to what language says but ultimately refers to a practical consensus.

This basis is not only a ground (a "bedrock", in Wittgenstein's words), but also functions as a limit. It is the outer limit of a chain of justifications, beyond which one can only appeal to the factual. The form of life points to the given, which does not mean that it is necessary, neither on logical grounds nor according to natural or historical laws. For Wittgenstein, the ultimate foundation is neither an essence nor a first cause. It is the limit beyond which we can no longer give explanations but only show how things work or just themselves. Meaning, then, is not an ultimate truth, but an open set of actions and reactions in which human beings tend to coincide insofar as they participate in a way of life.

Understanding friendship as a form of life places it in this liminal and fundamental terrain from which the possibilities of concrete agreement and disagreement between those who recognise themselves as such are established. More than a model of virtue or a utopian ideal, friendship is an open-ended set of ways of doing that make sense through the ways in which those who participate in it bring it to life. If friendship relationships have not been given their own institution, it is not because they are too pure or marginal, but because they are in the background of all other relationships. From this idea we can move on to the idea of "way of life" that the philosopher Simone Kotva updates, drawing on the long tradition between philosophy and spirituality. One of the most interesting elements of her approach is the way in which she relates activity and passivity, without opposing them, as inseparable conditions of attention. The consequence of this is "to avoid falling on the side of either pure activity -voluntarism- or pure passivity -determinism; is the desire of the middle" (Kotva, 2020, 136). I propose to understand this "middle" not as an Aristotelian middle term, but as a medium or a field of relations, in which we can understand ourselves as agents among others, both active and passive, linked through our interactions with beings participating in a more-than-human world. This world is not the result of the sum of individuals nor of the systematisation of functions. Not being reducible to either, it can be considered a common world.

Animals and Things

This common world is full of many beings, elements, and materials that we call "natural" because they are not produced or created by humans, but also of objects that are born of our imagination and our hands. These we call artificial. All of them, natural and artificial, are part of an inseparable human and non-human reality with which we establish relationships of closeness, affection, empathy and, in a way, friendship. In thinking about friendship, we need to consider the extent to which these relationships modify the notions of reciprocity and responsibility that we assume to be the symmetry of uniquely human relationships.

As an example of a practical reflection, Mexican researcher, and artist Shaday Larios presents herself and her fellow adventurers as "object detectives" (Larios, 2019). She has explored this transitional dimension between the human and the non-human from multiple perspectives. She shows how "the non-human narrates and transfigures itself or through the situated connections it establishes with humans in a place of political enunciation" (Larios, 2022). She thus invokes an animistic community that may have a sacred or remote origin but is in fact part of the material culture of any society. Not only do we project memories or affections onto objects such as a childhood toy, a piece of jewellery given to us by a loved one, the suitcase of an exile, or an old photograph, but the objects and presences of the "natural" world (a tree, a river, the profile of a mountain, the birds that visit us on the balcony, etc.) have their own agency that demands a way of responding and approaching us in each case, because the "in-between" that is created is not an empty or neutral space.

As a background to these practices, we can invoke the philosophy of Merleau-Ponty, that is also concerned with this "in-between", which is not the void that separates a subject from an object, but the place where the relation between the two takes place. Towards the end of his work, he proposes the concept of "chiasme" as a way of dynamically approaching this medium of relationships that are reciprocal, but not in an external way, based on a balanced idea of exchange, but by their mutual implication.

> Le chiasme, la réversibilité, c'est l'idée que toute perception est doublée d'une contre-perception, est acte à deux faces, on ne sait plus qui parle et qui écoute. Circularité parler-écouter, voir-être vu, percevoir-être perçu (c'est elle qui fait qu'il nous semble que la perception se fait dans les choses mêmes) – Activité = passivité.
>
> (Merleau-Ponty, 1964, 318)

It is where the encounter becomes possible, not as a limit or boundary, but as a transition between the ways of being and dispositions of each party, again both active and passive. From these ontologies of material and sensitive being, such as Merleau-Ponty's and the detectives of objects, the meaning of encounters, their dynamics, and their effects are expanded. It also expands the ways of understanding and engagements and affects in our societies are and how they are given.

Friendship is a very specific kind of affect, with a very restricted field of application in the Western philosophical tradition. For example, the concept of friendship was taboo in the field of animal behaviour studies. However, since 1985, with the work of the American anthropologist Barbara Smuts *Sex and friendship in Baboons* (Smuts, 1999) the term has been used, and since the 2000s it has appeared frequently and with wide acceptance in the scientific literature. The key to this shift in perspective has been a change in focus: these studies are not concerned with what makes friendship uniquely human and how to differentiate it, but with identifying the behaviours by which both animals and humans show preferences for particular individuals in the group and shape them into recognisable attitudes (physical proximity, play, attention, mutual support, grief, joy, etc.). Contrary to what was believed for centuries, these parameters have been found in several species. Contemporary science has also been able to study the physiological basis of these bonds of preference, loyalty and trust (basically the presence of oxytocin and vasopressin), which reinforces the evidence of the existence of elements of continuity between the human world and other species, even in the delicate field of friendly relations. Friendship relationships, that is, relationships of preferential affection that are not linked to food or reproduction, are sought after and beneficial for a wide range of species and broaden the contours of each species.

For the authors of an outstanding research on animal friendship, the only difference between the human world and the rest of what we know so far is the degree of strangeness built into communal living:

> We think it is not so interesting anymore to examine how humans may differ from other animals in their motivations for interact with friends, since, as we showed, they probably do not differ that much. Instead, of much greater interest are how humans interact with strangers, especially in modern societies (…). Many species lack frequent interactions with strangers and, thus, have no need for elaborate cognitive mechanisms to deal with such encounters.
>
> (Massen, Stercj, de Vos, 2010, 1.402)

How much strangeness is necessary and at the same time bearable in a specific common world? This question can have a biological or a political meaning, or both. The answer moves between different ontological realities, objects, animals and people, and puts all kinds of entities in relation to each other, without scales or hierarchies. Friendship has to do with the ways in which we make this relationship possible, not from abstract categories, but from concrete experiences of mutual knowledge and affection as the basis for an ethics of responsibility, attentive to the strangeness of the other, human, and non-human. The affective relationship between this widened sense of strangers is not a romantic mania or a whim for the bizarre and unknown, but a practice of expanding worlds that has deep ethical implications.

From the perspective offered by the observation of non-human bonds, it becomes clearer that the division between a human world capable of ethically aspiring to unconditionality and a purely instrumental non-human world does not hold. Friendship is a non-instrumental motivation that binds certain individuals and at the same time is beneficial for the species and its relationship with the environment, as it allows for the expansion of trusting relationships with the stranger. As the philosopher Brian Bannon puts forward,

> To have a friendship with nature is to acknowledge that the natural environment we live in contributes to our agency, identity, and well-being, but also to care for nature because the relationship itself is of value. Recognizing this contribution, we are called upon to express gratitude, concern, compassion, etc. in our actions so as to preserve (or restore) the health, integrity, and beauty of nature.
>
> (Bannon, 2017, 54)

Friendship is a dimension of experience that makes it possible to combine utility and unconditionality, as well as the particular and the general. From this broadening of the concept of friendship, which goes beyond the limits of the strictly human world, one of the great paradoxes left to us by classical reflection on friendship can be approached. It is the paradox according to which a relationship that has no utility or purpose is at the same time so beneficial. Aristotle began his reflections in the *Nicomachean Ethics* by affirming that no one would choose to live without friends, but at the same time he condemned friendships guided by an external purpose. Epicurus disagreed with this view and defended the usefulness of living among friends as indispensable to a philosophical life free from servitude and fear. Can this confidence on mutual support and reciprocity be understood from a more-than-human perspective? Only from an environmental phenomenology that focuses on non-intentional relationships among strange agents, preceding voluntary bonds, can we arrive to a non-utilitarian concept of utility. It is the basis for developing an ethic of reciprocity that does not depend on voluntaristic concepts such as benevolence or altruism.

Don't Talk to Strangers

If differences in the field of friendship are not substantial but differences of degree in terms of the strangeness of those involved, we must go a step further in our understanding of this strangeness and analyse how it occurs in human experience, not in the abstract but socially situated. Nothing and no one are strange in itself. Strangeness is not a characteristic, but an impression that one feels or manifests before a presence that one does not know where to place or how to interpret. It can occur with human presences, objects, or animate beings. Anything can be strange as long as someone reacts with discomfort, literally confusion.

One of the most important rules of childhood socialisation is "don't talk to strangers". We receive this instruction from a very early age, with no other clues than the examples given to us by mothers, fathers, or other guardians as to who deserves this warning. Children may cry out in terror at a grandparent or other relative, but approach unfamiliar people, animals, or objects without caution. Younger children may open their eyes without blushing and hold their gaze, ask about their injuries or approach strangers, while at the same time rejecting the proximity of their closest relatives. This causes unease and rejection, even discomfort and offence, because it shows that the boundaries of what is familiar and safe are not defined.

How do you learn to recognise a stranger? Strangeness is not the difference. There are people or groups who are very different from you, whether for physical, cultural, or social reasons, who are not perceived as strangers. It is enough that their difference has been properly identified and classified and that they appear, as expected, in the place that corresponds to them. Every day, in any city, we meet dozens and hundreds of strangers who do not necessarily seem strange to us. But it can happen that one day someone very close to us starts behaving so strangely that we think "I don't recognise you, it's not you". This shift towards the limits of the familiar and the known is what we call the stranger. What makes the stranger strange, then, is neither unfamiliarity nor difference, but its appearance as something out of place.

Anyway, the slogan "don't talk to strangers" teaches us to situate and recognise what is in its place, and to act in accordance with those expectations. However, would we make friends if we never talked to strangers? It is often the case that when adolescents stop playing with classmates, neighbours or family friends, they start going out or bringing home friends who raise the concerns of parents and guardians: where do they come from, do we know their families, what kind of environment do they move in? It is not just a question of avoiding dangers or bad influences. The underlying question goes further: how far can we extend the margin of trust into the unknown without endangering our core, without ceasing to be who we are? Making new friends makes us strangers to ourselves. To learn to make friends is also to begin to discover our own strangeness, which has been hidden from us until now.

Marriage is a fundamental institution of social order because it makes the other one's own. This operation can be conflictual, when it involves "others" unacceptable to the group for reasons of religion, age, or social status, or it can be harmonious, when it reinforces the identity and expectations of the families that unite their offspring. This operation is the great tragedy of love and the one that gives rise to all the stories of impossible encounters. Friendship performs the opposite operation: it makes what is one's own strange, starting with oneself. Like the characters of E.T. in the film or Enkidu in the ancient *Poem of Gilgamesh*, the friend does not come from a known place, nor does he come to stay. In the literature of friendship, there is always a moment when the friend leaves or dies: its truth is that of an encounter in separation. This is why friendship exists as a permanent invention of routines, rituals, calendars, languages, and motifs of encounter between people who, however close as friends, will remain strangers.

It is the *Poem of Gilgamesh*, the most ancient of human writings known to us, which tells the story of the friendship between the prince of the city of Uruk, Gilgamesh, and a being we would today call natural or savage, that is, uncivilised, whose name is Enkidu. He does not belong to any human group, neither superior nor inferior, but has not been socialised. It is therefore an encounter between the civilised and the savage, which is not presented as an opposition. Nor do they recognise each other. Rather, they love each other, join forces, and become necessary to each other, to the point where Enkidu's death makes Gilgamesh discover his own finitude. Can we today think of an encounter with the "natural" in the same way? Can we move from opposition to the discovery and care of shared finitude? To do so, we need to approach the idea that friendship is not a relationship between beings who recognise each other as identical, but who start from the possibility of loving and treating each other well out of mutual otherness.

The Democracy of Strangers

Friendship with the stranger is not only an ethical virtue that opens us to others, human, and non-human, but also has a political dimension. One way of defining democracy is as a social and political system in which it is possible to trust strangers. Democracy depends on conversation between strangers and, if properly conducted, should dissolve the divisions that block it. These are Danielle Allen's introductory statements to her book *Talking to Strangers* (Allen, 2004), in which she makes the very point that the injunction "don't talk to strangers" blocks the way to truly democratic societies because it makes mistrust the central element of community life. Unlike the family home or other forms of association and coexistence, the political home is based on interaction between strangers. For Allen, democratic citizenship would be one in which we can relate to strangers without harming anyone.

At least two conditions are necessary for political coexistence between strangers. One is that there are no processes of expulsion, segregation, or extermination aimed at creating identical communities, or "communities of similarity", in the words of Achille Mbembe (Mbembe, 2016). This identity-based concept of similarity is the antithesis of the political concepts of equality and fraternity. In addition to this negative condition, there must be another, perhaps less obvious, condition: that strangeness is not reduced to a classification of pre-established differences. A society based on diversity is closer to making foreignness possible than one based on a single identity, but it does not necessarily guarantee it. On the contrary, it can make it more difficult to understand, since everything has to be catalogued. Other species, its characteristics, benefits and dangers, too. The classifications of disciplinary society (class, level of education, marital status, nationality, criminal record, etc.) have not disappeared, but are superimposed on the new categories to form a mosaic in which no piece, however small, is left loose. From now on, any strange presence, human and non-human, that does not find its place in the classification will be understood as a dangerous presence.

The Society of Enemies

To live among strangers and strange beings is also to live among potential enemies. Even nature itself can be faced as a potential enemy. Achille Mbembe, in his analysis of contemporary democracies and their origins in war and colonialism, shows how the democratic language of pacifying societies is based on two principles: the first is the artificial separation of a community of equals (the hallucinatory dream of a community without strangers) and the rest. This "rest" is not the different, but the dangerous, that is, the enemies insofar as they represent a threat: the spectre of extermination.

Carl Schmitt had already defined the concept of the enemy in terms of an antagonism that is not circumstantial but fundamental. Schmitt's main thesis is that as long as a people exists in the political sphere, it must decide for itself who is a friend and who is an enemy.

> The political is the most intense and extreme antagonism, and every concrete antagonism becomes that much more political the closer it approaches the most extreme point, that of the friend-enemy grouping. In its entirety the state as an organized political entity decides for itself the friend-enemy distinction.
>
> (Schmitt, 1998, 30)

Without this decision, there can be more or less violent relations of exchange between groups and between individuals, as liberalism maintains, but there is no sovereignty and therefore no politics. The existence of the enemy is not a metaphor, but the real possibility of being eliminated by another sovereign group. The enemy is not the effect of disagreement or difference, but the political embodiment of the possibility of non-being.

Today, the threat of extinction, as the possibility of non-being, is also projected onto the human species in its relationship with what we call nature, in two ways. On the one hand, human activity causes the extinction of other species and makes our own species thinkable. On the other hand, non-human agents, such as a virus or a natural disaster, are seen as enemies to be fought with the same concepts and techniques as those of war. This is how, for example, the war against COVID-19 was approached in 2020, and how planetary decisions in the face of the pandemic were framed as a "war against the virus".

Hence the second principle of the society of enemies analysed by Mbembe, which he sees as a planetary generalisation of Schmitt's world: the permanent state of war, which can be either formal war or war as a permanent state of exception. For Mbembe, the generalisation of war is the *phármakon* of the global world created by colonisation, the evil circle of its death and life. Democracies are therefore "circles of separation" based on the struggle against the enemy. Necropolitics, which organises its power around the double possibility of dying at the hands of others or being exterminated by them, is the essence of globalised democracy.

The difference between the current state of the world and the realities of Hobbes or Schmitt is that the field of antagonisms has exploded. The separation between inside and outside, between peace and war, between equal and enemy, cannot be located in a single boundary or marker of identity. Separation is omnipresent, as is war. The enemy is everywhere, everyone can be the enemy.

So, who is a friend, and how do we know a friend? What sense of friendship is possible in a society of enemies? Basically, the possibilities are reduced to three: to deny it, to privatise it and to displace it. The negation follows the position of Carl Schmitt, for whom friendship is not a directly positive value, nor is it sought for its own sake, but is only activated when an enemy threatens existence. Thus, friends are those who are threatened by the same enemy. From this paradigm, friendship remains a good associated with peace and the security of the state, but unlike Greek and Roman concord, it is a good that is only sought and achieved under the pressure of an enemy. Mbembe goes beyond Schmitt to show that what really dominates in a society based on enmity is not so much hatred of the enemy, but the "desire for the enemy". This desire is at the root of political life and is expressed out of necessity and hatred. In order to exist, the enemy is needed and desired, feared and hated at the same time.

The privatisation of friendship is linked to the neoliberal drift of these same societies. Although they seem to be contradictory, in reality they coexist and feed each other. In a context of war and generalised hostility as a reality of politics, friendship relations are understood as a pre-political or anti-political sphere in which individuals manage their affective capital, whether as a space of comfort and protection or as capital to make their own lives profitable. They are not only relations of utility or pleasure, as Aristotle would say, but also serve to maintain a depoliticised life. This kind of friendship does not confront the enemy but presupposes it and places it on another level. It is no coincidence that today's exhibitionist subject, who lives in networks, demonstrates his detachment from the world by displaying his contacts, his relationships, his social capital, and his private happiness full of friends.

Neither the exteriority of the enemy nor the privatised identity of the friend allows us to understand what lies at the heart of the experience of friendship as constitutive of a common world: the affective bond with the strangeness of the other and other beings. Friendship is a relationship without clear contours or places of its own, because it is precisely this form of interest, love and inclination towards another that will be part of our lives and shape our common worlds. The "we" of friendship is not a proprietary "we", neither a restricted human "we", but an intimacy that displaces the sense of ownership, among human relationships but also towards other species and beings. We can trace this concept of friendship through history in some of its key figures and concepts. The purpose of what follows in this chapter is not to seek a historical foundation, but to broaden the framework for understanding our present by drawing on the threads of continuity with ideas and experiences from different experiences of the past.

Life in Common: Two Views

Epicurus, originally from Samos, founded a community on the outskirts of Athens in 306 BC, at a time of social and political crisis. It was characterised by the coexistence of men and women from different social classes and backgrounds, including prostitutes and slaves, which caused controversy and scandal in Greece at the time. The openness and importance of this community is reflected in the inscription written on its door: "Stranger, your time here will be pleasant. In this place the greatest good is pleasure". The importance of pleasure and the meaning of this term in modern, consumerist societies have given Epicureanism a false slant. For Epicurus, pleasure meant the absence of disturbance or pain in a life threatened by the fear of death and the gods.

On the basis of materialistic atomism and the conviction of the mortality of the soul, Epicurus and his companions fought against the fear of punishment and the afterlife, and they justified the search for the pleasures of the soul and the body in this life. For them, friendship is a natural and indispensable good for this healing of the soul, a fundamental element of the good life, because only by trusting in the help of a friend can these fears be extinguished. All friendship must be sought for its own sake, but its origin lies in usefulness. The fact that the root of friendship is usefulness does not mean that the friend is an instrument for one's own good. Rather, it indicates that the need for help is inherent in man, who is aware of his fragility and therefore exposed to fear.

Epicurus thus departs from the Greek ideal of virtue, which was based on the notion of self-sufficiency. Like the earlier philosophers, Epicurus seeks to overcome the turmoil and disease of the soul caused by boundless passions (i.e., those passions that make us dependent because nothing can satisfy them), but the difference is that he recognises the insufficiency of the individual. Individual self-sufficiency is an aristocratic ideal. Only a real and concrete community can respond to the anxiety inherent in human nature. From this perspective, the practice of friendship is part of the natural and necessary pursuit of happiness, along with philosophy: trust in the help of the friend and an understanding of nature are the two conditions for the healing (*phármakon*) of both suffering and bondage. "28) The same conviction which inspires confidence that nothing terrible lasts forever, or even for long, also enables us to see that in the midst of life's limited evils, nothing enhances our security so much as friendship" (Epicurus, source online, n.d.). Friendship is not confined to a strictly human world but is inscribed in the uncertainty of nature to share fears and contribute to a better life.

This collective and practical idea of friendship, linked to the continuum of natural bonds, as a fundamental practice for the construction of a common world was to be developed later. Étienne de La Boétie, known for his *Discourse on Voluntary Servitude* and for being the best friend of Montaigne, puts forward in this treatise an idea of friendship as resistance to tyranny. For La Boétie, servitude is an unnatural state.

> The very beasts, God help me! if men are not too deaf, cry out to them, "Long live Liberty!" Many among them die as soon as captured: just as the fish loses life as soon as he leaves the water, so do these creatures close their eyes upon the light and have no desire to survive the loss of their natural freedom.
>
> (La Boétie, 1975, 52)

If servitude is unnatural, freedom is natural. For La Boétie, freedom is not an individual attribute, but a shared sense of existence among peers. Nature is such that the common duties of friendship occupy a good part of our lives:

> Our nature is such that the common duties of human relationship occupy a great part of the course of our life. It is reasonable to love virtue, to esteem good deeds, to be grateful for good from whatever source we may receive it, and, often, to give up some of our comfort in order to increase the honour and advantage of some man whom we love and who deserves it.
>
> (La Boétie, 1975, 41)

Friendship, argues La Boétie, is a natural disposition towards the care and welfare of others. Its condition is freedom, but he does not understand it as the classical Greeks did. For La Boétie, freedom is not self-sufficiency or the absence of needs, but the recognition of the importance of the companionship that human beings find in their equals. We are free because we feel that we are companions. The Epicurean practice of friendship, which was ethical and philosophical, finds a political dimension in this approach.

In La Boétie's text, friendship takes on a directly political meaning, not as a dimension of justice in the *polis*, but as the antithesis of a tyrannical society. Under the power of what La Boétie calls "the One", subjects tend to merge with his figure or seek complicity in the plot. There can be no friendship where there is cruelty, disloyalty, and injustice. And when the wicked come together, there is plotting, not companionship; they do not love each other, but fear each other; they are not friends, but accomplices. La Boétie's text poses a riddle: is friendship the alternative to tyranny? In other words, would the opposite of a tyrannical society be a society of friends? At no point does La Boétie offer a model of society or how to build it. Rather, he makes a gesture that, despite its anachronism, could be called phenomenological: he takes a step back, rather than projecting a hereafter. He takes an observant approach to what he calls "nature", which includes the everyday, the habitual, the factual, and proposes a politics based on description rather than prescription. That is, he describes what would be a natural tendency in order to show how it has been corrupted and how it might be rectified. Instead of returning to things themselves, as phenomenology did at the beginning of the 20th century in the face of the abuses of positivism, it would be a matter of returning to friendship itself in the face of the abuses of a society of fear and power.

Can this gesture be repeated today? Tyranny changes forms and names, but today it is visible in the renewed concentration of political and economic power in the globalised world. Power is concentrated in a few hands, and fear and suspicion are spreading. As a result, paranoia and conspiracy become common sense. Friendship, or what remains of it, is the opposite of what La Boétie calls "companionship". It is synonymous with a private circle of contacts or privileges. To have friends is to have capital. La Boétie's phenomenological and political action consists in undoing paranoia, denaturalising domination and discovering the political and emancipatory bases of natural affection. What is relevant to La Boétie's perspective on the current challenges of an environmental ethic is that nature is not understood as a single essence, but as an open set of relations and practices that link us affectively to a diversity of agents.

Brothers, Sisters, and Comrades

This concrete and practical idea of friendship in a common world continues throughout modernity to the present day through the concepts of comradeship, fraternity, and *sorority*. They may seem to be strictly political concepts, constructed from the artificiality of futuristic utopias. They carry with them a modern imaginary based on the nature-culture opposition. However, their evolution over the last two centuries also places them as imaginaries for thinking about ethics and politics that encourage commitment to a world that is more than human.

Comradeship is the political relationship between individuals organised around the goal of radical transformation. From this concept of comradeship, this concept of political friendship defines a sphere limited to human action of a voluntaristic type, rather than to active and passive openness and complicity. But let us look at the concept more closely, the comrade not only fulfils an organisational function but also makes commitments of solidarity and loyalty to other comrades and is faithful to a truth in both discourse and practice. He therefore makes comradeship a condition of life and its conditions.

As researcher and activist Jodi Dean discusses in her studies of comradeship, comradeship also functions as an "ideal of the self" (Dean, 2019) as a perspective that members of a revolutionary party or organisation take on themselves. This is because the truth that gives meaning to comradeship is not theoretical or ideological but is tested through the practices of life and action. It implies a radical transformation of the very identity of the person who cannot leave anything out. It therefore works as a conversion from the conscience of our bonds and commitments to others. Jodi Dean traces the Latin origin of the term "comrade". It comes from *camera*, which means "space" or "room". Comrades, then, would be those who share a shelter, a roof, a place. They are guided not only by the horizon of the transformation of society, but also by the possibility of taking shelter together. In times of climate emergency, what are the conditions for sharing shelter? What does it mean to have a "self" in relation to others, human and non-human?

Comradeship is not only a project. Is collective action, but also trust, mutual support and solidarity in the rearguard. The genealogy of this kind of relationship goes back a long way. It can be linked to the comradeship between warriors, mutual support between women, loyalty between soldiers, etc. In order to understand its intrinsically revolutionary meaning, linked to the practical idea of the radical transformation of the existing order, it is interesting to follow the path proposed by the philosopher Michael Walzer in his research on radical politics and community (Walzer, 1965). Walzer locates the origin of the idea that there could be specially appointed and organised groups of men who could play a creative role in the political world, destroy the established order, and rebuild society according to the word of God or the plans of their peers in Calvinism and its forms of organisation.

Feminist thinkers and organisations question comradely practices and expose their patriarchal and androcentric limits. It is a critical turn that links the revolutionary horizon and direct action with the activity of care in the rearguard and everything related to the maintenance of life. There is no liberation that is not also the liberation of life in all its forms of oppression, including the most invisible and mundane. Sorority is currently a fashionable term that can be found in traditionally "feminine" (and very little feminist) magazines or as the main value of mainstream novels and series aimed at a female audience. Although a Spanish intellectual such as Miguel de Unamuno had already expressed the need to introduce the term in 1921, it only entered the Spanish dictionary in 2018, when it was declared the neologism of the year. Before that, it had gone through a long history of terminological and political struggles, waged mainly by North American and Latin American feminists since the 1960s.

Although it is often presented as the feminine transposition of the concept of fraternity (frater means brother in the masculine), it is closer to the sense of comradeship. It denotes the link between women in the community of struggle, with a double aspect: as a network of mutual support and as a front of antagonism against patriarchy and its forms of oppression. The anthropologist Marcela Lagarde defines it as "the ethical, political and practical dimension of contemporary feminism" (Lagarde, 2009, 126). Lagarde emphasises that sorority undermines the traditional enmity or rivalry between women, an imaginary confrontation that the paternalistic gaze of patriarchy has cast a shadow of doubt and division over relations between girls and women. The practices of the sisterhood would thus be a journey from supposed enmity to the conquest of their friendship. This does not mean being satisfied with a friendship that consists of friends hugging, helping, and supporting each other without questioning their role, which was also accepted by the patriarchy as long as it was about helping with the laundry, tea cups, or sexual confessions in the big city, depending on the time or social class. The term "sisterhood" indicates that this care is not possible without a common struggle against the subaltern position of women.

Sorority is thus a bond that becomes political because it subverts power relations through the apprenticeship of friendship. It is a strange friendship because sorority is not a private friendship, but a personal bond that is not privatised. It is an alliance between strangers that is not devoid of affection. The writer Gabriela Wiener sums it up in words that could not be more appropriate: "it is the friendship between women who are not even friends" (El País, 18 March 2021). Far beyond an ethics of care, the sisterhood redefines the spaces and times of comradeship, inscribing it in a continuum of vital situations that do not depend solely on the formal organisation of political struggle. For this reason, contemporary feminism has converged with environmental ethics through the theoretical and practical current known as ecofeminism. As the Spanish thinker and activist Yayo Herrero argues, "the architecture of our current societies, built on patriarchal, anthropocentric and capitalist foundations, endangers the ecological balances that make human life possible, hinders the interdependent relationships that sustain us as humanity, and threatens to cause a true anthropological shipwreck" (Herrero, 2023, 81). It is therefore essential today to combine the ethical ideals that have defined humanity, including the concept of friendship, with an experience of interdependence in all spheres of life.

Conclusion

A phenomenology of friendship, such as we have briefly presented in this chapter, is a necessary starting point for liberating the concept of friendship both from its capture as a private affection between individuals and from its reduction to a space of protection in a society of enemies. This twofold danger, which today threatens our ways of relating and bonding with others, has great consequences for our ability to think of ourselves, both in theory and in practice, as beings who give meaning to a common world. This meaning is not only a symbolic and human construction but is elaborated through the forms of life that communicate with humans and non-humans, other living beings or artificial objects. These forms of life do not constitute closed or identical worlds but are elaborated through encounters with the strangeness of others and the commitment to its equal value. The possibility of creating common worlds with the strangers, human and non-human, is the fundamental practice of friendship.

Note

1 This chapter is an advance on my research about friendship *La pasión de los extraños* (Galaxia Gutenberg, 2025).

References

Agamben, Giorgio. 1998. *Homo Sacer. Sovereign Power and Bare Life*, Standford: Stanford University Press.

Allen, Danielle. 2004. *Talking to Strangers*, Chicago: University of Chicago Press.

Bannon, Bryan E. 2017. Being a Friend to Nature: Environmental Virtues and Ethical Ideals, *Ethics, Policy & Environment*, vol. 20, no. 1, pp. 44–58, DOI: 10.1080/21550085.2017.1291824

Dean, Jodi. 2019. *Comrade: An Essay on Political Belonging*, London: Verso Books.

Epicurus, Principal doctrines (n.d.), source: https://epicurism.info/etexts/PD.html

Foucault, Michel. 1981. «De l'amitié comme mode de vie» (entretien avec R. de Ceccaty, J. Danet et J. Le Bitoux), Gai Pied, no 25, April 1981, pp. 38–39. Also in: *Dits et Ecrits* tome IV texte n°293.
Herrero, Yayo. Summer 2023. "La vida humana en un mundo justo y sostenible", en La economía española ante una mirada crítica, Dossieres, *EsF*, no. 50: 81–85.
Kotva, Simone. 2020. *Effort and Grace. On the Spiritual Exercise of Philosophy*, London: Bloomsbury Academic.
La Boétie, Etienne. 1975. *The Discourse of Voluntary Servitude*, trad.by Harry Kurz, Auburn, Alabama: Ludwig von Mises Institute.
Lagarde, Marcela. 2009. Pacto entre mujeres y sororidad, *en Aportes para el debate*, 25: 123–135.
Larios, Shaday. 2019. *Detectives de objetos*, Segovia: La uÑa RoTa.
Larios, Shaday. 2022. *Teatro de objetos documentales*, Segovia: La uÑa RoTa.
Marina, Garcés. 2025. *La Pasión de los Extraños: Una filosofía de la amistad.* Barcelona: Galaxia Gutenberg.
Massen, J., E. Stercj, and H. de Vos. 2010. Close social associations in animals and humans: functions and mechanisms of friendship. *Behaviour*, vol. 147, no. 11, pp. 1379–1412.
Mbembe, Achille. 2016. *Politiques de l'inimitié*, Paris: La Découverte, 2016.
Merleau-Ponty, Maurice. 1964. *Le visible et l'invisible*, Paris: Gallimard.
Schmitt, Carl. 1998. *The Concept of the Political*, Chicago: University Chicago Press.
Smuts, Barbara. 1999. *Sex and Friendship in Baboons*, Cambridge: Harvard University Press.
Walzer, Michael. 1965. *The Revolution of the Saints. A Study in the Origins of Radical Politics*, Cambridge: Harvard University Press.
Wittgenstein, Ludwig. 2009. *Philosophical Investigations*, New Jersey: Wiley & Sons.

SECTION 5

Vegetal and Animal Life

17
EMPATHY WITH PLANTS

Hayden Kee

Introduction

Can we empathize with plants? The question confounds us on first hearing, or even provokes laughter, for at least two reasons. For one, the term "empathy" is currently used in numerous diverse and sometimes contrasting senses in common parlance and in academic research in different fields and traditions. This is certainly true of classical and contemporary phenomenology where, as we will see, the term signals something not entirely synonymous with its common acceptation in modern English. For two, we are not entirely sure what plants *are*, and hence whether they would be appropriate targets for empathy. Plants play third fiddle in our lives (if they get a fiddle at all!) behind humans and other animals. Little wonder, then, that they have received scant attention from the major figures of the classical phenomenological tradition.

In this contribution, I survey the views of classical phenomenologists on whether we can empathize with plants. I then turn to the ongoing contemporary debate. I close by suggesting some possible implications and paths for further research. Plant empathy constitutes a sort of limit problem for phenomenology. Pursuing it, I propose, has implications not only for how we think about plants and empathy but also for our very understanding of the biological and embodied character of consciousness itself.

Historical Sources

I will begin this historical overview with the work of Husserl and Stein, for they were primarily responsible for establishing the now standard usage of the term "empathy" (*Einfühlung*) in phenomenology, and for developing what has now been streamlined into the dominant contemporary phenomenological account of the phenomenon. I will then turn to Scheler and the Munich psychologists and phenomenologists, who had a rather more expansive use of the term and understanding of empathy and related phenomena, including their application beyond the animal domain.

 DOI: 10.4324/9781003538097-22

Husserl

Ideas II (Husserl 1989), which Husserl began in 1912 but did not publish in his lifetime, provides a canonical phenomenological treatment of empathy. Husserl views empathy as a perceptual or perception-like act in which one subject experiences another being as not merely a material thing, but as a two-fold unity of "thing and subject, along with the subject's psychic life" (170/162). While the other thing-subject is experienced as a genuine unity, there is also a "fundamental difference" between the way we experience the thingly aspect of another embodied subject and the way we experience its subjective, psychic life (171/163). While the thingly aspect of the other is experienced in "primal presence," just as other perceptible things are, the other's subjective, psychic, and bodily sensory experience (the *Leib*, or "lived body") is "co-presented," or given in "appresence."

Though Husserl doesn't elaborate the point in any detail, and may be ambiguous about it in certain passages, in *Ideas II*, the basic and paradigm target of empathy is not in fact another human being as such, but another animal – or, as Husserl often calls them, using the Latin, other *animalia*. Indeed, the concepts of empathy and animalia appear to be very nearly mutually defining. Empathy is precisely the act of consciousness through which animalia are given, and animalia are just those objects that are (perceived as) "primally present Corporeal bodies [*Leibkörper*] with appresented interiority" (1989, 171/163). Such a determination might seem to preclude from the outset any possibility of empathy with plants. However, it must be stressed that "animal" is a technical phenomenological concept, not a biologistic one. It is not synonymous with the biological concept of the animal, and it is not necessarily coextensive with the biological-taxonomic animal kingdom. It may turn out that some members of the biological animal kingdom, such as placozoa and sponges, which have no nervous system, may not have the relevant kind of "appresented interiority" to count as animalia. Meanwhile, nothing precludes the possibility that some entities that are not members of the animal kingdom are two-fold unities of things and subjects, and hence *animalia* in the relevant phenomenological sense. Candidates might include sentient robots, extraterrestrial species, or even, indeed, plants, if they possess the relevant kind of psychic-subjective life and hence are appropriate targets for empathy. Curiously, then, plants could turn out to be "animalia."

In his notes on the foundation of the sciences of zoology and botany, in the manuscript for *Ideas III* building upon the work of *Ideas II*, Husserl directly addresses the question of empathy with plants (1980, 7ff./7ff.). Husserl has claimed that the foundations for zoological science (including the human sciences) rest on the researcher's ability to experience or interpret (*Eindeutung*) animals as having sensory experiences distributed across bodily sense organs. It rests, in brief, on the ability to empathize with the lived bodies of other animals, which would not be possible if the researcher herself were not likewise an animal, lived body. Husserl then considers whether botany should be founded in a similar way. Here is his answer:

> The obvious kinship between animal and plant [...] does not go so far that a determinate interpretation of the plant as a lived body [*Leib*] would be possible. [...] The *general and entirely indeterminately performed empathy* that permits the analogy does not suffice for the researcher; he requires concrete experience of concrete sensitivities [*Empfindsamkeiten*], related to concrete organs, whereby the analogy of plant organs with animal organs must be sufficiently great to ground the probability of the interpretation. If this is lacking, then the treatment of botany as a material natural science suffices – or rather, no other treatment then possible. This would not exclude plants'

having sensitivities after all; it only means that we would be incapable of recognizing them, because we are lacking any bridge of empathy and of mediately determinate analysis. Nevertheless, I here only wanted to be as accommodating as possible to the prevailing field of physiological botany and biology in general, and I leave open the question whether interpretive experience cannot play – or whether in fact it is not playing – its fruitful role after all[.][1]

A few comments are in order regarding this remarkable passage.

First, a great deal hangs on what Husserl means by "sensitivities" (*Empfindsamkeiten*). After all, in their responsiveness to a wide range of environmental stimuli, one might think that plants are in fact sensitive. I take it to be clear from the context of *Ideas II/III*, though, that the relevant sense of "sensitivity" here is not mere responsiveness to stimuli. Rather, it includes sensation, or *sentience*, in the sense of what we would today call *phenomenal consciousness*. I will return to this point and its significance for plant empathy in the following section.

Second, note that Husserl is not strictly excluding the possibility of empathy with plants altogether. The context here is the foundation of the sciences of zoology and botany. Husserl has simply stated that the analogy between plant and animal organisms is not sufficient to warrant an interpretation of plant organs as sensory organs in the relevant sense. However, Husserl claims that the very fact that we can make such an analogy, albeit one insufficient for scientific purposes, presupposes a "general and entirely indeterminately performed empathy" with plants. While Husserl does not elaborate the point, the suggestion seems to be that we experience the plant not as a mere material thing, like the stone, but also not as anything like a typical, determinate experience of another animal as a sensing, self-moving locus of embodied subjectivity. The plant, it seems, inhabits a liminal space between thing and animal that neither our typical human experiential categories nor our phenomenological conceptual framework is well equipped to capture.

Third, while Husserl has claimed that our (pseudo-)empathy with plants is insufficient to ground a scientific study of plants understood as phenomenally conscious subjects, he adds two important qualifications (1980, 8f./10). For one, he is not outright denying that plants may in fact have "sensitivities" in the relevant sense. He is simply denying that we would be able to recognize them and study them in a rigorously scientific way. Second, he stresses the provisional character of these comments, pertaining as they do to the field of "physiological botany" as it existed at the time. This seems to leave open the possibility that empathy and the serious interpretation of plants as lived bodies could come to play a role in plant science, should future developments motivate such a transformation of the science's theoretical framework.[2]

Stein

Edith Stein's work *On the Problem of Empathy* (1989 [1917]) contains a section on "The Phenomena of Life" (68–71/77–80) that portrays a more "determinate" empathy with plants than Husserl acknowledged. Stein maintains that certain general expressions of vitality, such as vigour or sluggishness, can be seen in plants in a way appropriately analogous to how we see these in humans and animals through empathy (69/77). Where for Husserl the analogies between animal and plant sense organs were insufficient to allow for empathy with plants, Stein holds that more general phenomena of vitality evident in both animals and plants present a case of "genuine analogy" (70/79). Nonetheless, she remains undecided on the question of whether such phenomena of life are indicative of anything "essentially psychic" or only provide an "essential basis for psychic existence," such as we find it in animals.[3]

Lipps and Geiger

Husserl's and Stein's understanding of empathy was shaped by their critical engagement with authors associated with the Munich school of phenomenology. In the early 1900s, Theodor Lipps first cast the problem of how we understand other subjects as a question of empathy (Debes 2015; Zahavi 2014). Previously, the term had been employed in aesthetics to designate a sort of projective "feeling into" the objects of aesthetic appreciation, a usage which does not imply the attribution of anything like a subjective, experiential self to the target of empathy. Lipps himself used the term in this sense in his writings on aesthetics (Lipps 1903) before also specifying empathy with other subjects as one species of empathy broadly construed. While the projective aspect of Lipps' account was later critiqued by Scheler (and Husserl and Stein), the Munich phenomenologists' writings sometimes use the term "empathy" or similar terms in something like the broader scope associated with the earlier aesthetic discussions, a usage that Husserl and Stein eschew. It was in this context that, in the 1910s, Moritz Geiger explored the question of "empathy" with the non-animal natural world (Geiger 1911; Schloßberger 2019). Geiger located a "feeling of oneness" (*Gefühl des Einsseins*) with nature that is the culmination of empathy (*Einfühlung*), or perhaps even its covert, forgotten foundation (1911, 37f.). These descriptions, however, were promptly criticized (Schilder 1914) and are yet to be been taken up by contemporary phenomenologists interested in the possibility of empathy with plants.

Scheler

Max Scheler is one phenomenologist who did take up this theme from Geiger. Indeed, Scheler's account of the experience of others is intimately bound up with his understanding of the experience of nature (Schloßberger 2019). I will here focus on three related issues: the expressivity of perceptual experience, the phenomenon of *Ein<u>s</u>fühlung*, and the ontology of plant life.

A traditional way of formulating the problem of other minds is to ask how it is that we come to ascribe mind to some bodies we perceive, given that all we really perceive are mere physical bodies. For Scheler, however, this way of posing the problem is fundamentally backwards. What we primarily perceive, Scheler maintains, are not mere bodies, but rather expressive psychophysical unities. And this applies not only to our experience of humans and other animals. Rather, the entire perceptual world is initially perceived in this way. Artefacts and natural objects are also initially experienced as expressive. Perception is basically animistic for Scheler. As such, the problem of understanding other minds is not how we come to attribute mind to some bodies, but rather how we come to see that some bodies do not have mind, or to refine within our experience the specific ways in which different beings are minded. Learning, as Scheler puts it, "is not *animation* [*Be-seelung*], but a continual 'de-animation' [*Ent-seelung*]" (1954, 239).

Scheler, like Lipps and Geiger, and unlike Stein and Husserl, recognizes a critical role for something called *Ein<u>s</u>fühlung* (with an "s" between "Ein-" and "*-fühlung*") in the account of empathy. The term is typically translated as "identification," but more literally and in analogy with the English coinage of "empathy," itself modelled after its counterpart German term

(Debes 2015), we might translate this as "uni-pathy" or "one-pathy." To avoid confusion in what follows, I will leave the term untranslated. Scheler only recognized the significance of *Einsfühlung* in the second, 1923 edition of his monumental work on empathy. The insight led to major revision of the work and Scheler's overall understanding of the broad phenomena of empathy (Scheler 1954, xlvi). What exactly Scheler understood by *Einsfühlung*, however, is more difficult to pin down. Scheler introduces the concept by way of a range of seemingly marginal examples including totemism in "primitive" thought, the religious mysteries of antiquity, hypnotism, the mental life of children, the experience of "truly loving sexual intercourse," and even a somewhat gruesome case from insect behaviour (19–31). In cultural traditions where the world is viewed as an organic whole (i.e., nearly all with the exception of the Judeo-Christian traditions and the recent mechanistic-scientistic worldview), *Einsfühlung* with the natural cosmos as a whole is possible (77ff.).

Scheler distinguishes empathy from emotional contagion, and most contemporary phenomenologists follow him in this. Yet he puzzlingly describes *Einsfühlung* as a "heightened form" and "limiting case" of contagion (1954, 18). He views *Einsfühlung* as an involuntary process taking place below the level of fully wakeful consciousness (96f.). It is a "vital causality" operating in a region between rational purposiveness and mere mechanical efficacy (35). It is a process that occurs only when the spheres of bodily sensation and cognitive rationality are "almost wholly empty of particular content" (35), thus making space for *Einsfühlung* to occur. One might be tempted to regard such experiences of *Einsfühlung* as interesting yet abnormal, barely within or just beyond the scope of what we typically think of as empathy. Scheler, however, regards the capacity for *Einsfühlung* as foundational for anything else we call empathy, at least in evolutionary and developmental respects and even though, in any given episode of normal adult empathy, no immediately preceding or undergirding act of *Einsfühlung* is required. *Einsfühlung* thus plays a central though little understood and little appreciated role in Scheler's understanding of empathy.

Scheler also held original views, expressed in the context of his philosophical anthropology (2008, 7–10), on the ontology of plant life. Drawing on then recent scientific studies and the revelations of timelapse photography, Scheler maintains that plants exhibit *psyche*, a term whose extension for Scheler is, in an Aristotelian sense, coextensive with life. In particular, plants exhibit the most basic level of *psyche*, impulsion (*Drang*), exhibited in their striving tendency towards light. While Scheler claims that all living things exhibit "being-for-themselves," an "*inwardness* through which they also are aware of themselves [*sie sich selber inne werden*]" (7), he is unequivocal that plants do not have anything that could properly be called sensation or consciousness. Nonetheless, Scheler does recognize in plants the "primordial phenomenon of expression": "There is a certain physiognomy of their internal states, of the states of their impulsion as the inwardness of their life. The expression of such states may reveal weakness, strength, abundance, or deficiency" (9).

This last remark is not entirely surprising, given that, as we have seen, for Scheler the entire perceptual world exhibits a kind of animistic expressivity. In this passage, however, Scheler appears to be attempting to identify a special expressivity characteristic of the most basic vital functions as opposed to those of the inorganic world, one that could help us develop the specificity of the *Eins-* or *Einfühlung* directed towards plants in contrast to inorganic objects. Nonetheless, while phenomenologists interested in the possibility of empathy with plants may find much inspiration in Scheler's work and that of other Munich phenomenologists, a degree of caution is also required. The notion of *Einsfühlung* and its alleged foundational role with respect to empathy as it is now typically conceived remain underdeveloped and problematic.

In fairness, the phenomena themselves are somewhat obscure, and Scheler indicates at times that experiences of *Einsfühlung* may be difficult to access for the typical modern, adult, Western subject. Challenges are compounded by somewhat infelicitous terminological choices on Scheler's part, and the lack of a contemporary translation of his most important work on empathy. Finally, along with his expansive understanding of the entire perceptual world as an expressive psychophysical unity, some critics have worried that Scheler overstates the directness of our experience of others, perhaps even collapsing the distinction between self and other.

Conrad-Martius

There is one final figure from the cast of early phenomenology who must be mentioned in our discussion: Hedwig Conrad-Martius (Bello 2002). In the context of a staunchly realist ontology, and informed by both personal experience and the most recent research in theoretical botany, Conrad-Martius developed a nuanced account of plant being in works such as *Metaphysical Conversations* (now translated in English – Conrad-Martius 2023 [1921]) and *Die "Seele" der Pflanze* (Conrad-Martius 1934 - untranslated). While there has been a slight resurgence of interest in Conrad-Martius' work from phenomenologists in recent years, those involved in the so-called "vegetal turn" in current continental philosophy and environmental philosophy are thus far yet to take notice.

A number of other figures in the classical phenomenological tradition have discussed experiences of nature, including encounters with the vegetal world. There are famous trees, for example, in the works of Buber, Heidegger, and Sartre, as objects of reverence, wonder, or existential horror. Different phenomenologists' various views on the questions of empathy (broadly construed) and organic life more generally will naturally have consequences for the question of plant empathy. As a complete survey of such literature would be impossible, and as to my knowledge no other figure apart from those already surveyed has directly discussed the question of plant empathy (or plant ontology in ways that are original and relevant for the question of empathy[4]), I will now turn to the contemporary debate.

Contemporary Debates

In recent years, a number of distinct developments have converged to raise anew the question of plant empathy: The so-called "vegetal turn" in the environmental humanities (Di Paola 2024); the increased interest in the question of plant ontology, intelligence, and even sentience in botany and theoretical biology; and the extensive research on empathy and social cognition in phenomenology, philosophy of mind, and cognitive science.

A natural starting point for the contemporary plant empathy discussion is Michael Marder's (2012) paper "The Life of Plants and the Limits of Empathy." Marder maintains that attempts to empathize with plants do not recognize the radical ontological differences between human and plant life and thus end up projecting "constructs and expectations of the human empathizer onto the object of empathy" (259). This conclusion follows from both Marder's views on plant ontology and his understanding of empathy. For Marder, plants are too radically different from humans and other animals to be suitable targets for empathy. They "live without psychic interiority," "defying all notions of individuality," in a kind of "unimaginable passivity." They are unfeeling, or at least they do not feel themselves feel, and they "do not suffer in the same way as sentient beings" (263). As for empathy, Marder understands it as an attempt to probe into the other's psychic interiority "by means of projective imagination" (262). Such an attempt is naturally bound to fail, given that plants, for Marder, do not possess any kind of psychic interiority. Not only that, but the attempt to empathize with plants, when taken as a foundation for a vegetal ethic, reveals the anthropocentrism and even the narcissism of empathy and its "endemic problems" in all ethics: "The inaccessibility of the other's psychic interiority and of experience forces the empathizer to project her own feelings onto the other" (262).

Matthew Hall has rejected both Marder's account of plant being (Hall 2021) and his understanding of empathy (Hall 2022). Hall points out that Marder's account of empathy assumes that empathy is projective, a view that, as we have already seen, phenomenologists deny. In its place, Hall substitutes a phenomenological understanding of embodied empathy drawing largely on Zahavi's (2014) account. Hall's understanding of plant life and embodiment is heavily inspired by Scheler, for whom, as we saw above, plants exhibit a kind of expressivity that, Hall believes, "brings plants within the orbit of Scheler's theory of empathy" (128). The phenomenological understanding of the expressive embodiment of the psyche also entails rejecting Marder's assumption of the "inaccessibility" of the other's experience (Marder 2012, 262). For Hall, plants, appropriately understood, may indeed be suitable targets for empathy, appropriately understood, though cultivating this capacity may take some training.

In response to Hall's proposal, Kee has drawn attention to issues with Scheler's views (discussed above), and Hall's assimilation of them to Zahavi's contemporary, somewhat streamlined account on empathy (Kee 2025). On the latter view, sentience (in the sense of phenomenal consciousness) is central to the question of empathy. Hence, if plants are not sentient, a default view that Kee provisionally accepts, then they are not, *prima facie*, suitable targets for empathy. Thus, attempts to empathize with them will in fact, as Marder had argued, often result in projective misunderstanding, whether zoomorphic or anthropomorphic, of plant being. This observation suggests a further qualification to the standard phenomenological view of empathy and sentience. Following Merleau-Ponty, Kee views sentience as intertwined with the organic, sensible body. Thus, even though sentience is central in the phenomenological account of empathy and focal in most typical empathic encounters, such empathy always implies an organic body, itself not fully transparent to sentience, as the "dark lining" of sentience. Insofar as we have a minimal, heavily qualified experiential access to our own sensible, organic body, as the dark other and counterpart of sentience, to the same extent we may be able to empathize with plants.

Such empathizing, Key argues, would imply a reverse direction of the movement that empathy with animals and persons usually entails. Husserl often describes such empathy in terms of a building up of stratified layers: first a material and lived body (*Leibkörper*) experienced as such, then a psyche (*Seele*) localized in and built upon the perceived lived body, and, where relevant, a person or spirit (*Geist*) built upon the lived body and soul. By contrast, empathizing with plants would require a downward movement, a descent or de-building of the sentient lived body that is appresented when we empathize with animals. Rather than *co-presentation* of such "appresented interiority" (1989, 171/163), we would have to *de-present* the sentient layer to appropriately empathize with plants. We would have to descend into the basement, as it were, of our organic embodiment. But since the dimension of embodied vital functioning that we share with plants is usually not experienced by us, qua sentient subject (or, if it is, it is experienced sentiently) such a "descent" appears to be blocked to us. The attempt to perform it, then, borders on an "impossible phenomenology" (Toadvine 2003), and empathy with plants thus construed turns out to be a paradoxical and problematic notion, a limit problem for phenomenology.

Open Questions

Rethinking Empathy

Encountering a limit phenomenon may lead us to revise our understanding of the typical or paradigm case. Taking seriously the question of plant empathy and surveying the relevant historical and contemporary discussions of it may lead us to rethink empathy as such. Most contemporary authors advancing a phenomenological account of empathy regard it as a *sui generis* mode of experience, not to be confounded with other adjacent experiences. They thus tend to ignore the question of whether there are genetically prior experiences are capacities that found empathy. As we have seen, Scheler maintained a founding relationship between *Einsfühlung* and *Einfühlung*. While the former notion, as articulated by Lipps, was critiqued by Stein (2008, 16), her critique may have been unjust (Stueber 2006, 8), and Scheler's 1923 variant appears not to have received a full hearing in the context of recent research on empathy. What both Scheler's (1954) and Kee's (2025) discussions of empathy point towards is the possibility of an origin or rooting of empathy in something older, deeper, and darker that is not entirely understood within current dominant treatments of the topic in phenomenology.

Sources of Inspiration

Such a rethinking of empathy might motivate us to consider alternative sources of inspiration than the typical examples that have informed the conceptualization of empathy thus far. Kee has suggested that interoception, perhaps cultivated through contemplative practice, might heighten our access to the organic dimension of animal embodiment that is partially shared with plants (Kee 2025). We have seen that Scheler draws insight from a wide and varied range of examples in developing his views on *Einsfühlung*. This was necessary since, he claimed, the capacity had largely atrophied in "the average adult product of modern civilization" (1954, 31). Alternative experiences *and traditions* thus may need to be invoked if we are to avoid WEIRD (Western, Educated, Industrialized, Rich, and Democratic) biases in our understanding of empathy, its foundations and origins, and its possible targets. The arts, too, may help us express and nourish an experience that is largely inexplicable to the intellect (Kee 2025).

Ethics

Even if some qualified form of empathy with plants is possible, the role of empathy in plant ethics is as debatable as it is in animal or human ethics. While Marder (2012) maintained that there are endemic problems of empathy in all ethics, many would reply that it depends on what one understands empathy to be, and the role it is meant to play in ethics (e.g., Aaltola 2018; Gruen 2015). Marder's critique may only hold if we assumed the projective account of empathy that most others in the discussion reject. Hall (2022) hopes that empathy will lead "to a sustained recognition of plant lives, and the foundation for an intersubjective understanding of our relationships with plants" (134).

Life Sciences

We have seen that it was precisely in his concern for the foundations of the life sciences in *Ideas III* that Husserl addressed the question of empathy with plants. There, he denied that any kind of determinate empathy with plants is possible, and hence also denied that such empathy plays a meaningful role in the foundation of botany. This was in contrast to zoology, where empathy does play a foundational role. There is a later unpublished note from Husserl, from around the time of the *Crisis*, that also addresses empathy and the foundations of biology (Husserl 2013). The material is less developed than the *Ideas III* discussion, but some commentators have interpreted it as evidence that Husserl later came to view empathy as central to *all* biology (Meacham 2013; cf. Merleau-Ponty 2022, 40ff.). This would be in line with recent enactivist discussions of the role of phenomenology in the life sciences. Evan Thompson, for example, claims that empathy is foundational for biology as such. He connects this claim to Jonas' dictum that "life can be known only by life."[5] In the appeal to Husserl here, as with the appeal to Merleau-Ponty (Kee 2018, 2020), enactivists may unwittingly be expanding claims made by earlier authors about animals (or *animalia*) specifically to apply to all organisms.[6] Interestingly, the paradigm organism enactivists often reference when elaborating their understanding of life is a motile unicellular organism. With its real-time (albeit microscopic) locomotion, such organisms certainly solicit our empathy in much the same way that animals do, whether or not such empathy is ultimately appropriate. But in the major works on enactivism, we find almost no mention of plants, their distinctive ontology, and the appropriateness of empathy directed towards them.[7] That is, we find in enactivism only accounts of the foundations of biology as a whole, and nothing comparable to Husserl's nuanced consideration of the proper experiential foundation of botany as such. Pending such an account, we may worry that contemporary enactivists are simply projecting zoological or anthropological forms upon plant life.

Rethinking Plant Ontology

There is considerable interest today in questions of plant behaviour, cognition, and intelligence. More researchers are seriously entertaining the possibility that plants are sentient, and some even speak of plant neurobiology. These remain highly speculative fields of inquiry. Nonetheless, phenomenologists should recall Husserl's provisional attitude towards the possibility of plant sentience and remain open but critical with respect to new findings from the sciences. In particular, our view on the possibility and nature of empathy with plants will likely require revision if we have good reasons to believe that plants are sentient.

Acknowledgements

Funding was provided by an Early Career Scheme grant from the University Grants Committee of Hong Kong SAR (CUHK 24609822).

Notes

1 Husserl (1980, 8f./9f. – translation modified, italics added). Cairns (1976, 74) records a quite inconclusive discussion with Husserl on the question of empathy with plants. We also find a mention of the connection between empathy and biology in a late addendum to the *Crisis* (Husserl 2013), which I will return to below.

2 This possibility has been explored by Gaitsch and Vörös (2016).

3 Stein (1989, 69/78). Husserl held the latter view (e.g., 1980, 104/117).

4 This is true even of Hans Jonas, whose views on vegetal life, while not necessarily incorrect, may strike us as "somewhat inadequate" in light of more recent discoveries, as one recent commentator has argued (Coyne 2020, 67ff.).

5 Thompson (2007, 162–65). Thompson cites Husserl's *Ideas II* and *III* (1989, 170–180; 1980, 94–98), yet these passages, in line with those from *Ideas III* discussed in detail above, claim that empathy is required for the constitution of animals and the founding of zoological science, not for all living beings and life science as such. Likewise, the appeal to Jonas does not necessarily imply a role for empathy, as Jonas developed his view without overtly invoking it (Bogotá 2024).

6 Here again, issues with and inconsistencies across translation may be causing part of the confusion. The English translation of *Ideas III* renders "Leib" as "animate organism." While as we have seen Husserl precisely distinguishes the study of plants from that of animals, with only animals being *Leiber*/animate organisms, the choice of English term could lead one to believe that Husserl is thinking of *all* organisms when he is only in fact speaking of *animalia*.

7 For example, Varela et al. (1991); Thompson (2007); however, see Thompson (2022) for a recent, critical and insightful enactivist discussion of plant sentience, and Gaitsch and Vörös (2016) for a proposed revision to Husserl's position from *Ideas III*.

Works Cited

Aaltola, Elisa. 2018. *Varieties of Empathy: Moral Psychology and Animal Ethics*. Lanham: Rowman & Littlefield Publishers.

Bello, Angela Ales. 2002. "Hedwig Conrad-Martius and the Phenomenology of Nature." In *Phenomenology World-Wide: Foundations — Expanding Dynamics — Life-Engagements. A Guide for Research and Study*, edited by A. Ales Bello, M. Antonelli, G. Backhaus, O. Balaban, G. Baptist, J. Bengtsson, J. Benoist, et al., 210–32. Dordrecht: Springer Netherlands. https://doi.org/10.1007/978-94-007-0473-2_20

Bogotá, Juan Diego. 2024. "Life, Sense-Making, and Subjectivity. Why the Enactive Conception of Life and Mind Requires Phenomenology." *Synthese* 204 (3): 101. https://doi.org/10.1007/s11229-024-04746-1

Cairns, Dorion. 1976. *Conversations with Husserl and Fink*. The Hague: Martinus Nijhoff.

Conrad-Martius, Hedwig. 1934. *Die "Seele" der Pflanze: biologisch-ontologische Betrachtungen*. Breslau: Franke.

Conrad-Martius, Hedwig. 2023. *Metaphysical Conversations and Phenomenological Essays*. Edited by Christina M. Gschwandtner. Boston: Walter de Gruyter.

Coyne, Lewis. 2020. *Hans Jonas: Life, Technology and the Horizons of Responsibility*. London; New York: Bloomsbury Academic.

Debes, Remy. 2015. "From Einfühlung to Empathy: Sympathy in Early Phenomenology and Psychology." In *Sympathy: A History*, edited by Eric Schliesser. Oxford University Press. https://doi.org/10.1093/acprof:oso/9780199928873.003.0015

Di Paola, Marcello, ed. 2024. *The Vegetal Turn: History, Concepts, Applications*. Cham: Springer.

Gaitsch, Peter, and Sebastian Vörös. 2016. "Husserl's Somatology Reconsidered: Leib as a Methodological Guide for the Explication of (Plant) Life." *Phainomena* 25 (98–99): 203–28.

Geiger, Moritz. 1911. "Zum Problem Der Stimmungseinfühlung." *Zeitschrift Für Ästhetik Und Allgemeine Kunstwissenschaft* 6: 1–42.

Gruen, Lori. 2015. *Entangled Empathy: An Alternative Ethic for Our Relationships with Animals*. New York: Lantern Publishing & Media.

Hall, Matthew. 2021. "How Plants Live: Individuality, Activity, and Self." *Environmental Philosophy*, January. https://doi.org/10.5840/envirophil20211797

Hall, Matthew. 2022. "Empathy for Plants." *Environmental Ethics* 44 (2): 121–36. https://doi.org/10.5840/enviroethics20225237

Husserl, Edmund. 1980. *Ideas Pertaining to a Pure Phenomenology and to a Phenomenological Philosophy: Third Book: Phenomenology and the Foundation of the Sciences*. Translated by Ted E. Klein and William E. Pohl. Dordrecht: Martinus Nijhoff Publishers.

Husserl, Edmund. 1989. *Ideas Pertaining to a Pure Phenomenology and to a Phenomenological Philosophy: Second Book: Studies in the Phenomenology of Constitution*. Translated by R. Rojcewicz and A. Schuwer. The Hague: Springer.

Husserl, Edmund. 2013. "Addendum XXIII of The Crisis of European Sciences and Transcendental Phenomenology." *Journal of the British Society for Phenomenology* 44 (1): 6–9. https://doi.org/10.1080/00071773.2013.11006784

Kee, Hayden. 2018. "Phenomenology and Naturalism in Autopoietic and Radical Enactivism: Exploring Sense-Making and Continuity from the Top Down." *Synthese* 198 (9): 2323–43. https://doi.org/10.1007/s11229-018-1851-3

Kee, Hayden. 2020. "The 'Surplus of Signification': Merleau-Ponty and Enactivism on the Continuity of Life, Mind, and Culture." *Journal of French and Francophone Philosophy* 28 (1): 27–52. https://doi.org/10.5195/jffp.2020.919

Kee, Hayden. 2025. "Roots of Sentience: An Alternative Phenomenological Approach to Empathy with Plants." *Environmental Philosophy* 21 (2). https://doi.org/10.5840/envirophil2025115142

Lipps, Theodor. 1903. *Ästhetik: Psychologie Des Schönen Und Der Kunst*. Hamburg: Verlag von Leopold Voss.

Marder, Michael. 2012. "The Life of Plants and the Limits of Empathy." *Dialogue: Canadian Philosophical Review / Revue Canadienne de Philosophie* 51 (2): 259–73. https://doi.org/10.1017/S0012217312000431

Meacham, Darian. 2013. "Biology, the Empathic Science: Husserl's Addendum XXIII of the Crisis of European Sciences and Transcendental Phenomenology." *Journal of the British Society for Phenomenology* 44 (1): 10–24. https://doi.org/10.1080/00071773.2013.11006785

Merleau-Ponty, Maurice. 2022. *The Possibility of Philosophy: Course Notes from the Collège de France, 1959–1961*. Translated by Keith Whitmoyer. Evanston: Northwestern University Press.

Scheler, Max. 1954. *The Nature of Sympathy*. Translated by Peter Heath. London: Routledge.

Scheler, Max. 2008. *The Human Place in the Cosmos*. Translated by Manfred S. Frings. Evanston, Illinois: Northwestern University Press.

Schilder, Paul. 1914. *Selbstbewusstsein Und Persönlichkeitsbewusstsein: Eine Psychopathologische Studie*. Berlin: Springer.

Schloßberger, Matthias. 2019. "Phänomenologie Der Naturerfahrung. Einfühlung Und Einsfühlung Mit Der Natur Bei Geiger Und Scheler." In *Natur Und Kosmos. Entwürfe Der Frühen Phänomenologie*, edited by Hans Reiner Sepp, 95–115. Traugott Bautz.

Stein, Edith. 1989. *On the Problem of Empathy*. Translated by Waltraut Stein. 3rd Revised edition. Vol. 3. The Collected Works of Edith Stein. Washington, DC: ICS Publications.

Stueber, Karsten. 2006. *Rediscovering Empathy: Agency, Folk Psychology, and the Human Sciences*. MIT Press. https://direct.mit.edu/books/monograph/2793/Rediscovering-EmpathyAgency-Folk-Psychology-and

Thompson, Evan. 2007. *Mind in Life: Biology, Phenomenology, and the Sciences of Mind*. Cambridge, Mass: Harvard University Press.

Thompson, Evan. 2022. "Could All Life Be Sentient?" *Journal of Consciousness Studies* 29 (3–4): 229–65. https://doi.org/10.53765/20512201.29.3.229

Toadvine, Ted. 2003. "The Primacy of Desire and Its Ecological Consequences." In *Eco-Phenomenology: Back to the Earth Itself*, edited by Charles S. Brown and Ted Toadvine, 139–53. Albany: SUNY Press.

Varela, Francisco J., Evan Thompson, and Eleanor Rosch. 1991. *The Embodied Mind: Cognitive Science and Human Experience*. Cambridge, MA: The MIT Press.

Zahavi, Dan. 2014. *Self and Other: Exploring Subjectivity, Empathy, and Shame*. Oxford: Oxford University Press.

18

PRACTICING PASSIVITY

Through Mystical and Vegetal Being

Simone Kotva

In this chapter, I will be offering a guide to passivity, weaving together arts of noticing born in the Great Acceleration with arts of attentiveness developed at the time of the early modern "death of nature."[1] In particular, I will be spending time with passivity in the work of the French Catholic mystic Jeanne Guyon (1648–1717). My reading of Guyon arose in the context of training in the tradition she represents: the Western contemplative heritage. I am not sure I would have arrived at my understanding of passivity had I not experienced what Guyon describes in her work; nor would I have found it of interest to ecological thinking. Practice deepened the way I came to write about the nature of passivity – and the "passivity" of nature. For this reason, the chapter entwines scholarship with personal reflection, situating my reading of Guyon in a winding story of cultivating spiritual exercises toward an ecology of practices.[2]

Beginning to Practice

My draw to practicing passivity began with my interest, as a graduate student, in the relationship between philosophy, mysticism, and ecology. I came across Pierre Hadot's *Philosophy as a Way of Life: Spiritual Exercises from Socrates to Foucault* and found that both nature and passivity were important in Hadot's work. Throughout his published work, culminating in *The Veil of Isis: An Essay on the History of the Idea of Nature*, Hadot mentions states of passivity as ways of communing with the cosmos.[3] I was interested in testing this claim for myself but did not come around to practicing the spiritual exercises in question until many years later. I was then already knee-deep in a new research project on mysticism and ecology, had read widely on the topic and was preoccupied with the preponderance of *vegetal* life in mysticism. The Buddha experienced enlightenment while seated under the Bodhi Tree (a *Ficus religiosa*) and many figures in Western Christian mysticism received inspiration in vegetal milieus, notably gardens. I knew that mystical visions in historical texts often took the form of vegetal allegories and was familiar with the countless examples of spiritual doctrines represented by the anatomy of plants. What I did not know was whether these vegetal metaphors were connected, in any way, to spiritual practices themselves and the kinds of passivity described by Hadot. I wanted to see whether passivity was metaphorical or experiential and I had a hunch that vegetal life might help me understand the practice of passivity better.

DOI: 10.4324/9781003538097-23

I decided to begin practicing by using historical handbooks of spiritual direction from the Western Christian mystical tradition, with which I was most familiar. I set to work with enthusiasm but soon experienced difficulties. At the core of the mystical tradition I was engaging with is the *via negativa*, or "path of negations." Practicing the *via negativa* involves letting go of things that shape ordinary awareness: first images, then language, then concepts, then the idea of self and, finally, beliefs. Through this process a person is supposed to move from meditation – where the mind is directed actively toward God – to contemplation – where the mind becomes God's passive recipient.

This was a process I found difficult and for the longest time I felt I was not doing anything very useful or *ecological*. Instead of letting go of my "self," and drawing closer to the cosmos, I was becoming more engrossed in myself. I struggled to make space for the practice, detesting the daily commitment. I endured for nine months, then put the practice aside, thinking I might return to it but not knowing when that return would happen, if at all. I continued with my research, but the scholarship was no longer as satisfactory as it once had been. Six months went by. Then, one day, without quite knowing why, I decided to pick up the practice again. I still experienced difficulties, but they seemed to be decreasing. It took me about another nine months to become skilled at letting go, and when I did, I began having the experiences characteristic of contemplative prayer in the Western Christian tradition. At their most intense, these experiences resulted in my skin becoming hypersensitive. An intense feeling of bliss would descend into my body from the crown of my head, bringing with it copious amounts of tears. The experience was erotic except it was not sexual and in fact quite distinct from sexual arousal. Odd things also happened to my sense of self-perception. I was left with a nearly continuous, background sensitivity in which my "self" was not focused to a single place but spread out over the whole surface of my body. It was as if self was skin, and "I" was dispersed to the sensations received by the cutaneous covering holding my body together. According to the texts I was working from, these were all signs that I had begun to contemplate God with "passive loving attention."[4] I would not have claimed, though, that I was experiencing God, nor did I feel entirely passive, or not exactly. It was rather as if I had discovered, for the first time, what was animate about my body's apparent inertia. I could feel myself sensing; I sensed my sensations.

Then something else happened. When the contemplative state abated, it would be succeeded by a heightened awareness to everything around me; and – this was unexpected – I began sensing *plants* with especial vividness. Everything around me and within me was more vivid, but I felt myself attuned to vegetal life with especial force, as if I had become, momentarily, plantlike myself. Plant life seemed to solicit my attention. Trees stood out, their presences but also personhood suddenly undeniable. This vegetal attunement would arrive during my contemplative practice and eventually began seeping into everyday awareness. After a while, this attunement became a new normalcy for me. It would fluctuate in intensity but would be present to some degree throughout the day. It still is.

Vegetal Mysticism

Experiencing myself *as* a plant was something I had not at all expected from practicing passivity. It is not mentioned by Pierre Hadot, the philosopher who set me onto the path of passivity as an ecological practice, and I could see no reference to it in the literature on contemplation in Western Christian mysticism. Through my research I had amassed accounts of mystics experiencing one-ness with the world and sympathy with all creatures (including plants) but nothing about *becoming* plantlike. More promising was the scholarship on gardens in Western Christian mysticism and the role of vegetal allegory, which is extensive. However, I could find nothing about contemplation as a vegetal state. The only detailed discussion of the topic I could find was in scholarship on Yoga. Historical manuals sometimes describe the practitioner, or *yogin*, becoming plantlike in the deeper stages of meditation.[5] This spoke to my experience, but I failed to find any similar discussion about the Western contemplative heritage in which I had learnt to practice passivity.

I then went back to the early sources in the Christian tradition, and when I did, I noticed something. I found that comparisons between contemplative states and vegetal life were, if not exactly common, then certainly not unheard-of. I also found that, although most modern scholarship makes barely any mention of these metaphors, several feminist historians of mysticism – such as Grace Jantzen, Barbara Newman and Amy Hollywood – had touched on closely related questions in mediaeval women's spirituality. I was especially struck by Jantzen's *Becoming Divine: Towards a Feminist Philosophy of Religion*, where the vegetal is introduced as an alternative symbol for the divine life, one that foregrounds flourishing, receptivity and natality.[6] I decided to write a genealogy of what I had begun calling "vegetal mysticism," focusing on the history of Christian mysticism up to the eighteenth century. Soon I was tracing an alternative history of practicing passivity, one that was told through descriptions of vegetal life.

Subsequent stages of my research took an increasingly vegetal turn. I studied the way pioneering biologists like Anthony Trewavas, Daniel Chamovitz and others described plant intelligence and was intrigued to discover that the way mystics describe passivity before God bears more than a passing resemblance to the way scientists studying plant behavior today understand how plants sense their environments and respond to external stimuli.[7] Setting out to flesh out this correspondence, I found myself in good company. While I had been practicing vegetal attunement through experiments in Christian contemplation and discovering the historical significance of passivity as a vegetal activity in Christian mysticism, others had been making parallel discoveries in philosophy.

In recent years, several philosophers – among them Emanuele Coccia, Luce Irigaray, Michael Marder and Elaine Miller – have labored to put into philosophical language the recent scientific discoveries regarding plant intelligence.[8] What I was calling "vegetal mysticism" finds a parallel in critical plant theory, and I began relating this theory to the epistemic role of *passiveness* in contemplation. Marder is the most observant of the parallel between plant intelligence and contemplation, suggesting that what he calls "plant-thinking" is evidenced not only in the way mystics so often choose to celebrate trees and flowers, but, more importantly, in their way of relating to the world, which (whether conscious or no) closely emulates vegetal modes of interacting with environments. "Unjustly considered insensitive by philosophers and public opinion alike," writes Marder, "plants by far exceed animals and human beings in their attentiveness to what is going on around them with regard to the levels of light, heat, moisture, vibration…They are constantly in touch with the elements."[9] For Marder, plant-thinking has "ethical offshoots" that find their closest equivalence in mindfulness and the exercises of mystical spirituality. Plants show an exemplary form of mindfulness by relating to the world with "unconditional […] hospitality," "absolute openness" and continuous attentiveness.[10] Plants do not single out some objects over others but open themselves to their environments, "becoming themselves" in "contiguity" with the elements rather than in differentiation from "air, moisture, soil, warmth, sunlight."[11] The idea of plants as mindful, and of mindfulness as a plant-like capacity, connects the mystical practices I am discussing here to an emergent discourse in the philosophy and science of vegetal minds.

Marder's concept of "vegetal mindfulness" has been interpreted, plausibly but not without question, as a retreat "into a private utopia...tantamount either to misanthropy or to the hatred of animals."[12] One of the risks of speaking about spiritual exercises in terms of the vegetal is that one thinks it is a matter of comparing such exercises exclusively to creatures with roots in the ground. A major point in Marder's argument, however, has been to show the way in which "vegetal mindfulness" is not restricted to plants:

> If we wish to discover vegetal mindfulness in ourselves, we should look no further than our bodies' unconscious attention to the surroundings...Without knowing it, we pay attention to the world on the surface, from the porous, essentially open, cutaneous membrane that keeps our living bodies intact, and, at the same time, communicates with whatever lies beyond them.[13]

For Marder, studying trees, flowers and other plants corrects ideas regarding the assumed superiority of humans *qua* animals, but also reveals how much of human animality is in fact already vegetal: how much of our awareness is mixed up with a basic sensitivity to inner and outer environments. Skin, for instance, practices a similar, incessant awareness of the world. Marder argues that when we shift our awareness to our skin, we may also be learning how to be attentive in the manner of plants.

Skin and its sensitivity had also been a part in my own experience practicing passivity in the context of Christian mysticism. In *Green Mass: The Ecological Theology of St. Hildegard of Bingen*, Marder even plays with the idea that the *via negativa* of Christian mysticism may be, at a deeper level, a *via vegetativa*, "way of vegetation."[14] Marder creates a graft of mystical and vegetal language, studying the way plants and mysticism form a relationship that is symbiotic as well as symbolic, made possible through the paradox of passive activity that plants – as well as mystics – express in their ways of sensing the world.

For my own part, the experience of practicing passivity has been impossible to separate from the promiscuity of crossing species. Like speculative fiction, indigenous knowledge practices, and other arts of noticing, the practice of passivity creates symbiosis that opens zones of contact between creatures. Whether this is the experience also of the historical mystic I will be turning to now, I will never know. But the significance of vegetal metaphors when it comes to the practice of passivity suggests that, at the very least, it is an educated guess. If so, we have in mysticism a rich store of practical know-how with strong bearing on the current planetary emergency. Amidst all the writing *about* passivity, there is very little advice on how to practice passivity. Perhaps there still lingers around the term too much of the romance of doing nothing. It is my hope that the practice outlined below will gesture toward an antidote.

Jeanne Guyon's Practice of Passivity

Passivity is often pitted, unhelpfully, against action, implying a dualistic paradigm. Nothing could be further from the truth. Action implies passivity, just as any muscular effort depends on a rhythmical exchange between tension and relaxation. Even so, it can sometimes be surprisingly difficult to convey the matter of action's passiveness. This was part of the reason I wanted to try *practicing* passivity for myself. I did work very hard on my exercises, even though the exertion was, as the mystic and philosopher Simone Weil liked to call it, in the end a "negative effort."[15] I was trying *not* to try – not to think, not to reason, not to judge, not to be reactive, not to desire. I was trying to be open and receptive to what might emerge. I did so every day, for several months, until my awareness began to shift. At stake was not passivity as such but rather a way of acting that draws attention to the already passive nature of any action, allowing me to cultivate and foreground that passivity.

Passivity in Christian mysticism is an old practice, referred to variously as contemplation, prayer of stillness, prayer of the heart, or simply the prayer of passivity. It is characterized by silence, simplicity and the absence of clerical supervision; one striking feature of Christian contemplation is its informal nature. The practice has origins in early Christian monasticism but, although it is ancient, historically it was subject to repeated censure from Church authorities, especially in the early modern period. During the seventeenth century the work of the French Catholic Jeanne Guyon (1648–1717) became emblematic of an escalating crisis of mysticism.[16] Guyon defended the practice of contemplation, at great risk to her personal safety: because of her teachings, Guyon spent ten years in house arrest and imprisonment.[17] One reason Guyon came under attack was the language she used to describe contemplation. Guyon argued that contemplation was a passive activity, but also that it was a *natural* activity. In the early modern period, one common way to control the practice of Christian spirituality was to stipulate the need for spiritual directors, sacraments and supernatural graces. Guyon insisted that contemplation was easy and did not require either priests or sacraments. She wrote that a person attempting to contemplate God should imitate involuntary bodily functions, such as breathing, tasting and digestion; or else that they should look at how the elements move and how animals and plants behave in relation to the elements. For Guyon, not only was contemplation natural, but it was also nature's own activity.

A brief glance at Guyon's influential, but contentious, *A Short and Easy Method of Prayer*, first circulated in 1685, gives us a sense for the connection between passivity and vegetality in Guyon's mysticism. Guyon challenges the idea that prayer requires special learning. In Guyon's words, "Nothing is easier than experiencing God and tasting him…The only way to look for him is as easy and natural as breathing the air and is no more than that."[18] She also compares prayer to breathing, breastfeeding, photosynthesis, the elements (the movement of water is a favorite image), and bird's flight, among many other analogies drawn. I describe one of these – breastfeeding – in more detail below. As I read Guyon, the principal imagery with which Guyon deals in her mysticism is akin to what Marder defines as the vegetal: an activity that lacks conceptual organization yet remains intentional, such as one finds in plants but also in humans in the regulatory activities of the gut, internal organs and the sensitivity of the skin. The same examples used by Marder are used by Guyon to describe passivity before God.

Guyon, though, also goes further than Marder, offering a method for cultivating as well as theorizing passivity. In *Short and Easy Method*, Guyon suggests setting aside thirty minutes every day for practicing. One begins by addressing God as divine love without the use of formal prayers but simply by arousing what Guyon calls the "affections," that is, one's sensitivity or, better put, ability to *feel*. After a while (it could be minutes or months, depending on the person's state of openness), one will experience, Guyon writes, a brief but overwhelming flow of "affection" such that speech but also thought becomes extremely difficult. In this state it is important not attempt to grasp at the experience but to pause and enjoy it, arousing the affections once more as the state begins to ebb away. Contemplation, for Guyon, begins with these moments of enjoyment, which, cultivated over time at each successive ebbing of affectivity are intensified and prolonged, and expanded into everyday awareness: "outside prayer, we ought to be such as we should wish to be during it."[19] In *Short and Easy Method* contemplation does not lead to a climax event but is supposed to transform ordinary, everyday perception, leading to a new disposition. Guyon writes extensively about how her life changed because of cultivating passivity. She explains how she became happier and at peace, able to deal with life's vicissitudes and withstand years of ruthless interrogation. She also reflects on an increase in her cognitive abilities but also, interestingly, her ability to notice the suffering of others. According to Guyon, once she had begun to enjoy the fruits of contemplation, she was able to respond to the needs of others more readily, the needs of humans but also animals and plants. Far from reducing Guyon to a state of inertia, the practice of passivity allowed Guyon to become more and more engaged with her surroundings.[20]

Part of this engagement with her surroundings is also reflected in the analogies Guyon uses to describe the practice of passivity. Breastfeeding is one of Guyon's favored images. She writes that contemplation should be pictured in terms of eating or taking nourishment, "as a small child attached to his [sic] mother's breast shows us sensibly."[21] A nursing child will begin by making efforts to stimulate the flow of milk, "but when the milk comes in abundance, he [sic] is content to drink it with no more movement," and will continue in this way, even though it scarcely seems possible "that he [sic] is getting nourishment [...] gently in peace without moving."[22] In contemplation, one need not do nothing; one need simply know when nothing more needs to be done in order for there to be satisfaction; that is, when to let go in order to let things be.

What is interesting about this and other descriptions is that they show the *activity* implicit in passivity. Guyon is emphatic about the velocity of what she calls passivity explaining, in *Short and Easy Method*, that "the more the soul is at peace, the more it moves with greater speed," citing the example of a body falling through water.[23] At stake is a cosmology in which nature's liveliness – the model, in all these examples, for contemplation – is viewed as passive *and* active, at the same time although not in the same way. The English Quaker mystic and natural philosopher Anne Conway, a contemporary of Guyon, uses a related argument when insisting that in the "region of perfect tranquility" there is "*the strongest and swiftest motion*."[24] In Conway's picture, tranquility is used to describe a *state* of matter and not matter's liveliness (or lack thereof). Mystical tranquility seems like rest, Conway explains, but this is not so because nothing is happening. Rather, it seems like nothing is happening because the movement being performed is not a struggle, it is "without any disturbance."[25] Like Guyon, Conway cites examples of bodies in free-fall, arguing that there are many cases in which matter appears to be at rest not because it is inert but because it is not struggling to stay in motion.

Another way in which passivity is pictured as an activity is by thinking about it in terms of the relationship of earthlings to the sun: "The creature may, indeed, open the window; but it is the sun [itself] that must give the light."[26] In this analogy, where the sun represents God and the window is the "heart" of the contemplator, the point is to scrutinize the idea that prayer causes God to appear, but it is also a critique of the idea that spiritual exercises are activities separate from the ordinary ways of mattering. Does this mean that a creature should not attempt, by their own effort, to "open the window," as Guyon puts it?[27] No, she reasons, reflecting that they key is to realize that while an effort is required, it is not an effort contrary to the body's ordinary way of being. Rather, contemplation should be modelled on photosynthesis, a passive activity in which solar energy is both received passively and processed actively.

If contemplation is more like nursing and photosynthesis than it is akin to a disembodied, cognitive exercise, then an old image from Christian mystical theology is put into question: the "spiritual combat," in which the soul is imaged as a fortress that must be defended against outside influences by mortifications and other heroic efforts to withstand the temptations of the flesh. Guyon adopts but also critique this central metaphor from asceticism: "the interior is not a stronghold taken by the cannon, but instead by love [...] God does not ask anything extraordinary or too difficult."[28] Alternatively, the "combat" can be refigured in non-violent terms:

> if we try to combat distractions directly, we will in fact irritate the situation, making it worse. Instead, we should focus more deeply on a vision of the present God and very simply recollect. Therefore, we combat these distractions indirectly and without thinking, which is a very effective method.[29]

Here, if there is an "enemy" to be defeated, it is the notion that matter is, in and of itself, evil and a source of temptation. As Guyon's were translated into English, Guyon's ideas became influential among Protestant dissident groups working for social change, especially the Quakers, in whom we see contemplative impulses already present and now refigured as means of increasing equality between the sexes, promoting political peace and also peace between species: Quakers took "inspiration for eco-justice imperatives and communitarianism *from mystic trance*."[30]

Passivity as Tactics

In looking to the early modern period for practices of passivity, I was influenced not only by its writers but also its historians. The French philosopher and historian of mysticism, Michel de Certeau, argued that in the late sixteenth and early seventeenth centuries ecstatic experiences became the subject of a distinct discourse, which its French advocates called *la mystique*, "mystics." One distinctive feature of mystics was a reaction to formal methods. Many writers argued – most notably Jeanne Guyon – that formal methods were not necessary in order to experience God. Instead, they emphasized minimal procedures, using words like "simple," "easy," and "natural" to describe spiritual exercises, comparing contemplation to bodily activities like breathing, digestion and even to photosynthesis.

Certeau's work on mystics was also the basis for a comprehensive critique of technique and practice from the perspective of embodiment, in which he sought to show how "[the ordinary] introduces itself into our techniques – in the way in which the sea flows back into pockets and crevices in beaches – and how it can reorganize the place from which discourse is produced."[31] Certeau pictured the ordinary not as a lack of technique but as an unconscious technique or habit, and argued that one way of rethinking technique was through a renewed attention to the small and largely unnoticeable repetitions through which bodies live, move and sustain themselves. He thought that mystics, *la mystique*, had attempted to do just this, and one key insight from Certeau's analysis, which I developed as I began writing about my research, is the prominence, in the spiritual exercises of early modern mystics, of passivity in the sense of *affectability*; that is, the importance of movements so small they almost escape notice, and yet of such importance that without them life would not be possible.[32]

Certeau wrote about affectability mostly from the perspective of human bodies, and several problems with his approach to mysticism have been noted, not least his emphasis on interiority.[33] Yet his interest in mysticism as a practice of the "ordinary" anticipates the ecological perspectives developed by feminist philosophers and historians. One scholar to recognize this idea, with explicit reference to the contemplative practices and early modern mysticism, was Elizabeth Potter. She puts it as follows: "[Many mystics during the seventeenth century] held a kind of materialist pantheism which identified God with the created world and so placed the spirit of life and cause of motion within terrestrial and celestial bodies themselves." "Materialist pantheism" is implicit in practices of contemplation which approach natural activities as models for contemplation, and see in the lifeways of animals and plants the figuration of the spiritual life, meaning that all that was needed in order to experience God was to look "at the world around them [...] because all of nature is full of God."[34]

Pantheism here is "materialist" because matter to many early modern mystics was not viewed as dead, inert or devoid of agency, in contrast to the mechanistic models developed during the early modern period by René Descartes, Isaac Newton, Robert Boyle, and Francis Bacon. Early modern forms of Christianity that advocated contemplative approaches to spirituality presented alternatives to religious but also to scientific orthodoxy, drawing instead on vitalist notions that Potter traces back to the natural magic tradition of Renaissance Neoplatonism. This connection was also shown by Carolyn Merchant in *The Death of Nature: Women, Ecology and the Scientific Revolution*. Merchant demonstrated the way in which ideas from the natural magic tradition and from Christian Kabbalah converged with mystical spirituality, the latter "[emphasizing] the inward presence of God, the living Christ within each individual, and the vitality of the living word."[35] When mystics during the early modern period insist that contemplation demanded no more effort than a plant's, saying – as Guyon did – that it was involuntary and passive, what was being rejected was the idea that spiritual practices were different to creaturely life.

Some scholars of mysticism have proposed that rather than see passiveness as central to the spiritual exercises of mystics, though, we should appreciate it instead as a rhetorical device. In the case of texts written by women like Guyon it is important to recognize that the passiveness we see described and endorsed "is not primarily about letting go of power, ego, and will, as it is passing through women's gendered 'nothingness' to finally gain perspective and wisdom over their own lives."[36] In order not to risk equating mystical passiveness with a negation of agency, however, it is not necessary to proceed to the other extreme and equate it with the acquisition of a will-to-power. As Beverly Lanzetta continues, "women mystics discovered, as did many of their male counterparts, that healing the wound of inferiority comes about not from the assimilation or imitation of the independent, autonomous, individualistic self – what some call the 'patriarchal self'."[37]

The reclaiming of mystical passivity and "nothingness" has been an important theme in several recent readings not only of women mystics but also of mystical theology in relation to philosophy more broadly. Writing from the perspective of queer theory, Kent Brintnall takes this, the claim that in mysticism "the subject's relation to itself, to its sense of its power and prerogatives" is "radically challenged," or, as Brintnall puts it, that there is "something to be gained by adopting practices that...will 'ruin in [us] that which is opposed to ruin'," to be the actual content of mysticism.[38] It has also been claimed that mystical "nothingness" resonates with the "ontological quandary" of the colonized, where nothingness "is the space or state of ultimate indeterminacy," in which the self is experienced not as "internally undifferentiated" but "*displaced*...internally incoherent, fractured, contradictory and always in a state of becoming."[39]

"Displacement" is also the literal meaning of "ecstasy,"[40] a concept connected intimately with contemplation, and in Christian mysticism it leads not to a negation of agency nor to a glorification of inferiority but to a refiguring of subjectivity in terms of becoming, dependency and vulnerability (the sense that one is not self-sufficient author of one's own existence). Mystical passiveness may be understood as an attempt to reclaim agency in a non-possessive mode that would resist the cultures of violence that identify agency narrowly with the possession and acquisition of power.

One reason for reading the early modern texts that give instructions for how to practice passivity is the detailed way in which they suggest a way of resisting the charge of nihilism without endorsing agency *qua* acquisition of power. It promises to do so through a clarification of the nature of passivity. I have so far characterized passivity as a creaturely or multi-species disposition, and it became important, both to my research and my practice, to view passivity as a *habit*, that is to say, to see passivity as passive in the way an unconscious or spontaneous action is passive, meaning that the passiveness in question is less like inertia and more like sensing, a comparison we also find frequently in Guyon's texts. If we are to read Guyon today, however, it also becomes important to work with a concept of "life" that is not based on a new dualism separating activity from passivity, recognizing instead that what we call life is always mixed with passiveness, modes of passiveness that compose every action, even as they compose the spontaneous generation of life from itself.

Guyon's practices, however, are also paradoxes, paradoxical calls to silence, indecision and non-conformity as a means of active resistance to what Certeau would call the "strategies" of dominant regimes. In *The Practice of Ordinary Life*, Certeau imagined the ordinary as "tactics" of opposition affording resistance to what Certeau called the "strategies" imposed by systems and institutions. "Strategies" are defined as a "Cartesian attitude…it is an effort to delimit one's own place in a world bewitched by the indivisible powers of the Other," an attitude Certeau associated with modern science, politics, and military strategy.[41] By contrast, tactics were seen by Certeau as precisely embracing the Other: "the space of a tactic is the space of the other."[42] Like "mystics," "tactics" in Certeau's description are open to the Other. Again, tactics shares with mystics an orientation to the moment; tactics are "limited by the possibilities of the moment"; they are also "determined by the *absence of power* just as a strategy is organized by the postulation of power."[43] Like every contemplative practice, Guyon's is a practice of attentiveness as well as of attention, aimed at cultivating concentration but also its opposite: a relaxed yet alert awareness in which an ability to notice and to respond spontaneously to suffering and demands for solidarity might be possible. Emerging from a time not only of widespread misogyny but also of violent ideological clashes between warring factions of Christian religion following the Reformation, Guyon was acutely attuned to the need for *common ground*. Scarred by the coercive methods of Catholic clergy, she found herself unable to enjoin practices that failed to let others be. In the mounting tensions of post-Reformation Europe, when Catholic but also Protestant missions were intensifying, Guyon chose to articulate her "tactics" in terms of listening, waiting and letting-be. Opposed to the control exercised by clergy, religious rituals and even doctrinal belief, Guyon looked instead to the spontaneous activities of the human body, but also to those of animals and plants, water, sunlight and air. Her notion of contemplation "as easy and natural as breathing," arose as a protest against the control of thought, speech and sensation by systems of power. By articulating the spiritual life in terms of non-discursive life, Guyon found a common ground that might resist such control, not in order thereby to escape human society, but rather to disarm its strategies of violence.

Contemplative Ecology Redux

Reading Guyon and practicing her methods of passivity, my modest suggestion is that contemplation makes tangible the lines between passivity and noticing, but also the lines between passivity and the crossing of species. Freedom in the passivity practiced by Guyon is realized in the relinquishing of power-over and the recognition of radical dependency and a fundamental affectability; it comes into force at the moment when forcefulness becomes weak and identifies itself as indistinct from sensation and from the life of sensing matter.

There is, in other words, a recognition of a structural homology between, on the one hand, the passivity of the soul as it relates to God in contemplative prayer, and, on the other hand, the passivity of life as the living relate to environments. Following this structural homology to its terminus we arrive at an ecological understanding of contemplation but also at a "contemplative ecology" such as was introduced recently by Douglas Christie. Protesting modern individualism, contemporary culture finds itself in a "reciprocal and deepening affinity between the language of spirituality and the language of the natural world."[44] Such a view does not entail the naturalization of spirituality nor the spiritualization of nature, because the idea of a "contemplative ecology" – of a spiritual practice that is also participating in the "natural" attitude of creaturely life – is about refusing to choose between a spiritual and a naturalistic paradigm when expressing intimacy with environments.

To contemplate is to co-labor alongside myriads of critters contemplating the elements as they contract, passively, oxygen into carbon dioxide. Ecstasy is an experience that reveals to the one who experiences it the passive activities with which life nourishes itself from itself, and ecologies are the condition of that revelation, even as they frequently become its content. The ecological condition of ecstasy explains the close connection between persons engaged in spiritual exercises and eco-justice and how it is that mystical revelations so frequently are concerned with the plight of earthlings. But even a minimal definition of contemplation as a practice of passiveness points to an ability to be attentive by one's surrounding. The heart of attentiveness is an alert receptivity more like patience than persistence. Contemplation, or something very much like it, becomes unavoidable in a philosophy that wants to think carefully about the habits that make care possible. Yves Citton's *Ecology of Attention* argues that attention needs to relate to the world the way a contemplative relates to icons: as "a presence that opens me further."[45] Citton proposes that it is because we fail to be attentive in ways that narrow rather than open the field of perception, that we do not see how "all our representations…are woven into the material realities" of our lives. Instead, we rely on "enquiry, research and calculation before we can become properly aware of [material realities]," eventually rendering awareness of material realities inseparable from enquiry, research and calculation.[46] The problem today, with the narrowing of attention and a concomitant focus on mental efforts to fathom the secrets of matter, is the same as that faced by mystics in the early modern period. This is why it is important, today, to think with their approaches to the practice of passivity.

Notes

1 Carolyn Merchant, *The Death of Nature: Women, Ecology and the Scientific Revolution* (San Francisco: Harper and Row, 1980).

2 Isabelle Stengers, "Introductory Notes on an Ecology of Practices," *Cultural Studies Review*, Vol. 11, No. 1 (2005): 183–196.

3 On Hadot, nature and contemplation, see Pierre Hadot, *The Selected Writings of Pierre Hadot: Philosophy as Practice*, eds. and trans. Federico Testa and Matthew Sharpe (London: Bloomsbury, 2020), 163–182.

4 John of the Cross, *The Collected Works of St. John of the Cross*, Rev. ed., trans. Kieran Kavanaugh and Otilio Rodriguez (Washington, DC: ICS Publications, 1991), 690 (*Living Flame of Love* 3.34). *The Living Flame of Love* describes, in allegory, the different stages of heightened affectivity characteristic of contemplative passivity.

5 Mircea Eliade, *Yoga: Immortality and Freedom*, trans. W. Trask (New York: Pantheon, 1958), 66.

6 Grace Jantzen, *Becoming Divine: Toward a Feminist Philosophy of Religion* (Manchester University Press, 1998), 156–170.
7 E.g. Daniel Chamovitz, *What a Plant Knows: A Field Guide to the Senses of Your Garden – And Beyond* (Scientific American, 2012); Anthony Trewavas, *Plant Behaviour and Intelligence* (Oxford: Oxford University Press, 2014); Suzanne Simard, *Finding the Mother Tree: Discovering the Wisdom of the Forest* (Knopf, 2021).
8 Elaine P. Miller, *The Vegetative Soul: From Philosophy of Nature to Subjectivity in the Feminine* (State University of New York, 2002); Michael Marder, *Plant-Thinking: A Philosophy of Vegetal Life* (Columbia University Press, 2013); Jeffrey Nealon, *Plant Theory: Biopower and Vegetable Life* (Stanford University Press, 2016); Luce Irigaray and Michael Marder, *Through Vegetal Being: Two Philosophical Perspectives* (Columbia University Press, 2016); Emanuele Coccia, *The Life of Plants: A Metaphysics of Mixture*, trans. Dylan Montanari (Polity, 2019).
9 Irigaray and Marder, *Through Vegetal Being*, 155.
10 Irigaray and Marder, *Through Vegetal Being*, 157–158.
11 Irigaray and Marder, *Through Vegetal Being*, 158.
12 Irigaray and Marder, *Through Vegetal Being*, 166.
13 Irigaray and Marder, *Through Vegetal Being*, 158–159.
14 Michael Marder, *Green Mass: The Ecological Theology of St Hildegard of Bingen* (Stanford University Press, 2021).
15 Simone Weil, *Waiting on God*, ed. Joseph-Marie Perrin and trans. Emma Craufurd (Routledge and Kegan Paul, 1951), 55.
16 Bernard McGinn, *The Crisis of Mysticism: Quietism in Seventeenth-Century Spain, Italy, and France* (Crossroad, 2021).
17 Jeanne Guyon, *The Prison Narratives of Jeanne Guyon*, trans. Dianne Guenin-Lelle and Ronney Mourad (Oxford University Press, 2011).
18 Jeanne Guyon, *Selected Writings*, ed. and trans. Dianne Guenin-Lelle and Ronney Mourad (Paulist Press, 2012), 59.
19 Guyon, *Selected Writings*, 28.
20 On Guyon's diaries and the effects of contemplative prayer on her mental health and psychic abilities, see the fascinating discussion in Marie-Florine Bruneau, *Women Mystics Confront the Modern World: Marie De L'Incarnation (1599–1672) and Madame Guyon (1648-1717)* (State University of New York Press, 1998).
21 Guyon, *Selected Writings*, 72.
22 Guyon, *Selected Writings*, 72.
23 Guyon, *Selected Writings*, 82.
24 Anne Conway, *The Principles of the Most Ancient and Modern Philosophy*, trans. Allison P. Coudert and Taylor Corse (Cambridge: Cambridge University Press, 1996), 27. My emphasis.
25 Conway, *Principles*, 27.
26 Guyon, *Selected Writings*, 97.
27 Guyon, *Selected Writings*, 97.
28 Guyon, *Selected Writings*, 73.
29 Guyon, *Selected Writings*, 61.
30 S. M. Komashin, "How Ecology and Economics Brought Winstanley and Nitobe to Quakerism," in *Quakers and Mysticism: Comparative and Syncretic Approaches to Mysticism*, ed. John Kershner (Cham: Palgrave Macmillan, 2014), 63–84 (69). Emphasis in the original.
31 Michel de Certeau, *The Practice of Everyday Life*, trans. Steven Rendall (Berkeley: University of California Press, 1984), 5.
32 Michel de Certeau, *The Mystic Fable, Vol. I: The Sixteenth and Seventeenth Centuries*, trans. Michael B. Smith (Chicago: University of Chicago Press, 1982), 10.
33 See, for example, Bernard McGinn, *The Foundations of Mysticism: Origins to the Fifth Century* (London: SCM Press, 1992), 310–313 ("Appendix: Theoretical Foundations"); Leigh Schmidt, "The Making of Modern 'Mysticism'," *Journal of the American Academy of Religion*, Vol. 71, No. 2 (2003): 273–302.
34 Elizabeth Potter, "Modeling the Gender Politics in Science," *Hypatia*, Vol. 3, No. 1, (1988): 19–33 (22).
35 Merchant, *Death of Nature*, 256.

36 Beverly Lanzetta, *Radical Wisdom: A Feminist Mystical Theology* (Minneapolis: Fortress Press, 2005), 96.
37 Lanzetta, *Radical Wisdom*, 96.
38 Kent Brintnall, "Erotic Ruination: Embracing the 'Savage Spirituality' of Barebacking," in *Negative Ecstasies: Georges Bataille and the Study of Religion*, eds. Jeremy Biles and Kent Brintnall (New York: Fordham University Press, 2015), 51–67 (66–67).
39 An Yountae, *The Decolonial Abyss: Mysticism and Cosmopolitics from the Ruins* (New York: Fordham University Press, 2018), 11–13.
40 Kevin Corrigan, "Ecstasy and Ecstasy in Some Early Pagan and Christian Mystical Writings," in *Greek and Medieval Studies in Honor of Leo Sweeney, S.J.*, ed. William J. Carroll and John J. Furlong (New York: Peter Lang, 1994), 27–38.
41 Certeau, *Everyday Life*, 32.
42 Certeau, *Everyday Life*, 37.
43 Certeau, *Everyday Life*, 38.
44 Douglas Christie, *The Blue Sapphire of the Mind: Notes for a Contemplative Ecology* (Oxford: Oxford University Press, 2013), 5.
45 Yves Citton, *The Ecology of Attention*, trans. Barnaby Norman (Cambridge: Polity Press, 2017), 181, citing Jean-Yves Leloup, *Un art de l'attention* (Paris: Albin Michel, 2002), 76.
46 Citton, *Ecology of Attention*, 82.

Bibliography

Brintnall, Kent. "Erotic Ruination: Embracing the 'Savage Spirituality' of Barebacking." In *Negative Ecstasies: Georges Bataille and the Study of Religion*. Eds. Jeremy Biles and Kent Brintnall, 51–67. New York: Fordham University Press, 2015.

Bruneau, Marie-Florine. *Women Mystics Confront the Modern World: Marie De L'Incarnation (1599-1672) and Madame Guyon (1648–1717)*. Albany: State University of New York Press, 1998.

Certeau, Michel de. *The Mystic Fable, Vol. I: The Sixteenth and Seventeenth Centuries*. Trans. Michael B. Smith. Chicago: University of Chicago Press, 1982.

Certeau, Michel de. *The Practice of Everyday Life*. Trans. Steven Rendall. Berkeley: University of California Press, 1984.

Chamovitz, Daniel. *What a Plant Knows: A Field Guide to the Senses of Your Garden – And Beyond*. New York: Scientific American/Farrar, Straus and Giroux, 2012.

Christie, Douglas. *The Blue Sapphire of the Mind: Notes for a Contemplative Ecology*. Oxford: Oxford University Press, 2013.

Citton, Yves. *The Ecology of Attention*. Trans. Barnaby Norman. Cambridge: Polity Press, 2017.

Coccia, Emanuele. *The Life of Plants: A Metaphysics of Mixture*. Trans. Dylan Montanari. Cambridge: Polity, 2019.

Conway, Anne. *The Principles of the Most Ancient and Modern Philosophy*. Trans. Allison P. Coudert and Taylor Corse. Cambridge: Cambridge University Press, 1996.

Corrigan, Kevin. "Ecstasy and Ecstasy in Some Early Pagan and Christian Mystical Writings." In *Greek and Medieval Studies in Honor of Leo Sweeney, S.J.*. Ed. William J. Carroll and John J. Furlong, 27–38. New York: Peter Lang, 1994.

Eliade, Mircea. *Yoga: Immortality and Freedom*. Trans. W. Trask. New York: Pantheon, 1958.

Guyon, Jeanne. *The Prison Narratives of Jeanne Guyon*. Trans. Dianne Guenin-Lelle and Ronney Mourad. New York: Oxford University Press, 2011.

Guyon, Jeanne. *Selected Writings*. Ed. and trans. Dianne Guenin-Lelle and Ronney Mourad. Paulist Press, 2012.

Hadot, Pierre. *The Selected Writings of Pierre Hadot: Philosophy as Practice*. Eds. and trans. Federico Testa and Matthew Sharpe. London: Bloomsbury, 2020.

Irigaray, Luce and Michael Marder. *Through Vegetal Being: Two Philosophical Perspectives*. New York: Columbia University Press, 2016.

Jantzen, Grace. *Becoming Divine: Toward a Feminist Philosophy of Religion*. Manchester: Manchester University Press, 1998.

John of the Cross. *The Collected Works of St. John of the Cross*. Trans. Kieran Kavanaugh and Otilio Rodriguez. Washington, DC: ICS Publications, 1991.

Komashin, S. M. "How Ecology and Economics Brought Winstanley and Nitobe to Quakerism." In *Quakers and Mysticism: Comparative and Syncretic Approaches to Mysticism*. Ed. John Kershner, 63–84. Cham: Palgrave Macmillan, 2014.

Kotva, Simone. "The Astonishment of Philosophy: William Desmond and Isabelle Stengers." In *Astonishment and Science: Engagements with William Desmond*. Ed. Paul Tyson, 258–270. Eugene, OR: Wipf and Stock, 2022.

Lanzetta, Beverly. *Radical Wisdom: A Feminist Mystical Theology*. Minneapolis: Fortress Press, 2005.

Marder, Michael. *Plant-Thinking: A Philosophy of Vegetal Life*. New York: Columbia University Press, 2013.

Marder, Michael. *Green Mass: The Ecological Theology of St Hildegard of Bingen*. Stanford: Stanford University Press, 2021.

McGinn, Bernard. *The Foundations of Mysticism: Origins to the Fifth Century*. London: SCM Press, 1992.

McGinn, Bernard. *The Crisis of Mysticism: Quietism in Seventeenth-Century Spain, Italy, and France*. New York: Crossroad, 2021.

Merchant, Carolyn. *The Death of Nature: Women, Ecology and the Scientific Revolution*. San Francisco: Harper and Row, 1980.

Miller, Elaine P. *The Vegetative Soul: From Philosophy of Nature to Subjectivity in the Feminine*. Albany, NY; State University of New York, 2002.

Nealon, Jeffrey. *Plant Theory: Biopower and Vegetable Life*. Stanford: Stanford University Press, 2016.

Potter, Elizabeth. "Modeling the Gender Politics in Science," *Hypatia*, Vol. 3, No. 1 (1988): 19–33.

Schmidt, Leigh. "The Making of Modern 'Mysticism'." *Journal of the American Academy of Religion* 71, No. 2 (2003): 273–302.

Simard, Suzanne. *Finding the Mother Tree: Discovering the Wisdom of the Forest*. New York: Knopf, 2021.

Trewavas, Anthony. *Plant Behaviour and Intelligence*. Oxford: Oxford University Press, 2014.

Ward, Patricia. "Madame Guyon and Experiential Theology in America." *Church History*, Vol. 67, No. 3 (1998): 484–498.

Ward, Patricia. *Experimental Theology in America: Madame Guyon, Fénelon, and Their Readers*. Waco, TX: Baylor University Press, 2009.

Weil, Simone. *Waiting on God*. Ed. Joseph-Marie Perrin and trans. Emma Craufurd. London: Routledge and Kegan Paul, 1951.

Yountae, An. *The Decolonial Abyss: Mysticism and Cosmopolitics from the Ruins*. New York: Fordham University Press, 2018.

19

A PHENOMENOLOGY OF SEXUAL DIFFERENCE AND PLANT LIFE

Michael Marder

19.1 Vegetality and Sexual Difference

Reflections on eco-phenomenology would be incomplete without a sustained discussion of sexual difference. First, following the insights of psychoanalysis and of a later generation of phenomenologists (notably, Jacques Derrida and my colleague-coauthor Luce Irigaray), from its apparently secondary position this difference shapes the ontology of human beings. More than a feature of embodiment, sexual difference is coextensive with embodiment as such, albeit never from a total or totalizing position. Second, in keeping with key figures in German *Naturphilosophie*, sexual difference is inherent to life itself, even when the organisms in question do not resort to sexual reproduction. As Friedrich Schelling notes in his *First Outline*, "throughout the whole of nature, absolute sexlessness is nowhere demonstrable, and an *a priori* regulative principle requires that sexual difference be taken as point of departure everywhere in organic nature [*organischen Natur auf Geschlechtsverscheidenheit auszugehen*]."[1]

Taken together, the two arguments on the fundamental character of sexual difference in human ontology/embodiment and in organic nature have a strong bearing on phenomenology. There are no phenomenological perspectives on the world free of this difference, nor are there any phenomenological worlds uncolored by it. It affects every aspect of human existence largely through the apparatus of what the early Heidegger calls *preunderstanding*, a non-thematized, implicit interpretation of ourselves and others, a lived relation to our facticity. Which is not to say that this interpretation may not become explicit, thematized in relation to the most diverse elements of being-in-the-world, as it does, for example, in gender theory or in critical phenomenology. And, as the transition to critical phenomenology suggests, in becoming thematic, sexual difference is also opened to questioning, to a critique that unsettles its outlines—above all, its imprisonment in a binary system.

Schelling's approach, consonant as it is with cosmological thought outside the Western fold, charts yet another promising, more obviously eco-phenomenological, path toward the thinking of sexual difference. The sexuation of organic nature as a whole means that it is not only a cornerstone of human ontology, but of the ontology of organic existence *tout court*.

 DOI: 10.4324/9781003538097-24

Here, however, a more nuanced treatment of the subject is in order, seeing that, by definition, we are not dealing with an abstract and homogeneous foundation (that would take us back to metaphysical grounds) but with a fractured beginning, eschewing totalization.

Although sexual difference begins most exuberantly (after the initial experimentation of bacteria with quasi-sexual conjugation and gene transfer) in the vegetal world, it does not find its pure origin in that world but cuts right through the plant kingdom, whose members rely both on sexual and on asexual modes of reproduction. The difference that prevails there is between difference and its absence, and it is this versatile difference between difference and indifference that stamps vegetal ontology with its unique character. Perhaps, this character cannot be really said, but only written; writing is, after all, a vegetal field where words grow silently and grow silent. Plants do not exactly provide "us" with a meta-language for expressing ourselves: to say sexual difference—or anything else for that matter—with plants is not to produce vibrations in the vocal cords, the sound passing through the resonators of the throat, mouth cavity and nose, so as to finally undergo articulation by the tongue, soft palate, and lips. Saying it with plants is much more passive than speaking; it is conveying "our" entire bodily extension, flesh or corporeity where distinctions between individual bodies and body politic do not yet, or no longer, matter. Suffused with sexual difference, the corporeity called human bears such a difference as a trace of vegetality.

Yet, it is precisely here, in the suffusion of animal and human bodies with sexuality, that Hegel records their divergence from the sexualization of plants. According to his *Philosophy of Nature*, sexual difference is "only quite partial" in the vegetal kingdom ("*der Unterschied ist so nur ganz partiell*")[2] for two reasons. First, plant sexuality is indeterminate, and "the differences are very often changeable while plants are growing." Second, and more importantly, this sexuality is concentrated in the flower, a detachable and superfluous part of plant, *ein abgeschiedener Teil*. "The different individuals," Hegel writes,

> cannot therefore be regarded as of different sexes because they have not been completely imbued with the *principle* of their opposition [*sie nicht in daß* Prinzip *ihrer Eingegensetzung ganz eingetaucht find*]—because this does not completely pervade them [*nicht ganz durchbringt*], is not a universal moment of the entire individual…

In keeping with Hegelian dialectics, "we" cannot include plants in saying *we* or *us*: not because they cannot say it themselves (though voice, which initially resounds in animality, is a crucial factor of subjectivity for Hegel) but because they are too indeterminate, undecided between sexual and asexual modes of being and, within sexuality, between the masculine and the feminine sides. Vegetal sexual difference oscillates between the disjunction and the conjunction of the polarities it interrelates, on the one hand, and their irrelevance to asexual reproduction, on the other. Its refinement and determination will require a transition to animal existence, where the entire flesh is awash in sexuality, arranged in oppositional formations of the masculine against the feminine, *us* versus *them*. What Hegel postulates, then, is the animalization of sexual difference through "the principle of opposition," as a sign of the animal's engagement, interest, and non-indifference in itself and in its other. And he interprets the precarious situation of the plant as that of an indifferent difference: uncommitted, utterly malleable, plastic, fluid due to its presumed lack of involvement, the incapacity to negate itself and its other on a path to the flowering of subjectivity. The breath-taking freedom of the plant flips, on Hegel's prompting, into a nearly mechanistic individuality, a moment of technicity of and in nature, dovetailing in a passive manner with technocracy's *active* indifference that neutralizes all meaningful differences.

To stay with technocracy for a moment: what this regime entails is the governance of people, as well as of inanimate objects, plants, and animals on the basis of technical rationality: not even logic but logistics, rules shorn of any substance, yet, despite their self-professed neutrality and allergy to any kind of sovereignty, subservient to global capital. (In a dark parody of Hegel's spirit, it is neither substance nor subject!) Technocracy is the political ramification of commodity fetishism that catches all living and non-living beings in the nets of the indifference it flaunts as the perfection of an even-handed, non-corrupt, well-oiled political and politically correct mechanism. In response to this implacable force that makes everything and everyone equally governable, predictable, and subject to manipulation, insistence on difference (notably, on sexual difference) is of paramount significance. Irigaray has recently acknowledged and begun elaborating the ecologico-political framework of sexual difference for the age of neutralization that by far exceeds human beings. For my part, I attend to the shape of this difference in the vegetal world, from which it hails. So, how about a crisp schema: plants against technocracy; sexual difference against abstract indifference; the finite, limited-gendered-growing-decaying *us* against the disembodied, gender-neutral but actually masculine, metaphysical *us* who invariably stand against *them*?

To the polymorphous sexuality of plants, to their non-oppositionality, Hegel opposes "the principle of opposition," that is, the dialectical principle *par excellence*. The implications of this move are broad. "We" can no longer say *we* or *us* without, in the same breath, standing over and against "them," who are not just "not-us" but ontologically opposed to us. With this proviso: so long as "we" suppress "our" vegetal heritage. Being oneself is being against the other, not thriving in contiguity with alterity in the manner of plants; when all is said and done, it is being an animal, animalizing all difference, including sexual, in saying (roaring, screeching, growling, chirping, croaking, bellowing, hissing, barking, squeeking…) *we, nosotros o nosotras* in opposition to others who are not "us." From Hegel's standpoint, not being against the other, the plant is also not itself. Immersed in the immediacy of an affirmation toward the outside, often indistinguishable from the inside, the plant is not bathed in, soaked, permeated (*eingetaucht, durchbringt*) with sexual difference as the principle of opposition. Or, at least, not totally (*ganz*) so, for this is a matter of partiality and totality: the totality of difference *or* its detachable, easily differentiated expression in a relatively free-standing part. The plant's immersion in pure immanence separates its sexual organs, namely the flowers, from the rest of its body that disobeys the principle of opposition, in and through which difference is registered, organized, and incorporated into organic being. The Hegelian flower is cut apriori, culled from the vegetal extension to which it belongs, as it were, by accident, as a playful and not entirely necessary supplement.

19.2 Sexual Ontology, Vegetal Ontology

Plant sexual ontology condenses in itself all the promises and paradoxes of plant ontology as such. Their slipping away from the principle of opposition liberates plants from two things at once: the dominion of a principle and the logic of oppositionality. Growing and reproducing regardless of a single governing rule (say, the "law" of sexual reproduction) and without beginning, an-archaically, plants do not invest their *we* with a predetermined meaning. Rather, meaning itself, and above all the sense of vegetal collectivity, blossoms and comes to fruition at the pace of their life of lives, full of surprising twists and turns, of differences that are "very changeable," mutable, alterable: *sehr wandelbar*. Plant world is home to different differences, that is, differences that preclude sameness both at the level of their form and at the level of content, whereas Hegel strives to set up a dialectic of difference in content and formal sameness, which is the principle of negation and contradiction. Sexual difference is not an exception from this rule. By being *too* different in its vegetal version, it reverts back, in Hegel's eyes, to indifference and non-differentiation, affecting the "genus-process," *Gattungs-Prozeß*, and the life-process of each plant.

What is the vegetal *we*, then, quilted together as it is from different differences, absent a foothold, or a roothold, in sameness? (I note in passing that radical difference is a quality of vegetal subjectivity, which cannot be observed within the knowledge context of biology. Evidently, plant cells and DNA do not display that kind of difference, even if they support it with remarkable phenotypic plasticity and, more pertinently, fluid sex determination). In the case of animal and human communities, a *we* consists of a number of individuals; it names a grouping or an assembly comprised of members who are distinct from the collectivity and who, participating *in* it, maintain their identity *outside* it. But the Hegelian point is that, in order to occupy this position inside-out or outside-in, the individual must be imbued with the principle oppositionality, starting with the oppositional organization of sexual difference. In its entirety, as a whole, the individual is determined by the possibility of a total negation by the other, with sexuality and death, *Eros* and *Thanatos*, fleshing out the totality of existential negation. The individuals' participation in a community is a return, a conscious taking-hold of, or, in totalitarian systems, delivering themselves to, the forces responsible for their individuation. First and foremost, their sexual individuation, seeing that "the entire habit (*habitus*) of the individual must be bound up with its sex [*Der ganze Habitus des Individuums muß mit seinem Geschlecht verbunden sein*]," that is to say, with the principle of oppositionality that defines dialectical sexual ontology and dialectics as such, itself an expression of oppositionally organized sexual difference in thought.

Plants, for their part, are and think differently, indicating an escape route from dialectics within dialectics and putting into question the suffocating totality of community.

> The difference reached by the plant, which is a difference of one vegetative self [*einem vegetativen Selbst*] from another vegetative self, each of which has the urge to identify itself with the other—this determination exists only as an analogue of the sexual relation [*ein Analogon des Geschlechts-Verhältnisse*]. For the sides of the relation are not two individuals.

The vegetal *we* is made not of individuals but of super- or infra-individual differences, because the "self" of a plant is not an individual, either. In the flora, sexual difference is a relation involving something more or something less than two individuals, which, for Hegel, is no longer or not yet a relation, but its "analogue"—literally, the putting into proportion of differences. The analogy of the sexual relation slots plants above and below *logos*, the realm of principles and oppositions summed up in the principle of oppositionality. Circumventing the mediations of *logos*, the vegetal *we* is unsayable, uncontainable in the medium of the voice and inappropriate to a way of assembling thoughts in the formations of contra-diction or non-contra-diction. Rather than inside-out or outside-in a community, the plant is completely outside, which, if we follow Hegel, comes to being completely within, based on the indifference of radical difference.

19.3 The Superficiality of Plant Existence and Sexuality

Besides the extra-individual nature of vegetal existence, which contributes to the misalignment of communal boundaries between this existence and that of animals and humans, the "analogy of the sexual relation" explains the essential superficiality of plant life. Sexualization grants the dialectical subject its form, as well as its specific character. At its crest, only "when the inner, generative forces have reached complete penetration and saturation [*die ganze Durchbringung und Sättigung*], does the individual possess the sexual impulse, and only then is it awakened." The subjective interiority of the animal and the human is the product of the withholding of the sexual impulse within a totality forged by generative forces in a state of "complete penetration and saturation." Before the act of penetration, the individual is penetrated by sexual difference, to the point of repletion and satiation, triggering the sexual impulse. Its subjectivity is nothing other than a force (here, the drive, "generative forces") kept at bay and inexhaustible in its occasional exteriorizations. Any *we* that may emerge from the tensions and bursts of that accumulated force will happen as a meeting of two or more interiorities, of dark recesses and reserves where each subject will hide, despite and in the course of its exposure to the other.

Conversely, the vegetal *we* will be superficial, craving nothing more than a caress of sunrays or of water, of a gust of wind or butterfly wings spreading pollen. In its exteriority, the "analogy of the sexual relation" is a brushing of two or more surfaces that do not occlude a hidden reserve but open unto to the other unreservedly. There might be friction but no opposition there, where in place of the drive, pulsion of force, "an impulse," one finds "the development of its [the plant's] organs [*daß Bildende seiner Organe*]," the formation and opening of flowers. Flowering toward the other (outside the vegetal world, femininity in general has been frequently associated with such flowering) is in stark contrast to penetrating or being penetrated by the other. How to imagine the *we* of this flowering and the flowered-toward…? How to say *we* or *us* with this other *we*, the one constitutively turned toward the other and other even to itself?

Hegel, for one, is incapable of doing, saying, or writing so. Devoid of opposition, vegetal difference becomes indistinguishable from indifference, otherness from sameness, unconditional openness to the other from absolute closure. So much so that Hegel disclaims the sexual difference of plants: "The plant therefore is asexual [*Geschlechtlos*], even the *Dioecia*, because the sexual parts form a closed, separate circle [*eine abgeschlossenen, besonderen Kreis*] apart from their individuality."[3] Beyond dialectical being, concretized across its self-negations and self-determinations that commence with its abstract opposition to nothing, there is but non-being; outside the confines of dialectical thought—only non-thought. To be sure, Hegel could not have known about phytoestrogens and complex hormonal networks traversing the bodies of plants and carrying these and other hormones well beyond the flowers. But this is not a good excuse. Behind his use of botanical data at the cutting edge of nineteenth-century science is a double methodological thrust I have already pointed out: (1) to cull the flower from the rest of the vegetal body, whose asexual and deathless non-individuated limits it transgresses and (2) by culling the flower, to cut the plant from relationality and a possible model of community—something "we" must resist today.

Why then read *Philosophy of Nature*, Part II of the *Encyclopaedia of the Philosophical Sciences* in "our" twenty-first century, given that the author of this treatise seems to preclude the more just, inclusive, ecologically and socially sensitive ways of articulating the common? Because dialectical reason works behind Hegel's back: in addition to co-opting and neutralizing, it promotes the vectors of liberation, among them the vegetal and the communal. Take, for instance, the early formulation of sexual difference in plants in Paragraph 348:

> But the plant does not attain to a relationship between individuals as such [*als solcher*] but only to a difference, whose sides are not at the same time in themselves whole individuals…; therefore, the difference, too, does not go beyond a beginning [*einem Beginn*] and an adumbration of the genus-process.[4]

A relationship between individual terms recedes to the background in the measure that the terms are individuated and foregrounded, obstructing with their definite outlines the very relation they participate in; in other words, they leave no room for relationality or difference as such. Plants, however, attain to a difference underlying any relation, a silent word underwriting all speech. They have already articulated their *we* before "us" and as a precondition for "our" saying anything whatsoever. That articulation "does not go beyond a beginning"—in fact, it may even precede the beginning—but its commencement is one "we" must always return to in every aspiration to a beyond. The uniqueness of vegetal difference is that it is generic albeit not abstract, different *as such* and *not as such*, essentially and existentially. For this reason, it admits diverse genres and genders into its midst, even as it withdraws from the relational grounds it fertilizes.

I wonder, at the same time, what would have happened, were Hegel's *Philosophy of Nature* to transition not only from vegetality to animality but also from animality ("the animal organism," with which it culminates) to humanity, and were it to do so precisely as a philosophy of *nature*, rather than a phenomenology of *spirit*. How would the manifold differences in sexual difference have appeared? In which light would they have been cast? A possible conjecture is that, once humanized, sexual difference will be wrapped in tragedy. Already an animal experiences sexuality as the embodied principle of opposition, permeating the entire organism, and, therefore, as "the negative power [*die negative Macht*] over the individual, realized through the sacrifice of this individual which it replaces by another."[5] The very negative power that produces the individual by turning the self against itself and against its other consumes its product in the species, to which the organism is subordinated. The *we* gains an upper hand over the *I*, if only by adding another *I* meant to replace the first. Resolving oppositionally configured sexual difference in progeny is not a happy synthesis but, on the contrary, intergenerational sacrifice, continuing to fuel dialectical frictions between the singular and the universal.

As for the human, sexual difference reaches over into the legal and political spheres, where the opposition between the masculine and the feminine comes to a head. In *Phenomenology of Spirit*, Creon and Antigone enact the tragic dimension that exacerbates the power of negativity palpable in animal sexuality. Sublimated, the principle of oppositionality that saturates a sexed body spills over into distinct conceptions and practices of communal being with irreconcilable demands laid on the subject: the *we* of the family and a living tradition versus the *we* of the state and a deadening law. The gendered assignment of roles and modalities of communal belonging yield ultimately incompatible and discordant tonalities of saying *we* or *us*, depending on who it is that is speaking. But the tragedy, *stricto sensu*, is that, while the human is both, the interplay of sexual difference leads to a sacrifice of one term in the relation (here, Antigone) to the other (Creon). That is a pyrrhic victory of humanity, its most intimate difference and relationality in ruins, over which presides a neutral and neutralizing "human law," as opposed to the "divine law" that dies with Antigone. Negative power migrates from the species-organism to the law-legal subject nexus, making Creon one of technocracy's pioneers.

Because Hegel takes the generic character of sexual difference in plants to be "formal," and so immune to internal negation and to permeation by the spirit of opposition, he concedes that the element of sacrifice is absent from vegetal sexuality. Instead of a negative power, the plant incarnates "this positive side," *diese positive Seite*, of the genus-process, which is one and the same as its "relationship to the outer world."[6] Its *I* is immediately a *we*, and its *we* an *I* oriented toward exteriority. The self-production of a vegetal living being is its reproduction in the other, and its replacement by another is an affirmation of the heteronomous self. The positive side that knows no sacrifice is, in the end, "the only necessary side of the negation [*allein nötige Seite der Negation*]"[7] oblivious to dialectical mediations. *Too* different, it lapses into sameness. So, whereas the animal or animalized iterations of sexual difference are tragic, the vegetal varieties are comic, full of sudden reversals, mishmashes, and confusions. ("In *Phenomenology of Spirit*, Hegel asserts that plants were the living enablers of the "mingling" or "blending," *Vermischung*, of natural forms and forms of thought—hence, what I term plant-thinking.")[8] The comic element of vegetal sexuality resides in these confusions or immediate identifications of extremes as much as in "playful superfluity"[9] and contingency, given that the plant can reproduce itself without resorting to sexual difference.

To the problem of the *we*, to the question concerning "us," sexual difference in plants silently responds that existence *is* coexistence. No doubt, this insight holds for human beings, too, and Heidegger put it with utmost lucidity when he described Dasein as Mitdasein, being-there as being-with. But, although playful inversions and confusions are abundant in human life, they are nowhere near the plasticity of vegetal existence, sexual or otherwise. At this point, again, the significance of Hegel is glaring. For all that one may learn from plants, direct mimesis (a non-representational imitation of their being) is a non-starter. Hegel's take-home message is that mediation is ineluctable, not the least for ecologically different sexual differences. Deepening the tragic standoff of animality, human sexuality has not yet become human; if anything—and much like family, society, law, politics and other ingredients of the Hegelian right (*Recht*)—it has been hyperanimalized. Becoming human does not imply rejecting "our" rich non-human heritage, but, on the contrary, drawing on more of this heritage, especially on the repressed vegetal dimension of sexual difference. The challenge is to embrace the plant and the animal, surface and depth, non-oppositionality and oppositionality in a condition I can only call tragicomic. To combine, in resisting global technocracy, what Nietzsche derided as "bearish seriousness" with flowery levity. To say *us* or *we* with "them," who need not be either conflated with or segregated from "us"…

19.4 L'écriture végétale

I have mentioned at the outset of this text that writing is a medium most propitious to vegetal expression, be it a self-expression or one that is collaborative with other lifeforms, including the humans "we" are yet to become. Now, if it is indeed the case that plant-thinking drastically changes how we view both plants and ourselves (in relation to them, each to her- or himself, and to each other), then a qualitatively new mode of writing is required to express such seismic shifts of perspective. For my part, I have not tried deliberately to cultivate a writing style that, in the footsteps of last century's experiments by philosophers including

Hélène Cixous, Julia Kristeva, and Luce Irigaray with *l'écriture féminine* ("feminine writing"), I would like to term *l'écriture végétale* ("vegetal writing"). But the absence of a deliberate plan did not mean that the seeds of vegetal writing remained dormant and have not germinated in something recognizable only in retrospect. In effect, the unconscious manner in which this non-project unfolded was probably more faithful to plants than all the painstaking attempts to emulate their growth in conscious, self-conscious, and representation-bound human thought.

The rapid extension of my initial project from *Plant-Thinking* (2013) took me by surprise. I often joke that my philosophy of vegetal life itself grew like a plant, tending in all directions at once: toward the sciences and the fine arts, ethics and politics, intellectual biography and autobiography. There is more than a grain of truth in this statement. The content, form, and method of plant-thinking cannot be easily parsed out into its independent components, its *how*, *where*, and *when* influencing the *what*. Its life and its text are as much about the milieu and the context as they are about whatever is conveyed by means of the formal expression all too swiftly inserted into a venerable lineage of "critical" and "academic" discourses.

I started with a theoretical exploration of the place of plants in metaphysics and of the role they could play in a postmetaphysical transition. From there, I embarked on a journey through the fields, groves, forests, and gardens of philosophy, seeking guidance for an alternative reading of the tradition (and of certain philosophical biographies) from the vegetation that proliferated in the canon's blind spots. The results of this exercise are gathered in *The Philosopher's Plant* (2014). Soon thereafter, together with Irigaray, I delved into a collaborative auto-hetero-phyto-bio-graphy, the sharing of a life or of lives among the co-authors, the readers, and plants themselves. Thus *Through Vegetal Being* (2016) was born. Another collaborative attempt (this time with Anaïs Tondeur) sifted through autobiographic, aesthetic, historical, and phytological prisms and also came to fruition that same year in *The Chernobyl Herbarium* (2016b). And, finally, the collection *Grafts* (Marder 2016a) appeared, as I say in the preface to that book, "in the in-between space where previously unimagined forms of thinking eventually come to life" thanks to the interventions, interactions, interpretations and interviews focused on the thinking *of*—about but also proper to—plants.

A definite pattern comes through here, in this path I have briefly sketched out: moving from the universal to the singular and, in an immediate reflux, back again from a more or less solitary work to thinking with and growing with others, human or not. The movement thus followed a trajectory from "us" through *I* to another *we*. But a loose pattern is not the same thing as a method—itself a way, which is universally prescribed or at least universalizable and therefore valid to the extent that it lends itself to being trodden by many feet over and over again. If I am correct in teasing out from my itinerary a tendency toward the singular *we*, then the path I have taken thus far is an anti-method, which, on its own terms, prohibits any following-after, though readers are absolutely free to let the above prohibition fall on deaf ears. In the ambiguous interplay of its singular and universal moments, in its exemplary non-exemplariness, I discern the outlines of *l'écriture végétale* that replicates the singular universality of the plants themselves. While plants are parts of the natural world (understood in the Greek sense of *phusis*: overall growth, emergence, coming-to-appearance, or coming-out-of-itself), they embody, in all their singularity, the universality of nature. While *l'écriture végétale* departs from and circles back to the vegetal, it sweeps the entire world into its midst, not sparing "us" either.

I am not going to dwell at length on the convergences and divergences between *l'écriture féminine* and *l'écriture végétale*. Such a comparison deserves much more care and attention than I can give it here. One thing ought to be mentioned, however: both kinds of writing invite misinterpretation based on what amounts to the seduction of classification. It is assumed that the feminine or the vegetal are classes of beings like any other, that is to say, discrete kinds or immaterial boxes within which women and plants respectively belong. Yet, nothing is further from the truth. Not only do vegetal and feminine ontologies embrace unfathomably more than the ontic realities of plants and women, but they are, in themselves, plastic enough to accommodate their other (the human, the masculine…) who has traditionally denigrated and abused them. In their singularity, *l'écriture féminine* and *l'écriture végétale* gesture toward universality.

In conclusion, another word on writing and its relation to plants is in order. Note that I am not saying (or writing—and that is not a minor nuance) that the experimental modes of expression discussed here break the silence imposed on certain living beings as a curse or a prison. Instead, these modes of expression dispense with the deafening noise of *logos* and put their hopes in the voice-less supplement of speech, namely writing, which, according to Jacques Derrida, comes first, preceding that which it supplements. Slipping below the incessant "talk-show" (I hear this term with a literal, as well as a phenomenological, ear) that speech is or has become, *l'écriture féminine* and *l'écriture végétale* touch upon something of arche-writing, the practice of tracing and inscribing inseparable from existence as such. That they do so is testimony to the fact that neither of them is easily classifiable, and so is not detainable within the empirical confines of a writing style, let alone of identity politics everywhere seeking the "voices" of those who seem to represent previously oppressed groups.

L'écriture végétale is further distinguished by two imperfectly overlapping frames, their edges unaligned with one another. On the one hand—the writing or archi-writing of plants and of the vegetal aspect in creatures normally not classified as plants; on the other—the writing about vegetal archi-writing, retracing its traces, or letting these traces come to the fore, finally liberated from a long history of their erasure, not least in the domain of sexual difference, which encompasses nothing less than ontological difference. The gap between the two frames is far from a debilitating condition; it opens a breathing space that guarantees some measure of autonomy to plants, even as it foments our intimate relation to them. Supplemented by plant-writing (and the supplement always comes first!), plant-thinking operates within the enabling limits of this double framing that determines it without imposing the suffocating closure of classificatory thought.

Notes

1 F.W.J. Schelling, *First Outline of a System of the Philosophy of Nature*, translated by Keith R. Peterson (Albany, NY: SUNY Press, 2004), p. 36.
2 G.W.F. Hegel, *Philosophy of Nature: Encyclopedia of the Philosophical Sciences, Part II*, translated by A.V. Miller (Oxford: Oxford University Press, 2004), p. 344. Until the next indication, all subsequent quotations are from this page of the text.
3 Hegel, *Philosophy of Nature*, p. 345.
4 Hegel, *Philosophy of Nature*, p. 343.
5 Hegel, *Philosophy of Nature*, p. 346.
6 Hegel, *Philosophy of Nature*, p. 346.
7 Hegel, *Philosophy of Nature*, p. 347.
8 G.W.F. Hegel, *Hegel's Phenomenology of Spirit*, translated by A.V. Miller (Oxford & New York: Oxford University Press, 1977), Paragraph 706.
9 Hegel, *Philosophy of Nature*, p. 345.

References

Hegel, G.W.F. *Philosophy of Nature. In Encyclopaedia of the Philosophical Sciences in Basic Outline, Part II*. Translated by A.V. Miller. Oxford: Clarendon Press, 1970.

Hegel, G.W.F. *Phenomenology of Spirit*. Translated by A.V. Miller. Oxford: Oxford University Press, 1977.

Heidegger, Martin. *Being and Time*. Translated by John Macquarrie and Edward Robinson. New York: Harper & Row, 1962.

Irigaray, Luce, and Michael Marder. *Through Vegetal Being: Two Philosophical Perspectives*. New York: Columbia University Press, 2016.

Marder, Michael. *Plant-Thinking: A Philosophy of Vegetal Life*. New York: Columbia University Press, 2013.

Marder, Michael. *The Philosopher's Plant: An Intellectual Herbarium*. New York: Columbia University Press, 2014.

Marder, Michael. *Grafts: Writings on Plants*. Minneapolis: Univocal Publishing, 2016a.

Marder, Michael. *The Chernobyl Herbarium: Fragments of an Exploded Consciousness*. With Anaïs Tondeur. London: Open Humanities Press, 2016b.

Schelling, Friedrich Wilhelm Joseph von. *First Outline of a System of the Philosophy of Nature*. Translated by Keith R. Peterson. Albany: State University of New York Press, 2004 [1799].

20

APPROXIMATE RELATIONS

Chimpanzee Paintings and the Performance of the Human

Louise Green

Encounter

I have never met a chimpanzee. They are not native to South Africa although some do dwell here in the Jane Goodall Chimpanzee Eden, a part of the Umhloti Nature Reserve in the northern part of the country. I have had an encounter with one of their primate relatives. It took place in a section of rainforest outside a small town on the south coast of South Africa. Walking alone along a narrow jeep track through dense forest, I observed ahead of me the dark furred back of something big. Although there are no wild bear in South Africa, the size of it, the shock of finding myself in company, brought the word 'bear' into my head. The next instant I recognized the foolishness of this thought. The big male baboon walking towards me down the track was followed by an entire troupe made up of baboons of all ages, all walking with their characteristically purposeful, stalking, slightly sideways gait except for the little ones which skittered along in a more playful manner. In this encounter, retreat, my first impulse, was impossible. Behind me the trail crossed a river mouth fordable only at low tide then ran along the lagoon with dense forest on the other side. We approached each other each on our own side of the track. In that moment of vulnerability, I felt compelled to express something through my body and gesture – containment, intention, purpose, a lack of confrontation. I felt subject to the demand of the baboon to inhabit my body expressively. We moved towards each other until at an invisible signal the entire troop disappeared into the undergrowth on the far side of the track, and I was allowed to pass.

My initial misrecognition of the creature as a bear is one that is inscribed in the Cape or Chacma Baboon's Latin name, *Papio Ursinus*, the bearlike baboon. It is one of the largest of the monkey species. Even after the figure on the track resolved itself into something familiar, it confronted me initially only as a large threatening body. Yet as we drew closer to each other, I recognized something else – that we were both embodied creatures whose feet carried us across the ground and whose eyes looked out at the world. In his 'Introduction to Eco-phenomenology,' David Wood, invites us to consider the implications of such a shared embodiment. He writes: 'Seeing (and hearing and touching) is a phenomenon of the differentiation of the world into discrete bodies, including ourselves, that occupy distinct places at particular times, bodies endowed with a mobility that reflects their needs and desires' (79).

 DOI: 10.4324/9781003538097-25

In another context, it might have been possible for me to merely observe the baboon, but as we walked towards each other I felt compelled to respond. Its purposeful walk was a question which I could only answer with the movement of my own body.

In her book *The Logos of the Living World: Merleau-Ponty, Animals and Language*, Louise Westling (2014) proposes that the work of phenomenologist Maurice Merleau-Ponty provides foundational insights for reconsidering human relations with animals. Merleau-Ponty's thought, she argues, 'anticipates present debates in critical animal studies and understands literature and other human arts as part of a continuum of meaningful communications and aesthetic behaviors throughout the living world' (2014, 3). By insisting on perception as an embodied activity and emphasizing the biological continuity between human and other forms of life, he invites us to think about animal and human bodies not merely as biological entities but as complex, expressive forms. In this essay I focus on a particular set of animal bodies, ones which bear a haunting similarity to human bodies. Chimpanzees and non-human primates have historically performed a significant role in philosophical reflections on the relation between humans and animals. I have used the term 'approximate' in the title of this essay both to suggest the nearness of the relation between people and apes but also to draw attention to the inexactness that characterizes attempts to define this connection. Evolutionary science proposes that humans and chimpanzees share a common ancestor. Yet this abstract knowledge does nothing to clarify the nature of the relation between these two species co-existing in the present. Eco-phenomenology offers a different way of conceptualizing the connection. Merleau-Ponty uses the phrase 'strange kinship' to describe human's relation with animals, to place the emphasis, Kelly Oliver (2009) explains, 'not on descendants or on generation but on a shared embodiment in a shared world, even if the style of body and the style of inhabiting that world are radically different' (222). My phrase 'approximate relations' echoes Merleau-Ponty's formulation but proposes something rougher and more improvized. It opens the way for thinking both about moments of mutual recognition and shared kinesthetic perception while maintaining an awareness of the limits of this connection. Knowledge across species remains of necessity approximate, constrained by different styles of embodiment and modes of inhabiting the world. What follows then is not a discussion about the latest developments in scientific research into the cognitive abilities and the forms of symbolization of apes. Instead, drawing on recent eco-phenomenological insights, it focuses on primates as occupying a special place in debates about what it means to be animate, in philosophical reflections on the nature and quality of life, and discussions about freedom and different forms of captivity and captivation.

I begin this essay with a reflection on one well-known iteration of this debate, philosopher Martin Heidegger's argument with the poet Rainer Maria Rilke, over the meaning and implications of the concept of 'the open.' Rilke introduces the idea of the open in the eighth of his *Duino Elegies*, first published in 1923. Heidegger's refers to the concept in a series of lectures in 1935–1936 but responds more fully in the essay 'What are poets for?' published in 1947. I take as my starting point, Georgio Agamben's extended critical reflection on the concept published in his book *The Open: Man and Animal* in 2002. Although some specific types of animals are called upon in this exchange to act as examples, for the most part this discussion is concerned with the categories of animal and human in the abstract. In this essay, I explore the effect of interrupting this philosophical discussion with three encounters with specific animals. The first is my own brief meeting with the baboon, which I will return to at later. The second is a fictional encounter, described by Franz Kafka in his story 'Report to the Academy.' In this story, the chimpanzee Red Peter addresses the unspecified 'academy' charged with the task of giving an account of his transformation from ape to human. The final encounter takes place in the London Zoo and involves the biologist and painter Desmond Morris and the chimpanzee, Congo. Congo and Morris in their own way confront the academy, both scientific and artistic, as Morris attempts to theorize a similar evolutionary trajectory for the history of modern painting. In the second two cases a chimpanzee is assigned the position of animal guide, mediating not only the obvious divide between humans and animal but also between human and human, between past and present. In all three encounters, the first experiential, the second imagined and literary and the third real and at least ostensibly scientific, the human and the animal are not abstract entities but rather confront each other in that contradictory and imperfect space of the social and historical.

The Nature of Captivity

In his book *The Open*, Italian philosopher, Georgio Agamben (2004 [2002]) notes a number of ways in which animals have been represented within the Western philosophical tradition. In one tradition, which finds expression in the work of Martin Heidegger, animals are regarded as mute and captivated by their environment. They are in Heidegger's terms 'poor in world.'[1] This means they are unable to escape the limits of their instinctual engagement with objects. They are not free and their world is reduced to those markers of significance or 'disinhibitors' that have relevance to their survival. Agamben explains Heidegger's position in the following way:

> The mode of being proper to the animal, which defines its relation with the disinhibitor, is captivation (*Benommenheit*)...In so far as it is essentially captivated and wholly absorbed in its own disinhibitor, the animal cannot truly act (*handeln*) or comport itself (*sich verhalten*) in relation to it: it can only behave (*sich benehmen*).
>
> (Agamben 2004, 52)

Heidegger makes an important distinction between two modes of being in the world, the one, *Benehmen*, translated into English by the term 'behaviour' and the other, *Verhalten*, by the term 'comportment.' The Oxford English dictionary definitions of these terms reveal that they are closely related. The term 'comportment' includes within its definition, behaviour, and the term 'behaviour' shares with comportment a relation to the notions of conduct and bearing.[2] Yet behaviour also has a more specialized meaning associated with it, a meaning which emerges in conjunction with a particular field of scientific research. Behaviour in its scientific sense, emerges in relation to the study of animals (including human animals), as part of the physical evidence through which a creature can be known. Behaviour, in this sense, could be described as the automatic performance of the body in a particular situation divided in some way from consciousness and choice. Agamben describes how Heidegger's thoughts about animals emerge in response to contemporary biological and zoological studies. It is in relation to these that he develops the term *Benehmen* or behaviour to describe the mode in which animals inhabit the world.

> Since the animal is ceaselessly driven in its manifold instinctual activities on the basis of its captivation and the totality of its capacities, the animal fundamentally lacks the possibility of entering into relation either with the being that it itself is or with beings other than itself.
>
> (Heidegger, quoted in Agamben 2004, 54)

The animal relates only to things within its 'disinhibiting ring' or its environment in this narrow sense of the word. It has no apprehension of or access to the world beyond those specific markers of significance which address its capabilities. Heidegger explains:

> Plant and animal depend on something outside of themselves without ever 'seeing' either the outside or the inside, i.e., without ever seeing their being unconcealed in the free of being. It would never be possible for a stone, any more than for an airplane, to elevate itself toward the sun in jubilation and to stir like the lark, and yet not even the lark sees the open.
>
> (Quoted in Agamben 2004, 58)

It is interesting to consider how Heidegger positions the lark. The sentence does not make it clear whether the lark is permitted 'jubilation' or whether it is permitted only mobility, the ability to 'stir' which distinguishes it from the stone and the aeroplane but does not gain its entry into an apprehension of the 'open.'

Heidegger's argument about the open is formulated in response to a fascination and an anxiety provoked by the use of 'the open' by poet Rainer Maria Rilke. Agamben sums up Heidegger anxiety in these terms:

> And precisely because the 'essential border between the mystery of the living being (plant or animal) and the mystery of what is historical' is neither experienced or thematized in Rilke's poetry, the poetic word falls short of a 'decision capable of founding history,' and is constantly exposed to the risk of 'an unlimited and groundless anthropomorphization of the animal,' which even places the animal above man and in a certain way makes a 'super-man' of it.
>
> (Agamben 2004, 59)

'Anthropomorphization' is a term of judgement often used to shut down reflections on the inner life and experiential world of animals. The term raises the problem of the proper mode of knowing animals–the extent to which it is acceptable to recognize in animals elements that have been identified as human. It draws attention to the fact that the animal stands as a problem for interpretation. Viewed from the outside, the actions of animals may in certain contexts be read as responses to external 'markers of significance' or they may appear enigmatic, as responses to a world which is not available to human apprehension.

Rilke employs 'the open' to describe the animal's relation to the world in *The Eighth Duino Elegy*. It begins:

> With all its eyes the natural world looks out
> into the Open. Only *our* eyes are turned
> backward, and surround plant, animal, child
> like traps, as they emerge into their freedom.
>
> (Rilke 1995, 377)

Viewed from Rilke's perspective it is human adults who are captivated by a mode of looking he describes as having eyes 'turned backwards.' Their experience of the world is 'mediated through consciousness and self-consciousness.'[3] Forced to see 'objects,' they cannot access what is really out there. The animal, on the other hand, is free to inhabit 'the open.' As a poet,

Rilke is not required to define his terms. 'The Open' thus remains open, indicating toward a utopian and unmediated world of experience, without performing the impossible task of defining it. He does this through the mediation of animals.

> We know what is really out there only from
> The animal's gaze; for we take the very young
> Child and force it around, so that it sees
> Objects – not the Open, which is so
> Deep in the animals' faces.
>
> (Rilke 1995, 377)

It is the opacity of animals, their enigmatic quality that makes it possible for them to hold this meaning. It is the impossibility of looking with the animal's eyes that makes what they see such a compelling metaphor for an unthought relation with the world. Yet considering the gaze of the animal provokes a sense of partial recognition, the sense of being aware of something which occupies peripheral vision, but which cannot be brought into focus.

What is at stake in this disjointed conversation begun by Rilke and continued by Heidegger and Agamben about the pleasure and freedom of the animal? Derrida and others have, of course, criticized the philosophical tradition which takes the concept of the 'animal in general' and opposes it to the concept of the 'human' (Derrida 2002).[4] In drawing his conclusion that animals are 'poor in world' Heidegger refers to the work of German biologist Jakob von Uexküll and in particular to his concept of *umwelt*. *Umwelt* describes an animal's environment from its own perspective. For each species, certain elements within the space that surrounds them are charged with meaning while others disappear. To exemplify the way animals are bound by certain markers of significance in their environment, Uexküll uses the example of a tick. Agamben devotes an entire short chapter to 'The Tick' and the limited way it interacts with its environment, moved only, according to Uexküll, by the smell of butyric acid, the temperature of 37 degrees, and the type of skin characteristic of mammals (Agamben 2004 46). Heidegger also introduces an example of a partially dismembered bee which continues to suck honey despite the fact that its abdomen has been cut away, unable according to Heidegger to recognize even its own bodily integrity. There is a gap which threatens to derail the conversation altogether between Uexkull's tick, Heidegger's partially dismembered bee[5] and the 'alert, warm animal' Rilke refers to in the Eighth Elegy (Rilke 1995, 379). Yet the term 'animal' persists in having an almost fetishistic hold over the imagination. Contained in the word is the root 'anima,' meaning according to the Oxford English Dictionary 'breath,' 'life,' and 'soul' and it is within this constellation of ideas that the hold of the animal becomes apparent. To talk about the animal is to talk about the possibilities of life and 'animation,' a mobility which is not merely behaviour, a mode of inhabiting space which exceeds the simple biology of habits. The 'animal' calls into being the open in this sense since it requires a space in which this mobility, this breathing takes place. In this sense 'animal' is not about outward form but rather about a quality of being in the world. It presents itself in contrast with the human not in absolute terms but at this particular historical moment when the question of human modes of life, the management of life, and its restrictions are particularly pressing.[6]

It is unsurprising then that Rilke's approach, which continues a particular tradition of thinking about animals in which they are conceptualized as having knowledge from which humans have been alienated, is particularly suggestive in the increasingly 'administered world' of the twenty-first century.[7] Alongside developments that marginalize animals even more extremely from everyday life and the increasing industrialization of farming, there has emerged an almost obsessive representation of wild animals, which have become the site of a peculiar kind of value.[8] Yet Rilke's 'open' like Heidegger's presupposes two perceptual worlds, which only partially overlap. Recent scholarship in eco-phenomenology, however, shows that Heidegger's clear distinction between human world and animal environment represents a misreading of Uexkull's concept of umwelt. In their article, 'The Phenomenology of Animal Life' cognitive ethologist Dominique Lestel, feline sociologist Jeffrey Bussolini, and animal studies scholar, Matthew Chrulew comment on the overlapping nature of sensory environments:

> Depending upon their particular sensory orientations and ecophysiological grounds, some animals will perceive each other and interact, while others will remain ignorant of each other's existences. No animal, humans included, possesses the capability to perceive and take account of each and every animal present.
>
> (Lestel 2014, 134–135)

Ethologist and philosopher Roberto Marchesini, they suggest, offers an alternative less rigid conceptualization of umwelt. He describes how 'an Umwelt, properly understood, is as much a field of possibilities as a constraining limit. Animals, including humans, exercise creativity in transcending their species-specific immersions and forming interspecies associations and bonds' (Lestel 2014, 135).

To make their argument, they offer two examples. These two examples are not long term ethological studies but rather fragments, points of contact in which animal behave in surprising ways: The first is the case of orangutang, Wattana, who learns to tie knots; the second, a group of cats in a New York apartment who develop a passion for chillies, deploying a number of un-catlike behaviours such as opening the fridge despite additional tape and child-locks in pursuit of their goal. Although frequently stigmatized as unscientific, anecdotes, Lestel et al. argue, are a recognized part of everyday ethological practices and can be valuable for elucidating 'complex questions of behaviour and culture' (129). Drawing on the anecdotal evidence of these New York cats and the knot-tying orangutang, they make two recommendations that are valuable for thinking about interactions with baboons and chimpanzees. The first is has to do with claims made about animal behaviour. To understand surprising behaviour they suggest, 'we must jettison the habits of thinking species in general in order to think instead the singularity of animals' (126). Second, they emphasize 'the importance of the principles of surprise and politeness that leave a maximum latitude to animals in interaction and research' (126). By this they mean keeping open the possibility of being surprised, of not shutting down the possibility for creative interaction. Drawing on these insights, this essay considers the open not as the space of the encounter with pure being which Heidegger suggests can be arrived at through the experience of boredom nor as the enigmatic and inaccessible realm which animals see beyond us when they look past us but rather the admittedly limited space of encounter between species in which the outcome is not established in advance.[9]

In *The Open*, Agamben reflects on the divisions of 'life' proposed by Aristotle and later Bichat noting the distinction they make between 'organic life,' the unconscious bodily functions and 'animal life' which is defined by its relation to the external world. This initial division of life within man, Agamben suggests, is what makes possible the later distinction between man and animal. He suggests that:

> In our culture, man has always been thought of as the articulation and conjunction of a body and a soul, of a living thing and a *logos*, of a natural (or animal) element and a supernatural, social or divine element.[10]
>
> (2004, 16)

Agamben's point is that this conjunction or articulation should more accurately be read as an incongruity, the 'practical and political mystery of separation' (16). Yet if life were divided up in another way 'animal life' might be defined as something at once corporeal and social, engaged in a form of expressivity which is not language but not silence either. In contrast with politically and socially alienated human life, which to use Rilke's formulation is confined by the act of self-reflection, which sees only objects, animal life (in humans and animals) might be considered as that which offers the potential for interactivity.

In Kafka's story, the chimpanzee Red Peter occupies an ambiguous position between ape and human, physically an ape but no longer possessing his ape nature. He is captivated, not by his environment, but by human language and so provides a useful figure for reconsidering Heidegger's distinction between behaviour and comportment and for suggesting another mode of being, the mode of performance.

Behaviour, Comportment, and Performance

Franz Kafka's short story, 'Report to the Academy', originally published in 1917, is one of a number of texts which address in various ways the problem of this partial recognition, the enigmatic being of the ape, and how it engages the subjectivity of others. Kafka's story is narrated by Red Peter, who is captured in Africa and taken to Europe on a ship. On the journey he transforms himself into a human to avoid the fate of captivity.

It is interesting to consider the process of becoming human described in the story. The ape first imitates the gestures of the sailors provoking amusement and interest. But in order to become human imitation is not sufficient. He is required to overcome the tremendous revulsion he feels for schnapps and to actually take a drink before he is able to speak his first word in human language. In Heidegger's terms he has to make the move from behaviour to comportment, to wrest himself out of his disinhibiting ring in order to inhabit the world. Yet for the ape to consider this transformation, he must already possess an awareness of the world. Kafka's ape is never captivated. He is merely captured, and the only possible freedom lies in abandoning his ape nature. Red Peter explains that

> my ape nature fled out of me head over heels and away, so that my first teacher was himself almost turned into an ape by it, had soon to give up teaching and was taken away to a mental hospital.
>
> (Kafka 1983, 258)

Red Peter's ape nature leaves him, entering the world at large as a free floating and potentially disruptive force, but his ape body remains. And it is partly because of his physical abilities that he is able to acquire the education necessary 'to reach the cultural level of the average European' (258). He describes how he appointed five teachers, placed them in communicating rooms and 'took lessons from them all at once by dint of leaping from one room to another' (258). There is a typically Kafkaesque dark humour in this fantastic image of the chimpanzee's education, undertaken not through quiet study but in very animal-like leaps and bounds.

The ape's animal life, once his attachment to his ape nature has been broken, exceeds human life to such an extent that he can consume European culture with apparent ease.

Red Peter overcomes with difficulty and determination his revulsion at human behaviour but the physical freedom he gains as a result of his transformation has certain costs as he explains in his report. Becoming human means the determined giving up of his origins and as he enters further and further into the human world, his memory of his past diminishes. Becoming human, the story suggests, is a process of forgetting. Humanness is marked by the violence it does to other forms of life. They become, the narrative implies, literally unthinkable.

In response to the Academy's request that he give them an 'account of the life [he] formerly led as an ape,' he explains that

> the opening in the distance...through which I once came myself, has grown so small that, even if my strength and my will power sufficed to get me back to it, I should have to scrape my very skin from my body to crawl through.
>
> (1983, 250)

Although Red Peter frames this return as a movement through time, he describes it terms of an actual, physical process – one which would involve the scraping off of his skin. It is as if the life he led as an ape remains inside him but it can only be accessed through a process which physically removes his skin, an image suggestive of the process of dissection. His life history can only be read in his dissected body, after he has been transformed from a speaking subject into an object of scientific knowledge.

Kafka's story involves an act of mimicry, a measuring or mapping of animal life against human life and human life against animal life. It comments on different degrees of separation and connection. In a complex network of partial recognitions and misrecognitions, human life emerges as an uncanny, disturbing approximation of animal life. Asked to speak to the academy, the ape in Kafka's story finds that in the medium of human language he is a stranger to himself. Language does not make possible full disclosure.

The philosopher, Ludwig Wittgenstein draws attention to this aspect of language – it makes possible communication but does necessarily reveal the nature of the other. In his brief discussion of animals, Wittgenstein having made his famous comment about lions ('If a lion could speak we would not understand him') hastens to qualify what might appear as a sweeping claim about the absolute difference between animal and human by insisting that language does not necessarily enable understanding.

> It is however, important as regards this observation that one human being can be a complete enigma to another. We learn this when we come to a strange country with entirely strange traditions; and what is more, even given a mastery of the country's language. We do not understand the people. (And not because of not knowing what they are saying to themselves). We cannot find our feet with them.
>
> (Wittgenstein 1958, 223)

Wittgenstein's observation is useful for two reasons. First, because it suggests that language does not exist in isolation from a particular form of life. Language is not a neutral conceptual tool. To describe the defining characteristic of humans as language misses the situatedness of acts of communication in particular games or encounters. The speaking animals in fiction and the apes in captivity taught human language are both captured by human life, both literally and symbolically. Second, it introduces the idea that the enigmatic quality of animal behaviour, which resembles but does not match or confirm our own is similar to the enigmatic quality of human behaviour which resembles but does not match or confirm our own. In other words, the enigmatic quality of animal life is not what separates it absolutely from human life. Instead, it is the notion that the animal's life is entirely available to us that marks it off as something absolutely other.

Morris's exhibition of chimpanzee paintings is interesting because although ostensibly the result of a scientific undertaking, the data collected are, in themselves, enigmatic. The chimpanzee paintings offer not definitive knowledge of the species life of the ape but rather suggestive parallels, partially familiar forms.

Play and Creativity

The exhibition of chimpanzee painting organized by Desmond Morris is very much part of the intellectual culture of the 1950s. It formed part of the Institute for Contemporary Arts 1957 programme entitled *Primitivism*. The first part of the programme was devoted to drawings done by Australian Aboriginal children. The second part, which ran from 17 September to 12 October, included twenty-four paintings by Congo and twelve by another chimpanzee housed at the Baltimore Zoo called Betsy (Lenain 1997, 93). A quotation from the exhibition catalogue makes clear the way in which chimpanzees are being enlisted to mediate and confirm a particular relation between Europe and the pre-modern world of Australian Aboriginal culture. The exhibition catalogue, written by Morris, states that:

> The patterns [in chimpanzee paintings] give us an important new source of information to help us trace the origins of human art. Much useful knowledge has been gathered in recent years from the study of the work of children, the art of primitive people, the pictures of the mentally unstable or deficient, and the remnants of prehistoric cultures. We now have drawings and paintings by our closest relatives the chimpanzees as further guides.

A particular relationship to time marks the narrative that makes possible this particular grouping of subjects – children, 'primitive' people, prehistoric cultures, chimpanzees, and the insane. All of them are, from Morris' point of view, outside of historical time, and contain a trace of that connection with nature which has been lost in modern society (Morris n.d.). Drawing on a somewhat crude and unquestionably racist evolutionary narrative, Morris articulates a particular understanding of the way in which animal life and human life might be correlated.

It is with this end in mind that Morris in his role as biologist for the London Zoo conducts a series of experiments with the resident chimpanzee, Congo. In reading an early version of this paper to a Ugandan colleague she was immediately disturbed by the chimpanzee's name – the constantly spoken yet invisible reiteration of some non-specific yet potent association with Africa. The name refers to but also partly obscures the lurking ideological content in the apparently scientific approach to chimpanzee painting. The chimpanzee always inhabits a space in the popular imagination at the same time as it inhabits the serious world of scientific investigation.[11]

The site of the production of these paintings is also significant. Unlike the partially open space of the reserve mentioned earlier, the zoo is a space in which the encounter between animal and human is highly structured. Congo's painting took place as part of Morris' investigation into the origin of the aesthetic impulse. Congo is invited, encouraged, manipulated into what Morris calls 'painting play' and learns over time to mimic the behaviour of a painter. He learns and becomes absorbed in the process of making visible marks.

Morris attempts to interpret Congo's paintings by seeing in their similarity to modern styles of abstract painting the suggestion that there is a biological origin to art. Modern art, he suggests, becomes more and more abstract as it loses it function as communication or rather as this function is taken over by photography.[12] Abstract art becomes then 'pure aesthetic experiment' and in this way comes to resemble the painting of primates. However, the abstractness of the categories that Morris is able to come up with suggests that Congo's painting represent a problem for interpretation which cannot be solved through relying on their superficial resemblances to modern abstract art. After mentioning the six basic principles of artistic self-expression, 'self-rewarding activation,' 'compositional control,' 'calligraphic differentiation,' 'thematic variation,' 'optimum heterogeneity,' and 'universal imagery' he comments that:

> Any rules that are basic enough to apply to several related species, rather than to one species or (as is more often the case in art theory) to more than one epoch of one species, must indeed be fundamental to the activity concerned.
>
> (1962, 168)

The ambiguity, however, lies in what precisely the 'activity concerned' is. To establish these categories, Morris has to elide one very significant aspect of their production. At no point do the paintings exist for Congo as products. It is up to the human facilitator both to provide Congo with paint, one colour at a time, and to remove the painting at a certain point before the painting is discarded and destroyed. The paintings are not abstract in the same sense as modern art is abstract because they are not a deviation from a previously representational form of articulation. What Congo is doing when he accepts the paintbrush and marks the page, remains enigmatic.

In the encounter between Morris and Congo, Congo learns to imitate not human 'behaviour,' since painting is not a habitual and functional action, but rather something which appears to more closely approximate Heidegger's idea of comportment. Congo learns to engage in a human cultural practice which exceeds the functional. Semiotician, Thomas Sebeok (1981), in his discussion of various artistic activities in animals (including Congo's paintings), notes that to ascribe an aesthetic sense to animals troubles deeply held convictions about the nature of animal life. He comments that:

> We find it difficult to conceive of art as a coherent part of animal life and can scarcely imagine it as an adornment of the creatures' leisure. All researchers in this field are stamped by a tension between a deeply felt conviction on the part of many distinguished and sensitive biologists that artistic activity indeed exists in the animal world and the inability to face its presumed lack of importance, even uselessness.
>
> (232)

The fact that animals engage in what to human researchers appear to be artistic activities suggests that they are moved to respond to their environment in ways that go beyond the kind of captivation ascribed to them by Heidegger. In Congo's case, he does not undertake this action spontaneously, but he nonetheless responds to Morris' prompts and extends the range of his interactions with those around him. He does this in the necessarily confined space of the zoo, but the very form of the 'experiment' requires that Morris allow Congo the freedom to assert something of his own subjectivity. In his discussion of paintings by apes and by Congo in particular, ethologist and primatologist, Frans de Waal, points to evidence that 'suggests that apes have a sense of colour and of when a painting is complete, becoming quite annoyed if their painting is interrupted before they are finished' (171). After reviewing a number of cases de Waal concludes that 'painting apes have a sense of both balance and completeness, enjoy the visual effect of what they do, and create regularities and patterns but are not out to produce a lasting product' (173).While it remains unclear what motivates their engagement in the activity of painting, they bring to it a sense of purpose. Although it is not possible to determine exactly what that purpose is, it seems likely that painting offers relief from the boredom of life in the zoo. The paint, paper and paint brushes become markers of significance in Congo's *umwelt* given the paucity of other possibilities. Morris is required by the logic of his own argument to create an opening in the condition of Congo's captivity in which his animal life might be asserted.[13]

This immediate, constrained but also creative encounter between Morris and Congo in the zoo is followed by a different kind of encounter between Morris and the product of Congo's play and the public. The exhibition represents, in part, Morris' attempt to intervene in the entirely cultural debate about the status of modern abstract art.[14] Yet in bringing science out of the laboratory into the public domain, what happens instead is almost the opposite. The scientific artefacts, the paintings, acquire a sudden and for Morris at least unexpected value as something else. In *The Biology of Art* (1962) Morris explains that:

> The Institute for Contemporary Arts was bombarded with requests to put the Congo paintings on sale and, in a weak moment, it agreed to do so. A rather high price was placed on them, so that only a few would go, to the I.C.A.'s most insistent visitors. A few days later it was decided that this was a mistake and that the selling of pictures would start a trend that would lead away from the scientific value of the work. To our horror, we found that it was too late, for, in those few days, practically every Congo picture in the exhibition (twenty-four Congos were on show) had been sold.
>
> (1962, 27)

It is this fact, this convergence of desire around Congo's paintings that make them interesting in a way entirely separate from the way envisaged by Morris. The public's desire to possess paintings by Congo, even at a 'rather high price,' suggests that people attributed a value to these paintings which went beyond their value as 'scientific artefacts.' The paintings by Congo and Betsy, moved from one context to another, were transformed from the results of a 'scientific' experiment to cultural artefacts within the complex value system of modern art. The exhibition, taken up as a manifestation of surrealism, became less a rational exposition of a scientific principle than a fascinating example of a kind of 'pure psychic automatism,' pure expression without the deforming intervention of human reason.[15]

Morris' self-confessed 'horror' at the sale of the paintings seems to indicate a desire to fix the meaning of the paintings within the framework of interpretation that he had established. The paintings by chimpanzees, like the paintings by the Aboriginal children, were to be read as exemplary of a common 'primitive' or essential aesthetic sense. The value of the paintings, for Morris, lies in their formal qualities, their formal resemblance to modern abstract art. The conditions of their production need to be bracketed off in order for these formal qualities to be made comparable to the work of contemporary painters. What Morris excludes from his consideration of the paintings is precisely that aspect which Congo might consider to be primary – the aspect of play. Although Morris recognizes this is how Congo regards the activity of painting, he does not consider it significant to the interpretation of his data – the paintings themselves.

In discussing the relationship between children and animals, Theodor Adorno suggests that what children and animals share is not a relation to some imagined developmental or evolutionary origin, but rather a common orientation to the world of things. In play, both children and young chimpanzees see objects but do not take them entirely seriously. Congo recognizes his paintings but is happy to destroy them once the pleasure of making them has come to an end. In *Minima Moralia*, under the heading 'Toy shop,' Adorno suggests that children perceive in animals something that eludes the reduction of things to commodities. Animals confirm children's suspicion that everything cannot be reduced to a relation of equivalence (1978, 228). Animals represent for children, Adorno claims, the promise of a mode of being not yet captivated by the world of commodities.

Adorno's comments suggest a way of reconsidering the opposition described earlier in the paper between the mode in which animals and humans inhabit the world. If for Heidegger animals are captivated by their disinhibiting ring or environment, for Adorno humans are captivated by a mode of perception schooled by the practical demands of consumer capitalism. Like the disillusioned Red Peter whose bleak account recognizes the limits to his hard won 'human' freedom, Adorno describes the nature of human captivity as a kind of forgetting, an inability to apprehend the world outside the categories imposed on it by the instrumental logic of the 'academy.'

Children, Adorno argues, perceive the 'contradiction between phenomenon and fungibility that the resigned adult no longer sees' (1978, 228). In other words, they recognize that the quality of things cannot be contained within their value as articles of exchange.[16] Play, Adorno suggests, is the way in which children refuse to observe the rules of commercial society. He writes:

> In his purposeless activity the child, by subterfuge, sides with use-value against exchange value. Just because he deprives the things with which he plays of their mediated usefulness, he seeks to rescue in them what is benign towards men and not what subserves the exchange relation that equally deforms men and things.
>
> (1978, 228)

Play permits the child to evade the demand that things are real and useful only in so far as they fulfil a particular social purpose. Adorno uses the example of a little truck which is moved around in imitation of the actual movement of vehicles yet remains outside any real economy, a sort of allegorical fragment of the real. He writes: 'The unreality of games gives notice that reality is not yet real. Unconsciously they[children] rehearse the right life' (1978, 228). The utopian promise of animals is precisely of this kind – a stubborn existence outside of mediated usefulness. This is not to deny that animals have been made useful, incorporated in innumerable different ways into the practices of advanced industrial society, as product or image or research object. Yet in actual encounters such as the encounter between Morris and Congo they cannot be reduced simply to this, assimilated entirely to the purposes of man.

In the conclusion to his book, *Monkey Painting*, Thierry Lenain comments on the fact that while paintings by chimpanzees are not art we cannot dismiss the aesthetic effect they create. They create the 'illusion of a work of art.' He continues: 'It is useless trying to remind ourselves that the forms are nothing more than gestures made visible, and that from the animal's point of view they are devoid of interest' (1997, 184). De Waal (2001) suggests something similar. In describing the difference between human and ape painters, he suggests that 'instead of worrying about the cumulative impact of an entire series of brush strokes and dabs, apes give the impression of taking a kinesthetic and visual pleasure in each separate action' (170). The paintings resemble art but they are not art. Yet what they are, is perhaps more interesting. What are 'gestures made visible' if not traces of the life of the body, marks which express kinaesthetic and visual pleasure?[17] Following Lestel's injunction to 'think…the singularity of animals' (126) and to maintain a disposition of 'surprise and politeness' in animal encounters, it might be possible to see the paintings as Congo's address to the academy from his place of captivity. He has not escaped like Kafka's Red Peter to live a human life, and perhaps as a result retains a different mode of engaging with the world. His body remains mobile and expressive, and it is this kinaesthetic perception that is revealed in the enigmatic fan-like strokes of the brush on the page.

The three encounters described in this paper invite a reconsideration of questions of captivity. Heidegger's claim that animals are captivated by their environment and cannot see beyond certain biologically determined markers of significance cannot adequately account for Congo's use of paintbrushes, paint, and paper. Merleau-Ponty offers a compelling metaphor for describing a more nuanced understanding of environment. 'For Merleau-Ponty,' Oliver writes, 'the Umwelt is not a narrow band chained around the organism but a melody that haunts it. Animals, like humans, are haunted and fascinated by their environments and fellow beings, making their captivation more awestruck than mechanical' (215). Although Congo's precise intentions remain enigmatic, it is clear that he has, in collaboration with Morris, expanded his sensory engagement with his environment in creative ways. Merleau-Ponty's insights are valuable for reframing the debate I described at the beginning of this essay about the different ways in which human and animals relate to the open. Yet they do not take into account the fact that many humans and animals exist in different kinds of physical captivity tied to environments (such as zoos) which are unlikely to inspire awe. Shared embodiment makes both humans and animals vulnerable to the production of umwelts depleted of all markers of significance, all reasons to find expressive pleasure in the body.

In Kafka's story, Red Peter makes this case quite emphatically in his gloomy address to the human academy. For him becoming human involves a reduction of the possibilities of his animal life. It is his animal exuberance that allows him to acquire European culture in leaps and bounds, yet at the same time it is precisely this exuberance that he needs to crush for him to occupy a place in the limiting human *umwelt*. Red Peter escapes from literal captivity into the bleak and limited freedom of the human world; Congo remains in captivity but asserts within the frame of the 'scientific experiment' a limited freedom. Following Rilke's understanding of animals, Congo's paintings might be supposed to contain a representation of what animals see, a glimpse of the unthinkable 'open.' But if following the insights of eco-phenomenology we acknowledge that chimpanzees, baboons, and humans share an embodied animal life, these paintings might contain instead something more intriguing, traces of the expressive life of the body. The popularity of these paintings, I suggest, indicates an emerging nostalgia for this dimension of animal life, the reminder of the body as expressive, the medium for a kind of 'language game' we can only play with animals.

The 'open' in this sense might be precisely the social space opened up by the form of expression which some animals expect of each other but which humans, uncertain and divided about their status as animals, have largely forgotten.

The encounter with the baboon I described at the beginning of this essay, neither hostile nor friendly, was unusual. It took place in a narrow margin of uncontrolled space. Although part of a reserve, it was nonetheless for that moment open in the sense of being outside of any official structure for mediating the relation between human and animals. This openness permitted me to experience the encounter as one in which I was, or felt myself to be, subject temporarily to the demand of the baboon. I was reminded of myself as partially, incompletely, possibly inadequately animate, possessing an animal life insufficient to match that of the baboon. In trying to describe the sense one has of a baboon's demeanour in such an encounter the word which presents itself is 'affront.' A baboon appears to regard the world and those around him or her with affront – a word which seems to describe the degree of challenge, the relentless directness and levelness of a baboon's gaze. To encounter a baboon is always to be made aware of its refusal. They engender a feeling of alarm. You feel at risk not simply from violence but also from a kind of foolishness. To be in the presence of the baboon is to be on the verge of becoming a figure of fun, to be startled out of yourself. This is partly because of the way in which they take control of the space of the encounter, taking whatever they wish without the usual wariness of wild animals, but it also has to do with the way in which they appear to recognize and yet dismiss the subjectivity of humans. Instead of submitting themselves to human culture, or ignoring it, they assert their own. Yet this assertion also involves a mutual recognition. One of the key insights of eco-phenomenology is that despite differences in 'the style of bodies and the style of inhabiting th[e] world,' species are connected by a 'shared embodiment in a shared world' (Oliver 222). For Merleau-Ponty animal life persists in humans through the world of sensory perception. Ted Toadvine (2014) explains Merleau-Ponty's point in this way: 'the animals within us are the lives of "my eyes, hands, and ears, which are so many natural selves"' (112). Although extremely brief my encounter with the baboon invited me to transcend my species-specific immersion and instead to reconnect with an animal self that lives on in the body though it seldom rises to the level of consciousness. Caught by surprise and unable to avoid the encounter, I experienced momentarily the creative possibilities and kinaesthetic pleasure of a world of sensory experience expanded by the presence of a baboon.

Notes

1 In their complex discussion of the distinction between animal and human Horkheimer and Adorno represent this tradition as defining animal life as dreary and insubstantial because of animals lack concepts and language (Adorno & Horkheimer 2002, 205).

2 The distinction between *Benehmen* and *Verhalten* like the distinction between behaviour and comportment is subtle. The two can, in some contexts, be used interchangeably.

3 In describing Rilke's concept of the open, Eric Santner explains that for Rilke: 'man's relation to things is crossed with borders, articulated within a matrix of representations that position him, qua subject, over against the world, qua object of desire and mastery' (Santner 2006, 1–2).

4 See also the discussion in Cary Wolfe's 'Humanist and Posthumanist Antispeciesism' (Wolfe 2009, 46).

5 Agamben discusses Uexküll's work on ticks and mentions Heidegger's interest both in Uexküll's research and research done on bees whose abdomens had been partially cut away (Agamben 2004, 45–47; 52–53).

6 I am thinking of Foucault's claim that modernity can be characterized by the biopolitical management of life being but also of Adorno's 'administered world,' in which he suggests it is almost impossible to breath (Adorno 1998, 193).
7 See John Berger's discussion in 'Why Look at Animals?' (Berger 1980, 14–17).
8 Nicole Shukin describes how animals are transformed into 'capital' in a number of different ways, both as material resources and as symbolic values (2009, 7).
9 In a note Haraway sums up Heidegger's position on the distinction between man and animal in the following way:

> Precisely what differentiates man and animal, what puts them into opposite and unbridgeable singularities, is the possibility of 'profound boredom,' utter disconnection from function, for man, and the animal's inescapable poverty of the world through an unbreakable tie to function and determined attachment.
>
> (2008, 368)

10 Agamben traces some of the history of the concept of life and discusses the way Bichat makes a distinction between animal life and organic life (2004, 13–16).
11 This is particularly true for Congo since before his exhibition, he had already become the mascot of a weekly television programme called *Zootime* (Lenain 1997, 78–79).
12 Desmond Morris describes both the details of his 'experiment' with Congo and his method of interpretation in *The Biology of Art* (Morris 1962).
13 This space, however, does not remain open for long. Lenain comments that Morris' project with Congo 'had to be halted at the end of 1958 because Congo became too assertive and energetic to sit quietly for painting lessons' (1997, 93).
14 The exhibition apparently met with mixed response. Lenain describes how the president of the Royal Academy wished to take legal action against the exhibitors (1997, 93).
15 In the Surrealist Manifesto, André Breton defines surrealism as follows: 'Pure psychic automatism, by which it is intended to express, whether verbally or in writing, or in any other way, the real processes of thought. Thought's dictation, free from any control by the reason, independent of any aesthetic or moral preoccupation' (quoted in Read 1974, 132). In an interview, Salvador Dali claimed authoritatively to have paintings by sea urchins: 'I now have paintings by sea urchins. I shall soon be showing some of their work in New York. By nature they are all artists. Only some Chimpanzees can paint, but *all* sea urchins can. They paint with their tube feet and their art is genuinely informal. The hand of the chimpanzee is quasi-human – the hand of Jackson Pollock is almost animal (Dali 1961, 84).
16 It is noteworthy that this element in children can itself be drawn into the realm of advertising. In a Master card advertisement on South African television, a list of children's toys and prices is read out while a child is shown ignoring the toys and playing with the cardboard box they came in. The final line '...watching her play with the box instead, priceless. Some things money can't buy, for everything else, there's Master card' shows the ability of advertising to attach itself to anything which might still be considered a sign of value.
17 Lenain discusses the work of Austrian artist Arnulf Rainer who tries to imitate the work of the ape painters. An image from Lenain's book shows him sitting slightly behind a chimpanzee attempting to curl his hand and adjust his body to the posture of the ape. Rainer then incorporated the chimpanzee's painting as segment within his own imitation of the chimpanzee painting. What is revealed, Lenain claims, is the impossibility of the artist producing a 'pure mark' one which actually resembles the pure gestural mark of the chimpanzee (1997, 180–181).

References

Adorno, Theodor. 1978. *Minima Moralia: Reflections from Damaged Life*. Translated by E. F. N. Jephcott. London: Verso.

Adorno, Theodor. 1998. "Education after Auschwitz." In *Critical Models: Interventions and Catchwords*, 191–204. New York: Columbia University Press.

Adorno, Theodor, and Max Horkheimer. 2002. *Dialectic of Enlightenment*. Stanford: Stanford University Press.

Agamben, Giorgio. 2004. *The Open: Man and Animal*. Translated by Kevin Attell. Stanford: Stanford University Press.
Berger, John. 1980. "Why Look at Animals?" In *About Looking*, 1–27. New York: Pantheon Books.
Dali, Salvador. 1961. "The Quantum Gun at Port Lligat." Interview with Mervyn Levy, *Studio* 162, 82–85.
De Waal, Frans. 2001. *The Ape and the Sushi Master: Cultural Reflections of a Primatologist*. New York: Basic Books.
Derrida, Jacques. 2002. "The Animal That Therefore I Am." *Critical Inquiry* 28 (2): 369–418.
Haraway, Donna. 2008. *When Species Meet*. Minneapolis: University of Minnesota Press.
Kafka, Franz. 1983. *The Penguin Complete Short Stories of Franz Kafka*. Translated by T. Tania and J. Stern. London: Penguin.
Lenain, Thierry. 1997. *Monkey Paintings*. London: Reaktion Books.
Lestel, Dominique, Jonathan Bussolini, and Matthew Chrulew. 2014. "The Phenomenology of Animal Life." *Environmental Humanities* 5: 125–148.
Morris, Desmond. 1962. *The Biology of Art*. London: Methuen.
Morris, Desmond. n.d. "Paintings by Chimpanzees." In *Exhibition Catalogue*. London: Tate Gallery Archives.
Oliver, Kelly. 2009. *Animal Lessons: How They Teach Us to Be Human*. New York: Columbia University Press.
Read, Herbert. 1974. *A Concise History of Modern Painting*. London: Thames & Hudson.
Rilke, Rainer Maria. 1995. *Ahead of All Parting: Selected Poetry and Prose of Rainer Maria Rilke*. Translated by Stephen Mitchell. New York: The Modern Library.
Santner, Eric L. 2006. *On Creaturely Life*. Chicago: University of Chicago Press.
Sebeok, Thomas A. 1981. *The Play of Musement*. Bloomington: Indiana University Press.
Shukin, Nicole. 2009. *Animal Capital: Rendering Life in Biopolitical Times*. Minneapolis: University of Minnesota Press.
Toadvine, Ted. 2014. "The Time of Animal Voices." *Environmental Philosophy* 11 (1): 109–124.
Westling, Louise. 2014. *The Logos of the Living World: Merleau-Ponty, Animals, and Language*. New York: Fordham University Press.
Wittgenstein, Ludwig. 1958. *Philosophical Investigations*. Oxford: Basil Blackwell.
Wolfe, Cary. 2009. "Humanist and Posthumanist Antispeciesism." In *The Death of the Animal*, edited by Paola Cavalieri, 45–58. New York: Columbia University Press.

21

ECO-PHENOMENOLOGY AND PASSIVITY IN ANNA BARBAULD'S "THE CATERPILLAR" AND JOHN CLARE'S "THE MOUSE'S NEST"

Lisa Vargo

In their dialogue "Eyes and No Eyes or, the Art of Seeing" from *Evenings at Home*, Anna Barbauld and John Aikin offer a primer for children about the importance of perception. A teacher asks two students about a walk along a heath and in a meadow. While one boy complains the walk is dull and declares that his fellow student "is so tedious, always stopping to look at this thing and that," the other describes with wonder what he notices (Aikin and Barbauld 1794, 93–4). This leads the adult to conclude "so it is one person walks through the world with his eyes open, and another with them shut; and upon this difference depends all the superiority of knowledge the one acquires above the other" and the unwilling student is told to "learn that eyes were given you to use" (Aikin and Barbauld 1794, 108–9). This simple narrative suggests that the matter of how one might use one's eyes to understand the natural world captures a preoccupation of British Romantic writing that has come to be associated with ecological and green matters, despite predating the creation of the word "ecology" in the 1870s (Oxford English Dictionary, OED).[1] Industrialization and the occurrence of climate change brought on by the eruption of Krakatoa in 1815 helped lead writers to a recognition of what we now call the Anthropocene, and an awareness of the kinds of potential for disaster that might exist. An alternative path within these debates is anticipated by eco-phenomenology, and in particular those discussions which recognize that passivity offers the possibility of a responsiveness that challenges notions of perception and our ethical interaction with the natural world. Mary Shelley's 1818 novel *Frankenstein*, composed during the summer of climate change following Krakatoa, reflecting contemporary interest in polar exploration and experiments by Sir Humphrey Davy and Luigi Galvani with electric currents, offers a well-known example of debates about human control of nature. While the creature that Victor Frankenstein assembles is constructed of human and animal parts that tragically do not synthesize, two lesser-known poems embody a lack of synthesis as meaningful actions. What is exceptional is how the poets' everyday encounters with nature might lead to a recognition of passivity and a sense of "do-nothing," anticipating Merleau-Ponty's suggestion that what "we meant by passive synthesis was that we make our way into multiplicity, but that we do not

 DOI: 10.4324/9781003538097-26

synthesise it" (Merleau-Ponty 1962, 496). John Clare's "The Mouse's Nest" (written circa 1835–7) and Anna Barbauld's "The Caterpillar" (published posthumously in 1825)[2] ultimately resist notions of an easy and sentimentalized kinship between humans and nature, anticipating Ted Toadvine's contention that "an ethical response to nature becomes possible only when we are faced with the *impossibility* of reducing it to the homogeneous, the continuous, the predictable, the perceivable, the thematizable" (Toadvine 2003, 140). How passivity leads to a recognition of the impossibility of reduction within an "art of seeing" in "The Mouse's Nest" and "The Caterpillar" forms the subject of this essay.

Clare and Barbauld present exceptions to other writers of the Romantic period with respect to their marginal places in society. The resistance of these writers to notions of perception that assume mastery and domination over nature might be related to their own social positions. Barbauld and Clare were outsiders to mainstream British society—John Clare, an agricultural labourer and so-called "peasant poet" dispossessed by enclosure of common lands who spent the latter years of his life in an asylum, and Anna Barbauld, a member of nonconformist intellectuals associated with the Warrington Academy, a group denied their civil rights via the Test Acts and regarded as enemies of the state for their support of the French Revolution and advocacy for rights. It is a commonplace in commentary on their writings that their social positions influenced their perception of the world.[3] Their writings are exceptional in a number of ways: Clare for his resistance to the conventions of contemporary literary culture and Barbauld for the political positions exemplified in her "Epistle to William Wilberforce" and writings opposing the Test Acts and her poetic jeremiad "Eighteen Hundred and Eleven." Clare's class position and Barbauld's nonconformist faith enable more subtle and nuanced ways to look at the natural world. Clare's accidental encounter with a mouse and her young juxtaposed with the image of water running over stones and reflecting the sun and Barbauld's declaration that sparing a caterpillar crawling on her arm represents "the weakness of a virtuous mind" each describe a moment that "opens up and develops an access to Nature and the natural that is both independent of the conceptuality of the natural sciences and of traditional metaphysics" (Wood 2001, 94). Their exceptional independence of perspective is present in their writings.

"The Mouse's Nest" and "The Caterpillar" each present an embodied experience by a first-person narrator who seems to be a version of the poet and describes an encounter with a life form through which the impulse towards mastery described in *Frankenstein* as the desire to "penetrate into the recesses of nature and show how she works in her hiding places" (Shelley 1996, 32) is overturned with a surprising move to passivity and the recognition of states of being that resist synthesis. A reading of the poems via eco-phenomenology and passivity uncovers just how radical their conceptions are presented in the poems. Ted Toadvine's argument can be applied to demonstrate how, in their narratives, the poems come to reject the "homogeneous, the continuous, the predictable, the perceivable, the thematizable" (Toadvine 2003, 140). The result is to offer examples of lived experience via a re-conception of the relation between human and animal through passiveness. Both poems have received a critical reception of looking beyond the matter of perception, but a resistance to a hermeneutical impulse offers new possibilities for reading the poems.

Clare's "The Mouse's Nest" presents an account of seeing with "open eyes" that recognizes multiplicity, which is not synthesized by the presence of the perceiving subject. In fact, it offers a narration that, in the end, rejects forcing nature into a narrative. A first-person narrator (presumably a version of Clare) finds a ball of grass, expecting it to be a bird's nest—Clare wrote a series of beautiful poems describing birds' nests—but instead discovers a mouse nursing her young. A variation on a sonnet of 14 lines, its seven rhyming couplets have been called "a strange poem, and that it is hard to put one's finger on what makes it so wonderful" (Hickman 2012, 81) or, perhaps drawing on Clare's use of Northamptonshire dialect, it does not unfold easily once "proged":

I found a ball of grass among the hay
And proged it as I passed and went away
And when I looked I fancied something stirred
And turned again and hoped to catch the bird
When out an old mouse bolted in the wheat
With all her young ones hanging at her teats
She looked so odd & so grotesque to me
I ran and wondered what the thing could be
And pushed the knapweed bunches where I stood
When the mouse hurried from the crawling brood
The young ones squeaked and when I went away
She found her nest again among the hay
The water oer the pebbles scarce could run
And broad old cesspools glittered in the sun.[4]
(Clare 1984, 263)

Part of its strangeness is tied to the specifics of the poem. John Barrell calls Clare a poet of particularity and immediacy crucial to an understanding of his poetry (Barrell 1972, 114, 118). With respect to his choice of the dialect word "proged," Seamus Heaney remarks that he might have used "poked" or "prodded," but "had he done either of these things, both he and his readers would have been distanced in a minimal yet essential way from the here-and-nowness, or there-and-thenness, of what happened" (Heaney 1995, 66). The "nowness" seems particularly significant to the invocation of passivity that focuses on the moment.

Another example of Clare's focus on nowness and particularity is the mention of knapweed bunches, locating the landscape in common meadow or marginal land. Knapweed is a perennial flowering plant associated with meadows and wasteland that does not flourish on cultivated land. While it is deemed a nuisance by farmers, the plant is noted for its ability to produce in abundance nectar for butterflies, bees, and the goldfinch. Accordingly, it offers a reference to a focus on wildness that exists on the margins. Clare is writing at a time when changes in farming practices through enclosure meant the decline of such plants and the accompanying decline and even extinction of species of bees throughout the nineteenth century (Hicks et al. 2016). With respect to farming, the bunches of knapweed are a weed, but within a landscape of common land, they represent biodiversity, a variety and number of species central to a sense of multiplicity which is not synthesized, as well as providing a cover for the mouse and her brood. It is part of the *oikos*—house or economy—that is managed by nature, and while it might be noticed by the eye, it can only exist if it is left alone and is not subject to human control.

The knapweed that when pushed aside leads to the disruption of the nest and its occupants, along with the narrator's surprise that it is not the bird's nest he anticipates, as well as his wonder at the oddness and grotesqueness of what he sees, would seemingly be followed by a response that moves beyond the immediacy of the moment. To read the poem as a narrative requiring a conclusion is seemingly reinforced by the sonnet form, whose sestet or final couplet, depending on the type (here the poem is closest to the English sonnet form with a final couplet), is intended to resolve the issue or problem that has been explored in the preceding 12 lines.[5] But Clare's poem resists resolution and ultimately a narrative as the observer turns from the vision of the mouse with her young who return to the nest after he has disrupted and separated the mouse and her brood to contemplate water barely flowing over stones in cesspools glittering in the sun.[6]

Why the poem moves in a seemingly *non sequitur* from the nest to the pools offers a crux illuminated by the presence of dialect, which once again lends itself to immediacy and away from some moralizing gesture or restoration that includes the human. A lead from Alan Vardy might be taken here with his suggestion that Clare refuses "to redirect meaning into the poetic self" via "a resistance to recuperation": "The human figure in the poem gives us no clues about what this event might mean" (Vardy 2003, 23). Vardy distinguishes an inability on the part of the reader to "construct a moral out of it," wishing to "preserve the truthfulness of that visceral moment's inexplicability" and in opposition to a poet like Wordsworth, who "would mediate on the event and eventually secure its meaning" (Vardy 2003, 23). What Vardy calls "vivid perceptual moments" (Vardy 2003, 24) are synonymous with particularity and passivity. Clare purposefully leaves an unresolved description of the mouse's nest and the cesspool to focus on an even more radical form of resolution that nevertheless rejects recuperation. The poet Seamus Heaney famously described Clare's ability to capture the "one-thing-after-anotherness of the world" and that the final couplet of the poem achieves "not a self-conscious effect but a complete absorption" (Heaney 1995, 67).

But Heaney's notion of "absorption" requires some parsing. If this absorption means that while he accepts what he sees, he does not intend to invoke kinship or moral reflection but to illustrate an ethics that resists the thematizable for a "one-thing-after-another-ness," thus expanding the scope of what is perceived while he does not become part of it. In the final lines of the poem, Clare changes his perspective from the up-close view of the circular nest to a distanced circle of the pools as he employs another dialect term to define what he is describing. Robert Macfarlane defines "cesspools" as "water that gathers on the 'cess,' or land between a river and its bank, when the river is low fenland" (Macfarlane 2015, 124).[7] The particularity of dialect offers another clue that he is situated on common land that is used for grazing and for gathering wood and turf for fuel, and has not been improved by enclosure. Chapman and Seeliger note about common lands that contrary to enclosure, "as a broad generalisation they were relatively well protected from erosion, and offered sanctuary to a far more diverse range of wildlife species" (Chapman and Seeliger 2001, 140). The range of what Clare's observer notices and the suggestion that this is land that has not been "improved" by advancements in agricultural methods are significant with respect to the role that passivity plays in the poem. Absorption signifies a passive recognition that resists the thematizing impulses of incorporation, commodification, or ownership.

Clare's resistance to co-option within a landscape that portrays pre-enclosure common land is illuminated by the writings of Brown and Toadvine. Clare rejects beliefs that "sanction, and even encourage the domination and technological control of nature" (Brown and Toadvine 2003, xvi). Clare's publisher, John Taylor, instructed him to "Speak of the Appearances of Nature more philosophically" (Hickman 2011, 10), but rather than rejecting philosophy, he is aware of an alternative stance that anticipates eco-phenomenology. Clare's poem teaches us how to perceive in a manner that does not dominate via a moral or attempt to control (think of Robert Burns's "best laid plans of mice and men"), and as some criticism on the poem suggests, the powerful urge to turn the poem into a narrative. An eco-phenomenological perspective reads the poem anew, enabling a perspective that a different sense of the ethical might arise via a passivity which resists, as Toadvine suggests, the thematizable. But that resistance is the point of the poem. Its irresolution is the resolution.

John Taylor's instruction to Clare of a need to "Speak of the Appearances of Nature more philosophically" might be said to miss the point of Clare's poetry, but cannot be said of Anna Barbauld. "The Caterpillar" bears resemblance to eco-phenomenology's critique of the reductionism of science and of those beliefs, as I have already mentioned, quoting Brown and Toadvine, that "sanction, legitimate, and even encourage the domination and technological control of nature" (Brown and Toadvine 2003, xiii). The scientist Joseph Priestley was a family friend (and the subject of some of her poems),[8] and Barbauld was aware of science and its complex interactions with nonconformist theology. The dialogue "Eyes and No Eyes or, the Art of Seeing" is but one example of the many essays on natural history and science that comprise the volumes of *Evenings at Home*. It is important to use one's eyes, but for what purpose is certainly more complex for Barbauld than a means to control nature.

While Clare adapts the sonnet form, it is telling that Barbauld employs the mock heroic in blank verse (Smith 2015, 552–4) to describe a European tent caterpillar, identified by Alice Den Otter (Den Otter 2004, 214) as *Order Lepidoptera, Family Lasiocampidae, Species Malacosome Neustria* (first described by Carl Linnaeus in his 1758 10th edition of *Systema Naturae*) and more commonly known as the caterpillar of the Lackey moth. The tent caterpillar is a familiar garden pest, feeding on oaks, willows, hawthorn, and blackthorn, and notorious for damage to fruit crops of cherries, plums, and apples. Despite her "sworn perdition to thy race" and "recent slaughter" having poisoned and "crushed whole families beneath my foot," she spares the "single sufferer" who is "grown human," reflecting on the seeming irony of "capricious Pity, / Which would not stir for thousands, melts for one."

> No, helpless thing, I cannot harm thee now;
> Depart in peace, thy little life is safe,
> For I have scanned thy form with curious eye,
> Noted the silver line that streaks thy back,
> The azure and the orange that divide
> Thy velvet sides; thee, houseless wanderer,
> My garment has enfolded, and my arm
> Felt the light pressure of thy hairy feet;
> Thou hast curled round my finger; from its tip,
> Precipitous descent! with stretched out neck,
> Bending thy head in airy vacancy,
> This way and that, inquiring, thou hast seemed
> To ask protection; now, I cannot kill thee.
> Yet I have sworn perdition to thy race,
> And recent from the slaughter am I come
> Of tribes and embryo nations: I have sought
> With sharpened eye and persecuting zeal,
> Where, folded in their silken webs they lay
> Thriving and happy; swept them from the tree
> And crushed whole families beneath my foot;
> Or, sudden, poured on their devoted heads
> The vials of destruction.-This I've done

Nor felt the touch of pity: but when thou,-
A single wretch, escaped the general doom,
Making me feel and clearly recognise
Thine individual existence, life,
And fellowship of sense with all that breathes,-
Present'st thyself before me, I relent,
And cannot hurt thy weakness.-So the storm
Of horrid war, o'erwhelming cities, fields,
And peaceful villages, rolls dreadful on:
The victor shouts triumphant; he enjoys
The roar of cannon and the clang of arms,
And urges, by no soft relentings stopped,
The work of death and carnage. Yet should one,
A single sufferer from the field escaped,
Panting and pale, and bleeding at his feet,
Lift his imploring eyes,-the hero weeps;
He is grown human, and capricious Pity,
Which would not stir for thousands, melts for one
With sympathy spontaneous: 'Tis not Virtue,
Yet 'tis the weakness of a virtuous mind.
(Barbauld 2019, 274–6)

Her recourse to natural history requires attentive scrutiny. Her brother John Aikin wrote a treatise on the subject, *An Essay on the Application of Natural History to* Poetry, taking poetry to task for "too cursory and general survey of objects, without exploring their minuter relations; and is only to be rectified by accurate and attentive observation, conducted on somewhat of a scientific plan" (Aikin 1777, 10). Barbauld's poetry follows her brother's union of natural history and poetry, at the same time holding with Brown and Toadvine that "experience must be treated as the starting point and ultimate court of appeal for all philosophical experience" (Brown and Toadvine 2003, xi). Many of her poems include science as a subject for poetry; Barbauld connects the imagination of the poet with the inventors of the hot air balloon via their contemplations of soap bubbles in another mock-heroic poem, "Washing Day." She was aware of Hume's theory of associationism of knowledge gained via sensory experience. Den Otter summarizes how the poem is typically read:

> By the end, the reader is left to ponder the whole system of moral virtue: is it virtuous to protect pests and parasites as necessary parts of this system or is such protection merely an indulgent weakness of a virtuous mind?[9]
>
> (Den Otter 2004, 211)

The reading presented here suggests a different path with respect to its rather enigmatic conclusion.

In a subtle and closely argued essay Inhye Ha argues that "eighteenth-century posthumanist formulation of self and community, one that arises from the poet's ability to register the beauty and vitality in nonhuman species" might "epitomize a two-way street where a nonhuman tames a human subjectivity in profoundly ethical ways" (Ha 2018, 455) and create community arising from an aesthetic response to the beauty of the creature that inspires sympathy (Ha 2018, 461). The result is an "idealized notion of political community between humans and nonhumans" (Ha 2018, 466) in which the "speaker's decision to release the caterpillar, despite her previous history of killing other insects without much self-consciousness, indicates that a sensorial recognition made possible by the encounter can actually result in an act of attention and care" (Ha 2018, 475). While this is a very persuasive argument, I think it represents a step towards where Barbauld ultimately arrives in her contemplation of the caterpillar.

Sharon Smith also confirms Barbauld's belief in the need for connection between observer and observed (Smith 2015, 557); at the same time, she recognizes that it is accompanied by an awareness of the potential for violence that might come from the scientific gaze. As Smith explains, "the dissecting knife that must cut, open, and ultimately destroy in order to bring the hidden elements of nature into view" (Smith 2015, 558). The poet satirist likewise has this potential and must include the satirist herself within the scope of, as Smith puts it, "her own mock-heroic gaze" (Smith 2015, 559). To forget this self-inclusion is to miss the poem's subtle argument; if Clare's poem is characterized as difficult to account for what makes it so wonderful, Barbauld is often a difficult poet whose subtle arguments are easily misunderstood. Her poem enacts looking with one's eyes open: the poem's narrative is achieved through sensory experience of careful description and a direct address to the caterpillar; she explains she "scanned thy form with curious eye," carefully noting its colour, texture, and physical motion (Barbauld 2019, lines 3–12). A philosophical progression here can be traced as has been outlined. Knowledge gained via sensory experience leads to violent destruction of the tent caterpillars, hyperbolically described in terms of a mock-heroic extended metaphor of war: "The victor shouts triumphant; he enjoys / The roar of cannon and the clang of arms, / And urges, by no soft relentings stopped, / The work of death and carnage" (Barbauld 2019, lines 32–5). Barbauld includes herself in her satiric gaze as her past actions are subjected to ethical discussion and her sparing of the single caterpillar she addresses in her poem.

But to stop there is to miss the subtlety of Barbauld's thinking that leaves out her own recognition shared with Ted Toadvine that "an ethical response to nature becomes possible only when we are faced with the *impossibility* of reducing it to the homogeneous, the continuous, the predictable, the perceivable, the thematizable" (Toadvine 2003, 140). Barbauld's self-satire does not end with a recognition of fellow feeling, which she rejects in the final two lines of the poem. To view the lines as reconciling humans and nature is to miss her point. It is significant that the description and decision for non-action (passivity) occur via the caterpillar's physical contact with her arm and finger: "My garment has enfolded, and my arm / Felt the light pressure of thy hairy feet; / Thou hast curled round my finger" (Barbauld 2019, lines 7–9). Merleau-Ponty's statement in the *Phenomenology of Perception* has a rather uncanny connection:

> The body is the vehicle of being in the world, and having a body is, for a living creature, to be intervolved in a definite environment, to identify oneself with certain projects and be continually committed to them… .if it is true that I am conscious of my body via the world, that it is the unperceived term in the centre of the world towards which all objects turn their face, it is true for the same reason that my body is the pivot of the world.
>
> (Merleau-Ponty 2002, 432)

A sense of experience lived in the body leads to a bodily understanding of perception on Barbauld's part that takes the poem beyond the difficulties of morality that are so unsatisfactory and even glib. Barbauld often ends her poems with an ironic twist or reversal, and this poem is no exception as it concludes with a cryptic observation: "'Tis not Virtue, / Yet 'tis the weakness of a virtuous mind" (Barbauld 2019, lines 41–2).[10] She overturns the abstract ethical arguments that she has been exploring via the mock-heroic form for another understanding arising from perception, resembling Levinas' contention that ethical arguments which take freedom as their basis are self-serving, as well as Merleau-Ponty's consciousness of the body. Barbauld's declaration that sparing the caterpillar crawling on her arm represents "the weakness of a virtuous mind" captures a sense of the self-serving, and might be more properly seen as a further moment of the poem's self-satire, describing a moment David Wood observes in his definition of eco-phenomenology that "opens up and develops an access to Nature and the natural that is both independent of the conceptuality of the natural sciences and of traditional metaphysics" (Wood 2001, 94). The observer of the caterpillar would see anew.

Barbauld offers an example of lived experience through which a re-conception of the relation between human and animal takes the poem into the more startling recognition of the sense of "radical passivity" envisioned by Levinas, in which we pre-consciously understand our inherent responsibility through others, summarized in Benda Hofmeyr's introduction to the collection *Radical Passivity*:

> ethical action is made possible by a primordial, an-archical responsibility that is pre-consciously felt as the 'Other- within-the-Same'. We are passively obligated before we can actively choose to help. Levinas therefore argues that the needy other disturbs and incapacitates our customary egocentric ways, and that this 'radical passivity' enables us to recognize our inherent responsibility towards others in need.
>
> (Hofmeyr 2009, 1)

Barbauld's gentle self-mockery recognizes how perception leading to her sparing the caterpillar needs to be recognized in new terms. She describes an act of egocentrism and weakness that must be rejected, whereas viewing the caterpillar with a passive response leads to a more meaningful sense of ethical action beyond the "virtuous mind." Barbauld's poem enacts Ted Toadvine's radical sense that nature "is, if anything, the refusal of the hegemony of perception, language, and thought; it is the 'pocket of resistance' and the unpredictable par excellence" (Toadvine 2003, 150). Barbauld's passivity is unexpected as she rejects the violence of human hegemony towards nature for a different understanding of the relation.

The readings I have offered will likely not overturn the more traditional ways of understanding Clare's and Barbauld's poems, readings, I have argued that they maintain an assumption of human dominance and overlook how the works are so remarkable and atypical of Romantic era poems about human encounters with nature. Understanding how they anticipate an eco-phenomenological perspective means that they are intended ultimately to abandon a human-centred view in favour of open-endedness and irresolution, or, it might be argued, a rejection of the ego of the poet for a resolution that paradoxically recognizes the value of passivity. The poets' rejection of easy anthropomorphic morals is a measure of their poems' greatness as works of art. "The Mouse's Nest" and "The Caterpillar" are poems of quiet yet determined activism that deserve our thoughtful attention for what they might teach us.

Notes

1 Some notable examples include writings by Jonathan Bate, Mary Ellen Bellanca, Greg Gerrard, Richard Kerridge, James McKusick, and Timothy Morton.
2 Den Otter (2004, 212n9) argues that it was written June 1815.
3 See Jonathan Bate, *John Clare: A Biography*. Farrar, Straus, and Giroux, 2003 and William McCarthy, *Anna Letitia Barbauld: Voice of the Enlightenment*. Johns Hopkins University Press, 2008.
4 As I will discuss below, "cesspools" takes the form of the colloquial "sexpools" in the manuscript version.
5 Opinion is divided about how to read the end of the sonnet. John Lucas contends that the description of the sun on the pools represents the restoration of the natural order following its disruption. This is to place an optimistic reading of the poem within the mainstream of Romantic nature writing.
6 Hickman suggests that Clare simply "looks away" and rejects critics' attempts to impose meaning on the incident as openness forms "the logic of the narrative" which is fragmentary and inconclusive" (Hickman 2011, 8–9).
7 See OED: A dialect term means a turf or peat bog or "A space of ground lying between a drain or river and the foot of its bank." Timothy Morton suggests that the term itself offers a crux as scholars including John Goodridge and Robert Hayes prefer Clare's original spelling "sexpools" which are "small pools formed in the hole left by turf cutting" (Morton 2008, 192n).
8 See "The Mouse's Petition" and "An Inventory of the Furniture in Dr. Priestley's Study."
9 See also Weldon who suggests that the poem asks "readers to consider that accepting the Rational Dissenting imperative to activism and social change prompted by moral change can begin with the simple yet vital act of accepting the viewpoint of another living being—human or animal" (Weldon 2002, 13). Caroline Dauphin (2021) views the poem through an ecofeminist reading of georgic.
10 Consider the ending of her mock-heroic "Washing Day" where contemplation of how soap bubbles inspired the Mongolfier brothers' invention of the balloon: "Earth, air, and sky, and ocean, hath its bubbles, / And verse is one of them—this most of all (Barbauld 2019, lines 85–6).

References

Aikin, John. 1777. "An Essay on the Application of Natural History to Poetry. J. Johnson." In *Eighteenth Century Collections Online*. http://library.usask.ca/scripts/remote?URL=http://find.galegroup.com/ecco/infomark.do?contentSet=ECCOArticles&do?cType=ECCOArticles&bookId=0188301.00&type=getFullCitation&tabID=T001&prodId=ECCO&do?cLevel=TEXT_GRAPHICS&version=1.0&source=library.userGroupName=usaskmain

Aikin, John and Anna Letitia Barbauld. 1794. "Evenings at Home; or, the Juvenile Budget Opened. Consisting of a Variety of Miscellaneous Pieces, for the Instruction and Amusement of Young Persons." In *J. Johnson. Eighteenth Century Collections Online*, vol. 4. Accessed September 19, 2025. https://link-gale-com.cyber.usask.ca/apps/doc/CW0115794858/ECCO?u=usaskmain&sid=bookmark-ECCO&xid=579cc537&pg=1

Barbauld, Anna. 1825. *The Works of Anna Lætitia Barbauld*, vol. I. Edited by Lucy Aikin, Longman, Hurst, Rees, Orme, Brown, and Green, pp. 278–80.

Barbauld, Anna, 2019. "*The Collected Works of Anna Letitia Barbauld.*" In *Volume I: The Poems, Revised*. Edited by William McCarthy. Oxford University Press.

Barrell, John. 1972. *The Idea of Landscape and the Sense of Place, 1730-1840: An Approach to the Poetry of John Clare*. Cambridge University Press.

Bate, Jonathan. 2003. *John Clare: A Biography*. Farrar, Straus, and Giroux.

Brown, Charles S. and Ted Toadvine, eds. 2003. *Eco-Phenomenology: Back to the Earth Itself*. State University of New York Press.

Cataldi, Suzanne and William S. Hamrick eds. 2007. *Merleau-Ponty and Environmental Philosophy: Dwelling on the Landscapes of Thought*. State University of New York.

Chapman, John and Seeliger, Sharon. 2001. *Enclosure, Environment & Landscape in Southern England*. Tempus Publishing.

Clare, John John Clare. 1984. *Ed. Eric Robinson and David Powell*. Oxford University Press.

Dauphin, Caroline. 2021."From Suzanne Verdier to Anna Barbauld: An Ecofeminist Revolution of the Georgics." *Ecozon@*, vol. 12, no. 2: 101–16.

Den Otter, Alice G. 2004. "Pests, Parasites, and Positionality: Anna Letitia Barbauld and 'The Caterpillar'." *Studies in Romanticism*, vol. 43, no. 2: 209–30.
Ha, Inhye. 2018. "The "Fellowship of Sense": Anna Letitia Barbauld and Interspecies Community." *Studies in Romanticism*, vol. 57, no. 3: 453–78.
Heaney, Seamus. 1995. "John Clare's Prog." In *The Redress of Poetry*. Faber.
Hickman, Ben. 2011. "John Clare and the End of Description." *John Clare Society Journal*, vol. 30: 5–22.
Hickman, Ben. 2012."A Response to John Lucas's Note." *John Clare Society Journal*, vol. 31: 81–82.
Hicks, Damien M. et al. 2016. "Food for Pollinators: Quantifying the Nectar and Pollen Resources of Urban Flower Meadows." *PLoS One*, vol. 11, no. 6: e0158117. 24 Jun. 10.1371/journal.pone.0158117
Hofmeyr, Benda. 2009. *Radical Passivity: Rethinking Ethical Agency in Levinas. of Library of Ethics and Applied Philosophy*. Springer.
Lucas, John. 2012. "'Mouses Nest'—Again: A Note." *John Clare Society Journal*, vol. 31: 77–79.
Macfarlane, Robert. 2015. *Landmarks*. Hamish Hamilton.
McCarthy, William. 2008. *Anna Letitia Barbauld: Voice of the Enlightenment*. Johns Hopkins University Press.
Merleau-Ponty, Maurice. 1962. *Phenomenology of Perception*. Trans. Colin Smith. Routledge.
Merleau-Ponty, Maurice. 2002. *Embodied Perception, in The Phenomenology Reader*. Edited by Moran, Dermot and Timothy Mooney. Routledge, pp. 423–59.
Morton, Timothy. 2008. "John Clare's Dark Ecology." *Studies in Romanticism*, vol. 47, no. 2: 179–93.
Shelley, Mary Wollstonecraft. 1996. "Frankenstein." In *The Novels and Selected Works of Mary Shelley*, vol. 1. Edited by Nora Crook. W. Pickering.
Smith, Sharon. 2015. "'I Cannot Harm Thee Now': The Ethic of Satire in Anna Barbauld's Mock-Heroic Poetry." *European Romantic Review*, vol. 26, no. 5: 551–73.
Toadvine, Ted. 2003. "The Primacy of Desire and its Ecological Consequences." In *Brown and Toadvine*, pp. 139–53.
Vardy, Alan. 2003. *John Clare, Politics and Poetry*. Palgrave Macmillan.
Weldon, Amy. 2002. "'The Common Gifts of Heaven': Animal Rights and Moral Education in Anna Letitia Barbauld's 'The Mouse's Petition' and 'The Caterpillar'." *Cardiff Corvey: Reading the Romantic Text*, vol. 8. https://www.romtext.org.uk/articles/cc08_n02/
Wood, David. 2001. "What Is Ecophenomenology?" *Research in Phenomenology*, vol. 31, no. 1: 78–95.

SECTION 6

Literary Intersections

22

WISE PASSIVENESS

Responding to Nature with Wordsworth and Merleau-Ponty

Cassandra Falke

Eco-phenomenologist Ted Toadvine writes that any philosophy of nature faces two main questions: first, "how can we think nature starting from our situation within it?" and second, "is there a means of thinking nature that can take into account its excess over our projections and cultural stereotypes concerning it" (2009, 7). This chapter draws on Maurice Merleau-Ponty's concept of the chiasm to address these questions arguing that, because in Merleau-Ponty's thinking we are constantly "bound to the world through all its parts, up against it," our every action has ecological impacts (1968, 131, fn. 1). In explaining the chiasm, Merleau-Ponty refers to the reversibility of sensation. "[W]hen my right hand touches my left hand," I can focus on my right hand touching or my left hand being touched (1968, 133–34). Correspondingly, since we live in and among the "flesh of the world" (1968, 144), we can shift focus from our impacting touch of the world to that which is being impacted or to how we are being impacted by the world. There is a continuity between us and others that enables discovery and reflection. In our chiasmic encounters with human and non-human others, however, we never have full access to the receiving side. We sometimes cannot perceive the impact of our "touch" at all. Much has been written about Merleau-Ponty's chiasm in relation to nature,[1] but I think considering the concept in light of the philosopher's near-contemporary work on passivity can produce more concrete eco-phenomenological insights relevant for Toadvine's questions.[2]

I explore passivity in the essay's second half by turning from phenomenology to poetry. In his 1798 poem "Expostulation and Reply," William Wordsworth advocates for "wise passiveness" in relation to the natural world as sensually given. Explicated in light of Merleau-Ponty's chiasm, the concept of "wise passiveness" describes an orientation toward the natural world that is receptive, epistemologically humble, and, I think, extremely useful for grounding an eco-phenomenological ethics that is meaningful normative. It may seem counterintuitive to address a twenty-first-century question by pairing a nineteenth-century poet and a twentieth-century phenomenologist, but each of these thinkers engages the question of our sensual reception in a way that subverts a strong self/nature division. Furthermore, the Romantics were profoundly concerned with the pre-conscious impact the natural world has on individuals and communities because, like us, they began to recognize as historically contingent certain ways of being among non-human nature. Facing an unprecedented rate of

 DOI: 10.4324/9781003538097-28

technological change and the attendant cultural disruptions, poets working in the 1790s–1820s looked on the natural world with a dawning sense of its precarity, its vulnerability to human destruction.

Two hundred years later, we see writ large the consequences of industrialization and global capitalist expansion that the Romantics could only imagine. Every step, every word, all the wasting and saving implicates us in unseen human and more-than-human relations in ways that we now know we cannot know at the time of impact. We now know, for instance, that carbon dioxide waste impacts the atmosphere for hundreds of years after its production (Boyle and MIT, n.d.). We know that automotive and airplane engineers did not know that in the mid-twentieth century. In another lifetime, more will be known about the damage we do now. While science works to understand more about the consequences of our collective and individual actions, eco-phenomenology works to articulate modes of orientation toward uncertainty itself that account for the indivisibility of humans and/as nature and the excess of beyond human givenness. "Wise passiveness" is my name for such an orientation.

Merleau-Ponty's Chiasm

In the notes for his Spring 1955 course on "Passivity" at the Collège de France, Merleau-Ponty asserts that the goal of his phenomenological project writ large is "to affirm the identity with being of the perceived world as such" (Merleau-Ponty 2010, 124). He sees Husserl and Heidegger united in the quest to explain how this being arrives for us in the immediacy of lived experience (Merleau-Ponty 2005, viii). In phenomenology, he writes, "the world is always 'already there' before reflection begins." As a "rigorous science," phenomenology's "efforts are concentrated upon re-achieving a direct and primitive contact with the world, and endowing that contact with a philosophical status" (Merleau-Ponty 2005, vii).

Perception, for Merleau-Ponty, is neither simple nor solitary. The "problem" of explaining it, he thinks, is that philosophers lose their poise between the conflicting forces of activity and passivity, the perceived world and the perceiving subject. There are so many factors in relation to which we are both active and passive. On the "world side," the flow of time belies any attempt to take a still frame perception of what is around us and treat that as an object of analysis. Time carries us along as passive participants, offering events that we construe retroactively as things. What pretends to be an object is actually "full of gaps, ellipses, allusions; objects are 'physiognomies,' 'behaviors,'—[there] is anthropological space and physical space" (Merleau-Ponty 2010, 124). Every instant of experience in the natural world is ready to unspool a flow of these spontaneous gaps and connections. The branch outside my window is, allusively, Wallace Stevens's "The bough of summer and the winter branch;" the cycle of seasons the branch evokes is among "Things to be cherished" (Stevens 1915, II: lines 14, 7). I see from its movement that a bird has just left. There is the ellipsis. I realize the movement drew my attention, but know, I have not seen the bird. The lady on the sidewalk did not see a branch or a bird. The sidewalk predisposed the lady to be there. Anthropological space and physical space overlap.

My son's experience, looking out a parallel window facing the same tree, will have been significantly different because there are also complexities on the "subject side" of phenomenological thinking. On the "subject side," Merleau-Ponty chides himself to not consider "only the 'natural' body." One must rather

> consider everything that is sedimented above and describe the subject resolutely, not as consciousness, i.e., coincidence of being and knowledge, or pure negativity, but as the X to which fields (practical no less than sensory) are open.—In particular, it is necessary to introduce imaginary fields, ideological fields, mythical fields—linguistics and not only [the] repletion of sensing.
>
> (Merleau-Ponty 2010, 124)

All the sedimentations that conducted me, as a bank conducts a river, toward Stevens, bird, lady and sidewalk will have a radically different shape for my son as he looks out, and he will have been the only site of their passing cohesiveness. Imagination, ideology, myth and language are not separate sediments added to the gaps and spaces, but elements already within the moment's composition as the moment occurs for him. But for all the uniqueness of his experience or mine, the myth, the language, the moment; ideology and imagination (as a capacity); bodies and birds all precede and exceed our moment of engagement with them. We cannot pretend agentic control over the possibility of our engagement with these any more than we can take credit for making our hearts beat.[3]

Although, as subjects, each of us comes forth at the junction of this chiasm, Merleau-Ponty does not want to explore the subject psychologically, but rather to describe philosophically how "[the] perceptual consciousness [is] introduced by the body from the world, which is not a world for me alone, but also for other bodies and in this way common history" (2010, 125). Common history is not only human history, but the provision of the earth itself that allows that history to occur. The dynamic order that includes all the world-side and subject-side considerations listed above originates from and within the earth. "The 'earth' is *archē*: [it] bears the possibility of all being above the nothingness, above the flood—seed of the threatened world, on the basis of which everything blooms again" (2010, 129–30).

Until his death in 1961, Merleau-Ponty returned to this X where fields open around a perceiver. He discovers more and more sedimentary layers and more forces meeting. The X becomes the figure of the chiasm. His most extensive examination of the chiasm is an essay found among his papers at his death and published in English as "The Intertwining – The Chiasm" in *The Visible and the Invisible*. It is here that he uses the figure of a person's left and right hands touching to think through the X at the center of his ontology. "When one of my hands touches the other, the world of each opens upon that of the other because the operation is reversible at will, because they both belong (as we say) to one sole space of consciousness" (1968, 141). This reversibility is the "fundamental phenomenon" that characterizes "seeing and the visible," "speech and what it signifies" (1968, 155, 154). It is the model through which Merleau-Ponty thinks about a moment my hand or eye reaches something that is not me. But, beyond the space of a single consciousness, reversibility becomes more difficult to understand. He acknowledges that if I touch someone else's hand, I cannot experience the inverse sensation of being touched, but he also wants to argue that there are not, in the moment of contact, two separable phenomena: first, my touch, and second, the experience of being touched by me. We can think of those two as separate phenomena, but that is secondary to an originary unity in the act. When experiencing binocular vision or two hands touching or two eyes looking at the same time as two hands touch, say, a cat, one does not have left eye vision plus right eye, plus left hand feeling plus right compiled as a little "bouquet" of separable experiences of "consciousness of" cat (1968, 141). With my cat, it is all black fluffiness together. Similarly, while there are things I do not know about my cat's experience of being looked at and petted, her being looked at and petted is simultaneously part of my experience of doing it. It shares a temporality: I anticipate whether she is in the mood, I adjust to what she seems to like at the moment. I can think about what it must be like to enjoy a good scratch behind the ears, but that would be a reflection that comes secondary to my more immediate experience of her enjoyment. There is much to be said about how the experience of my cat's enjoyment may or may not resemble the experience of another person's enjoyment of being touched, but the particulars of that difference are not as important for my argument as the fact that the (cat or human) other's enjoyment is instantaneously already part of my enjoyment.

If I give some cryptogamic soil a good scratch, its enjoyment or displeasure is no part of my experience. I do know, based on research and the wisdom of people who have watched desert landscapes for decades, that it takes years for a spot to regrow its cryptogamic crust if I have walked on it.[4] I know that during those years, the now-exposed soil will be subjected to wind and water erosion and that nearby plants will not benefit from the nitrogen conversion of soil-dwelling bacteria and lichen. Because the time scales of soil growth are far different than the time scale that my cat and I at least partially share, I cannot experience lichen or bacteria death with a phenomenologically describable immediacy. The fact that I immediately look for somewhere else to step when I see cryptogamic crust arises from the experience I have had before and beyond that moment. The fact that I am killing things is not derivable from the moment itself. Walking in some soils, people are useful seed spreaders, so the fact that walking would do damage is not a given. What the difference between my touch of another person's hand, scratching my cat, or walking on cryptogamic soil reveals is that although each of these encounters has the chiasmic immediacy of impacting/being impacted, the degree of what can be accessed from my side of the encounter varies. We would therefore do well to be mindful of how much we do not know about the effects our actions have on the more-than-human natural world.

Question One: How can we think of Nature from our Position Within it?

All possibilities – of taking perspective, receiving birdsong, making decisions, expecting the future – all depend on the finite earth that bears and blooms. That being the case, what insights can we now bring to Toadvine's first question, bearing in mind both the X that each of us is at the moment and the earth as the "seed of the threatened world?" How can we think about nature starting from our situation within it? For one thing, we see that every encounter is ecological because all action impacts the earth. Acts like throwing out a milk carton and driving to the store for more are as ecologically impactful as walking across a wolf's path. But, we are much more likely to think about the wolf's reception of our presence than we are about the earth's reception of waste, both because our personal waste joins a depersonalized mass of human waste and because we do not live at a time scale that allows us to see the cumulative effects of our life's waste. If the healthy atmosphere flew away like the bird on a branch, or broke like the cryptogamic soil, I might notice that my touch, up against the world, had two sides in this instance, too, impacting and impacted. But I cannot see or hear where the earth is touched by my carbon.

My impact is not abstract; my own "movements incorporate themselves into the universe they interrogate, are recorded on the same map as it" (1968, 133). As "universal flesh," "the world," which is to say the experiential complement to the given earth that bears and blooms, registers every movement as surely as my left hand feels the touch of my right, but my ability to inhabit or imagine the receiving side is limited by the smallness of my perceptual access, knowledge and attention. Even acts that have no immediate material manifestation, such as naming a place we walk to an uncareful walker, have an impact. That place is now more vulnerable. There are states of awareness in which we act with caution even without knowing the consequences of our actions. We do not want to wake a sleeper. We do not want to crack the ice. Phenomenology may not have the resources to habituate people to anticipate the impact of our constant contact with earth, but it has the resources to describe the contact, and I think Merleau-Ponty's chiasm offers a useful figure for consolidating all the fields that both shape and register our impact. Perhaps knowing about that potential impact could create the same latent awareness we have with sleepers or ice.

Question Two: How to think Nature's Excess over our Projections and Cultural Stereotypes

The impossibility of thinking through all the world-side and subject-side fields that we impact and are impacted by leads us into Toadvine's second question: how can we think nature's excess? Ideology, language, myth – some of the factors Merleau-Ponty mentions are not natural, per se, but they do not unfold anywhere other than the earth, so it is in our earthly dwelling that their consequences are felt. Also, we could try to list all the natural factors and lives that coexist at one moment and find that list interminable. The bird's history with that tree. The tree's history with that sidewalk. The germs that lady carries. The weather they and I all share. Because, as Merleau-Ponty says, "absolute plentitude is the result of isolating analysis" (2010, 124), with all the world-side and subject-side variables for chiasmic encounters, the most ecologically beneficial reaction to a moment's perception cannot be pre-determined. If I walk at dusk through a field near a wolf reintroduction area and see tracks, my thoughts about whether or not to tell the rancher I work for will be inflected by my impression of him, the tension between animals raised for food and predatory animal agency, the tension between humans and wolves as predators, etc. Many factors beyond my control enabled me to notice the tracks. A less demanding shift at work and the cold combine to produce wakefulness. Having heard that wolves were released nearby and knowing what wolf tracks look like, encourages me to watch the ground. A light rain that afternoon meant their paws sank in. My understanding of the past that led to the event of my seeing tracks is quite limited, as is my ability to anticipate the effects of saying I saw them. Absolute plentitude may stretch around me in the moment, but complexity only increases as I try to take in the past and the future. I can relativize the moment in relation to any number of world-side fields, knowing I have left my own tracks and my own smell on the receiving side of my chiasmic encounter. I know that, subject-side, I have been affected and will approach the field more alert next time. But the complex conditions similar to those that facilitated the event of seeing the tracks will condition the reception of news that wolves are on the property, exceeding the possibility of knowledge.

Recognizing inter-relationality that extends beyond our own perceptive power produces ethical questions but not ethical norms. Even so, I am responsible for trying to understand how my touch will be received. Two forms of relativization seem consistently beneficial. First, I could think of the impact of my acts on non-human life. If the wolves have been using that field as a place to move unnoticed, and if they are nearby when I pass, then tomorrow they will not count on a field without humans. But that is not enough to ensure they would not come back and be found by the rancher. Second, I can try to know what I cannot know. Why were they crossing the field? Other than the threat to cattle we ranch workers had heard so much about, what difference would their presence in the landscape make? In the late nineties, when wolves were reintroduced to the Greater Yellowstone ecosystem, wildlife biologists chose to admit what they did not know and wait. In the end, the project has "triggered a still-unfolding cascade effect among animals and plants—one that will take decades of research to understand" (Farquar). Beavers are back, elk populations are healthier. Willows are regrowing to secure the banks of rivers. Not everyone thinks this recovery has been worth it. Although the impact on livestock was less than expected, the impact on elk populations, in an area where hunting provides a lot of food and profit, was devastating. In 1995, when wolf reintroductions began, the elk population had reached a peak of around 18,000 animals. The population dropped to 4,000 by 2013. It is now growing again, and lots of species have benefitted from the wolf kills in the meantime, but waiting for that to happen required that scientists, locals and policymakers admit that they could not know the long-term consequences of the reintroduction (National Park Service 2024). For all the counting and mapping, they never will. In their commitment to the good of that ecosystem, they continue producing partial knowledge in a chiasmic and never fully transparent relationship with the dynamism of the place.

Nourishing the Seed of the Threatened World, or Living in a Wise Passiveness

Several Romantic period poets describe forms of receptive attention that I think can be elaborated to function as valuable eco-phenomenological concepts. The most famous of these is probably John Keats's "negative capability," which he defines as the capacity to be "in uncertainties, Mysteries, doubts, without any irritable reaching after fact & reason" (2001, 370). Keats is particularly interested in artists' ability to tolerate the uncertainties of inspiration. But the biologists following the Yellowstone wolf packs with more hope than certainty can also be described as exercising negative capability. Jonathan Mulrooney says negative capability marks "the space of fissure or temporal gap between human consciousness and the world" (2019, 261). Being wisely passive in the face of this gap requires that we accept mysteries and uncertainties about our impact on the natural world. But it is not a passiveness that dismisses the importance of considering that impact. Passiveness in the face of diminishing ecological health of the planet might be construed as relinquishing responsibility, saying we cannot know the environmental damage we are doing, and therefore we cannot be held responsible for that impact. Wise passiveness, in contrast, holds that because we can see some evidence of damaging or positive impact (walking on living soil or planting a tree) and because we see now the evidence of generations who were not wisely passive (atmospheric warming due to past production of carbon waste), we should recognize both that every action has an impact on the earth that we share as home with other human and non-human living things, and that we are responsible for trying to avoid unknowable damage.

Wise passiveness seems to me a legacy of the Romantic poets that has been underexplored and sometimes misinterpreted. Passivity is often approached in relation to action. The person sitting is about to stand (Swift 2018, 943). The one who does not act is resisting (Nersessian 2015, 3). Some scholars see the romantic commitment to passivity as a form of non-violent protest or a commitment to limit consumerism. I think these insights are important and correct, but only part of the story. In being essentially defined by its opposite, the nature and characterization of passivity itself slips away. Passivity is presented in several romantic period poems and letters as a commitment and a skill, something that can and should be honed based on the nature of what it means to be in the world. Much as we can take responsibility for cultivating wise passiveness, our capacity for it is given.

Wise passiveness includes negative capability, but Wordsworth's term goes further to explicitly suggest that the source of mystery arises from outside ourselves and comes to us unbidden. The phrase occurs in his 1798 poem "Expostulation and Reply." A character named William is sitting on a rock when a character named Matthew chides him for indolence. Here is the full poem.

"Why, William, on that old gray stone,
Thus for the length of half a day,
Why, William, sit you thus alone,
And dream your time away?"

"Where are your books?—that light bequeathed
To Beings else forlorn and blind!
Up! up! and drink the spirit breathed
From dead men to their kind.

"You look round on your Mother Earth,
As if she for no purpose bore you;
As if you were her first-born birth,
And none had lived before you!"

One morning thus, by Esthwaite lake,
When life was sweet, I knew not why,
To me my good friend Matthew spake,
And thus I made reply:—

"The eye—it cannot choose but see;
We cannot bid the ear be still;
Our bodies feel, where'er they be,
Against or with our will.

"Nor less I deem that there are Powers
Which of themselves our minds impress;
That we can feed this mind of ours
In a wise passiveness.

"Think you, 'mid all this mighty sum
Of things forever speaking,
That nothing of itself will come
But we must still be seeking?

"Then ask not wherefore, here, alone,
Conversing as I may,
I sit upon this old gray stone,
And dream my time away."

(2008, 129)

Since the poem begins from Matthew's point of view, no notice is given to where the speakers are until the fourth stanza. The focus is entirely on the affront of William's leisure, the element under human control. In the advertisement for the first edition of *Lyrical Ballads*, Wordsworth says the poem was based on "a conversation with a friend who was somewhat unreasonably attached to modern books of moral philosophy." This friend positions William as a subordinate – he is standing, William sitting; he asserts authority over how William spends his time. He positions William as a child of Mother Earth and not the oldest, most responsible one. His language of birth and death seems playful, but also emphasizes the limited time we mortals have to get things done. Natural elements in the scene are reduced to metaphors or functionality. The stone is for sitting; the light is for reading. And books, that man-made light in the fifth line, are where wisdom really lies. Matthew values light as a metaphor for human wisdom (line 5) and metaphorizes earth as a Mother (line 9), whose identity is defined by her human children. Light and earth are not independent entities, but figuratively or literally serve humanity.

William's tone in the fourth stanza immediately changes the feel of the poem. In the voice of the narrator, he emphasizes their emplacement in time – it's morning – and space – Esthwaite Water. Not just his life, but life in general is sweet. Apparently, he has considered asking why and dismissed the question as unimportant. Matthew is his "good friend." Good friend, good place, good morning; it is enough. The "I" position from which Matthew spoke is so confident that he does not even see himself. Matthew never uses "I." But William, in addition to seeing himself as emplaced, thinks his "I" in a shared way. This is at first puzzling because he mentions eyes, ears and bodies, which are irrevocably one's own and not another's, but he uses "we." "The eye," in general, "cannot choose but see." "*We* cannot bid the ear be still," and "*Our* bodies feel" (lines 17–20). The order here is interesting. Eyes are what one uses to read and look at lakes, so William starts with the sense Matthew also began with, the sense that allows most humans the greatest sense of mastery over their surroundings, but then he moves to sound and touch, senses that we more easily recognize as bringing things to us unawares. The organs of sense are not described as willful, but they are powerful, and in their will-less way, impart things to us that we have not chosen to receive.

In my exploration of the chiasm in the chapter's first half, I focused on how the world receives our touch, but here the focus is on William's reception. The speaker alludes to mysterious powers abroad that offer food for our minds, almost like a magic meal appearing. And importantly, wisdom is not something beyond us that we acquire from the sound of lakes any more than by the reading of books. Wise passiveness arrives as a gift, a banquet during which we can "feed this mind of ours", but what is wise is somehow the whole convivial event, not just the thing we would take in. Wise passiveness comes via our senses, which act before and beyond the will. It acts in concert with (capital P) Powers that move through but are not coextensive with the natural world. These Powers could be conceived secularly as the perpetuating energies involved in growth and decay or as something divinely generated.[5] The speaker takes more interest in their operation than in their source. This contrasts with Matthew's implication that wisdom is "a light bequeathed," "a spirit breathed / From dead men" that *can* be taken in (drunk). If wisdom, for Matthew, is both light and drink, then it nourishes us in and of itself but also shapes the way we perceive whatever we direct our attention to, the way the sun lights a room. This is not an unreasonable claim. Of the two positions on wisdom voiced in the poem, it is the more commonsensical. William's is definitely the odder of the two and not just because he is sitting, "alone" "Conversing." It is also odd because he characterizes "wise passiveness" as something we are "in." What does that mean?

One can be in a place, in a mood, or in a position. That position can be physical or social or a general metaphor for how we find ourselves positioned by forces and decisions beyond us. Being positioned involves relationality to others and to a broader pattern. It does not eliminate the act of will somewhere in the process, but whatever is *being* positioned is implicitly passive. Think of an actor, told by a director to move to a spot relative to another actor. The actor must then inhabit the behavior that proximity calls for. He can determine what that behavior is but only in relation to what the proximity would call for. In *The Prelude* (1805), Wordsworth abstracts the condition of our embodied positionality to a general orientation to the world: "by reason and by will" he writes, we acknowledge "dependency sublime" (1995, 332). By acknowledging his position in conversation with Matthew, on a morning, beside Esthwaite water, on an "old grey stone," embodied with eyes and ears, William contests the role the will plays in becoming who we are. He chooses to sit; he chooses to respond to Matthew. He does not suggest that we do not have a will, but he suggests that the role we think it plays in perception and learning is being exaggerated by Matthew. He is only able to sit on that rock with that view because it is already there. The things there are "for ever speaking" (line 26), linnets forever singing, lake forever lapping. Wordsworth cannot choose to initiate or halt that speaking. He can only choose to receive. "The eye cannot choose but see," but we are projecting a sort of Netflix-addicted orientation onto the world if we think it can choose *what* it sees.

Wise passiveness, as articulated here, enhances the concept of the chiasm by describing one element of chiasmic receptivity in a particular situation (sensual reception), by suggesting that engaging in receptivity is a choice, and by praising it as a wise choice. Merleau-Ponty repeatedly insists that the language of passive reception is haunted by the assumption of an already-existing subject that can receive.[6] Framing the poem as a dialogue between two named speakers further reinforces the illusion of stable, autonomous selves, but the fact that things are agentically speaking and William hearing, seeing, smelling whether he wills it or not, figures the self as porous, shaped by forces outside it. William's choice to inhabit a wise passiveness is expressed through verbs that are grammatically active but physically passive – sitting and dreaming (lines 31, 32). Conflating the receiver too fully with what surrounds him risks presuming too much access to that which we receive. Merleau-Ponty stresses that "There is no coinciding of the seer with the visible. But each borrows from the other, takes from or encroaches upon the other, intersects with the other, is in chiasm with the other ... The things touch me as I touch them and touch myself" (1968, 261). Although the "flesh of the world" is "distinct from my flesh," my reception is "not a 'psychic state,' but intra-corporeal" (1968, 261).

Wise passiveness describes an orientation to the world, in the sense that Sara Ahmed uses the term orientation. Orientations are about where we start, about the situatedness of our bodies, and "about the directions we take that put some things and not others in our reach."[7] We can choose to orient ourselves more fully toward certain "objects of thought, feeling, and judgment, and objects in the sense of aims, aspirations, and objectives" by altering our habits, but we do so within social and environmental fields (world side) that orient us prior to our willful interaction with them (Ahmed 2006; 553, 560). William locates wisdom in the attitude or posture of passive reception. We cannot control or predict or will in any way what appears in that magic banquet. The wisdom is in waiting for it.

In keeping with Merleau-Ponty's sense of incorporeal entwinement with the world, physical positioning matters for Wordsworth's sense of what it means to be "in a wise passiveness." The wisdom one can seek and get is associated with Mathew, who is walking up and starting conversations. He bids William get "Up! Up!" but we learn in the poem's penultimate line that

William does not get up. He sits. He maintains a posture of attuned receptiveness. Not every moment calls for the same kind of receptivity, however.

In the companion poem, which follows "Expostulation and Reply" in all editions of *Lyrical Ballads*, William actively instructs Matthew. He now counsels Matthew to get "Up! Up!" Both the physical movement and William's entreaty to Matthew are portrayed as responsive when the poems are read together. As finite bodies in time and space, they need to get going if they want to receive the abundance William sees all around them.

> Up! up! my Friend, and quit your books;
> Or surely you'll grow double:
> Up! up! my Friend, and clear your looks;
> Why all this toil and trouble?
>
> The sun above the mountain's head,
> A freshening lustre mellow,
> Through all the long green fields has spread,
> His first sweet evening yellow.
>
> (2008; 130, lines 1–8)

This is not just any yellow, but that first, passing "sweet evening yellow," which they do not want to miss. Wordsworth here engages in what Merleau-Ponty describes as "a passivity that bears an activity" (1968, 272), the reception of a call that invites a response. Activity is enabled by things giving themselves in a particular way. Although that is always a private gift, mediated, for example, by whether one can hear linnets or not, by whether a linnet is singing at that moment or not, these gifts are just spilling out around us all the time as what Wordsworth calls "the light of things."

"Come forth into the light of thing" William continues, "Let Nature by your teacher" (2008; 130, lines 15–16). These two lines, I think, arise from the same insight Merleau-Ponty has with the chiasm. If long green fields and linnets call or reach or teach or spill forth light, it implies a receptive capacity in both us and them. I am not anthropomorphizing them to say this because I don't say that all forms of receptivity are human, but equally, fields and linnets are not mere objects. Even rock, which seems the opposite of receptive when sleeping or walking on it, receives our impression. Look at the old castle steps. To attribute this reception to cause and effect at a mineralogical level misses the point. Whether we can explain eroding steps and hillsides with the physics of abrasion, we cannot stop them. We cannot start them if we go mountain walking, and this chiasmic interaction, I think, is the insight Wordsworth and Merleau-Ponty share.

Conclusion

In closing, I want to tease out four ways in which we are responsible to nature within wise passiveness. These are orientation, reception, remembering, and passive action. With "things forever speaking" the call is issued, and we are not only able to respond (response-able), but whatever we do or do not do with our bodies and voices in the world will be our response – the tree we planted, the land we cleared, the beauty we noticed, described and advocated for.

We cannot not respond. To use the language of ecology (from the Greek *oikos* meaning "home"), our earthly home has welcomed us. Have we been good guests?

Orientation: think of your body metonymically. It orients your attention to the world. Sit up straight, you pay more attention. Close your eyes to use your ears. The body is not the self, as anyone who has lived with pain or disability knows, but it is where we are up against the world. It is where the world touches us. William, in "Expostulation and Reply" and "The Tables Turned," exemplifies this orientation.

Reception: with things forever speaking, how do we know what to listen to most? What do they say? As Merleau-Ponty alludes to with his proliferation of "world-side" and "subject-side" factors pressing on us, there will always be more than we can possibly consider going on in a given moment. In keeping with Wordsworth's exhortation to wisdom, it seems wise to try to appreciate what we receive from the non-human world and what it receives from us, bearing in mind that we can never fully know. One day, my family and I were walking with a friend in Northern Norway. There were lots of animals, but few of the two-legged sort. A juvenile male fox became very interested in my son, who was eleven at the time. That day, he became a fox that had seen people. We shared food with him, which I would not do with an urban fox, but I wagered that if another person came to the area, he or she would also be benign as people go. But I cannot know. Our touch is upon him and his upon us, and I don't know what we did to him. To us, his visit was a gift. His beauty was so wonderfully superfluous. All five humans who met this fox will have received different things. For me, it was partially a promise that if someone so delicate could flourish in the Arctic, I could, too.

From reception to remembrance: The summer before I went to college, I lived mostly beside the Chattooga River. One morning while bathing, I looked at my hands beneath the water and found them alight with glittering water. The Chattooga has quite a lot of mica, a flaky kind of fool's gold, and just at that moment, the sun hit the water above my hands in a way that revealed it all. But it wasn't all; I looked, and it was on every side of me. The sun was striking just so. Warmth and light on a cold river morning. Ten years later, in a period of grief, I went walking by the River Eden in Northern England. I was greeted by deciduous trees, with a cold-flowing river coursing through limestone, and the smell of petrichor and the oils of rotting leaves, and I sailed back to all that mica shining. The gift, the uncontainable saturation of that moment, was returned to me, the light suddenly everywhere. That passive remembering came, as the first moment had come, as a gift. It was mine to receive and to discover why it was just what I needed. As Wordsworth writes elsewhere, such moments

> have no slight or trivial influence
> On that best portion of a good man's life,
> His little, nameless, unremembered, acts
> Of kindness and of love.
>
> (2008; 132, lines 33–36)

Passive action: To conclude, a chiasmic, wisely passive relationship to the natural world not only demands responsiveness and humility; it also imposes responsibility. We are placed in a bounded horizon, in limited bodies with precious few moments in the sun, but no one is placed exactly as and where we are. We are responsible for what can be seen and done where we are, and there is no norm for that any more than there is a norm for how to be a gracious guest. Thankfulness helps. Not being a burden is advisable. Don't trash the place. Leave a little something for the next guest. What we do and don't do will never be enough. We won't have the first, last or most important word, but it will be our voice – of prayer, praise, instrumentalization, reception, dominance or awe – that helps whoever's near us and listening to find their place in the longer song.

Notes

1 For an excellent overview, see Brian Bannon's "Flesh and Nature: Understanding Merleau-Ponty's Relational Ontology" *Research in Phenomenology* Vol. 41, No. 1 (2011): 327–57. For Toadvine's own speculations about "The Human-Nature Chiasm," in Toadvine, *Merleau-Ponty's Philosophy of Nature*, 107–30.

2 Merleau-Ponty's "Working Notes" from January 1959-March 1961 are published in *The Visible and the Invisible*, pp. 165–275. The note for 1 November 1959 bears the heading "The Chiasm." Just a few years prior, in 1954–55, Merleau-Ponty had taught courses on "Institution" and "Passivity" at the Collège de France. In thinking through passivity, he thematizes sleep, memory and the unconscious, themes which seem very indirectly related to engagement with non-human nature, but the overall question driving his inquiries into passivity and the chiasm relates to "transforming the dependence undergone into conscious and willed dependence" (*VI*, 117), a transformation imminently relevant for human's relation to non-human nature.

3 In describing the "passivity of our activity" in November 1959, Merleau-Ponty writes "It is not I who makes myself think any more than it is I who makes my heart beat" (1968, 221). Claude Lefort, who edited *The Visible and the Invisible* and wrote a forward for *Intuition and Passivity* also identifies this passage as a key indication that "the problem of passivity is not a regional problem" in Merleau-Ponty's late phenomenology but a very central concern (2010, "Foreword" xxi–xxii).

4 The National Forest Service has a good description of cryptogamic or cryptobiotic soil crusts in its description of Glen Canyon: "Cryptobiotic Soil Crusts" 17 June 2024. https://www.nps.gov/glca/learn/nature/soils.htm.

5 Scholarship by romanticists takes a range of positions on this question. James Chandler leaves the source of the Powers an open question in *Wordsworth's Second Nature* (University of Chicago, 1984), 150–55. For J. R. Watson, the powers are clearly divine: *Wordsworth's Vital Soul: The Sacred and Profane in Wordsworth's Poetry* (Macmillan, 1982), 156–59. Evan Gottleib reads the poem as "remarkably congruent with object-oriented philosophy's basic project" (29) *Romantic Realities: Speculative Realism and British Romanticism* (Edinburgh University Press, 2016).

6 For example, "the term 'receptivity' is improper because it evokes a Self distinct from the present and who *receives* it" (Merleau-Ponty 1968, 190).

7 Sara Ahmed. "Orientations: Toward a Queer Phenomenology." *GLQ: A Journal of Lesbian and Gay Studies*, vol. 12, no. 4 (2006): 543–74. Starting p. 545, embodied situatedness 550, directions we take 552.

Works Cited

Ahmed, Sara. "Orientations: Toward a Queer Phenomenology." *GLQ: A Journal of Lesbian and Gay Studies* 12, no. 4 (2006): 543–74.

Bannon, Brian. "Flesh and Nature: Understanding Merleau-Ponty's Relational Ontology." *Research in Phenomenology* Vol. 41, No. 1 (2011): 327–57.

Boyle, Ed and MIT Climate Writing Team. "How Do We Know How Long Carbon Dioxide Remains in the Atmosphere?." (n.d.). https://climate.mit.edu/ask-mit/how-do-we-know-how-long-carbon-dioxide-remains-atmosphere#:~:text=%E2%80%9CThe%20first%2010%25%20goes%20quickly,thousands%20of%20years%20to%20leave

Chandler, James. *Wordsworth's Second Nature*. Chicago: University of Chicago, 1984.

Farquar, Brodie. "Wolf Reintroduction Changes Ecosystem in Yellowstone." *Outside: Yellowstone National Park Trips*. https://www.yellowstonepark.com/things-to-do/wildlife/wolf-reintroduction-changes-ecosystem/

Gottleib, Evan. *Romantic Realities: Speculative Realism and British Romanticism*. Edinburgh University Press, 2016.
Keats, John. "To George and Tom Keats, 21-27 December 1817." In *John Keats: The Major Works*. Edited by Elizabeth Cook. Oxford University Press, 2001. 370.
Merleau-Ponty, Maurice, *The Visible and the Invisible*. Edited by Claude Lefort. Translated by Alphonso Lingis. Northwestern University Press, 1968.
Merleau-Ponty, Maurice. *Phenomenology of Perception*. Translated by Colin Smith. Routledge, 2005.
Merleau-Ponty, Maurice. *Intuition and Passivity: Course Notes from the Collège de France (1954-1955)*. Foreword by Claude Lefort. Translated by Leonard Lawlor and Heath Massey. Northwestern University Press, 2010.
Mulrooney, Jonathan. "Reading Keats's Negative Capability." In *Afterword for Keats's Negative Capability: New Origins and Afterlives*. Edited by Brian Rejack and Michael Theune. Liverpool University Press, 2019. 259–62.
National Forest Service. "Cryptobiotic Soil Crusts." 17 June 2024. https://www.nps.gov/glca/learn/nature/soils.htm
National Park Service. "Elk." *Yellowstone National Park*. Updated 31 July 2024. https://www.nps.gov/yell/learn/nature/elk.htm#:~:text=Yellowstone%20provides%20summer%20range%20for,through%20hunting%20outside%20the%20park
Nersessian, Anahid. *Utopia Limited: Romanticism and Adjustment*. Harvard University Press, 2015.
Stevens, Wallace. "Sunday Morning." *Poetry Foundation*, 1915. https://www.poetryfoundation.org/poetrymagazine/poems/13261/sunday-morning
Swift, Simon. "Wordsworth and the Poetry of Posture." *ELH* 85, no. 4 (2018): 941–72. https://doi.org/10.1353/elh.2018.0034
Toadvine, Ted. *Merleau-Ponty's Philosophy of Nature*. Northwestern UP, 2009.
Watson, J. R. *Wordsworth's Vital Soul: The Sacred and Profane in Wordsworth's Poetry*. Macmillan, 1982.
Wordsworth, William. *"The Prelude" in The Prelude: The Four Texts (1798, 1799, 1805, 1850)*. Edited by Jonathan Wordsworth. Penguin, 1995.
Wordsworth, William. *The Major Works*. Edited by Stephen Gill. 2008. Oxford University Press.

23

IN THE MOOD FOR NATURE WITH HEIDEGGER AND THOREAU

Robert A. Winkler

Chapter

It is 1854, and Henry David Thoreau's *Walden; or, Life in the Woods* goes into print. In the book, Thoreau bears witness to the simple existence he experienced during his two-year stay in a self-built cabin at Walden Pond near Concord, Massachusetts. Foundational for American nature writing, *Walden*'s narrator is both actively and passively immersed in nature – a form of life Lawrence Buell famously described as "a more 'ecocentric' way of being" (1995, 1).

It is 1927, and Martin Heidegger publishes his magnum opus *Sein und Zeit* (*Being and Time* 1996) – conceived in his cabin at Todtnauberg, Black Forrest – wherein he radically reconceptualizes human existence (*Da-sein*, which literally means being-there) as always already thrown into a world. Against the grain of the Western philosophical tradition, Heidegger's Da-sein is not a subject facing objects, but a being-in-the-world which transcends apparently familiar conceptions of activity and passivity.

It is 1944/45, and Heidegger writes the so-called *Feldweg-Gespräche* (*Country Path Conversations* 2010) in the midst of death and destruction. Consisting of three conversations on the road, that is, on the country path, the text elaborates one of the most influential – yet also most cryptic – notions of the later Heidegger: *Gelassenheit*. "[A]s a form of willing non-willing" *Gelassenheit*, which is often translated as releasement, equanimity, or serenity, "both releases us *from* our will to dominance and our will to represent (*vorstellen*) and releases us *to* the possibility of encountering entities on their own terms" (Merwin et al. 2019, 7, emphasis in original).

In the following, we will return to these three particular scenes, that is, rather to the philosophical insights tied to them, in order to better comprehend how the relation of human being and nature becomes reconceptualized in a revolutionary manner – and to go from there to develop this nexus further. Both Thoreau and Heidegger were and continue to be influential in the fields of ecocriticism in general and eco-phenomenology in particular (Buell 1995; Brown and Toadvine 2003; Marder 2018).[1] Scholars accordingly stress the former's "practical" and the latter's "theoretical" contributions to the field, with Heidegger, so

DOI: 10.4324/9781003538097-29

to speak conceptualizing the active-passive attunement that Thoreau finds in and describes as being-in-nature. But what if we were to reverse this apparently clear-cut hierarchy to bring these two figures into a fruitful dialogue by treating Thoreau also as a philosopher and Heidegger also as a nature writer? On this understanding, *Walden* becomes a text that not only describes but also thinks through the nexus of environment, body, and attunement, a relation that the early Heidegger thoroughly conceptualizes. The latter, in turn, emerges as a nature writer, whose notion of *Gelassenheit* concretely manifests in his often neglected phenomenological texts like "Why Do I Stay in The Provinces?" (*Warum bleiben wir in der Provinz?* 1934) or "The Pathway" (*Der Feldweg* 1949). This chapter first theorizes Heidegger's conceptualizations of "nature" and "attunement" before exploring the complex affective ecologies underlying both *Walden* and Heidegger's nature writing in order to gain more insights into the relation of being-in-nature and the active-passive attunement of *Gelassenheit*.

Nature in Early Heidegger

Ecocritical scholars have eagerly appropriated Heidegger's philosophy since the 1970s into the subfields of deep ecology, eco-phenomenology, or environmental ethics.[2] It is, however, as Kalpita Bhar Paul observes, a particular Heidegger that has been most attractive for ecocritical scholars:

> Environmental ethics borrowing from the secondary literature on Heidegger broadly argues that the critique of technology, the concept of the fourfold, earth, authentic dwelling, and homelessness together generate a strong ecocentric perspective. However, [...] the literature tends to concentrate primarily on the later Heidegger's works at the cost of overlooking the early Heidegger's phenomenological exploration of human *Dasein* as 'being-in-the-world.'
>
> (Bhar Paul 2017, 93, emphasis in original)

Tying in with Bhar Paul, we will, in the following, concentrate on the early Heidegger's phenomenological exploration of human existence, that is, Da-sein, as being-in-the-world, to delineate how he conceives the nexus of existence, nature, and mood. But let us first explore the "nature" of Heidegger's conception of nature.

It is commonly assumed that in *Being and Time*, Heidegger subsumes nature under objectively present beings (*Vorhandenes*) with their kind of being, namely, objective presence (*Vorhandenheit*) – and indeed he does so. He states:

> Being is always the being of a being. The totality of beings can, with respect to its various domains, become the field where particular areas of knowledge are exposed and delimited. These areas – for example, history, *nature*, space, life, human being, language, and so on – can in their turn become thematized as objects of scientific investigations.
>
> (Heidegger 1996, 7, emphasis mine)

As an "object of scientific investigation," however, nature is derivative of world and worldliness as conditions of the possibility of Da-sein's existence. Heidegger writes that "[n]ature is itself a being which is encountered within the world and is discoverable on various paths and stages" (Heidegger 1996, 60). So, within an always already given world, nature can be encountered as objectively present beings – but not exclusively so, although this encounter with nature is the most prominent one in Western epistemology:

> One look at traditional ontology shows us that one *skips over* the phenomenon of worldliness when one fails to see the constitution of Da-sein of being-in-the-world. Instead, one tries to interpret the world in terms of the being of the being which is objectively present within the world but has not, however, even been initially discovered – in terms of nature. Ontologically and categorially understood, nature is a boundary case of the being of possible innerworldly beings. Da-sein can discover beings as nature only in a definite mode of its being-in-the-world. This kind of knowledge has the character of a certain "de-worlding" of the world. As the categorial content of structures of being of a definite being encountered in the world, "nature" can never render *worldliness* intelligible. But even the phenomenon "nature," for instance in the sense of the Romantic concept of nature, is ontologically comprehensible only in terms of the concept of world; that is, in terms of an analytic of Da-sein.[3]
>
> (Heidegger 1996, 61, emphasis in original)

Encountering nature as objectively present beings is hence simultaneously founded upon and results in scientific knowledge, which, in turn, "has the character of a certain 'de-worlding' of the world" as within this omnipresent framework, world, in fact, "has not, however, even been initially discovered."

What, then, are other "paths and stages" on and through which nature is discoverable, or rather, has always already been discovered within a – or rather as a – world? Another lengthy passage sheds light on this clearing:

> Thus beings are accessible in the surrounding world which in themselves do not need to be produced and are always already at hand. Hammer, tongs, nails in themselves refer to – they consist of – steel, iron, metal, stone, wood. "Nature" is also discovered in the use of useful things, "nature" in the light of products of nature. But nature must not be understood here as what is merely objectively present, nor as the *power of nature*. The forest is a forest of timber, the mountain a quarry of rock, the river is water power, the wind is wind "in the sails." As the "surrounding world" is discovered, "nature" thus discovered is encountered along with it. We can abstract from nature's kind of being as handiness; we can discover and define it in its pure objective presence. But in this kind of discovery of nature, nature as what "stirs and strives," what overcomes us, entrances us as landscape, remains hidden. The botanist's plants are not the flowers of the hedgerow, the river's "source" ascertained by the geographer is not the "source in the ground."
>
> (Heidegger 1996, 66, emphasis in original)

Nature, accordingly, is an ontological and categorial "boundary case of the being of possible innerworldly beings" in three different ways – with three different kinds of being. First, nature as objectively present beings in their objective presence. Second, nature as a thing at hand (*Zuhandenes*) in its handiness (*Zuhandenheit*), that is, "'nature' in the light of products of nature." Heidegger implicitly references nature in this kind of being in one famous passage of his 1954 essay "The Question Concerning Technology" (*Die Frage nach der Technik*):

> In the context of the interlocking processes pertaining to the orderly disposition of electrical energy, even the Rhine itself appears as something at our command. The hydroelectric plant is not built into the Rhine River as was the old wooden bridge that joined bank with bank for hundreds of years. Rather the river is dammed up into the power plant. What the river is now, namely, a water power supplier, derives from out of the essence of the power station.
>
> (Heidegger 2013, 16)

It is imperative to distinguish, in turn, between two different ways of handling nature here. First, the use of nature in the prevalent contemporary mode in which being shows itself in our days, namely, through technology as "enframing" (*Gestell*), which reduces nature (and everything else) to a mere "standing reserve" (*Bestand*) where "the river is dammed up into the power plant."[4]

This (ab-)use of nature conceals the second – yet more primordial – use that discloses nature as useful things (*Zeug*) within a referential context and referential totality through which Da-sein's very own being in its possibilities is made intelligible.[5] Heidegger delineates that

> [i]n its familiarity with these relations, Da-sein "signifies" to itself. It primordially gives itself to understand its being and poteniality-of-being with regard to its being-in-the-world. [...] We shall call this relational totality of signification *significance*. It is what constitutes the structure of the world, of that in which Da-sein as such always already is.
>
> (Heidegger 1996, 81, emphasis in original)

To appropriate John Haugeland (2013), "nature matters to us" and "we give a damn about it" because, and when, it is discovered as significant, useful things,[6] as "old wooden bridge that joined bank with bank for hundreds of years" and which is therefore "built into the Rhine River." In this context only, "[t]he forest is a forest of timber, the mountain a quarry of rock, the river is water power, the wind is wind 'in the sails'" (see also Cameron 2004).

The third – most surprising and unexpected – conception of nature and its kind of being explodes the neat categorization of things at hand/useful things in contrast to objectively present beings: "nature as what 'stirs and strives,' what overcomes us, entrances us as landscape." In this boundary case, Da-sein discovers nature in its poetic kind of being, a disclosing neither revealing its handiness nor its objective presence, but as, to appropriate Steven Spielberg, *Close Encounters of the Third Kind* in and through which nature presences in a different way of significance. Andreas Luckner (2007, 42) notes in passing that Heidegger here hints at a non-categorial and existential relation of Da-sein and nature, which he will pick up in later essays such as "Poetically Man Dwells," "Building Dwelling Thinking," or "The Thing," which cultivate the seeds sown here.

Does Heidegger give us a hint as to why nature in this poetic sense matters to us and how we might disclose it? The answer is as simple as it is overlooked and can be found in some of the existentials Heidegger delineates in *Being and Time*: attunement (*Befindlichkeit*) and understanding and attunement *as* understanding. Da-sein is always already thrown into a world, and it always already finds itself in a mood in which this world is disclosed to it. In his 1929 essay "On the Essence of Ground" (*Vom Wesen des Grundes*), Heidegger expresses the primacy of attunement in Da-sein's encounter with nature:

> Nature is manifest primordially in Dasein because Dasein exists as attuned (*befindlich-gestimmtes*) *in the midst of* entities. Since disposedness (thrownness) belongs to the essence of Dasein and is expressed in the unity of the full concept of *care*, however, it is here alone that the *basis* for the *problem* of nature can first be attained.
>
> (qtd. in Carman 2003, 194–95, emphasis in original)

In the Mood for Nature? – It is Da-sein's attunement, we can already glimpse, that harbors the potential for ecocritical scholars to work out a non-categorial, existential, sustainable, and poetic relation of Da-sein and nature, a relation we will further develop with recourse to Thoreau's *Walden*. But let us first zoom in on Heidegger's moods.

In the Mood with Heidegger

Heidegger was among the first to take philosophically seriously the notions of disposedness/attunement (*Befindlichkeit*), mood (*Stimmung*), and fundamental mood / fundamental attunement (*Grundstimmung*). Already indicated here by the different translations of these key concepts, "Heidegger's treatment of affective phenomena is terminologically disparate and inconsistent" (Hadjioannou 2019, x). Let us therefore tread carefully here and further delineate and clarify Heidegger's conceptual notions of what would in current terminology be called affectivity or affective phenomena in order to eventually better grasp how Da-sein's being-in-nature is always already a disposed and attuned one.[7]

Da-sein is being-in-the-world. Heidegger first delineates the elements of "world" – referential context, referential totality, significance – before he turns to "being in" to make intelligible an apparent paradox: The individual Da-sein has its own subjective world, its own worlds, in which it lives (for instance, the world of music; the world of Heidegger Studies; the world of bonsai; the world of my family), while we simultaneously all share a world out there, that is, in the sense of a container of things at hand and objectively present beings. How, then, is Da-sein in the world, how does it find itself "there?" By way of *Befindlichkeit*, which Heidegger introduces as a "fundamental existential:" "What we indicate *ontologically* with the term *attunement* is *ontically* what is most familiar and an everyday kind of thing: mood, being in a mood" (Heidegger 1996, 126, emphasis in original). Being in a mood *existentielly* is hence only possible on the condition of Da-sein's always already being attuned *existentially*. Let us exemplify these phenomena by briefly pointing to Joseph von Eichendorff's classic 1837 poem "Moonlit Night" (*Mondnacht*):

> It was like Heaven's glimmer
> caressed the Earth within
> that in Her blossom's shimmer
> She had to think of Him.
>
> The breeze was gently walking
> through wheatfields near and far;
> the woods were softly talking
> so bright shone ev'ry star.

Whereat my soul extended
its wings towards skies to roam:
O'er quiet lands, suspended,
my soul was flying home.
(The LiederNet Archive 2008)

Attuned by and in nature, the soul of von Eichendorff's lyrical I "extended its wings towards skies to roam: O'er quiet lands, suspended." Against the Western philosophical tradition, moods are thus not something individual, hermetically sealed, and at the command of Leibniz's monad or Descartes' subject, but, as Heidegger writes, "[m]ood assails. It comes neither from 'without' nor from 'within,' but rises from being-in-the-world itself as a mode of that being" (Heidegger 1996, 129). Mood situates Da-sein amidst its environment and connects it with its fellow *Da-sein* when we speak of "the mood in a room," or "the mood of the nation." Von Eichendorff accordingly imagines a tender interplay between nature *and* nature – "like Heaven's glimmer / caressed the Earth within / that in Her blossom's shimmer / She had to think of Him" – in and through which the lyrical I becomes attuned and the mood "but rises from being-in-the-world itself as a mode of that being." In the ontic existentiell mood for nature, the "soul was flying home;" nature itself, then, is the ontological existential mode of attunement which prepares the serene and equanimous ground on the basis of which the mood arises as when "[t]he breeze was gently walking / through wheat-fields near and far."

Befindlichkeit literally translates as situatedness in the sense of being somewhere; we will stick with "attunement," as suggested by Joan Stambaugh in her 1996 translation of *Being and Time*, the version which is also referenced throughout this chapter. Attunement aptly relates to the German word for mood – *Stimmung* – in its musical connotation "as it evokes a spectrum of possible ways of being in tune (or out of tune) [*(nicht) gestimmt sein*] with one's surroundings" (Slaby 2021, 244).[8] But back to the paragraphs themselves:

> Beings of the character of Da-sein are their there in such a way that they find themselves in their thrownness, whether explicitly or not. In attunement, Da-sein is always already brought before itself, it has always already found itself, not as perceiving oneself to be there, but as one finds one's self in attunement.
>
> (Heidegger 1996, 127–28)

We did not choose to be born, we did not choose the respective circumstances of our Da-sein, we are thrown in our existence and find ourselves always already in tune, attuned, in a mood. Heidegger, accordingly, defines the first essential ontological characteristic of attunement the following: "*Attunement discloses Da-sein in its thrownness initially and for the most part in the mode of an evasive turning away*" (Heidegger 1996, 128, emphasis in original).

Before theorizing and knowing, even before hammering – to refer to the famous example – we are thrown into a world in which we find ourselves before becoming active in any way.[9] Attunement, thus, is the passive-receptive dimension of Da-sein's "openness to world" as Heidegger specifies: "*Mood has always already disclosed being-in-the-world as a whole and first makes possible directing oneself toward something*" (Heidegger 1996, 137, emphasis in original).

In his 1929/30 lecture course "The Fundamental Concepts of Metaphysics: World, Finitude, Solitude" (*Die Grundbegriffe der Metaphysik. Welt – Endlichkeit – Einsamkeit*), Heidegger further works out the significance of moods for our being-in-the-world by proposing that some "fundamental attunements" affect our very inner and outer experiences. Let's look at the central and oft-cited passage:

> Attunements are *not side-effects*, but are something which in advance determine our being with one another. It seems as though an attunement is in each case already there, so to speak, like an atmosphere in which we first immerse ourselves in each case and which then attunes us through and through. It does not merely seem so, it is so; and, faced with this fact, we must dismiss the psychology of feelings, experiences and consciousness. It is a matter of *seeing* and *saying* what is happening here.
>
> (Heidegger 1995, 67, emphasis in original; see also Slaby 2021, 243)

Certain fundamental attunements like profound boredom, anxiety, wonder, or terror confront us with our naked existence; always already given, they "then attune[] us through and through" by seizing us completely and permeating our very being. Attunements do, on the one hand, materialize and "petrify" in concrete emotions on the level of the individual – in her moods – on the other hand, they show themselves as the all-encompassing environmental spheres which envelope Da-sein. To apply and develop terminology from a different field, which Heidegger himself brings into position: Being-in-the-world is always already attuned by and in an atmosphere, a climate, a weather.

Nature and Attunement in Thoreau's *Walden*

After having delineated the significance of nature and affective phenomena in Heidegger, let us now go one step further and propose that *Gelassenheit* is the fundamental attunement for our being-in-nature.[10] We will do so by appropriating Edmund Husserl's phenomenological rallying cry "Back to the things themselves!" (Husserl 1983, 35)[11] – "Back to the texts themselves!," that is, interrogating one of the classics of nature writing in order to see what it has to tell us about the relation of human being and nature. In what ways, then, is attunement in its literal manifestation as the atmosphere, the climate, the weather depicted in *Walden*? In other words, how does Thoreau get into the mood for nature?[12]

At the beginning of the text, *Walden*'s narrator is not – yet – immersed in the natural environment around Walden Pond.[13] The first and longest chapter, "Economy," harshly and sarcastically critiques the state of mid-19th-century American culture along with those "who are said to live in New England" (Thoreau 2004, 4). Thoreau, in fact, feels for and has empathy with his fellow neighbors who would suffer from conformity, superficiality, materialism, alienation, and depression in this kind of society and economy; far from merely lamenting, Thoreau proposes another "economy of living" which he unfolds for the reader in the form of affective nature writing (Kelleter 2010). Already in the next chapter, "Where I Lived, and What I Lived For," Thoreau gives over his narrator, and the reader, to nature and in this very process un-conceals the interconnectedness of environment, body, attunement – and *Gelassenheit*. This happens when the narrator begins to spend his days and nights in his as-of-yet unfinished cabin, "which, by accident, was on Independence Day, or the fourth of July, 1845" (Thoreau 2004, 84).[14] So let us delve into the passage:

> This was an airy and unplastered cabin, fit to entertain a travelling god, and where a goddess might trail her garments. The winds which passed over my dwelling were such as sweep over the ridges of mountains, bearing the broken strains, or celestial parts only, of terrestrial music. [...] With this more substantial shelter about me, I had made

> some progress toward settling in the world. This frame, so slightly clad, was a sort of crystallization around me, and reacted on the builder. It was suggestive somewhat as a picture in outlines. I did not need to go out doors to take the air, for the atmosphere within had lost none of its freshness. It was not so much within doors as behind a door where I sat, even in the rainiest weather.
>
> (Thoreau 2004, 85)

It is the cabin – as "a sort of crystallization around" him – that affords the narrator's finding himself disposed passively-receptively in an environment which "reacted on the builder." As a passive-receptive dimension of Da-sein's "openness to world" (Heidegger 1996, 137), this *being-in-the-cabin* is the precondition for being-in-nature, for literally becoming attuned in and by the atmosphere, the climate, the weather. Only because the narrator has permitted this passive receptivity – only because he has opened to the world – the calm yet powerful winds are enabled to enwrap him in high spirits. Strikingly, Thoreau's use of language here plays on the musical connotation of Heidegger's attunement as being in tune with one's surroundings, as when the winds the narrator is experiencing are described as "bearing the broken strains, or celestial parts only, of terrestrial music."

Thoreau's narrator finds himself attuned in an environment; in and of itself, this is hardly remarkable because "Dasein is never *not* in the mode of disposedness – it is constantly attuned to its surroundings in certain ways and thus specifically disposed" as Jan Slaby remarks (2021, 243, emphasis in original). On a more fundamental level, however, the narrator's active acting upon nature is put on hold, and this very pausing enables him to perceive nature as nature in the first place. The narrator's "activities" after moving into the cabin reveal this becoming. He notices and takes in the flora and fauna around him, namely, the rich bird population and the geographical particularities of Walden Pond, with this passive receptivity expressed in phrases such as "whenever I looked out on the pond it impressed me" (Thoreau 2004, 86). The narrator's statement about his cabin – "I had made some progress toward settling in the world" – becomes more and more understandable as a cultivation of the faculties of perception which enable him to settle in the world, that is, to understand nature in Heidegger's third sense, namely, as that inexhaustible source, "as what 'stirs and strives,' what overcomes us, entrances us as landscape" (Heidegger 1996, 66).

Thoreau practices *Gelassenheit*, this notion which Heidegger in his 1955 "Memorial Address" to the composer Conradin Kreutzer explicitly set in relation to technology's enframing:

> We let technical devices enter our daily life, and at the same time leave them outside, that is, let them alone, as things which are nothing absolute but remain dependent upon something higher. I would call this comportment toward technology which expresses 'yes' and at the same time 'no,' by an old word, *releasement toward things*.
>
> [...]
>
> *The meaning pervading technology hides itself*. But if we explicitly and continuously heed the fact that such hidden meaning touches us everywhere in the world of technology, we stand at once within the realm of that which hides itself from us, and hides itself just in approaching us. That which shows itself and at the same time withdraws is the essential trait of what we call the mystery. I call the comportment which enables us to keep open to the meaning hidden in technology, *openness to the mystery*.
>
> (Heidegger 1966b, 54–5, emphasis in original)

"Releasement toward things" and "openness to the mystery" transcend the calculating and ordering ratio of enframing and consequently entail the possibility for us to position ourselves in a non-instrumental relation to being and beings – and thus also toward nature (see Winkler 2020). Correspondingly, whenever Thoreau's narrator has a major genuine insight in *Walden* – when he actually understands his being-in-nature on his body – this is only possible on the basis of his finding himself in a prior state of passive attunement and attuned passivity. And only then his natural environment attunes him through and through to make him understand the surrounding nature as a world in its own significance. Thoreau, in fact, finds *Gelassenheit* (releasement, equanimity, serenity) in the first place given in and through nature herself. Mind his description of the pond's own attunement:

> This small lake was of most value as a neighbor in the intervals of a gentle rain storm in August, when, both air and water being perfectly still, but the sky overcast, mid-afternoon had all the *serenity* of evening, and the wood-thrush sang around, and was heard from shore to shore. A lake like this is never smoother than at such a time; and the clear portion of the air above it being shallow and darkened by clouds, the water, full of light and reflections, becomes a lower heaven itself so much the more important.
>
> (Thoreau 2004, 86, emphasis mine)

In the course of his inner and outer journey and the concomitant progression of the book, the narrator becomes generally more attuned to the various attunements emerging through the actual climate at Walden Pond. The awareness of and inscription into this affective ecology, consequently, enables him to move – or rather be moved – from passive receptivity (looking, seeing, perceiving) to a more active being-in-nature. This process happens already in the progress of the very passage we are interpreting here:

> Every morning was a cheerful invitation to make my life of equal simplicity, and I may say innocence, with Nature herself. I have been as sincere a worshipper of Aurora as the Greeks. I got up early and bathed in the pond; that was a religious exercise, and one of the best things which I did.
>
> (Thoreau 2004, 88)

As we can see in Thoreau's *Walden*, it is Da-sein's being-in-nature and becoming one with nature that exemplifies the cultivation of *Gelassenheit* as a fundamental attunement. This active-passive encounter with the natural world attunes our narrator and lets him encounter beings as beings, that is, *nature as nature* and not as enframed through technology's *Gestell*. On the contrary, *Gelassenheit*, as Heidegger proposes in the *Country Path Conversations*, "awakens when our nature is let-in so as to have dealings with that which is not a willing" (Heidegger 1966a, 61).[15] Only on the basis of this willing non-willing does nature find Da-sein's nature, that is, reach us on its own terms. This very process brings about a reversal of the well-established subject-object constellation, according to which the human being is always in command of how to encounter and use nature. The opening of the chapter "The Pond in Winter" is a telling exemplification of this reversal (see also Cavell 2002, 112–13):

> AFTER a still winter night I awoke with the impression that some question had been put to me, which I had been endeavoring in vain to answer in my sleep, as what-how-when—where? But there was dawning Nature, in whom all creatures live, looking in at

my broad windows with serene and satisfied face, and no question on *her* lips. I awoke to an answered question, to Nature and daylight.

(Thoreau 2004, 282, emphasis in original)

The beginning of spring at the end of *Walden* brings full circle Thoreau's understanding of "attuned awakeness" in the sense of Heidegger's awakening *Gelassenheit* that permeates and connects beings across all plains of existence – be they microscopic, human, more-than-human, or cosmic. "Only that day dawns to which we are awake. There is more day to dawn. The sun is but a morning star" (Thoreau 2004, 333) as the "Conclusion" memorably and infinitely perpetuates this conception. Given the Moebius strip structure of the text, however, already the first pages of "Economy" cast the seeds of this holistic philosophy of being attuned by – to eventually become one with – nature, a philosophy *Walden* will unfold "[t]o anticipate, not the sunrise and the dawn merely, but, if possible, Nature herself!" (Thoreau 2004, 17).

In the Mood for Nature! – Thoreau works out Da-sein's non-categorial, existential, sustainable, and poetic relation to nature we have found hidden in the early Heidegger at the beginning of this chapter. *Walden* is not a philosophical treatise in the strict sense of the word, but through the apparently passive act of making it through the text – equal parts novel, memoir, nature writing, and critique of society – the reader becomes a philosopher who understands, not least, by immersing herself in the vivid descriptions of the natural environments. Thoreau shows and not merely tells about *Gelassenheit* as this fundamental attunement, both cause for and effect of our proper being-in-nature.

Coda: Nature and Attunement in Heidegger's Nature Writing

Thomas Sheehan keeps on insisting "that Heidegger remained a phenomenologist from beginning to end" (Sheehan 2014b, 249). If that's true, what implications does this have for Heidegger's own, often neglected, phenomenological texts? We mean phenomenological here not in the strictly philosophical sense, that is, we are *not* dealing with theoretical texts thinking about phenomenology as a method, field, or practice; rather, these texts are short autobiographical reflections about Heidegger's own being-in-the-world and as such they lend themselves to be read phenomenologically, that is, to be interpreted on the basis of their own descriptions and worldmakings. So let us round our exploration off by taking a brief look at two texts of this kind, which, comparable to Thoreau's *Walden*, not merely *tell* about but essentially *show* the relation of being-in-nature and the active-passive attunement of *Gelassenheit*.

Heidegger infamously rejected the chair of philosophy at the Humboldt University in Berlin – Hegel's chair – for the second time in 1933 to stay in Freiburg and Meßkirch. In "Why Do I Stay in The Provinces?" (*Warum bleiben wir in der Provinz?* 1934), which has the prior caption "Creative Landscape" (*Schöpferische Landschaft*), he lays bare his reasoning for doing so. His Melvillian "I would prefer not to" go to Berlin and rather continue being-in-nature also thrives on and expresses *Gelassenheit* as a fundamental attunement. At the beginning of the short text, Heidegger first describes his cabin in its surrounding area to then proclaim that "[s]trictly speaking I myself never observe the landscape" (Heidegger 1981b, 27). This landscape is his "work-world" whose sublime beauty he does not take in actively and consciously by framing the landscape or contemplating its natural beauty. On the contrary,

> all of this moves and flows through and penetrates daily existence up there, and not in forced moments of "aesthetic" immersion or artificial empathy, but only when one's own existence stands in its *work*. It is the work alone that *opens up* space for the reality that is these mountains. The course of the work remains embedded in what happens in the region.
>
> (Heidegger 1981b, 27, emphasis in original)

Just like Thoreau's narrator, Heidegger's thinker understands his being-in-nature as a *being-in-the-cabin* in and through which "work alone [...] *opens up*" the reality of nature. It is, in turn, only on the precondition of his passive attunement and attuned passivity that his world, that is, his natural environment, makes him understand his own existence, an existence that conflates work, life, and nature.

The power of nature to open up the world for a Da-sein in attunement is also beautifully exemplified in Heidegger's "The Pathway" (*Der Feldweg* 1949). Reflecting on the respective country path of his hometown, Meßkirch, Heidegger interweaves childhood memories with deep philosophical insights to playfully lay bare that actually the human being is in need of the natural world. Along its way, the country path passes by a bench on which

> [o]ccasionally there lay [...] one or another of the great thinkers' writings which youth's awkwardness attempted to decipher. When the puzzles ran together, and no way out presented itself, the pathway helped. For it escorts feet quietly along the winding trail through the expanse of barren land. Time and again, thinking from the same writings or from one's own attempts traverses the trail which the pathway draws through the fields. The pathway remains as close to the step of the thinker as to that of the farmer walking out to the mowing in early morning.[16]
>
> (Heidegger 1973, 33)

Writing here about his own frustration as a young thinker struggling with the works of the greats, it is the pathway that literally carries him away to enable him to think again by becoming as attuned to his natural surroundings as the farmer is while pursuing his work.

Just as Thoreau takes time and a large part of his book to let his narrator become properly attuned to nature in *Gelassenheit*, Heidegger's nature similarly shares its insights and attunements only over time. Dynamically becoming, not merely statically being as

> [m]eanwhile, the hardness and smell of oakwood began to speak more distinctly of the slowness and constancy with which the tree grew. The oak itself spoke: Only in such growth is grounded what lasts and fructifies; growing means this: to open oneself up to the expanse of heaven and at the same time to sink roots into the darkness of earth. Whatever is genuine thrives only if man is both in right measure: ready for the appeal of highest heaven and preserved in the protection of sustaining earth.[17]
>
> (Heidegger 1973, 35)

Da-sein's being-in-nature thus not only exemplifies *Gelassenheit* as a fundamental attunement. Rather, the active-passive encounter with nature needs to be understood philosophically; that is, this fundamental attunement requires patience, repetition, and attentiveness in our concrete lived experiences in order to "then attune[] us through and through" (Heidegger 1995, 67). *Gelassenheit*, eventually, might in a Nietzschean sense become what it is, namely, an actual eco-phenomenology, which lets us encounter beings as beings and nature as nature.

Notes

1 For philosophical explorations that explicitly bring both thinkers together, see Cavell (2002), Morrison (2003), and Benesch (2012).

2 The most important name here is Michael E. Zimmerman who published the first book on Heidegger and environmentalism entitled *Heidegger's Confrontation with Modernity: Technology, Politics, and Art* already in 1990. For an excellent overview of the field, see Casey Rentmeester, who acknowledges that "[a]ttempts to construe Heidegger as a potential asset to environmentalism have sometimes been met with skepticism or even outright denial. The three overarching obstacles with using Heidegger as a novel source in environmentalism are as follows: 1) the concern of anthropocentrism; 2) Heidegger's relationship with Nazism; and 3) Heidegger's rejection of normative ethics" (Rentmeester 2016, xix).

3 Heidegger specifies: "'Nature' in the Kantian concept in the sense of modern physics" (Heidegger 1996, 61).

4 This (ab-)use and mastery of nature within the (literal) framework of technology is famously problematized in Heidegger's 1966 interview with journalists from the renowned German weekly *Der Spiegel* who visited Heidegger's hut in Todtnauberg:

> SPIEGEL: But someone might object very naively: what must be mastered in this case? Everything is functioning. More and more electric power companies are being built. Production is up. In highly technologized parts of the earth, people are well cared for. We are living in a state of prosperity. What really is lacking to us?
>
> Heidegger: Everything is functioning. That is precisely what is uncanny, that everything functions, that the functioning propels everything more and more toward further functioning, and that technicity increasingly dislodges man and uproots him from the earth. [...] All our relationships have become merely technical ones. It is no longer upon an earth that man lives today.
>
> (Heidegger 1981a, 56, translation modified)

5 The wording might sound odd here, but given a thing's merging in a referential context and referential totality,

> [s]trictly speaking, there 'is' no such thing as *a* useful thing. There always belongs to the being of a useful thing a totality of useful things in which this useful thing can be what it is. A useful thing is essentially 'something in order to ...'. The different kinds of 'in order to' such as serviceability, helpfulness, usability, handiness, constitute a totality of useful things.
>
> (Heidegger 1996, 64, emphasis in original)

6 John Haugeland pinpoints Da-sein's taking-care structure in contrast to Artificial Intelligence in Tao Ruspoli's insightful 2010 documentary *Being in the World* which features the who's who of American Heidegger scholars:

> The claim was that you could build such and such a machine. And you could test that claim by seeing whether you could build such and such a machine. And it couldn't be done in the 1950s or the '60s and it hasn't been done even now. Perhaps I can reach to one of the most fundamental aspects about the difference between human beings and machines by pointing to something about each of us, with which we are all deeply familiar and that is that 'It matters to us what happens in the world! It matters to us what happens to us! It matters to us what happens to our friends! It matters to us the progress of science and philosophy!' All those are desiderata, those are things to build a life on that one can summarize in the phrase: '*Giving a damn!*' And if you have that phrase then you can say in a word what AI has so far failed to come up with by saying the trouble with computers is that they don't give a damn.
>
> (Ruspoli 2010, 19:18)

7 In *Walden* as well as in *Being and Time* and the *Country Path Conversations*, we can trace affects, which circulate between and thus connect humans, more-than-humans, and the environment. The emerging field of Affective Ecocriticism develops this holistic perspective by "acknowledg[ing] the powerful role environments themselves play in shaping affective experience" (Bladow and Ladino 2018, 3). For a comprehensive overview of Heidegger's conceptualization of these phenomena, see Withy 2024.

8 In the respective entry on "Disposedness (*Befindlichkeit*)" in Mark Wrathall's *The Cambridge Heidegger Lexicon*, Jan Slaby brings together different translations of the term in order to shed light on its manifold meanings: John Haugeland's "findingness" or "so-findingness;" simply "situatedness;" Hubert Dreyfus's "where we're at ness;" "state of mind," as put forward by John Macquarrie and Edward Robinson in the first translation of *Being and Time* (1962); Bill Blattner's "disposedness;" and Joan Stambaugh's "attunement" (Slaby 2021, 243–44).

9 As Taylor Carman pinpoints:

> We have a preconceptual, precognitive access to the real *as such*, an access that is mediated neither by our familiar practices nor by conceptual or theoretical content, but rather in virtue of our disposedness (*Befindlichkeit*) or mood (*Stimmung*). Disposedness or mood is the disclosure of Dasein's thrown-ness (*Geworfenheit*), which establishes our primordial temporal sense of pastness or 'having been' (*Gewesenheit*).
>
> (Carman 2003, 192, emphasis in original)

10 In relating *Gelassenheit* to *Being and Time*'s existentials, we follow Alfred Denker (2011) to bring together the thinking of the so-called early and the later Heideggers to overcome the unproductive polarization in Heidegger Studies between Thomas Sheehan (2014a, 2025) for whom "Heidegger's Being" is merely meaning on the one hand, and Richard Capobianco (2014, 2024) for whom "Heidegger's Being" is merely mystical manifestation on the other. For a groundbreaking thinking through of *Gelassenheit* as a fundamental attunement against the "multifaceted problem of the will in [Heidegger's] thought-path" (xi), see Davis (2007).

11 Here is the full quote in the original context: "But to judge rationally or scientifically about things signifies to conform *to the things themselves* or to go from words and opinions back to the things themselves, to consult them in their self-givenness and to set aside all prejudices alien to them" (emphasis in original).

12 For a more detailed exploration of the affective ecologies around atmosphere, weather, and climate – including the contemporaneous political and racial climate – underlying and structuring *Walden* in particular and Thoreau's oeuvre in general, see Winkler (2024). In a similar vein, Jane Bennett construes Thoreau's affective ecologies as affording the potential for grasping – and changing – the political climate:

> Thoreau is thus among a group of thinkers who target the affective register of politics; he seeks to alter the emotions, aesthetic judgements, and *dispositional moods* that shape political wills, programs, affiliations, ideological commitments, and policy preferences. This, then, is a second aspect of Thoreau's thinking that is particularly useful to contemporary cultural discourse: *his exploration of the political relevance of moods.*
>
> (Bennett 2002, xxii, emphasis mine)

13 Scholarship on *Walden* usually takes "Thoreau" for "the narrator" and vice versa. In the following, I similarly use both interchangeably.

14 Much has been made of this "accident" in the secondary literature as, for instance, Stanley Cavell highlights:

> Of course the central action of building his house is the general prophecy: the nation, and the nation's people, have yet to be well made. And that the day is at hand for it to depart from its present constructions is amply shown in its hero's beginning and ending his tale with departures from Walden.
>
> (Cavell 1992, 116)

15 An abstract of *Country Path Conversations* (*Feldweg-Gespräche*) had been published in English in the edited volume *Discourse on Thinking* in 1966 under the title "Conversation on a Country Path about Thinking" along with the "Memorial Address" to the composer Conradin Kreutzer. The entire work was only published in 2010 as *Country Path Conversations* in an excellent translation by Bret W. Davis. The quote here is taken from the *Discourse on Thinking* volume.

16 Heidegger concludes his 1946 "Letter on Humanism" (*Brief über den Humanismus*) similarly with a poetic conflation of thinker and farmer:

> Thinking is on the descent to the poverty of its provisional essence. Thinking gathers language into simple saying. In this way language is the language of Being, as clouds are the clouds of the

sky. With its saying, thinking lays inconspicuous furrows in language. They are still more inconspicuous than the furrows that the farmer, slow of step, draws through the field.

(Heidegger 1977, 242)

17 The overlap with Thoreau is almost uncanny as he shares the identical image at the beginning of *Walden*:

> The soil, it appears, is suited to the seed, for it has sent its radicle downward, and it may now send its shoot upward also with confidence. Why has man rooted himself thus firmly in the earth, but that he may rise in the same proportion into the heavens above?
>
> (Thoreau 2004, 15)

References

Benesch, Klaus. 2012. "Cultural Immobility: Thoreau, Heidegger, and the Modern Politics of Place." *Amerikastudien/American Studies* 57 (3): 403–18.

Bennett, Jane. 2002. *Thoreau's Nature: Ethics, Politics, and the Wild.* New Edition. New York: Rowman & Littlefield Publishers, Inc.

Bhar Paul, Kalpita. 2017. "The Import of Heidegger's Philosophy into Environmental Ethics: A Review." *Ethics and the Environment* 22 (2): 79–98.

Bladow, Kyle, and Jennifer Ladino, 2018. "Toward an Affective Ecocriticism: Placing Feeling in the Anthropocene." In *Affective Ecocriticism: Emotion, Embodiment, Environment*, edited by Kyle Bladow and Jennifer Ladino, 1–22. Lincoln and London: University of Nebraska Press.

Brown, Charles S., and Ted Toadvine, eds. 2003. *Eco-Phenomenology: Back to the Earth Itself.* Albany, NY: State University of New York Press.

Buell, Lawrence. 1995. *The Environmental Imagination: Thoreau, Nature Writing, and the Formation of American Culture*. Cambridge, MA: Harvard University Press.

Cameron, W. S. K. 2004. "Heidegger's Concept of the Environment in *Being and Time*." *Environmental Philosophy* 1 (1): 34–46.

Capobianco, Richard. 2014. *Heidegger's Way of Being*. Toronto: University of Toronto Press.

Capobianco, Richard. 2024. *In Heidegger's Vineyard: Reflections and Mystical Vignettes*. New York, NY: Angelico Press.

Carman, Taylor. 2003. *Heidegger's Analytic: Interpretation, Discourse and Authenticity in 'Being and Time'*. Cambridge, UK: Cambridge University Press.

Cavell, Stanley. 1992. *The Senses of Walden. An Expanded Edition*. Chicago: The University of Chicago Press.

Cavell, Stanley. 2002. "Night and Day: Heidegger and Thoreau." *Revue française d'études américaines* 91 (1): 110–25.

Davis, Bret W. 2007. *Heidegger and the Will: On the Way to Gelassenheit*. Evanston, Ill.: Northwestern University Press.

Denker, Alfred. 2011. *Unterwegs in Sein und Zeit. Einführung in das Leben und Denken von Martin Heidegger*. Stuttgart: Klett-Cotta.

Hadjioannou, Christos. 2019. "Introduction." In *Heidegger on Affect*, edited by Christos Hadjioannou, ix–xviii. London: Palgrave Macmillan.

Haugeland, John. 2013. *Dasein Disclosed: John Haugeland's Heidegger*, edited by Joseph Rouse. Cambridge, MA: Harvard University Press.

Heidegger, Martin. 1966a. "Conversation on a Country Path about Thinking." In *Discourse on Thinking*, translated by John M. Anderson and E. Hans Freund, 58–90. New York: Harper & Row.

Heidegger, Martin. 1966b. "Memorial Address." In *Discourse on Thinking*, translated by John M. Anderson and E. Hans Freund, 43–57. New York: Harper & Row.

Heidegger, Martin. 1973. "Der Feldweg / The Pathway." *Listening. Journal of Communication Ethics, Religion, and Culture* 8 (1/2/3): 32–39.

Heidegger, Martin. 1977. "Letter on Humanism." In *Basic Writings*, edited by David Farrell Krell, 189–242. New York: Harper & Row.

Heidegger, Martin. 1981a. "'Only a God Can Save Us': The *Spiegel* Interview (1966)." In *Heidegger: The Man and the Thinker*, edited by Thomas Sheehan, 45–72. Chicago: Precedent Publishing, Inc.

Heidegger, Martin. 1981b. "Why Do I Stay in The Provinces? (1934)." In *Heidegger: The Man and the Thinker*, edited by Thomas Sheehan, 27–30. Chicago: Precedent Publishing, Inc.
Heidegger, Martin. 1995. *The Fundamental Concepts of Metaphysics: World, Finitude, Solitude*. Translated by William McNeill and Nicholas Walker. Bloomington, IN: Indiana University Press.
Heidegger, Martin. 1996. *Being and Time. A Translation of Sein und Zeit by Martin Heidegger*. Translated by Joan Stambaugh. Albany, NY: State University of New York Press.
Heidegger, Martin. 2013. "The Question Concerning Technology." In *The Question Concerning Technology and Other Essays*, translated and with an introduction by William Lovitt, 3–35. New York: Harper Perennial Modern Thought.
Heidegger, Martin. 2010. *Country Path Conversations*. Translated by Bret W. Davis. Bloomington, IN: Indiana University Press.
Husserl, Edmund. 1983. *Ideas pertaining to a pure Phenomenology and to a Phenomenological Philosophy. First Book: General Introduction to a Pure Phenomenology*. The Hague/Boston/Lancaster: Kluwer Academic Publishers.
Kelleter, Frank. 2010. "Ecology/Economy: Henry David Thoreau geht spazieren." In *Ökologische Transformationen und literarische Repräsentationen*, edited by Maren Ermisch, Ulrike Kruse, and Urte Stobbe, 177–92. Göttingen: Universitätsverlag Göttingen.
Luckner, Andreas. 2007. *Martin Heidegger: Sein und Zeit. Ein einführender Kommentar. 2. Auflage*. Paderborn: Verlag Ferdinand Schöningh.
Marder, Michael. 2018. *Heidegger: Phenomenology, Ecology, Politics*. Minneapolis, MN: University of Minnesota Press.
Merwin, Christopher, Wendland, Aaron James, and Christos Hadjioannou. 2019. "Introduction: Heidegger's Thinking Through Technology." In *Heidegger on Technology*, edited by Aaron James Wendland, Christopher Merwin, and Christos Hadjioannou, 1–12. New York, NY: Routledge.
Morrison, Ronald P. 2003. "Wilderness and Clearing: Thoreau, Heidegger, and the Poetic." *Interdisciplinary Studies in Literature and Environment* 10 (l): 143–65.
Rentmeester, Casey. 2016. *Heidegger and the Environment*. Lanham, MD: Rowman & Littlefield International.
Ruspoli, Tao, dir. 2010. *Being in the World*. New York City, NY: Mangusta Productions, DVD.
Sheehan, Thomas. 2014a. *Making Sense of Heidegger: A Paradigm Shift*. Lanham, MD: Rowman & Littlefield International.
Sheehan, Thomas. 2014b. "What, After All, Was Heidegger About?" *Continental Philosophy Review* 47 (3–4): 249–74.
Sheehan, Thomas. 2025. *Heidegger's 'Being and Time': Paraphrased and Annotated, Volume 1*. Lanham, MD: Rowman & Littlefield International.
Slaby, Jan. 2021. "Disposedness (*Befindlichkeit*)." In *The Cambridge Heidegger Lexicon*, edited by Mark Wrathall, 242–49. Cambridge, UK: Cambridge University Press.
The LiederNet Archive. 2008. "Eichendorff, J. von. 'Mondnacht.' Translated by Walter A. Aue." Accessed June 10, 2025. https://www.lieder.net/lieder/get_text.html?TextId=38193
Thoreau, Henry David. 2004. *Walden; or, Life in the Woods. The Writings of Henry D. Thoreau*, edited by J. Lyndon Shanley. Princeton, NJ: Princeton University Press.
Winkler, Robert A. 2020. "'Let It Be.' Heidegger's Conception of Technology Between Enframing and Releasement." In *Review of Heidegger on Technology*, edited by Aaron James Wendland, Christopher Merwin, and Christos Hadjioannou. *KULT_online* 62 (November). https://journals.ub.uni-giessen.de/kult-online/article/view/1064/1229
Winkler, Robert A. 2024. "The Weather in Thoreau: Different Dimensions of Climate (Change) in *Walden* and Beyond." In *Entanglements, Narratives, and the Environment: Inter-American Perspectives*, edited by Nicole Haring, Roberta Maierhofer, and Eva Bauer, 37–54. Lanham, MD: Rowman & Littlefield: Lexington Books.
Withy, Katherine. 2024. *Heidegger on Being Affected*. Cambridge, UK: Cambridge University Press.
Zimmerman, Michael E. 1990. *Heidegger's Confrontation with Modernity: Technology, Politics, and Art*. Bloomington, IN: Indiana University Press.

24

THE FLESH OF A DROWNED WORLD

Climate Trauma and Porous Living in J.G. Ballard's *The Drowned World*

Zlatan Filipovic

In his contribution to *Eco-Phenomenology: Back to the Earth Itself*, David Wood (2003) identifies what may well be a point of intersection for an ever-growing field of environmental studies, ecopoetics and related theoretical disciplines. The field itself has faced a minorization of its overarching, collective claims, where differences in theoretical affiliations, strategies and priorities seem to multiply, cutting new escape routes from already established regimes of meaning and enabling new discursive assemblages to emerge. However, where most of the expressions of basic ecological literacy appear to converge, finding a common departure point for their diverse frameworks of validation, resides in the fundamental "recognition of the constitutive quality of relationality" that Wood has identified as "the central virtue" of "the ecological perspective" (Wood 2003, 225). The simple fact that "things are what they are by virtue of their relations to other things" (225). This relational ontology, which privileges heteronomy and dependency over autonomy and freedom, places new demands on our regimes of being, inviting us to reconsider our place within the knotty fabric of our shared existence. Focusing on "relationalities of worldly engagement, both human and those of other creatures" (Wood 2003, 213), can thus help us develop a deeper sense of responsibility for the ecological communities that sustain us.

Since this notion of relationality and the way it transforms our understanding of the non-human world will also be integral to my reading of new ecological realities represented in J. G. Ballard's novel, *The Drowned World*, I will first establish a theoretical framework within which the significance of the term for our perceptual life can be developed. Tracing its critical resonance across heterogeneous yet related discursive fields will also reveal the underlying convergence of their theoretical horizons on the notion of relationality and its epistemological significance for the way we understand the world and our place within it. What will become apparent, however, is not only its transformative potential for our ecological awareness but also the fact that its implications can be traced back to our prereflective experience of the world. Since this experience of the world that precedes our objective representations of it finds its most productive articulation in Merleau-Ponty's (1968) phenomenology, his notions of "chiasm," as reversible intertwining of subject and object, and "the flesh of the world," as embodied experience,[1] will be used to both further develop Wood's (2003) claims

 DOI: 10.4324/9781003538097-30

regarding relationality and to tether its implications to our perceptual life. Merleau-Ponty's phenomenology provides an important touchstone for understanding the embodied nature of human experience and the fact that we are riveted to the world we have otherwise considered separate and readily available for appropriation. His alternate understanding of subjectivity and his conceptual challenge to Cartesian dualism carves out a way for us to see how our perceptual life is meshed with the objective world over which it is assumed to preside and how our place in the world is determined by our embodied experience of it.

This will also provide a theoretical departure point for my consideration of Ballard's novel that, through its exploration of environmental trauma, exposes both the frailties of human existence and the limits of scientific knowledge while also revealing the significance of deep time for our understanding of embodiment. Deep time and its impact on Ballard's world are not only pervasive in the natural landscape, reclaimed by the rising sea and the increasing tropical heat of the primeval past, but also in the perceptual life of those immersed in it. It is articulated as a geological recession of temporal horizons, giving rise to an uncanny world that is both familiar and remote, as the incipient past violates and succeeds the future. The temporal chiasm of the new world, however, also affects the perceptual life, causing dreams of "archaeopsychic past" (Ballard 2013, 56) or deep time to merge with and disarticulate the lived experience of the present. As the world is thus being reclaimed by the feral agencies of the resurgent past, the narrative of progress, dependent on continual evisceration of life worlds, is also gradually revealed as a narrative of self-destruction. In its rejection of established regimes of being, the novel thus calls for a more receptive existence, a new embodied style of being that harbours deeper phenomenological intimacies and alternate postobjective imaginaries of kinship with the perceived world. This style of being, however, that I will call porous living and that evokes Merleau-Ponty's (1968) register of "carnal adherence"[2] and "[e]ntwining," remains open-ended, hinting at unknown horizons of our existence where *conatus* or perseverance-in-being itself may be redefined.

In order to solicit an alternate reading of Ballard's novel and contribute to our understanding of relationality and its significance for ecological production of subjectivity, I will now consider how relationality and its understanding have been disseminated across multiple forms of inquiry whose conceptual and epistemological itineraries may differ but whose ecological interests and related biocentric concerns intersect at the point of our entangled existence. Although the notion of relationality may be repurposed to accommodate new imaginaries, its articulation as the implicit predicate of our ecological existence still retains its significance, constituting a departure point for ecological literacy. Tracing its recent theoretical iterations will thus open up a productive critical terrain in which our relation with feral ontologies, the frailty of our existence and the limits of our dominant ontological regimes could be considered. Although the exteriority of its relations may differ, the notion of entanglement is native to our lived experience and will thus be established on our prereflective engagements and on Merleau-Ponty's phenomenological understanding of embodiment as the carnal relation between self and world.

There Is No Me in the Mesh: Relationality's Porous Lives

Recent decades have witnessed an increasing critique of dominant anthropocentric assumptions that dictate our relations to the non-human world in environmental humanities. Considering the exigency of climate disruption, horizons that continue to privilege our own needs and prerogatives in order to maintain the prosperity of our economic existence have become increasingly untenable. An ontological shift towards bicentric regimes of being that has gradually emerged in our discursive practice has also led to a renewed interest in our prevalent relations to the non-human world, questioning the historical legitimacy of our imaginaries, where its articulation has been predominantly shorn of agency or any ontological depth. By reconsidering the notions of entanglement, materiality and agency, an array of alternate yet related topologies has emerged to provide novel and engaging ways of understanding relationality and the mesh of new intimacies that now constitute our existence.

In eco-deconstruction, the distinctive characteristic of relationality that is native to our ecological existence is referred to as "'originary environmentality'" or the "constitutive ecological embeddedness of mortal life" (Fritsch, Lynes and Wood 2018, 12). Analogous to endless deferrals of the final signified that maintain the alterity of the text in deconstruction, it consigns existence to an endless imbrication of other, feral agencies that make it possible. The fact that "[n]othing… is anywhere ever simply present or absent," that "[t]here are only everywhere, differences and traces of traces," as Derrida (1982, 26) suggests, implies that nothing is in virtue of itself alone but emerges only as a trace of countless others whose persistence itself is never fully present (to) itself but remains incomplete and contingent upon the triumph of other deferred agencies. This ecological entanglement or "interweaving" (Derrida 1982, 26), far from limiting human agency, opens up new ethical and political possibilities for engagement in a shared existence.

In Bruno Latour's (2007) work, our incompletion that relationality implies is articulated in terms of actor-network theory (ANT). Everything that determines and constrains our realities, Latour suggests, can be considered in terms of assemblages of heterogeneous actors who can only emerge in the displacement of networks consisting of other actors that constitute them. The multiplicity of bodies through which our "action is dislocated" (Latour 2007, 46) includes both human and non-human actors that have an equal stake in the production of the world that sustains and constrains us. What is emphasized, in particular, however, is the agency of non-human actors, and the symmetrical ontologies that privilege neither, which also undermine our precious myths of authorship and domination. In *Down to Earth*, Latour (2018), in fact, calls for a posthuman understanding of our deeply entangled present that, instead of manifesting the historical value of the autonomous subject and its epic narratives of emancipation, manifests rather our "newly rediscovered value of dependency" (83). This paradigm shift in our understanding of the place we occupy in the world that now openly contests our authority over it is due to the "New Climatic Regime" (Latour 2018, 91) that constitutes our new ontological horizon, assuming also new and challenging imperatives, making our dependence on other non-human actors painfully apparent.

Jane Bennet (2004) goes even further and argues for a "vitality of things" (348), a new ontological regime of abiotic agencies that have been suppressed by the *cogito* and its colonizing projects, often referred to as narratives of progress. Bennett calls it the "thing-power" of a new materialism that "depict[s] the nonhumanity that flows around but also *through* humans" (349, emphasis added). Human existence, in other words, is irremissibly embodied in and consigned to the world it presumptuously towers above. The radical reconsideration of non-human realities as active interlocutors constitutive of our existence in new materialism is also further related by Diane Coole (2010) to Merleau-Ponty's phenomenology and its abiding "emphasis on corporeality as productive negativity" (93). Indeed, both *Phenomenology of Perception* and *The Visible and the Invisible*, in particular, testify to the fundamental relationality of existence and what Merleau-Ponty (1968) sees as its "entwining" (269) in the shreds of itself held by the objective world. Every "relation with being," he writes, "is *simultaneously* a taking and a being taken, the hold is held, it is *inscribed* and inscribed in the same being that it takes hold of" (266). This "simultaneous experience of the holding and the held *in all orders*," he defines as a "*chiasm*" of our being in the world, not "above," in our "active grasp and... possession" of it, but "beneath" (266, emphasis added). It is thus our immanence to corporeal constituencies of our existence that defines the relational nature of our present rather than our capacity to transcend them. Considered in these terms, relationality then also implies that we are always in a proximity of spectral ontologies and half-revealed interlocutors that constitute or, rather, haunt the possibility of our existence.

This hauntology, as Karen Barad (2010) suggests, is inherent in the very notion of "entanglement" (243), a term Barad uses to account for the relational assemblages constitutive of our world. Relying on Derrida's understanding of conceptual indeterminacy and the fact that all meaning is haunted by what it disavows in order to emerge, and on quantum theory's indeterminacy of matter, Barad radicalizes the notion of relationality while also further developing its significance for new materialist ontologies. "Quantum entanglements," they argue,

> are far more ghostly than the colloquial sense of 'entanglement' suggests. [They] are not [merely] the intertwining of two (or more) states/entities/events, but a calling into question of the very nature of two-ness, and ultimately of one-ness as well. Duality, unity, multiplicity, being are undone.
>
> (251)

I am not only riveted to but irremissibly dispersed across the entire plane of material existence that holds the grains of my being in its open hand, like sand. The self, for Barad, is always carried away by the other that "is irreducibly and materially bound to, *threaded through*, the 'self'" (265). Following Derrida, our existence is thus only ever manifested in traces of its own dispersals, where it is vitiated and affirmed, at the same time.

In Timothy Morton's (2013) terms, the vulnerability of our existence, or, which amounts to the same thing, its relationality, is only another articulation of our irremissible inscription in "the mesh" (83) of interobjective relations that *include* intersubjectivity as only one of its local expressions. For Morton, everything is connected and manifested by virtue of its inscription *in* and *by* other objects. Climate disruption itself is often revealed in traces and displacements of its deferred effects that are decoupled from their initial sources and disclosed by other signs that only partially index the change. Climate change, literally, reveals itself by withdrawal, through and by other signs that often conceal direct relations of force.

For The Dark Mountain Project (2017), founded by Paul Kingsnorth and Dougald Hine in 2009, the fact of our entanglement is barely visible, indeed, invisible in Merleau-Ponty's phenomenological terms, due to "the stories we have been telling ourselves," (23) where one that has dominated all others, is "the story of civilisation" and its toxic fantasy of "human centrality, of a species destined to be lord of all it surveys, unconfined by the limits that apply to other, lesser creatures" (15). To counter this story, they call for an alternate form of writing they refer to as "uncivilized writing" (17), which will not only challenge the anthropocentric myths that constitute our living history, "the myth of progress, the myth of human centrality and the myth of our separation from nature," but also provide a space for alternate epistemologies to emerge that "see us [only] as one strand of a web rather than as the first palanquin in a glorious procession" (18).

What tethers all these different articulations of environmental crisis[3] and what ecological literacy, broadly speaking, appears to depart from is the fact that our existence is radically embodied and articulated by our immanence to feral agencies whose ontological significance we have disavowed since the Enlightenment. As Martin Shaw (2017) suggests, however, these kinds of reimaginings of the human subject "involve submission" (149). A new imaginary, where

> you are not the sole master of your destiny, but in the *unruly* presence of something vaster. You may have to get used to spending a little time on one knee. May have to bend your head. Without a degree of submission, healing, ironically, cannot enter... That's the price tag.
>
> (149, emphasis added)

The prospect of healing, in other words, can only be prepared through an ethical audacity of submission, where our endless narratives of new frontiers give way to exigencies of care that exist on the limits of our current ontological bias.

For Rosi Braidotti (2019), this opens up new frameworks of legitimacies for the human subject that now "has to be understood as a relational, embodied and embedded, affective and accountable entity and not only as a transcendental consciousness" (31). Significantly, the emergence of this new posthuman subject also implies a radical disruption of the Cartesian binaries that have ontologically validated our instrumentalization of non-human existence. The arrogative claims of the *cogito* are disrupted through the emergence of two related terms that the new posthuman imaginaries call for, "firstly, mind-body continuum," which, as Braidotti suggests, entails both "the embrainment of the body and embodiment of the mind" and "secondly, the nature-culture continuum," which is considered in terms of "'naturecultural' and 'humanimal' transversal bonding" (31).

Both of these terms are phenomenologically significant and indebted to Merleau-Ponty's (1968) persistent attempts to reconsider the human subject as a constitutively embodied subject. For Merleau-Ponty (1968), my body is both phenomenological and objective, both "sentient" or "for itself" and "sensible" or "in itself" (135–37), at the same time and without contradiction. It is "made up of two leaves," as he further suggests, "of which the one, that of the 'sensible,' is bound up with the rest of the world" (137) that it, *at the same time*, takes hold of.[4] Merleau-Ponty's rearticulation of the body's originary significance in perception and experience can thus provide a productive framework for understanding the "mind-body continuum" and the "nature-culture continuum" (Braidotti 2019, 31) in critical posthumanities. Furthermore, since my body, in Merleau-Ponty's register, is a situated and relational being, constituted by its openness to the world it is both *a part of* and *apart from*, it also provides a foundation for understanding the entanglement of mind, body and environment that is critical to ecocentric perspectives.

This porosity of living by which my body is defined and to which Merleau-Ponty refers as the *chiasm* of its ensembles, the reversible encroachment of its perceptual and its objective qualities,[5] challenges also the dualism of Cartesian distinctions between mind and body that has historically legitimated both colonial aggression and ecocide, sanctioning the (ab)use of the natural world and the black body identified with it for the expansionist projects and capitalist ventures of Western imperialism. Focusing thus on the originary embodiment of the human subject, Merleau-Ponty's phenomenological approach provides a more integrated understanding of our existence that is not only coextensive with the demands integral to ecological perspectives but that could also be considered as foundational in terms of its emphasis on lived experience from which the awareness of our deeply entangled existence emerges. As

Coole (2010) suggests, when reconsidering the ontological significance of matter, Merleau-Ponty's phenomenology shows how our overreliance on categories and identities indebted to rationalist presuppositions only serves to "reify and separate processes that are irreducibly interwoven within the perceptual lifeworld" (96). Indeed, "the idea of chiasm" or encroachment, for Merleau-Ponty (1968), of "*Ineinander* is on the contrary the idea that every analysis that *disentangles* renders unintelligible" (268) the lifeworld or *Lebenswelt* it is supposed to manifest. It is thus "a question of creating a new type of intelligibility" (Merleau-Ponty 1968, 268) that can solicit a preobjective experience of the world, what Merleau-Ponty, in *Phenomenology of Perception*, refers to as "antepredicative unity of the perceived world" that is "the ground of... [both] verbal expression (*Darstellung*) and intellectual significance (*Bedeutung*)" (2002, 273). It is "the antepredicative life of consciousness" (xvii), distinguished by porosities, entanglements and zones of indiscernibilities, by "brute or wild Being" of the perceived world, which, "ontologically, is primary" (Merleau-Ponty 1968, 170, 200), that fundamentally underlies our being in the world. Our immanence and our relationality, the living wholeness of being, is thus manifested in attestations of our lived experience that reveal the wild meaning of our existence, "our bewildering proximity to the whole of being in syncretic experience" (Merleau-Ponty 2002, 400), that is older than our synthetic structures of objective knowledge. And, for Merleau-Ponty (1968), the task of philosophy resides "in restoring a power to signify" without synthesis "a birth of meaning... [as] an expression of experience by experience" (155).

This also clarifies the significance of descriptive language for phenomenology that is not supposed to impose forms or genres to reassemble our being in the world but to use language as "the very voice of the things, the waves, and the forests" (Merleau-Ponty 1968, 155) to articulate its carnal existence. However, the question is, as he further argues, whether we can reconquer the wild meaning of our experience that precedes our objective reductions of it by the resources philosophy harbours in its conceptual language, or whether we need alternate forms of expression that may be more appropriate for its task. This alternate form of language that sheds its conceptual certainties in order to reveal the uncertainties of our being, the "encroachment of everything upon everything, a being by promiscuity" (Merleau-Ponty 1968, 234) as the flesh of all expression, can be productively considered in terms of literary language since literature signifies only by virtue of the strength it draws from the ambiguities of our existence, its capacity to embody the openness of our lived experience that no certainty of objective knowledge can match. Departing from this notion of literature and its capacity to solicit a more sensuous and, by implication, more discernibly embodied experience of the world, I will now consider how J. G. Ballard's novel, *The Drowned World*, could be used to better understand the prereflective porosity of our existence that both establishes our immanence in the world and humbles our overarching ambitions.

Porous Living: A Piece of the Main

"SOON IT WOULD be too hot" (Ballard 2013, 17). The opening sentence of the novel is as prescient and suggestive of our present as a prophecy that foreshadows its bitter end. Although Ballard's novel does not distinguish global warming and its devastating impact on the biosphere as explicitly anthropogenic, it still provides an alternate terrain for imagining a present that falls well within the remit of our shared realities. Ballard's speculative world has been drowned for seventy years, and its physics irrevocably changed due to runaway global warming, even if the increasing heat is explicitly related to solar storms that precipitated the melting of ice caps and longstanding glaciers, leading to rising sea levels and thermal expansion of the oceans. "The solar disc," as the narrator reveals early on, "was no longer a well-defined sphere, but a wide expanding ellipse that fanned out across the eastern horizon like a colossal fireball, its reflection turning the dead leaden surface of the lagoon into a brilliant copper shield" (17). As the global temperatures rose and human life, cornered by unprecedented heat waves and geophysical upheavals, had colonized the polar circles to establish a stronghold, a UN contingent of scientists, led by Colonel Riggs, was sent as a biological testing station across the drowned cities of Europe. Under the UN, relocated to Camp Byrd, one of the newly established cities in Northern Greenland, where human life and "old categories of thought" (25) could still find refuge, the station was intended to register the impact of climate change on the environment and rescue any survivors in the inundated lagoons of submerged European cities. Having arrived to London, which has become a tropical wetland, bursting with prehistoric life, Dr. Kerans, however, has realized that "little now remained to be done" and that "the biological mapping had become a pointless game," merely reaffirming the anticipations of the initial reports that were "heralding the momentous return of the age of the great reptiles" (19). Their "northward odyssey through the drowned cities in the south" that was "soon to end with their return to Camp Byrd" (20) seemed only to reflect the vanities of their own Promethean fantasies to actively contain and, through scientific domination, reinstrumentalize the natural world for explicit human interests. The evident ecological and evolutionary changes in biotic life have now produced contesting, liminal life worlds, as the human ascendancy has been rolled back and its native habitat aggressively reclaimed by the marine and tropical terrestrial biomes, reinstating the "ascendancy of amphibian and reptile forms" (34) in a rising tide of vanishing human present:

> In the early morning light a strange mournful beauty hung over the lagoon; the somber green-black fronds of the gymnosperms, intruders from the Triassic past, and half-submerged white-faced buildings of the 20th century still reflected together in the dark mirror of the water, the two interlocking worlds apparently suspended at some junction in time, the illusion momentarily broken when a giant water-spider cleft the oily surface a hundred yards away.
>
> (21)

The entangled life worlds, Dr. Kerans perceives from the lavish balcony of his mould-ridden suite at the Ritz, testify to the spectrality of the past that is always buried alive within the present, signifying the persistence of the biophysical past or deep time as a ghost that, in this case, assumes corporeal and menacing visibility. The spatial disjunctions and elemental shifts of climate change manifested in the inundated landscape are also catenated by temporal displacements where the forays of the forgotten past have now gained ascendancy, seizing the present in its extending embrace. This embrace of deep time, evident in the novel as the "nightmare world of competing organic forms returning rapidly to their Paleozoic past" (31), makes the hauntology of every present visible as the fact of temporal entanglements that underlie our objective narratives of representation through which we establish dominion. However, as Barad (2010) suggests, the beginning, "like all beginnings,"

> is always already threaded through with anticipation of where it is going but will never simply reach and of a past that has yet to come. It is not merely that the future and the past are not "there" and never sit still, but that the present is not simply here-now.

> Multiply heterogeneous iterations all: past, present, and future, not in a relation of linear unfolding, but [as] threaded through one another in a nonlinear enfolding of spacetimemattering, a topology that defies any suggestion of a smooth continuous manifold.
> Time is out of joint. Dispersed. Diffracted. Time is diffracted through itself.
> (244)

Time, in the novel, is, indeed, "out of joint" and "diffracted through itself." The past has become the new horizon of the future, evident in the present. The physiological changes in the Earth's biosphere that Kerans and the rest of the crew witness, "the completely altered… shape and contours of the continents" (Ballard 2013, 34), the "giant forests" (31) and "the ancient impassive faces" of "the iguanas and basilisks that had taken over the city" and were "once again… the dominant form of life" (30), do not only represent the entanglements of spacetime but its "intra-action," in Barad's (2010) terms. "*Intra-action* not interaction" of the past within the present as "a felt sense of… intra-activity" (244, 245). The constitutive relationality of the present as an assemblage of vitiating traces is embodied in the novel as an active participation of the past in the present, its "intra-action," rescinding its linear progression towards the future whose horizons have already passed. This confounded temporality of the past to-come that frames the narrative manifests also the hauntology of climate change, where past elements of ecocide and climate disruption, coupled with slow violence and the erratic phenomenology of its effects, contribute to the global haemorrhaging of our horizons. Our vanishing present, according to Morton (2013), is already spectral: "The end of the world has already occurred… It was April 1784, when James Watt patented the steam engine, an act that commenced the depositing of carbon in Earth's crust" (7). The chronotopic overlap or, rather, the entanglement or "intra-action" (Barad 2010, 244) of the past in the present, helps Kerans better understand the relationality of *conatus* and its embodiment in phenomenological terms while also revealing the ontological significance of non-human agency and its claim on the world:

> Looking up at the ancient impassive faces [of iguanas and basilisks] … a[t] their seats in the one-time boardrooms… tak[ing] over the city… Kerans could understand the curious fear they roused, re-kindling archaic memories of the terrifying jungles of the Paleocene, when the reptiles had gone down before the emergent mammals, and sense the implacable hatred one zoological class feels towards another that usurps it.
> (Ballard 2013, 30)

The relationality of *conatus* and its ecological entanglement are also manifested in the "steady decline of mammalian fertility" (34). With the "inverted ecological balances" (34–35) of the new world, the fertility rates of the remaining human communities have also progressively dropped, reminding Kerans that the "genealogical tree of mankind was systematically pruning itself, apparently moving backwards in time" and, as the anthropocentric paradigms of human privilege shift, "a point might ultimately be reached," as he further reflects, "where a second Adam and Eve found themselves alone in a new Eden" (35), presumably, none the wiser.

The changing perceptual realities or life worlds of the crew members that openly contradict and reveal the vanity of their objective existence and empirical pursuits have also induced dissociations in their absconding affective life. Each member, except Colonel Riggs, who, in the new world, represents the senseless tenacity of its old validation frameworks, experiences perceptual and affective changes of the new ontological regime, what Kerans regards as "zone[s] of transit," those liminal realms of indiscernibility that, in fact, constitute our perceptual life:

> The growing isolation and self-containment, exhibited by the other crew members of the unit and from which only the buoyant Riggs seemed immune, reminded Kerans of the slackening metabolism and biological withdrawal of all animal forms about to undergo a major metamorphosis. Sometimes he wondered what zone of transit he himself was entering, sure that his own withdrawal was symptomatic… of a careful preparation for a radically new environment, with its own internal landscape and logic, where old categories of thought would merely be an encumbrance.
>
> (24–25)

As if climate change and the increasing biophysical evisceration of the human world are gradually peeling off the stratified crust of objective knowledge and abstraction to reveal the phenomenological veracity of the world and its subjectivities, constituted at the relational intersections of non-human multiplicities. Kerans, in the course of the narrative, is progressively emerging as the new "posthuman subject of knowledge" that Braidotti (2019) defines as "embedded, embodied and yet flowing in a web of relations with human and non-human others" (4). The fact that subjectivity is transversally (dis)assembled by the traces of vitiation and dispersals across the living crust of the world, that it is carnally constituted, will become one of the more significant yet implicit revelations articulated in the novel that both Kerans and the reader are gradually coming to understand. Although subjectivity's inscription in the mesh, the chiasm (*ineinander*) or porosity of its living, may be phenomenologically or prereflectively understood, it is nevertheless objectively, ethically and politically represented as distinguished from its carnal alliances with the world it is both within and beyond. However, as Braidotti (2019) further suggests, "[s]ubjectivity is not restricted to bound individuals, but is rather a *co-operative trans-species effort*… that takes place transversally, in-between nature/technology; male/female; black/ white; local/global; present/past — in assemblages that flow across and displace the binaries" (3, emphasis added).

One of the dominant articulations of subjectivity's co-operative (dis)assemblage in the novel is the persistent recurrence of "organic memories" (Ballard 2013, 56) in the dreams shared by the crew members, which Dr. Bodkin, a senior biologist and Kerans' assistant, has been monitoring. As Lieutenant Hardman suffers a deteriorating onset of dreams dominated by "the oldest memories on Earth" (56), the objective structures of his temporal existence are becoming less and less discernible from his lived experience of deep time. "Every step" in the evolution of life, as Bodkin relates when discussing Lieutenant Hardman's deteriorating condition with Kerans,

> is a milestone inscribed with organic memories… each is a record of a thousand decisions taken in the face of a sudden physico-chemical crisis. Just as psychoanalysis reconstructs the original traumatic situation in order to release the repressed material, so we are now being plunged back into the archaeo-psychic past… The brief span of an individual life is misleading. *Each one of us is as old as the entire biological kingdom, and our bloodstreams are tributaries of the great sea of its total memory*… nervous system is a coded time-scale, each nexus of neurons… marking a symbolic station, a unit of neuronic time.
>
> (56, emphasis added)

For Bodkin, and eventually for Kerans, the geophysical changes are carnally reiterated as neuronic changes. Entire "biological kingdom[s]" with their feral history of death throes and triumphs are thus implicated in the possibility of every human thought. This holistic understanding of subjectivity and of its originary embodiment, the inscription of its immanence in the systems that far exceed its naïve fantasies of authorship, reflects the chiasmic relation of reversible encroachments in the assemblage of subjectivity as "an exchange between me and the world, between the phenomenal body and the 'objective' body, between the perceiving and the perceived" (Merleau-Ponty 1968, 215). So that, as Merleau-Ponty further suggests, "what begins as a thing ends as consciousness of the thing [and] what begins as a 'state of consciousness' ends as a thing" (215). The perceptual world of Lieutenant Hardman, however, so far suppressed by the claims of the *cogito* to emerge only in his dreams, has become almost indiscernible from the objective structures that dictate his otherwise arrogative existence. "Sometimes," as he confesses to Dr. Bodkin, "I think I have the dreams continuously, every minute of the day. Perhaps we all do" (Ballard 2013, 50). Having taken a risk, "using the heater to simulate the sun," and having raised "the temperature well into the 120s" (57), Dr. Bodkin has found that Lieutenant Hardman's condition did not decline any further but has rather improved:

> For the previous three weeks his dreams were almost driving him out of his mind, but during the last few days he's been much less disturbed, almost as if he were accepting the dreams and allowing himself to be carried back without retaining any conscious control.
>
> (57)

Lieutenant Hardman's acceptance of the new ontological regime will also lead to his imminent disappearance, as he breaks away from the station and turns South, "entering the world of total neuronic time, where the massive intervals of the geological time-scale calibrated his existence" (62). Renouncing the arrogative axioms of the recent past and their obsolete narratives of legitimation, Hardman submits to the new perilous imaginaries that may redefine his place in the world. His audacity, however, is transformative; it requires "a total reorientation of the personality" (57), as Dr. Bodkin observes, foreshadowing also Kerans' increasing openness towards a receptive existence that harbours a deeper, integrative intimacy with the world and the new feral possibilities of porous living. In light of this, Dr Bodkin's caveat regarding the intestate horizons of human existence could then be considered as auspicious: "If we let these buried phantoms [of deep time] master us as they re-appear we'll be swept back helplessly in the flood-tide like pieces of flotsam" (57), or, rather, like pieces of the main that we are.

However, rather than submitting to the new imaginaries of an entangled present, the remaining pockets of humanity, concentrated on polar circles, still rely on their own hubris to actively manage the ecological trauma: "The tacit assumption," as Kerans relates, "made by the UN directorate… [was] that within the new perimeters described by the Arctic and Antarctic Circles life would continue much as before, with the same social and domestic relationships, by and large the same ambitions and satisfactions" (58). The tenacity of our exorbitant claims on the world, the persistent fantasy of our birthrights and our enterprising abilities to transcend the limits the world has ferociously asserted in order for Man and his myths to yield, would thus continue unabated, providing yet another chapter in his adventure of being. Although this possibility, according to Kerans, was "obviously fallacious, as the mounting flood-water and temperature would show when they reached the so-called polar redoubts" (58), the inquiring crew members, including Colonel Riggs, who were still representing the old continents of meaning that do not move in step with the geological realties of the new ones, were hard pressed to understand "why... Hardman [should] choose to travel South?" (71). After all, there is "nothing but solid jungle... open sea" and searing heat there, "he'll *fry*" (71). Kerans, however, knew that South and "the uncertain but curiously potent image [it] summed up" (59), was the only bearing left on the compass: "'Colonel," he replied, "there isn't any other direction'" (71), its "ghostly deltas" (58) being the only available horizon, "diffusing outwards... like the heady vapours of some spectral grail" (59). The symbolic use of the "grail" and its "spectral" ontology/hauntology to represent the chiasm of the entangled present, the diffraction of spacetime through itself, that the invasion of the surging South in the novel articulates, is phenomenologically significant since it provides the possibility for a radical transformation of perception, opening up a new prereflective way of seeing and being in the world. South is not merely the only horizon left in Ballard's world but is also a phenomenological key to unlocking a different perception, one that is less reliant on synthetic structures and more attuned to the holistic expressions of the world. "[W]ith all its dormant magic and mesmeric power" (59), South enunciates the encroaching horizon that resides beyond the limitations of the *cogito*, beyond "the encumbrance" of the "old categories of thought" (Ballard 2013, 25) and its blockages to "wild Being" (Merleau-Ponty 1968, 170) that is ontologically primary. However, like the Grail itself, it represents also an elusive yet overwhelming challenge, demanding a fundamental shift in human understanding of being in the world that is both intimate and wild, harbouring both possibilities and uncertainties of existence.

This preobjective or "antepredicative knowledge" of the world, in Merleau-Ponty's (2002) terms, "our bewildering proximity to the whole of being in syncretic experience" (82, 400), is also enunciated in Kerans' gradually intensifying dreams, revealing his carnal participation in "the flesh of the world" (Merleau-Ponty 1968, 84) and the perceptual unity of porous living:

> As the great sun drummed nearer... the dense vegetation... was flung back abruptly, to reveal the black and stone-grey heads of enormous Triassic lizards... Kerans felt, beating within him like his own pulse, the powerful mesmeric pulse of the baying reptiles... [in] the lake, whose waters now seemed an extension of his own bloodstream. As the dull pounding rose, he felt the barriers which divided his own cells from the surrounding medium dissolving, and he swam forwards, spreading outwards across the black thudding water...
>
> (Ballard 2013, 86)

The flesh of the present and the transversal cartilage of our being that have so far been repressed and articulated as abject realities of the objective assemblage of the world and "its common" or "fleshless style" (Merleau-Ponty 1968, 111), dominated by scientism whose significance in the novel grows increasingly symbolic, are now revealed as the prevailing expressions of felt experience. The "terrestrial and psychic landscapes, as Kerans informs us, "were now indistinguishable..." (Ballard 2013, 89). His colleague and paramour, Beatrice Dahl, having had the same recurrent dream as other members of the crew, the orders had come through for everyone to leave and proceed back North to the harbours of old and jealously maintained imaginaries. Colonel Riggs, however, "was still obeying reason and logic, buzzing around his diminished, unimportant world with his little parcels of instructions like a worker bee about to return to his nest" (90). Kerans, Beatrice and Dr. Bodkin also decide to return, but not to the insolvent fantasies of progress represented by Camp Byrd. Rather than further disavow the non-human world, they resolve, instead, to affirm its imperatives and find their place within the new porous ecologies of the present that depart from ontological heteronomy, shorn of human prerogatives: "[I]n so far as we are part of the planet… we too are returning," as Dr. Bodkin explains. "This is our zone of transit, here we are re-assimilating our own biological pasts. That's why we have chosen to remain here" (108).

The resistance to transformative ontological change towards immanence and disassemblage of human privilege required by climate change is represented in the novel by the arrival of Strangman, distinguished by his menacing, "white smiling face" (108) and his racialized crew of "human scavengers" (38). They were "driven northward" (38) from "the Equatorial lagoons and archipelagos… pillaging the drowned cities" (105) and reclaiming the relics of a perished imperialist culture that Kerans aptly identifies as "bones" (112). While racial distinctions seem to have fallen apart under the searing light of the sun, revealing their limit as a myth constructed to justify abuse, they reemerge as a residual set of legitimacies when Strangman and his crew take possession of the lagoon:

> Kerans' heavy sunburn, like that of Beatrice and Dr. Bodkin, made him virtually indistinguishable from [Strangman's] crew, and the subtle [racial] distinctions between… [his crew members] had vanished. Strangman alone retained his original paleness, the effect emphasised by the white suit he had chosen.
>
> (109)

Their reemergence and authority, however, are implicitly related to "the old categories of thought" (25), as it is later revealed that Strangman stands under the protection of the "Reclaimed Lands Act and the Dykes Maintenance Regulations" (179) issued by the UN at Camp Byrd. Old categories of thought that have no ontological purchase in the porous terrain of the new present are thus not only epistemological but also racial, reaffirming the abiding dominance of imperialist knowledges where whiteness is constructed as the universal signifier of authority with a privilege of plunder.

As Strangman prepares to retrieve what he believes is the treasure in the submerged London planetarium, Kerans enlists to help in order to save Beatrice and Dr. Bodkin from the continuous torments they have been subjected to. Having cleared the lagoon of the "reptilian evil" (105) and the ferocious caimans Strangman and his crew had restrained into submission, Kerans descended into the entangled depths of the sunken universe and

> felt the warm blood-filled nausea of the chamber flood in upon him… the soothing pressure of the water penetrating his suit so that the barriers between his own private blood-stream and that of the giant amnion seemed no longer to exist. The deep cradle of silt carried him gently… Far above him, as his consciousness faded, he could see the ancient nebulae and galaxies shining through the uterine night, but eventually even their light was dimmed and he was only aware of the faint glimmer of identity within the deepest recesses of his mind. Quietly he began to move towards it, floating slowly… through the darkness alone, like a blind fish in an endless forgotten sea, driven by an impulse whose identity he would never comprehend.
>
> (127–28)

Kerans' carnal relation with the world is manifested as his own "consciousness" and its primacy recedes in the upsurge of the world around him. The "barriers" that distinguish "his own private blood-stream," the thickness of the flesh that separates him from the world that cradles it, are also what make him belong to it. The skin is the mesh, the inside a threshold of the outside. Separated by "the thickness… of the body," Merleau-Ponty (1968) writes, the "distance is not the contrary of… proximity, it is deeply consonant with it… not an obstacle between… [but a] means of communication." The body itself "far from rivalling that of the world, is on the contrary the sole means I have to go unto the heart of the things, by making myself a world and by making them flesh" (135). As the flesh of the world is revealed to Kerans in a womb-like chamber of a dome that softly, "like an immense placenta" (Ballard 2013, 127), cradles his body, it implies also that porous living is nascent and originary in terms of its ontological significance. The in-difference of identity he experiences in relation to the world, "an identity without superposition… that constitutes its natal secret" (Merleau-Ponty 1968, 135–36), emerges, thus, as an originary articulation of being-in-the-world for Kerans. And it is this perceptual openness that finally provides meaning for the reversed trajectory of their temporal existence and its "immense tidal pulse he had heard in his dreams" (Ballard 2013, 123).

Unable to regain a hold on the old world and its treasure, revealed by climate change as a myth that has lost its currency, Strangman decides to drain the lagoon in a convulsive pursuit of old fantasies. As Leicester Square is "resurrected from its own sewers," revealing this time, beyond "the grace… [that] had once distinguished [its] streets," the "filthy mantle" (146) of its own past, Kerans finds himself "hemmed in by the shifting planes of dissonant realities millions of years apart" (148). With Dr. Bodkin executed for the attempt to blow up the levee that keeps the chiasmic present at bay and Beatrice harried into submission by Strangman as his consolation prize, Kerans is now ready to commit himself to "[t]he brute or wild Being… [of] the perceived world" (Merleau-Ponty 1968, 170) that the South in the novel represents:

> His time there [in the clutch of the old world] had outlived itself, and the air-sealed suite with its constant temperature and humidity, its supplies of fuel and food, were nothing more than an encapsulated form of his previous environment, to which he had clung like a reluctant embryo to its yoke sac. The shattering of this shell… was the necessary spur to action… Both the past, represented by Riggs, and the present contained within the demolished penthouse, no longer offered a viable existence. His commitment to the future… plagued by so many doubts and hesitations, was now absolute.
>
> (Ballard 2013, 167–68)

The difference between Kerans' perceptual life, open to the wild horizons of porous living, and the objective realities, still deeply entangled with the colonial fantasies of ownership, becomes even more apparent as Colonel Riggs reappears *ex machina*. Returning from Camp Byrd, Riggs descends with his crew to defuse the escalating violence and reconsider the possibility of reoccupying the city, commending rather than restraining Strangman and his territorial projects to reclaim it:

> The last thing I intend to do is re-flood the lagoon. If anybody tries I'll personally blow his head off. Reclaiming land, particularly an urban area like this right in the centre of a former capital city, is a Class A1 priority.
>
> (180)

Realizing "that even a wider gulf now divided Riggs and himself," Kerans finally understands that it "was Riggs, and not himself, who was the time traveller" (179), ironically trapped in the predatory imaginaries of the past and its fantasies of human exceptionalism.

Having flooded the lagoon and narrowly escaped the Colonel and his crew, Kerans now turns south, towards the new, transversal cartographies of destratified existence. After a brief encounter with Hardman, whose "personality was now submerged deep within his mind" and could barely discern anything but "the sun [as] the sole entity strong enough" (196) to dictate his passage, Kerans picks his own and forays deeper and out of sight, "a second Adam searching for the forgotten paradises of the reborn Sun" (198). With its closing statement, the novel remains ambiguous regarding Kerans' redefined relation to the world and his possibility to persevere in it. Evoking Yeats' "The Second Coming," where some "rough beast," yet to be revealed by the apocalypse of modernity, is dragging itself "towards Bethlehem to be born" (Yeats 2000, 76), Ballard's "beast," however, is not as ominous. While also revealed by the auto-immune logic of modernity, Ballard's "second Adam" retains a nascent promise of hope for an alternate ontological horizon that, from the depths of our resistance, summons up our originary openness to the world. In an interview with Ronnov-Jessen (1984), Ballard also hints at an alternate imaginary opened up by the climate disruption. Discussing *The Drowned World*, he suggests that "[a]ll the other characters… react as most ordinary people would: if the dam bursts, they run for the hills. It's only the central character [Kerans] who sees the system of imaginative possibilities represented by the disaster." The more Kerans ventures in "the less of him there is" (n.d.). Or rather, the *more* of him there is. In his "descent through total time" (Ballard 2013, 100), Kerans, as it were, eavesdrops on "the flesh of things" that, in Merleau-Ponty's (1968) terms, "speaks to us of our own flesh" (193), revealing our own embodied existence and originary intimacy with the world. Porous living testifies both to our vulnerability but also to our potential for transcendence, opening a new ontological horizon beyond the human kingdom.

In the context of eco-phenomenology, *The Drowned World* provides a compelling account of porous living, a receptive style of being that abdicates from the privilege of its birthrights to accept the constraints of its immanence in the transversal constituencies of the non-human world. Through his gradual disclosure of perceptual experience, Kerans decides to renounce the "old categories of thought" (Ballard 2013, 25), blocking his passage to the originary promiscuity of existence, and to reemerge as a relational intersection of entangled ecological realities, a posthuman Adam. Kerans' surrender, however, is not an abnegation of responsibility but an epiphany of our carnal relation with the world and our responsibility for it. It is a concession that our future relies on our ability to reimagine our place in the world and find an alternate style of being—an unglorified living that accepts the finitude of its own existence.

This ontological imperative is further developed by the novel's exploration of time. *The Drowned World* vividly depicts the insinuation of the past in the present as a lived experience, intensifying the abiding impact of embodied temporalities. The submerged cities, the resurgence of prehistoric life and the "archaeopsychic past" (56) that emerges in the dreams of the crew members all testify to the significance of deep time in ecological existence. This hauntology, however, is not represented as a nostalgic restitution of the past that may be related to the charisma of pristine wilderness in deep ecology, but as an admission of temporal encroachments that compel us to consider the spectres of climate change as our *intra-active* interlocutors.

In light of this, Ballard's novel provides a stark caveat against the persistence of human privilege that contributes to the ongoing dismantling of anthropocentric narratives. The attempts to reclaim the drowned cities in order to reassert our dominance only rearticulate our perverse colonial fantasies in a world that openly contests the legitimacy of their claims. The novel suggests, instead, the need for new ontological imaginaries that reveal the originary constitution of our existence as a piece of the main.

Notes

1 The flesh of the world, however, is not an object, as Merleau-Ponty indicates, "is not explained by the flesh of the body, nor the flesh of the body by the... self that inhabits it," it is the simultaneity of them, the fact that, by virtue of my body, I participate in the world that constitutes me, at the same time (Merleau-Ponty 1968, 250).

2 "Why would not the synergy," as Merleau-Ponty (1968) inquires,

> exist among different organisms... Their landscapes interweave, their actions... fit together exactly [i.e., nothing is external to anything else]: this is possible as soon as we no longer make belongingness to one same 'consciousness' the primordial definition of sensibility, and as soon as we rather understand it as... a carnal adherence of the sentient to the sensed and of the sensed to the sentient.
>
> (Merleau-Ponty 1968, 142)

3 There are others, to be sure, that are just as significant within literary studies and environmental humanities: Deep Ecology, Ecofeminism, Social Ecology and Eco-Marxism, Material Ecocriticism that departs from New Materialism, Postcolonial Ecocriticism, Blue Humanities, Petrocultures and Energy Humanities and still others. Although they all open up heterogenous aspects of climate crisis and, thus also, account for the non-linear articulation of its phenomenology and impact, there are still significant theoretical and material overlaps, exigences and concerns that are common to most, but that various theoretical fields approach and emphasize differently.

4 Already in *Phenomenology of Perception*, for instance, Merleau-Ponty (2002) uses lived experience in order to instantiate this continuum of corporeality that recasts the subject as entangled and embodied. "My body," he writes,

> is recognized by its power to give me 'double sensations:' when I touch my right hand with my left, my right hand, as an object, has the strange property of being able to feel too... In other words, in this bundle of bones and muscles which my right hand presents to my left, I can anticipate for an instant the integument or incarnation of that other right hand, alive and mobile, which I thrust towards things in order to explore them. The body catches itself from the outside engaged in a cognitive process; it tries to touch itself while being touched, and initiates 'a kind of reflection' which is sufficient to distinguish it from [mere] objects....
>
> (106–7)

5 Developing his notion of *chiasm*, Merleau-Ponty (1968) writes: "[W]e situate ourselves in ourselves *and* in the things, in ourselves *and* in the other, at the point where, by a sort of *chiasm*, we become the others and we become world" (160).

References

Ballard, J. G. 2013. *The Drowned World.* New York: Liveright Publishing Corporation.

Barad, Karen. 2010. "Quantum Entanglements and Hauntological Relations of Inheritance: Dis/Continuities, SpaceTime Enfoldings, and Justice-to-Come." *Derrida Today* 3 (2): 240–68.

Bennet, Jane. 2004. "The Force of Things: Steps toward an Ecology of Matter." *Political Theory* 32 (3): 347–72.

Braidotti, Rosi. 2019. "A Theoretical Framework for the Critical Posthumanities." *Theory, Culture & Society* 36 (6): 31–61.

Coole, Diana. 2010. "The Inertia of Matter and the Generativity of Flesh." In *New Materialisms: Ontology, Agency, and Politics*, edited by Diana Coole and Samantha Frost, 92–116. Durham, NC: Duke University Press.

Derrida, Jacques. 1982. *Positions*. Translated by Alan Bass. Chicago, Ill: University of Chicago Press.

Fritsch, Matthias, Philippe Lynes, and David Wood, eds. 2018. *Eco-Deconstruction: Derrida and Environmental Philosophy. Groundworks: Ecological Issues in Philosophy and Theology*. New York: Fordham University Press.

Latour, Bruno. 2007. *Reassembling the Social: An Introduction to Actor-Network-Theory*. Oxford: Oxford University Press.

Latour, Bruno. 2018. *Down to Earth: Politics in the New Climatic Regime*. Cambridge, UK: Polity.

Merleau-Ponty, Maurice. 1968. *The Visible and the Invisible*. Edited by Claude Lefort. Translated by Alphonso Lingis. Evanston, Ill: Northwestern University Press.

Merleau-Ponty, Maurice. 2002. *Phenomenology of Perception*. London: Routledge.

Morton, Timothy. 2013. *Hyperobjects: Philosophy and Ecology after the End of the World*. Minneapolis: University of Minnesota Press.

Ronnov-Jessen, Peter. 1984. "Against Entropy." *Literary Review*. Accessed January 13, 2025. https://literaryreview.co.uk/against-entropy-interview-with-j-g-ballard

Shaw, Martin. 2017. "Small Gods." In *The Dark Mountain Project, Walking on Lava: Selected Works for Uncivilised Times*, edited by Charlotte Du Cann, Dougald Hine, Nick Hunt, and Paul Kingsnorth, 148–50. White River Junction, VT: Chelsea Green Publishing.

The Dark Mountain Project. 2017. *Walking on Lava: Selected Works for Uncivilised Times*. Edited by Charlotte Du Cann, Dougald Hine, Nick Hunt, and Paul Kingsnorth. White River Junction, VT: Chelsea Green Publishing.

Wood, David. 2003. "What Is Eco-Phenomenology?" In *Eco-Phenomenology: Back to the Earth Itself*, edited by Charles S. Brown and Ted Toadvine, 211–35. Albany, NY: State University of New York Press.

Yeats, William Butler. 2000. *Yeats's Poetry, Drama, and Prose*. Edited by James Pethica, 76. New York: W. W. Norton & Company.

25

MOVING WATER

Aquatic Experience in Emmi Itäranta's Novel *Memory of Water*[*]

Alice Sundman

Opening Reflections—Shaping Environmental Consciousness?

When Clive Hamilton, philosopher of ethics, calls for "a different kind of orientation to the Earth" (2017, 152), he emphasizes that humans now have a responsibility toward nature, place, and the earth that goes beyond ethics and morality as normally conceived: rather than judging human actions "on a scale of good and evil," we need to look at "where they fall on a scale of *care and neglect*" (150; original emphasis). Care, as understood in the field of care ethics, frequently involves acts of concern for someone or something.[1] Resulting from conscious decision-making, such acts are primarily actively constituted. However, underlying these acts are *passively* constituted affects, attitudes, and relations. Exploring these kinds of fundamental pre-reflective perceptions, I believe, gives a fuller understanding of that which underlies and prompts active caring, for care, as Hamilton sees it, should not be understood through traditional ethics or morality, but rather as "a deep, pre-ethical sense of responsibility" (149). So, rather than through a focus on how acts of caring, neglect, and mastery are *performed*, such responsibility can be better understood through attention to passive pre-reflective experiences, attitudes, and relations that underlie deliberate acts, for, as Arne Naess has suggested, "ethics follows from how we experience the world" (qt. in Toadvine 2009, 134). Citing Naess, Ted Toadvine argues that "[a]n 'ontological shift' in our perception of the world can alter our ethos by shifting our sense of *what is* and how we experience and interpret our relations with things" (2009, 134; original emphasis). Bryan E. Bannon, in the same vein, claims that "we must alter the very way nature is conceived in order to inspire the necessary changes in affection that can motivate a healthier relationship with nature" (2014, 17).

Could such a shift in how we perceive nature, place, and the earth be set in motion through literature? Taking this overarching question as a point of departure, this essay brings together ecological concerns with phenomenological notions and cognitive literary criticism focusing on affect in an eco-phenomenological exploration of a contemporary novel thematizing climate change, namely Emmi Itäranta's *Memory of Water* (2014).[2] The essay discusses how the

* This research was supported by the Anna Ahlström and Ellen Terserus Foundation. Open Access for this chapter was funded by the Sven and Dagmar Salén Foundation.

DOI: 10.4324/9781003538097-31

experience of water is portrayed in the novel, and whether these depictions have the potential to evoke affective responses in readers—to *move* them—and thus to impact perceptions, attitudes, beliefs, and, by extension, ethics in the real world.

Literature's impact on the real world has been discussed by critics such as Erin James, who points to "storyworld models" as potentially "a powerful driver of real-world change" (2022, 20) and Alexa Weik von Mossner, who suggests that readers' as well as non-readers' "understanding ... can be shaped or at least influenced by environmental narrative" (2017, 8). This essay takes a cue from Weik von Mossner's reflections on "embodied simulation," a "pre-reflective" (Weik von Mossner 2017, 36) process in which mirror neurons, a type of brain cell, are affected not only by a person's encounters with other people but also, and more importantly in this context, by literary texts[3]; as Weik von Mossner claims, "readers will map the sensations, emotions, and movements of a character onto their own brains, thereby understanding, and literally *feeling*, their interaction with the character's environment, its pleasures, and its pain" (2017, 25; original emphasis). Furthermore, as psychologists Melanie C. Green and Timothy C. Brock have shown, not only readers' emotions and attitudes may be affected by fictional texts, but also "real-world beliefs" (2000, 707). Such change in beliefs, Green and Brock contend, is dependent on the degree of "transportation," that is, immersion in a text (2000, 707). This in turn is related to, among other things, "the quality and meaning of the text" (2000, 705) as well as "if a reader feels that she has experienced realistic narrative events" (Green 2004, 251). Obviously, immersion in a text depends on a variety of intratextual as well as extratextual features and factors, and on multifaceted interrelations between these. As Weik von Mossner notes, "individual 'transportability' into a narrative world can vary considerably among actual readers" (2017, 32) and "our affective response to a literary text is determined by emotional memories" as well as "by the *vivacity* of the things we imagine while reading" (2017, 24; original emphasis). The focus of this essay, however, is not on actual readers' experiences, memories, or associations,[4] but on *the text's potential to affect readers* through textual features such as narrative style, thematic foci, and narrative perspective.

Particularly potent for readers' experiences of literary environments relevant to ecologically focused criticism, Weik von Mossner holds, are descriptions of *characters' experiences*. These, she suggests, create a sense of "empathetic engagement" and "feeling along with" characters that affect how "readers experience and relate to a narrative environment" (Weik von Mossner 2020, 133). *Memory of Water*, with its (mainly) first-person[5] account of the protagonist's experiences in a climate-changed world, is germane to a discussion of how literary imaginings can partake in shaping our environmental consciousness. The novel depicts a future world in which scarcity of drinking water permeates life in a small village in the fictional Scandinavian Union, a region under occupation by the power state New Qian. At this point in time, the world's shorelines are radically changed as large parts of the continents are submerged. At the same time, the former water resources in Scandinavia are said to be contaminated because of an accident that happened during the oil wars. In this world of a shortage of fresh water, the protagonist, the young tea master Noria Kaitio, has a secret to guard: following the old tradition of tea masters, she is the watcher of a hidden natural spring. The novel thus thematizes not only the presence and absence of water, but also ethical questions relating to the scarcity of water, the secret spring, and the question of the allegedly contaminated water resources of the Lost Lands.

As a novel portraying a climate-changed world, *Memory of Water* can be characterized as climate fiction. Climate fiction, or cli-fi, has been described as literature "[c]haracterized most frequently by efforts to imagine the impact of drastic climatological change on human life and perceptions" (Irr 2017, 1), "centrally concerned with manifestations of a changing climate on its ecological, economic, and social consequences" (Mayer 2014, 27), or simply "explicitly focused on climate change" (Schneider-Mayerson 2018, 474). Part of the interest in this genre, as Matthew Schneider-Mayerson holds, is the expectation that such texts might impact "environmental beliefs and behavior" (2018, 474)[6]; as Maria Laakso, Toni Lahtinen, and Hanna Samola claim, climate fiction "aims to warn people of the effects of climate change" (2019, 200). While Itäranta's novel indeed fulfils these criteria for climate change novels, it has also been described as "genre-defying" and read as "ecological trauma fiction" (Jytilä 2022, 214). And, as Claudia Egerer submits, it "resists the categories critics assign to it—young adult fiction, cli-fi, sci-fi, dystopian fiction, speculative fiction. It is all of these and none, and *invents its own poetics of water*" (2022, 34; emphasis added). In this essay, my main concern is not the question of genre. Rather, I focus on the novel's "poetics of water" and, more precisely, its portrayal of *aquatic experiences* in and of a climate-changed world and its potential to influence that which may underlie real-world active "beliefs and behaviors," such as pre-reflective emotions and attitudes.

In exploring the novel's portrayal of aquatic experience, the phenomenological concepts of the lived body, passivity, and activity are helpful. In line with both Edmund Husserl and Maurice Merleau-Ponty, I view the lived body as fundamental to experience: the lived body is "constantly there, functioning as an organ of perception" (Husserl 2001, 50) and constitutes the access point to place and the world. This relation can be *passive*, or pre-reflective, or it can happen *actively*, or reflectively, as when we become "aware of a world" (Merleau-Ponty [1945] 2002, 100). I understand passivity through Husserl as a pre-reflective state in which "[the ego] simply perceives, when it is merely aware, apprehending what is there and what, of itself, is presented in experience by itself" (2001, 93).[7] Activity, on its part, involves "the accomplishment of the ego who in the strict sense makes judgments, makes decisions, and who actively appropriates and establishes its acquisition of knowledge" (Husserl 2001, 106). Activity and passivity are not to be seen as binary notions; rather, in line with Husserl, I view them as interrelated, since "all activity essentially presupposes a foundation of passivity" (Husserl 2001, 276). As Anthony Steinbock phrases it, passivity "is the *basis for* activity as the realm of cognitive and logical experience" (2001, xlii; emphasis in the original). Likewise, in my reading of *Memory of Water*, I view passive perceptions, experiences, and attitudes as underlying acts of reflection and acquisition of knowledge.

In what follows, I explore the portrayal of the *experience of water* in Itäranta's novel. I argue that three modes of experience are vital for the potential to evoke a sense of water's being and significance in the climate-changed world presented in the text: the presentation of *immediate experience* invites the reader to imaginatively experience direct perception of water through embodied simulation. The explicit juxtaposing of actively acquired *abstract knowledge*, on the one hand, and lack of concrete experience, on the other, invites active reflection on that which may be lost in a climate-changed world. And the portrayal of communal *sedimented experience*, passed on from generation to generation, points to our responsibility for future generations while also suggesting a view of water as not just a resource to exploit but an element to experience, reflect on, and care for.

Immediate Experience—The Spring, the Rain, and the Absence of Water

Memory of Water abounds with textual moments portraying the concrete experience of water through the lived body. This kind of immediate experience, understood through Husserl's thinking as involving "our experience as it is 'given'" and being "mainly passive" (Overgaard 2008, 290), allows the reader to, in words borrowed from Weik von Mossner, "imaginatively experience *what it is like*" (2017, 139; original emphasis) to perceive water in this future world. Noria's first visit to the secret spring is such a moment. This happens when her graduation ceremony approaches and her father, who is the present tea master and Noria's teacher, shows her the spring as part of an age-old tradition in which it is the tea master's task to watch over the spring. This is Noria's first encounter with free-flowing fresh water. It is an experience that engages her body and her senses as she, for the first time in her life, hears, sees, and feels such an abundance of water.

Sensory imagery is crucial in the portrayal of Noria's experience of the spring. Her first perception of it happens through her hearing: "slowly I began to distinguish a stretching, growing sound …, familiar and yet strange. I had never before heard it flowing free, entirely pushed by its own weight and will" (11). The co-existence of strangeness and familiarity is emphasized by Noria's comparison of the strange wildness of flowing water "not chained in man-made confines" (11) with "rain knuckling the windows or bathwater poured on the roots of the pine trees" (11). The sound takes on palpable qualities as Noria experiences it as a materiality that can be sensed tactilely: "It wrapped me and pulled me in" (11). The closer they get to the spring, the stronger the sound: it "thrummed loud," "the darkness sang" and "received us with a rumble. It was like the roar of heated water at the bottom of an iron cauldron, but more like the sound of a thousand or ten thousand cauldrons" (11). The sensory imagery evokes an overwhelming presence of something wild and roaring filling the underground space.

When they reach the spring, Noria experiences the water through her sight:

> Water rushed from inside the rock in strings and threads and strands of shimmer, in enormous sheets that shattered the surface of the pond at the bottom of the cave when they hit it. It twisted around the rocks and curled in spirals and whirls around itself, and churned and danced and unravelled again. The surface trembled under the force of the movement.
>
> (12)

Here, the description of how the water moves is combined with a textual form that creates a rhythm resembling similar movements. This is created through the rhetorical device polysyndeton, seen in the "strings and threads and strands" and the water that "twisted around the rocks and curled in spirals and whirls around itself, and churned and danced and unravelled again," evoking an intensity and a sense of both the water and the text moving in winding, dancing, meandering ways. Visual imagery, seen by Gabrielle Starr as "play[ing] perhaps the largest sensory role" and "often experienced as having the greatest vivacity" (Starr 2010, 277) in texts, is here combined with—and intensified by—imagery evoking tactile feeling as the water's forms hold both visual and tactile potential, and, moreover, as the textual form in itself creates a sense of flowing rhythm.

Noria then uses her hand to *feel* the water: "I dipped my fingers in the water and felt its strength. It moved against my hand like breathing, like an animal, like another person's skin. It was cold, far colder than anything I was used to" (12). The sensory imagery portrays the lived body's experience of the water through tactile feeling. Later on, Noria takes her friend Sanja to the spring. They step into the water: "I hardened myself against the piercing chill of the water, letting it settle on my skin, even as it cut into my limbs and pinched along my back" (167). These descriptions focus on the tactile feeling in the body—the face, the hand, the skin, the limbs—in adjectives, nouns, and verbs evoking the feeling of the element: "cool and moist," "strength," "animal," "breathing," "piercing chill," "cut," "pinched." The lived body experiences the unfamiliarity of the element as Noria's senses passively take in the feeling of water.

In these passages, sensory imagery relating to sound, vision, and tactile feeling, together with the form and rhythm of the text, creates a sense of immediate experience of the water in the spring. If Weik von Mossner is right in claiming that "textual imagery relating to vision activates the visual cortex whereas textual imagery relating to sound, smell, taste, or touch activates other relevant brain regions" (Weik von Mossner 2020, 133), Itäranta's focus on how Noria experiences the spring allows the reader to imaginatively experience the spring in similar ways as the protagonist does. This potential is further enhanced by the first-person perspective, through which the reader is given access to the protagonist's reactions and feelings, enabling a sense of "feeling along with" (Weik von Mossner 2020, 133) Noria.

Similarly, the portrayal of the experience of rainfall invites readers to feel the rain along with the protagonist. Here, however, the experience is not restricted solely to humans: "its sounds enclosed the landscape within their soft fingers, stroking the soil and grass and tree-roots" (168). The sensory imagery suggests that it is the ground and the vegetation that experience the tactile feeling of the rain's sounds. If Weik von Mossner is right in holding that "somatosensory and motor cortices in readers' brains" will respond to the descriptions "regardless of the fact that these actions are attributed to nonhuman agents" (2017, 35), the text, here, invites the reader to experience these feelings also along with nonhuman subjects.

These experiences of water play out against a background of scarcity of fresh water. Such a shortage is indicated through phenomena such as "water guards" controlling that water allowances are not exceeded (19) and "water crime" (22) for which the punishment is death (163). Yet it is Noria's first-person experience of water's absence in the landscape that creates a feeling of drought. Her descriptions of a "dried swamp" (8), a "faded landscape of burned-out grass and bare stone" (8), a "river that is now a dry scar in our landscape" (26), and "a long, ditch-like furrow, smooth-worn with stones and sand gathered at the bottom" which is a "channel of a dried stream" (87) create images of how the water that once was has left marks in the ground. Embedded in these descriptions is a contrast between the arid landscape and the same place during a time when water still existed in abundance. The description of the Dead Forest displays such a contrast:

> The Dead Forest had once been called Mosswood, a name that recalled deep-green leaves moving in the wind and verdancy so lush and moist that you could feel it on your skin. Even longer ago, when words for such greenness were not needed yet, because it was a given in these lands, the forest had not had a name at all, so my father had told me. Now its bare trunks and branches twisted towards the sky sand-dry and colourless like a cobweb woven across the landscape, or the empty husks of insects caught in it. Life no longer circulated in them, their veins were brittled and broken, their skins frozen into letters of a forgotten language, near-incomprehensible marks of what had once been. Some trunks had wrung themselves into the ground, where they lay speechless, still. …
>
> The only sounds were the slow sinking of the forest, the wind clutching leafless twigs and the faint creaking of the trunks turned towards the earth.
>
> (203–04)

Here, dead dryness is contrasted with lush vegetation. Through sensory imagery focusing on sight and tactile feeling, a verdant forest is evoked, albeit in imagination only. The description of the "bare trunks and branches" uses visual imagery evoking dryness: they are "sand-dry and colourless" and have "veins" that are "brittled and broken." Note also how stillness and death are intertwined with a lack of language as trunks are "speechless." The dryness is also experienced through "sounds" of "the slow sinking of the forest, the wind clutching leafless twigs and the faint creaking of the trunks." These descriptions juxtapose a world of dryness with the idea of a living, breathing forest in which verdancy, birdsong, and the sound of leaves rustling in the wind are present.

So, while the text describes drought and absence of water, it simultaneously evokes lush vegetation and water flowing in rivers and streams, thus inviting the reader to experience, through embodied simulation, a landscape in which water exists in abundance, but also to feel, along with Noria, the loss of such a world.

Abstract Knowledge—Ice and Snow

There is a moment in Noria's story in which she reflects on the difference between abstract knowledge and actual experience:

> my mother's books did not tell everything. I wondered how snow would feel on my palm just before melting into water, or what ice would look like on a winter's day in a sun-glazed landscape … I was disappointed with the tall bookshelf and its contents, which promised so much and yet ignored what was most important. What good was it to know the composition of a snow crystal, if one couldn't resurrect the sensation of its coldness against one's skin and the sight of its glimmer?
>
> (42)

What Noria misses is the immediate experience, the lived body's passive, pre-reflective, and concrete experience, of a state of a natural element that no longer exists in its natural form. She has a theoretical understanding of what ice and snow are: "they were water that had taken a different form in low temperatures" (39). This knowledge is the result of an act in which she, to use the words of Husserl, "actively appropriates and establishes … [her] acquisition of knowledge" (Husserl 2001, 106). Producing ice in her freezer in an attempt to experience the feeling of winter does not provide her with the experience she misses; she "was no closer to past-world winters" (41). While her experiment indeed enables her to feel ice through her hand as she "pushed" it "into the icy wrap of water" (40), the multifarious experience of perceiving ice and snow in a natural environment cannot be recreated.

It is worth noting that the novel is not critical of abstract knowledge or of science; rather, it points to, in Noria's mother's words, "different ways of knowing" of which it "sometimes … was impossible to say which was the most reliable" (42). The juxtaposing of abstract knowledge and immediate experience invites the reader to actively reflect on the significance of different kinds of knowledge. Such reflection, moreover, invites further pondering of a world in which ice and snow no longer exist in natural form. This thematizing of the tension between knowledge and experience, while inviting active reflection, also invites readers to imaginatively experience the ice and snow that no longer exist. When Noria imagines snow on her palm and when she thinks of what a landscape would look like after snowfall, these images invite the reader to imagine these experiences along with her, and thus to envisage a world in which this experience can only be imagined.

Another experience of ice at a remove from immediacy is that of the "enormous ice cap that had once rested on the topmost peak of the world" (41). Noria knows they once existed, while now they remain, if at all, in the form of "small islets of ice, floating across the deserted sea, quiet and lifeless, carrying the memories of the past-world locked within, slowly giving in to water and melting into its embrace" (41). For Noria, knowledge of the ice caps is abstract. Yet through her imagination, she creates an image of floating ice that has the potential to evoke a sense of the ice as there in the sea.

For Noria, the ice is something that only potentially may still exist, something that could hold memories of a past world, while for the reader, this is an image of a potential future. The reader is invited to experience ice in a world without ice. Similar kinds of defamiliarization, that is, of the text "disrupting our habitual perception of the world, enabling us to 'see' things afresh" (Baldick 2008), permeate the novel. Noria's immediate experience of water, as we saw, makes the reader experience water against a background of a radically changed world, characterized by drought and drastically altered shorelines, thus offering us the opportunity to experience water in a future climate-changed world.[8]

Toward the end of the novel, Noria finally experiences frozen water in its natural form. One morning, she notices that her garden is covered in a "delicate layer of snow" (257). She experiences this through her lived body: "Not only the colour was different, but also the *scent*, and the *silence*" (257; emphasis added). She takes off her shoes and socks to feel the materiality of the frozen layer: "I wanted to know how the icy grass would feel under my bare feet. The stalks rustled and their paper-crisp coldness cut through me" (257). She sees "[w]hite frost" and "white-veined ghosts of leaves" (257). Sensory imagery creating the sight, the sound, and the feel of frost and snow invites the reader to experience these along with Noria, for whom abstract knowledge is now combined with immediate experience. The reader is invited to experience ice and snow in a world marked by the absence of these forms of water, and thus to feel along with Noria how direct experience combines with abstract knowledge at the same time as experience of ice and snow is defamiliarized.

Sedimented Experience—The Sea, the Lost Lands, and Ideas of Water

The sea is never immediately present in Noria's experience, yet her perception of it differs from her knowledge of frozen water in two significant ways: first, the sea exists in abundance in her world and is thus possible to experience. And second, her knowledge of the sea exceeds the abstract understanding of scientific observations—it exists in her consciousness as an internalized presence. It is out of sight, yet there: "Yet further that way was the sea, but it couldn't be seen from here even on bright days" (8). Her understanding of the sea simultaneously holds qualities of abstract knowledge and immediate experience: it is based on facts about seas and oceans at the same time as the sea is perceived as undeniably there, although it is not directly experienced.

Let us consider the passage in which Noria ponders two maps from different times, one from "the past-world" and the other from "the present-world": "I stared at the outlines of the continents and oceans, changed, barely recognisable. So much lost to salt and water" (64). She notices that "the old coastlines were long gone." There are "[d]rowned islands, coastal plains, river deltas turned salt-bitten; and large cities, now silent ghosts of lives past in their shroud of sea, everywhere, everywhere" (64). While Noria is looking at abstract representations of geographical locations consisting of lines and symbols on sheets of paper, she nevertheless experiences these as images of actual places that once held life, but that are now underwater. And later on, she thinks of what she will never have the opportunity to see: "somewhere, under the waves, arching like an ever-clouded sky, the ghost cities of the past-world, sharp-edged and mute as memories" (81). Lacking direct experience of these kinds of places, she nevertheless sees them as actually there.

In exploring this kind of experience, Husserl's ideas of *sedimentation*[9] are helpful. "[E]very accomplishment of activity," Husserl contends, "sinks … into passivity, and is sedimented in the accomplishments of original passivity" (2001, 275–76). From initially discussing sedimentation as a process in which activities of the ego become part of a subject's passivity, he moves toward viewing it as a process in which "human subjectivity can take over experiences not only from its contemporaries, but also from its predecessors" (Geniusas 2024, 10). This is because "[e]xperiences become sedimented through a common or experiential history and common social references" (Depraz 1998, 36) and thus form a "sedimentation of a tradition" (Steinbock 1998, 180). In *Memory of Water*, Noria's awareness of the sea is the result of a sedimentation of knowledge passed down through generations. Her awareness of seas and oceans is part of her sedimented and passively constituted knowledge of their existence.

Knowledge of the sea enters language and becomes part of Noria's way of expressing herself in imagery: "The pale sand rippled around dark-grey boulders *like water surrounding abandoned islands*" (27; emphasis added). And, "the thin grey light weighed on the winter-worn grass *like sea on underwater landscape*" (114; emphasis added). Note how sand and light in these similes are likened to water, as if water in seas were the familiar experience that highlights the qualities of sand and air. This imagery, moreover, brings out the contrast between dry sand and masses of water. Furthermore, the water that surrounds islands evokes an image of how rising sea levels have created new islands. For Noria, the knowledge of submerged land has sunk into her consciousness through sedimentation and become an internalized image of what the world is. For the reader, however, this is unfamiliar, something that we think of as happening at some point in the future.

In Husserl's thinking, Saulius Geniusas holds, sedimentations of tradition, or "generative sedimentations," are passed on "through *language*" (2024, 11; original emphasis). While it is plausible that Noria's knowledge and ideas of the sea are the results of having read about it in books and heard about it through verbal accounts, what the text suggests is that these kinds of knowledge and beliefs also *enter* language, become part of it in the form of passively constituted expressions and sayings, which, in turn, may passively influence attitudes and beliefs.

The sedimented knowledge of water in the history of the world also entails beliefs that are the result of active falsification of history. As Noria's mother once told her daughter, "Most believe that [the world] changed on its own, simply claimed its due" (40). Noria thinks of the sea as the agent in the processes that changed the Earth. There was a time "long before sea began claiming islands and coasts" (94), and she ponders "the big cities that drowned when the oceans changed their shorelines" (67). In her perception, the oceans are the agents. However, as she begins to understand through reading and listening to old accounts, the change in the ocean levels was due to human action. Noria and her friend Sanja have found some silver-colored circular discs (which today's readers recognize as CDs), which, they find out, contain recorded accounts from a secret expedition to the Lost Lands. The Lost Lands, it turns out, most likely hold resources of drinkable water, despite the widespread belief that they are contaminated. It is implied that the prevailing views of the sea as the agent in changing the shorelines and the Lost Lands as contaminated are the results of active falsification of history. The account on the discs tells "of ruin and devastation, of oceans reaching towards the centres of the continents, swallowing land and fresh water. ... Then it turned into a tale of truths forged and lies told and history changed for ever" (253). Actively formed misleading accounts of history have thus sunk into a collective consciousness through processes of sedimentation, now forming passively constituted attitudes and ideas prevailing in the world.

But, the text suggests, this false version of history can be remedied. The account recorded on the discs radically changes the two young women's perception of the world and its history. It "changed what we knew, changed what we felt, changed like a sea that rises and swallows all streets and houses, will not withdraw, will not give back what it has claimed" (195). Before she is killed, Noria writes down her story in her tea master's book, hiding it under a floorboard in the tea house. Her words may potentially reach other people in an unknown future. And Sanja, having noticed the water crime mark on Noria's door, brings the discs to her friend's mother, who is now doing research at the university in Xinjing in New Qian. Knowledge, actively mediated through language, can now potentially be actively revealed and passed on, and then passively sedimented into a new, truer version of history, thus offering a sliver of hope for future generations.

There is also a kind of sedimented experience that springs from ideas of and stories about water. Reflections on water are expressed in formulations resembling old sayings and mythical words of wisdom. These describe water in terms of a living, organic being whose features are only partly known to humans. Water is indifferent to humans as it "doesn't care for human sorrows" (1). It has memory: "Now only the water remembers" (13). And it has agency connected to feelings as it "isn't afraid to burn in fire or fade into the sky" (221). While such features, presented through personification of water, are indeed formulated, they also entail something that cannot be fully grasped by humans: "Water is the most versatile of all elements" (5), suggesting a multifariousness that escapes human attempts to pin down certain features. "It exists beyond all beginnings and ends" (221), indicating ungraspable infinity. And as it "cannot be held in one place" (113), it resists human control. It has an existence out of reach of humans as it "flows without slowing or quickening its pace in the darkness of the earth, where only stones will hear" (1). And there are "secret waters running deep inside the still stone of the fells" (34), to some extent known by Noria, yet mostly existing in hiding from human life.

Water, the novel suggests, is a fundamental life-giving element that affects the earth in its entirety. It is connected to other natural elements, to the earth, and to celestial bodies. Consider the following quote, in which Noria tells her readers about an old story recorded in all the tea masters' books in her home, and thus passed on through generations of tea masters:

> The story tells that water has a consciousness, that it carries in its memory everything that's ever happened in this world, from the time before humans until this moment, which draws itself in its memory even as it passes. Water understands the movements of the world, it knows when it is sought and where it is needed.
>
> (90–91)

While this is a story within the narrative, it nevertheless suggests and evokes the idea and image of water as having consciousness, as holding memories, and as understanding the world. This passage, Riitta Jytilä suggests, "goes against the fantasy of mastery and domination of nature, since memory is not something we possess" (2022, 223). Moreover, as water "understands the movements of the world," it constitutes a being that, rather than being controlled by humans, has the capacity to comprehend the world, including humans. As Jytilä also claims, water holds "a capacity to connect the human and nonhuman past, present and future" (2022, 224). Indeed, as Egerer proposes, "Water is what ties and holds everything together, because the linking of human and planetary history suggests that while humans forget, the planet, and water, remember" (2022, 34–35).

The age-old tradition of the tea ceremony is vital to the novel's portrayal of water as well as to the sedimented ideas of water's being and existence. As Egerer notes, the tea ceremony "stands in stark contrast to the lived reality of fear, oppression and killing that characterise Noria's world" (2022, 34). Furthermore, I propose, the tradition of the tea ceremony suggests a view of and relation to water that forms an alternative to possession, control, and objectification of water. In contrast to the military's attempts to own and possess water, Noria and her tea master ancestors view water as impossible to own and control: it "belongs to no one" (45); it "doesn't belong to us, but we belong to water" (221).

These sedimented views of water, passed on through generations of tea masters, suggest that water holds properties that are not fully graspable or available to humans. Water has consciousness and memory, it leads a life we cannot fully perceive or understand, and it holds, as Egerer notes, "an understanding of planetary entanglements surpassing human comprehension" (2022, 35). Water, then, cannot be owned or controlled, and it resists objectification. It holds the secret that gives life, and it exists in relation to the center of the earth, to the air, and to the moon. Through these features, it entails something that is unknown to humans, and it exists in relation not only to humans but also to the rest of the earth and to the universe.

Concluding Remarks—Toward New Views of Water

If we follow Weik von Mossner's ideas of how readers can be affected by literary texts, Itäranta's novel contains ample textual moments potentially evoking readers' affects. The novel's first-person account and the use of sensory imagery invite readers to feel along with the protagonist's *immediate experience* of water, and thus to imaginatively experience water through the passive process of embodied simulation. If Weik von Mossner as well as Green and Brock are right in claiming that literary texts may not only impact emotions, but also affect attitudes and beliefs, *Memory of Water* has the potential to *move* readers, that is, to prompt affective response as well as attitudinal change, through its textual features in combination with its thematic focus on scarcity of water in a climate-changed future. This makes possible a new understanding of water as it invites us to open our eyes to water's sound, sight, and tactile feeling as if experienced for the first time. The juxtaposing of *abstract knowledge* of frozen water with immediate experience of ice and snow has the potential to spark a sense of water in its frozen forms: by describing what is *not* there, the text evokes a sense of ice and snow as they feel on the skin and look like in the landscape. At the same time, the text invites reflection on the non-existence of ice and snow in a climate-changed world. The novel's evocation of *sedimented experience* in processes of active telling of (true or false) versions of history and articulation of ideas and stories sinking into communal passivity suggests how knowledge and ideas are passed on from generation to generation. The sedimented ideas of water in the tradition of tea masters suggest new ways of viewing water, based not on mastery but on an openness to water's unknown qualities and its existence not as a resource to exploit, but as an element partaking in the constitution of nature.

In the novel's portrayal of the experience of water as well as in the text's potential to affect readers, *passivity* is central: it is a fundamental part of immediate experience. It is what constitutes sedimented experience, which, while sometimes originating in activity, makes activity sink into communal passivity. And it is part of the very process in which readers are affected by literature through the passive, pre-reflective process of embodied simulation.

While the fictional world in *Memory of Water* is characterized by drought and water shortage, it is nevertheless water's presence, absence, qualities, and consciousness that are central in the novel's descriptions of the future climate-changed earth. Through Noria's first-person account, and through textual features such as sensory imagery and personification, the text invites readers to imaginatively experience water in such a world. As Weik von Mossner notes, climate fiction frequently conjures dystopian worlds. Such literary representation, she suggests, "might overwhelm us rather than move us" (2017, 146). *Memory of Water*, while indeed portraying a dystopian world, also emphasizes the beauty, bliss, and delight in the aquatic experience. Such a focus, I suggest, has the potential to contribute to a *valuing* of water and, by extension, of our shared nature. Through this, *Memory of Water* holds the potential to partake in shaping our environmental consciousness in ways that promote a caring impulse in relation to the natural world.

Could this potential also be seen in the light of Toadvine's and Bannon's respective calls for new ways of conceiving nature? In order to form a "rejuvenated relationship with nature," Toadvine argues, it is imperative to "take seriously the reality of the perceived world and the sensuousness of what is" (2009, 134–135). *Memory of Water* does indeed take seriously the perception and sensuousness of water through its emphasis on the multifaceted ways of immediately experiencing water through the lived body. When Bannon suggests a view of nature based not on human domination of nature, place, and the earth, but on an understanding of nature as constituted by "interrelations between bodies" and as holding "a sense other than given to it by humanity" (2014, 194), he calls for an understanding of nature in which "the ideal of mastery might be challenged by embracing nature's mystery" (186), since "an ontology that makes mystery an ineliminable aspect of our relationship with the world supports dispositions and attitudes toward nature, such as openness, adaptability, and caring" (2014, 34).[10] *Memory of Water*, as we have seen, suggests ways of relating to water that resist forms of mastery such as control and objectification. Instead, it proposes views of water as holding qualities and relations beyond those graspable and perceivable for humans: water exists in relation to the center of the earth where "only stones will hear" (1) as well as to the universe, where it "walks with the moon" (5), thus suggesting ways of imagining an ontology of water that is open to that which is ungraspable to humans and to a view of water as partaking in the constitution of nature and the Earth. By illuminating such features of water, I propose, the novel has the potential, through passive processes of embodied simulation, to participate in changing our conception of water by *moving* readers. Hence, it also has the potential to partake in prompting new ways of relating to water, place, and the earth, ways that may, ultimately, spur the kind of "deep pre-ethical sense of responsibility" that Hamilton calls for (2017, 149). Eventually, then, it has the potential to contribute to orientations to the Earth characterized by acts of caring.

Notes

1 This involves a more practical and active undertaking than Martin Heidegger's notion of care (*Sorge*). For example, Joan Tronto and Berenice Fisher's frequently quoted definition of caring emphasises "everything that we *do to maintain, continue, and repair* our 'world'" (Tronto 2013, 19; emphasis added). Maria Puig de la Bellacasa, basing her discussion on Tronto and Fisher's definition, emphasises "a generic appeal of care that makes it unthinkable as something abstracted from its situatedness" (2017, 6). And Kyle Whyte and Chris Cuomo, building on Carol Gilligan, describe an ethics of care as "approaches to moral life and community that are grounded in *virtues, practices, and knowledges* associated with *appropriate caring and caretaking* of self and others" (2017, 234; emphases added). When anthropologist Karine Gagné investigates how people in Ladakh in the Himalayas relate to the landscape, she is concerned with not only "beliefs and ideas about the non-human world," but also "*practical activities* through which people engage with the environment" (2018, 7; emphasis added). And when geographer Anna Krzywoszynska explores care and attentiveness in relation to soil, she focuses on how "soils become *objects of concern and of care* for conventional farmers" (2019, 666; emphasis added).

2 Itäranta wrote the novel simultaneously in English and Finnish; it was published in Finnish in 2012, and in English in 2014.

3 As Weik von Mossner notes, mirror neuron research has shown that "when we see another person act, we map those actions onto our premotor cortex, the part of the brain that is also active when we engage in actual movement" (2020, 132). Weik von Mossner bases her discussion on research showing that reading may activate mirror neurons in similar ways as interaction with people in the real world does (2020, 132–33).

4 Such an empirical study of Itäranta's novel was recently carried out among upper secondary school students in Finland; see Helle et al. (2024).

5 Noria is the first-person narrator throughout the novel except for the epilogue in which a third-person narrator takes over.

6 Schneider-Mayerson's article addresses this question by surveying American readers of climate fiction and their responses to such texts.

7 As Anthony Steinbock notes, Husserl does not give a clear definition of passivity in his writings. In his "Translator's Introduction" to Husserl's *Analyses Concerning Active and Passive Synthesis*, Steinbock outlines five "distinctive, albeit interrelated meanings" of the term. For a concise summary of these, see Steinbock (2001, xxxviii–xliii).

8 Importantly, however, experiences of, for example, drought and submersion, often imagined to be part of a future climate-changed world, are already happening; as Pieter Vermeulen (2020) notes, "Because the consequences of planetary destabilization are unevenly distributed, some constituencies are already living—or indeed, already mourning—a reality that more privileged communities fear will become part of their future" (2020, 112).

9 As Saulius Geniusas (2024) notes, "Husserl is the philosopher who transformed this geological metaphor into a philosophical concept" (2), and his use of the notion, according to Geniusas, varies from understanding it as "a static concept, a genetic concept and a generative concept" (14). For a more comprehensive account of sedimentation in Husserl's thinking, see Geniusas.

10 Bannon's use of the term mystery builds on Heidegger's discussion of truth and involves the idea that "it is not possible to disclose all the possible relations to any given being in a single manifestation of the being" (Bannon 2014, 85).

References

Baldick, Chris. 2008. "Defamiliarization." In *The Oxford Dictionary of Literary Terms*. Oxford University Press. https://www-oxfordreference-com.ezp.sub.su.se/view/10.1093/acref/9780199208272.001.0001/acref-9780199208272-e-293

Bannon, Bryan E. 2014. *From Mastery to Mystery: A Phenomenological Foundation for an Environmental Ethic*. Athens, OH: Ohio University Press. ProQuest Ebook Central.

Bellacasa, Maria Puig de la. 2017. *Matters of Care: Speculative Ethics in More than Human Worlds*. Minneapolis: University of Minnesota Press. http://www.jstor.org/stable/10.5749/j.ctt1mmfspt

Depraz, Natalie. 1998. "Imagination and Passivity Husserl and Kant: A Cross-Relationship." In *Alterity and Facticity: New Perspectives on Husserl*, edited by Natalie Depraz and Dan Zahavi, 29–56. Phaenomenologica, vol. 148. Dordrecht: Springer. https://doi.org/10.1007/978-94-011-5064-4_2

Egerer, Claudia. 2022. "Imaginings of Water: Anthropocene Waters and the Entanglement of the Living." In *Water Lore: Practice, Place and Poetics* (1st ed.), edited by Camille Roulière and Claudia Egerer, 23–38. London: Routledge. https://doi.org/10.4324/9781003218272

Gagné, Karine. 2018. *Caring for Glaciers: Land, Animals, and Humanity in the Himalayas*. University of Washington Press. http://www.jstor.org/stable/j.ctvd1c7sj

Geniusas, Saulius. 2024. "The Origins of Sedimentation in Husserl's Phenomenology." *European Journal of Philosophy*: 1–17. https://doi.org/10.1111/ejop.12931

Green, Melanie C. 2004. "Transportation Into Narrative Worlds: The Role of Prior Knowledge and Perceived Realism." *Discourse Processes* 38 (2): 247–66. https://doi.org/10.1207/s15326950dp3802_5

Green, Melanie C., and Timothy C. Brock. 2000. "The Role of Transportation in the Persuasiveness of Public Narratives." *Journal of Personality and Social Psychology* 79 (5): 701–21. https://doi.org/10.1037/0022-3514.79.5.701 https://ezp.sub.su.se/login?url=https://www.proquest.com/scholarly-journals/role-transportation-persu.siveness-public/docview/614344768/se-2

Hamilton, Clive. 2017. *Defiant Earth: The Fate of Humans in the Anthropocene*. Cambridge: Polity Press.

Helle, Anna, Toni Lahtinen, Olli Löytty, and Panu Pihkala. 2024. "Lukiolaiset Teemestarin kirjaa lukemassa: Ilmastokirjallisuuden herättämiä ajatuksia ja tunteita." [Upper Secondary School Students Reading Memory of Water: Thoughts and Affects Evoked by Climate Fiction]. *Kulttuurintutkimus* 41 (3): 22–37.

Husserl, Edmund. 2001. *Analyses Concerning Passive and Active Synthesis: Lectures on Transcendental Logic*. Translated by Anthony J. Steinbock. Dordrecht: Kluwer.

Irr, Caren. 2017. "Climate Fiction in English." *Oxford Research Encyclopedia of Literature*. https://doi.org/10.1093/acrefore/9780190201098.013.4

Itäranta, Emmi. 2014. *Memory of Water*. New York: Harper Voyager.

James, Erin. 2022. *Narrative in the Anthropocene*. Columbus: The Ohio State University Press.

Jytilä, Riitta. 2022. "Remembering in the Age of Global Warming: Emmi Itäranta's Memory of Water as Ecological Trauma Fiction." In *Nordic Utopias and Dystopias: From Aniara to Allatta!*, edited by Pia Maria Ahlbäck, Jouni Teittinen, and Maria Lassén-Seger, 213–27. Amsterdam/Philadelphia: John Benjamins Publishing Company. ProQuest Ebook Central.

Krzywoszynska, Anna. 2019. "Caring for Soil Life in the Anthropocene: The Role of Attentiveness in More-than-human Ethics." *Transactions of the Institute of British Geographers* 44: 661–75. https://doi.org/10.1111/tran.12293

Laakso, Maria, Toni Lahtinen, and Hanna Samola. 2019. "Young Saviors and Agents of Change: Power, Environment, and Girlhood in Contemporary Finnish Young Adult Dystopias." *Utopian Studies* 30 (2): 193–213. https://doi.org/10.5325/utopianstudies.30.2.0193

Mayer, Sylvia. 2014. "Explorations of the Controversially Real: Risk, the Climate Change Novel, and the Narrative of Anticipation." In *The Anticipation of Catastrophe*, edited by Sylvia Mayer and Alexa Weik von Mossner, 21–38. Heidelberg: Universitätsverlag Winter.

Merleau-Ponty, Maurice. (1945) 2002. *Phenomenology of Perception*. Translated by Colin Smith. London: Routledge.

Overgaard, Søren. 2008. "How to Analyze Immediate Experience." *Metaphilosophy*, 39: 282–304. https://doi.org/10.1111/j.1467-9973.2008.00543.x

Schneider-Mayerson, Matthew. 2018. "The Influence of Climate Fiction: An Empirical Survey of Readers." *Environmental Humanities* 1 November 2018; 10 (2): 473–500. https://doi.org/10.1215/22011919-7156848

Starr, Gabrielle. 2010. "Multisensory Imagery." In *Introduction to Cognitive Cultural Studies*, edited by Lisa Zunshine, 275–90. Baltimore: Johns Hopkins University Press.

Steinbock, Anthony J. 1998. "Spirit and Generativity: The Role and Contribution of the Phenomenologist in Hegel and Husserl." In *Alterity and Facticity: New Perspectives on Husserl*, edited by Natalie Depraz and Dan Zahavi, 163–203. Phaenomenologica, vol. 148. Dordrecht: Springer. https://doi.org/10.1007/978-94-011-5064-4_2

Steinbock, Anthony J. 2001. "Translator's Introduction." In *Edmund Husserl, Analyses Concerning Passive and Active Synthesis: Lectures on Transcendental Logic*. Translated by Anthony J. Steinbock. Dordrecht: Kluwer Academic Publishers.

Toadvine, Ted. 2009. *Merleau-Ponty's Philosophy of Nature*. Evanston, Illinois: Northwestern University Press.

Tronto, Joan C. 2013. *Caring Democracy: Markets, Equality, and Justice*. New York University Press. https://doi.org/10.18574/nyu/9780814770450.001.0001

Vermeulen, Pieter. 2020. *Literature and the Anthropocene*. London: Routledge. https://doi.org/10.4324/9781351005425

Weik von Mossner, Alexa. 2017. *Affective Ecologies: Empathy, Emotion, and Environmental Narrative*. Ohio State University Press. https://doi.org/10.2307/j.ctv11hpszq

Weik von Mossner, Alexa. 2020. "Feeling Nature. Narrative Environments and Character Empathy." In *Environment and Narrative*, edited by Erin James and Eric Morel, 129–46. Columbus: Ohio State University Press.

Whyte, Kyle, and Chris Cuomo. 2017. "Ethics of Caring in Environmental Ethics: Indigenous and Feminist Philosophies." In *The Oxford Handbook of Environmental Ethics*, edited by Stephen M. Gardiner, and Allen Thompson, 234–247, Oxford Handbooks, (2017; online edn, Oxford Academic, 12 Nov. 2015), https://doi.org/10.1093/oxfordhb/9780199941339.013.22

26

POETRY AND THE FORGOTTEN FUTURE

Towards a New Onto-Phenomenology of Poetic Work

Patryk Szaj

For Forrest Clingerman, who awaited my text on poetry but never received it.

Let me begin with poetry instead of a conventional introduction. The understated prose poem *Oblężenie* by Polish poet Andrzej Sosnowski, published in the 2015 collection Dom ran [Home of Wounds], reads as follows:

> So they rightly say that life today is experiencing a "real siege," because at the present moment it truly pays off. People line up at night, in queues of dreams, and upon being torn from a stunned sleep, they immediately rush to our centers and cash registers, which are now melodic, singing, and so personalized that every paying customer hears their favorite music at the end. This way, their positive emotions are always guaranteed. Even if someone seemingly devoid of desires or dreams appears, a real explosion of unforgettable experiences awaits them.[1]
>
> (2015, 10)

This piece appears to be a quintessential example of Hubert Zapf's concept of literature as cultural ecology (2016). According to the German literary scholar, a literary text – due to its semantic openness, imaginative intensity, and aesthetic complexity – acts as a functional equivalent of biodiversity, an ecological force within the cultural system that promotes its balance. The core of Zapf's theory is the triadic functional model of literature as cultural ecology, which assigns literature the roles of culture-critical metadiscourse, imaginative counter-discourse, and reintegrative interdiscourse. Let me examine *Oblężenie* through this lens.

First, it is worth asking who or what life 'really pays off' for at the present moment. Certainly not life as such (understood either as social life, *bios*, or as vital force, *zoe*). Sosnowski engages in the demythologization characteristic of culture-critical metadiscourse here. Although it seems that 'people' benefit from the blessings of consumer capitalism, in reality, it is they – more precisely, their cognitive-emotional labor – that become the commodity. Melodic, singing cash registers appear as a truly dystopian image of the omnipotence of capital, which, under the guise of personalizing the content provided to customers, collects their

DOI: 10.4324/9781003538097-32

data, feeds its algorithms, and then, in a vicious cycle, capitalizes on people's desires. Sosnowski undoubtedly mocks – especially in the last two sentences filled with excessively ecstatic epithets – the language of spectacle, i.e., capital that 'accumulated to the point that it becomes images' (Debord 2005, 17). The Debordian 'hijacking,' or 'rerouting,' *détournement*, through which the discourse of capital begins to work to its own detriment, mixes here with uncovering of the Lacanian (rather than Freudian) 'dreamwork.' People torn from a stunned sleep never truly wake up; they immediately engage in the dreamwork – creating a coherent principle of reality based on the constitutive repression of the Real.[2]

In this manner, *Oblężenie* functions as a culture-critical metadiscourse, unveiling the core elements, hidden assumptions, normative abuses, naturalizations, and constitutive outsides of the capitalist system from within. As Zapf articulates, it 'responds to hegemonic discursive regimes by exposing petrifications, coercive pressures, and traumatizing effects of dominant civilizational reality-systems that are maintained and reinforced by those discursive regimes' (2016, 104). These aspects are readily discernible in Sosnowski's text. However, reconstructing its role as an imaginative counter-discourse, which involves articulating what is inaccessible, unimaginable, or unachievable within the prevailing regimes of cultural self-interpretation, proves more challenging. This function is not directly expressed but rather emerges from the specific intensity of *Oblężenie*. This intensity, understood in a Deleuzian manner,[3] is evident in at least two instances within the piece. The assertion that life 'pays off at the present moment' implies a short-term perspective on profitability. Capital seeks to commodify life (both human and potentially non-human, though the poem does not specify) in a context where it has exhausted its planetary 'externalities.'[4] In doing so, it must externalize them once again, so to speak, repressing the fact (the Real) that the exploitation of life is only profitable in the short term ('at the present moment'), while in the long term, it inevitably leads to a metabolic rift, a fundamental contradiction between the logic of accumulation and the laws governing biological processes.

This longer-term perspective, merely hinted at in Sosnowski's poem, suggests that 'life' encompasses what Anna Tsing refers to as 'latent commons,' which persist in the pericapitalist space, on the fringes of capital, within its exteriorities (2015, 255). Capitalism cannot function without these commons (as it needs them to transform into alienated commodities), yet they simultaneously pose a mortal threat to it (by revealing alternative ways of life and temporal rhythms). Life, as a latent common good, represents a potentially disruptive force, undermining the dreamwork and exposing the fissures of post-capitalist desire. This second residue of the imaginative intensity of *Oblężenie* is evident in the poem's final sentence. For the mere fact that capital needs to colonize or enclose our desire demonstrates that, fundamentally, this desire is other-than-capitalist.

Thus, Sosnowski's poem operates as a reintegrative interdiscourse, bridging existing cultural systems with the elements they exclude, thereby restoring connections between previously segregated discursive spaces. If the siege of life is achieved through a Cartesian cut that separates desiring subjects from their fundamental interdependence with the rest of *zoe*, then *Oblężenie* – without any explicit assertoric commentary – intimates that the pathways for non-capitalist desire lie in uncovering the relationality that forms both the constitutive outside (the repressed Real) and the essential condition (putting life to work) of capitalism.

A concise interpretation of *Oblężenie* reveals that while Zapf's theory convincingly posits that literature functions as an ecological force within cultural systems, it falls short of explaining *how* or *why* this is feasible. I propose that this multidimensional efficacy of literature – particularly poetry – derives directly from its onto-phenomenology. Poetry inherently embodies what a scholar of Anthropocene lyric Tom Bristow terms 'historico-ecophenomenology' (2015, 18). It does so by virtue of its ontology – semantic density, imaginative intensity, aesthetic complexity, and above all, de-pragmatization and formal autonomy. It penetrates those layers of the 'here and now' that evade other discourses or are actively suppressed by them, exposes the concealed foundations of hegemonic cultural practices, and sustains the dynamic process of becoming. Consequently, irrespective of a poet's intentional aims, poetry serves as 'a cultural counterpoint to the political right, to eco-modernist ideology, and to human domination' (Bristow 2015, 16), which are forces that perpetuate the status quo. In essence, poetry reveals that the world is always more intricate than hegemonic cultural discourses would have us believe, and it can never be reduced to a linear conception of the 'present moment.' The stakes here are nothing less than the future. I will endeavor to substantiate these claims through a dialogue with three authors who, to varying degrees, engage with and expand the phenomenological tradition: Timothy Morton, Michael Marder, and Karen Barad.[5]

Back to the Future

But first, let me, so to speak, return to the future. Poetry does not actualize a potential future; rather, it emergently brings it forth from within itself. The affective-cognitive intensity and density of poetic text reveal that what we call the present is, in reality, an entanglement of various pasts, presents, and futures, all reduced to a single hegemonic vision of the 'here and now.' One of the most famous statements in this context is Adrienne Rich's programmatic – and programmatically paradoxical – phrase about Muriel Rukeyser's poetry: the poem 'reminds us of something we are forbidden to see. A forgotten future' (2007, 36). Thus, the work of the poem dislocates time, disrupting simple linearity. The future, in a sense, has long been here, though 'we' are not yet (or no longer) in it. While reminding of a forgotten future may evoke Platonic anamnesis, I believe the stakes of poetry lie more in revealing the 'possible' (see: Deleuze and Guattari 1987). By that, I mean the potential inherent in any system or set of elements, such as the unrestrained life in Sosnowski's poem that is captured within the assemblage of capital, yet simultaneously subverts it from within. Let me try to explain this more clearly.

Rich is, of course, not the only author to emphasize the futurality of poetry. In his monograph on Anthropocene lyric, Bristow cites John Kinsella: 'poetry has been the tool by which we behave and confront behavior, memorize and challenge those memories' (Bristow 2015, 62), which seems to confirm two of the three functions Zapf attributes to literature in his triadic model – critical metadiscourse and imaginative counter-discourse. For similar reasons, Ted Hughes argued – as paradoxically as Rich – that 'even bad poetry is good:'

> Poetry, in every unstable or revolutionary situation, is on the side of eruption. It speaks for what is suppressed, or rejected, or denied; it speaks for whatever in a society is alive but undervalued, for whatever is shit in the dark and made to suffer. It speaks for everything, finally, which the status quo deems irrelevant or undesirable or, perhaps, unmanageable, and so it speaks for the future, the time when those undesirables shall have broken up the status quo and come into the open as an accepted part of life.
>
> (cited in: Solnick 2017, 94)

These words illustrate that the counter-hegemonic potential of poetry is a matter of its onto-phenomenology, not the poet's intention. Every poem, even the weakest, performs the work of defetishization, undermining what the status quo reifies. What hegemonic discourse stabilizes – the excluded, unwanted, undesirable, insignificant, uncontrollable elements – poetry reintroduces into the dynamic process of becoming, from which a future emerges that is not a mere extension of the present but an unexpected, eventful 'to come.' It achieves this not primarily through its 'message,' but through the dynamic interplay of 'form' and 'content' as entangled phenomena (which I will examine more closely in dialogue with Barad).

Most importantly, the unveiling of a forgotten future is not an arbitrary act. Life (Sosnowski), the future (Rich), the possible (Deleuze/Guattari) are always already embedded in what 'is' as its repressed alternative. Pieter Vermeulen recalls Raymond Williams's conviction that the task of critical thought is to capture 'the dynamic interrelations, at every point in the process, of historically varied and variable elements' (Williams 1977, 121). According to Williams, the present consists of three overlapping, simultaneous temporalities: the dominant (the hegemonic and homogeneous status quo), the residual (elements of the past that actively influence the present and are never fully absorbed by it), and the emergent (what transcends the dominant, is repressed by it, but simmers beneath the surface of stabilized discourses as 'new meanings and values, new practices, new relationships and kinds of relationship are continually being created' [Williams 1977, 124]). Both residual and emergent resist the false homogeneity of the dominant culture. Poetry – precisely because of its onto-phenomenology – seems particularly suited to sustaining this resistance: reminding us of residues and supporting heterogeneous emergencies.

The Refuge for the Arche-Lithic

In the 2020 collection *Cielenie lodowca* [Glacier Calving] by another Polish poet, Marcin Ostrychacz, there is a poem titled *Kroki* [Steps]. It is divided into three stanzas, each describing one of the titular steps – pivotal moments in human history. The first step is a kind of mythical anthropogenesis, a moment (spanning several million years) of the separation of the genus Homo from the rest of the animal kingdom. The third step can be identified with what Anthropocene scholars call the Great Acceleration (Steffen et al. 2015): from moon landings, through the hippie Era of Ergot Enlightenments, to the complete commodification of planetary life, where 'livestock constitutes 75% / of all living animals' (Ostrychacz 2020, 35). However, before this, humanity takes the

> Second step, where seeds were deliberately sown,
> …
> Fields sprouted from the seeds of wild millet.
> From the fields, harvested were crops.
> Crops were turned into flatbreads.
> Flatbreads fed an ever-growing family.
> Family outgrew the house.
> Houses were joined into settlements.
> Settlements were surrounded by palisades.
> Bricks were fired, palaces built,
> salaries introduced.
>
> (Ostrychacz 2020, 33)

The structure of this fragment (almost a model embodiment of a syntactic poem, where each line equals a short simple sentence; chain-like coupling of lines) suggests that the successive threads of the narrated history logically follow from one another. Moreover, in the last two lines, the text clearly accelerates and further emphasizes the causal chain (breaking the previous 'sentence-like' structure of the lines; introducing the euphony of palaces-salaries). From the seeds of wild millet, not only fields but human civilization itself springs forth. The capitalist Great Acceleration is merely a distant consequence of this breakthrough: the Neolithic Revolution.

It is that revolution, as Timothy Morton argues, that created the first industrial machine in history: agriculture. Since then, the development of human civilizations has been governed by an 'algorithm' that the author of *Dark Ecology* calls agrilogistics. Agrilogistics 'promises to eliminate fear, anxiety, and contradiction – social, physical, and ontological – by establishing thin rigid boundaries between human and nonhuman worlds and by reducing existence to sheer quantity' (Morton 2016, 43). The fulfillment of this promise is based on three undeniable axioms:

(1) The Law of Noncontradiction is inviolable.
(2) Existing means being constantly present.
(3) Existing is always better than any quality of existing

(Morton 2016, 47)

According to Morton, the contemporary climate crisis has been virtually embedded in the agrarian model of functioning from the very beginning, partly because the capitalist regime reproduces its patterns: if life today is experiencing a 'real siege,' then this 'today' has lasted for 12,500 years. Traces of the entanglement of material-discursive practices of agrilogistics and capitalism can be found in the language itself. Morton points out the etymological proximity of the words *chattel* (property, personal goods), *cattle* (livestock), and *capital* (2016, 27). The condensed temporalities of agrilogistics and capitalism – feeding on the broader temporality of the more-than-human – overlap in Jeremy Prynne's poem *Die a Millionaire* too: '… what starts as irrigation ends up / selling megawattage across the grid' (2015, 15). The British poet plays here with the common etymology of the words current and currency: a stream of water, transformed into an electric current, initiates a stream (circular movement, flow) of capital. As Solnick points out in his commentary on the poem, what used to move freely – people, their genes, water – is grounded by technology. The author of *Poetry and the Anthropocene* also situates the sources of this shift in the Neolithic Revolution: 'The instrumentalization of the river begins with "irrigation" – and irrigation itself denotes that shift from the nomadic to the agrarian' (Solnick 2017, 154).

However, the intentions of agrilogistics have been muddled from the very beginning by something Morton calls the 'arche-lithic' – a temporality from before the Neolithic, from before all binary oppositions, revealing the primordial kinship of humans and non-humans. It is not a 'past' abandoned due to the Neolithic Revolution, but a heterotemporality that haunts the 'now.' It shines through the cracks of agrilogistics and reveals fractures within agrilogos, as it does not adhere to its axioms. The arche-lithic operates like Jacques Derrida's arche-writing, to which the author of *Dark Ecology* directly refers. It does not precede agrilogistics in a linear sense but rather precedes it in a logical order as

> the opening of the first exteriority in general, the enigmatic relationship of the living to its other and of an inside to an outside … which must be thought before the opposition of nature and culture, animality and humanity …
>
> (Derrida 1998, 70)

If agrilogistics attempts to remove the fundamental ambiguity, the arche-lithic constitutes the condition of the impossibility of this endeavor. The arche-lithic mind 'is immersed in a non-totalizable host of patterns that cannot be bounded in advance' (Morton 2016, 84). Importantly, the arche-lithic, as a temporality from before the Neolithic inscribed in our 'now,' characterizes both the organic world and – shining through it – the algorithm of agrilogos. It is a kind of glitch, an internal error in the material-discursive practices of agrilogistics, disrupting it with a virus of non-identity. It does not come from outside agrilogos but emerges from the very immanence of the world in which agrilogos is embedded. In this way, it resists its modes of knowledge and action.

The arche-lithic reminds us of something we were forbidden to know – forgotten virtual futures emerging from what agrilogos seeks to eliminate: ambiguity, complexity, multi-scalarity, and the weirdness of the world. It is no coincidence that I allude to Rich's formula here, as I believe that poetic work is inherently non-agrilogistic or, in other words, poetry provides refuge for the arche-lithic. Poetry does not adhere to the principle of noncontradiction; it constantly speaks both *p* and *non-p* simultaneously. It expresses the arche-lithic in the only possible way, non-assertorically, figuratively, in an ontologically dense language of paradoxes. Therefore, we can read about 'the end of the world [*p*], / which goes on [*non-p*]' (Małgorzata Lebda), encounter an apostrophe to decomposers 'my oppressors, [*p*] / my friends [*non-p*]' (Julia Fiedorczuk), or trust a poet who says 'nothing here is mine [*p*] / and everything is mine [*non-p*]' (Ilona Witkowska).

The last quote, in particular, appears as an implicit critique – and simultaneously an exit from – agrilogistics. It can be read as both a questioning of agrilogistics' claims to possess the world and an opening to the arche-lithic multiplicity, where what is 'mine' is no longer private property but a shared common good. Witkowska destabilizes the petrified social structure based on 'once and for all' occupied positions. Instead, as Monika Glosowitz writes:

> She activates a series of associative images and shows how easily these positions can be shifted if we understand that they are the result of affective investments conditioned by various ideologies (national, racial, economic). At the same time, she demonstrates that community-building affects can be distributed across hierarchical axes, mobilizing capacities for integration against oppression or – in other words – within a new politics practiced on imperial peripheries.
>
> (2019, 223)

Moving beyond the first axiom of agrilogistics also means that poetry escapes the dominant agriontological categorical difference between substance (referred to by Morton as Easy Think Substance) and accident. This difference has practical consequences. If every being consists of attributes and accidents, it is easy to objectify it under the pretext of separating its essence from unnecessary manifestations, and then harness this *eidos* to work for agrilogos.

Meanwhile, poetry disrupts the rules embedded in the hegemonic order, leading us beyond the binary logic of agrilogos, the monolingualism of power, or the utilitarian regime of capital. It is fundamentally a/telic,[6] escaping any *telos* instrumentally imposed on texts. Poetry opposes the agrilogistic eidos, essence, and agro-nomy with its a/nomic non/essence and aim/lessness. In each of these terms, the slash suggests that poetry is not simply the opposite of other material-discursive regimes. Instead, it works from within to unsettle them, creating a space of ambiguity and play where clear distinctions begin to blur. While poetry may gesture toward law, essence, or purpose, it never fully aligns with any fixed form of them. It speaks in a language that predates binary divisions – between law and lawlessness, essence and accident, purpose and aimlessness. In this way, the metaphysics of presence – the second axiom of agrilogistics, secured by what is foundational, the *arche* – gives way to an/archic multiplicity. From the perspective of agrilogos, this certainly appears as nonsense. But it is precisely this 'nonsense of poetry,' as Derek Mahon argued, that becomes a source of resistance:

> to a world engineered for maximum 'efficiency', competitive 'growth', 'global excellence' and 'world-class' foolishness of all kinds. The poetry nonsense sets itself up against regulation, system, utility. It's a last ditch of sanity in a naff world of exploitation and lies. It has no function and no exchange value [...] It is indeterminate, marginal, unimportant; and therein lies its importance. What we need is a dimmer, dreamier universe.
>
> (cited in: Solnick 2017, 120)

The programmatic statement of the Northern Irish poet captures each of the features of poetry that seem to constitute the arche-lithic: a specific principle of indeterminacy, rendering poem an entity with an undecided onto-epistemological character, thus eluding the unifying tendencies of agrilogos; heterodiscursivity, which, so to speak, introduces a 'virus of health' into the neoliberal regime suffering from the disease of efficiency; un/importance, which on the one hand means minimal social impact as a paradoxical disruptive force (existing outside the fetish of utility), and on the other, an escape from the logic of Easy Think Substance (*eidos*); futurity as something other than the mere extension of the present.

Sublimation of the Plant Soul

The arche-lithic, as a temporality from before the Neolithic that haunts our 'now,' intersects with another forgotten heritage of humanity – the plant soul, which, according to Michael Marder, has never been abandoned by us. Just as the arche-lithic in Morton's thinking destabilizes agrilogos, the plant – the 'weed of metaphysics' – in Marder's Plant-Thinking deconstructs the Western metaphysical project. One could say that the time of plants *is* the time of the arche-lithic – a forgotten heritage that does not reside in the past but constitutes another temporality (and an Other of temporality), a temporality from before metaphysics that still, though repressed, resides within us.

It is not, of course, about equating, in a metaphysical (agrilogistic) gesture, the arche-lithic with the plant soul. Rather, the idea is that if – following Deleuze and Guattari's recommendation – we should 'follow the plants' (1987, 11), it is because they teach us to function within the patterns of the arche-lithic. Marder's project arrives at similar conclusions to Morton's dark ecology through a different path. The plant soul is described by him as 'life in its an-archic bareness [that is] ontologically vibrant' (Marder 2013, 22), reminiscent of the joyful queerness of the arche-lithic. In both cases, this untamed multiplicity arises from indeterminacy, primordial ex/propriation, or foundational contamination by absence. In the plant soul, as in the arche-lithic, everything is still virtually possible. The metaphysics of presence fails to stabilize its becoming, and an-archy here literally means the absence of a source, *arche*, that would subordinate the plant to *logos*. As Marder writes, the plant is non-identical with itself, lacks an 'inside,' depends on its other, and the possibilities of its soul, simultaneously barely alive and overly vital, not entirely distinct from the external environment, remain indeterminate. It is precisely from this indeterminacy – the division at the beginning and in the beginning, which, as we will see, connects Marder's, Morton's, and Barad's thinking – that the generosity of the plant soul arises:

> That which is most common [the plant soul existing in all beings – PS] is the most widely shared and the first, the origin always already divided, falling apart, and as a consequence supplanted or usurped by another origin (the principle of animality). Plant-soul is a concrete expression of such division of and at the origin—the kind of primordial generosity that gives itself to all other creatures, animates them with this gift …, parts against itself, and in this parting and falling apart invites the participation of beings in the acts of living.
>
> (Marder 2013, 46)

When Witkowska writes 'nothing here is mine / and everything is mine,' it is the plant soul speaking through her. What in agrilogos has the character of a disjunctive alternative ($p \neq q$), in the arche-lithic, plant thinking, and poetry becomes the law of the 'included' middle, provided we treat the middle – as opposed to the center – as a series of variable, accidental, overlapping points of intersection that determine the non-identity of each being with itself. For Marder, this is evident in the case of plants, which do not form a concentrated organism or system but rather a diverse, dis/organized multiplicity. The same logic operates in the arche-lithic or arche-writing: what is situated in the central position, as *logos* subordinating all other elements of the organism/system, has an aporetic character. It simultaneously belongs (as its center) and does not belong to the structure, for it does not adhere to its dynamics (it plays the assigned role of its 'stabilizer').[7]

Bottom line is that such an established organism/structure collapses under its own weight. The arche-lithic or the plant soul disrupts agrilogos (metaphysics) from within, from the very immanence of the discourse of the One. They are not transcendent forces but something that always operates within agrilogos (metaphysics), privileging what virtually simmers in the 'here and now' at the expense of what is already actualized. In other words, they work like latent commons in Tsing's view. It is no coincidence that Marder describes the plant as a 'non-resource,' nor that Morton uses the term *agri*logos. Agrarian logic, which also characterizes industrial capitalist agriculture, appropriates this non-resource, arrests its an/archic generosity, and harnesses it to serve its own interests. However, the non-resource nature of plant ontology does not disappear. Therefore, for Marder, practices of deforestation or monocultural cultivation constitute an ontic error.

'Even in our highest endeavors, we remain sublimated plants,' writes Marder (2013, 40), asserting that philosophy is merely the subtlest embodiment of the living legacy of the vegetative soul. However, it seems that the plant soul finds a more faithful sublimation in poetry than in metaphysical reflection – one that does not overshadow or subordinate it to its own *telos*. Let me look at another poem, *biec* [to run] by Krystyna Miłobędzka from the volume *gubione* [lost] (2008) (2010, 333):

I would only like to run
run for nothing
run to nothing
just run
to run
run

This piece can be read as a self-reflection on the aesthetic function of language, which – according to the structuralist paradigm – is privileged by poetry. In the latter, the message focuses on itself because of the self-referential highlighting of figurativeness that exhausts itself in its own counting on being visible. Running for nothing, to nothing, without any overarching goal, is akin to doing something for the sake of doing it, escaping the discourses of productivity, utility, and pragmatism. For Roman Jakobson, this was the essence of poetry. The author of *Linguistics and Poetics* saw this as the autotelic nature of poetry. However, this self-purposefulness inevitably leads to Hegelian bad infinity – it is eternally incomplete, unattainable, unrealizable, thus transforming into a/telicity. Running for nothing, to nothing, exhausting itself in itself, is also an infinite, endless, futile run devoid of intentionality or – better – expressing a longing for 'non-conscious intentionality' (Marder 2013, 153–62). With this concept, Marder reverses Levinas's 'non-intentional consciousness,' which allows us to live 'otherwise than being' by splitting or spilling beyond itself, gravitating towards the other, towards transcendence. Meanwhile, the non-conscious intentionality characteristic of plants expresses a certain uncontrolled striving, multiplication of being and meaning rooted in immanence, beyond consciousness and beyond conscious representation. Growth that grows for the sake of growing. Running that runs for the sake of running.

Such growth/running cannot be represented, as it constitutes 'a pure multiplicity of immanence' (Deleuze and Guattari 1987, 157), a collection of diversities, a hospitable intersection point of everything that happens in its course and does not submit to hierarchical organization. In other words, it refers to Deleuze and Guattari's Body without Organs. For Marder, the embodiment of this BwO is the plant, but poetry operates analogously. Running for nothing, to nothing, just running is, of course – as non-conscious intentionality – an internally contradictory action. Even if one does not always run *for* something, one, at least by assumption, runs *somewhere*.

Poetic creation thus fulfills the fundamental premise that, according to Aristotle, likens us to plants: it does not respect the principle of non-contradiction (which, let me recall, is one of the three axioms of agrilogistics). 'To follow the plants' in this context means to destroy the classical (agri)logos. Non/logicality, a/telicity, and non/assertoricity constitute a potential subversive force (though, like any latent commons, it can also be appropriated or instrumentalized). In both plants and poetry – each in its own way, thanks to their ontological characteristics – 'the passages, the outlets, or the media for the other … [that] let the other pass through them without detracting from the other's alterity' open up (Marder 2013, 42).

However, we still have not fully explained how poetry recovers the forgotten future, how it reveals the temporality of the arche-lithic, the plant soul, what is possible, hidden in the folds of the 'here and now.' To answer these questions, let us make one more conceptual move and turn towards Karen Barad's agential realism.

Poetry and the Indeterminacy Principle

Wolfgang Iser once asserted that every literary text – especially a poetic one, I would add – contains a certain degree of indeterminacy. The German phenomenologist referred to Edmund Husserl's student, Roman Ingarden, who believed that every literary work is constructed in a dual manner: as an artistic object, it functions only potentially, characterized by schematic structure and empty spaces that demand concretization (actualization) during the reading process. It is only through the aesthetic experience of the reader that it becomes an actualized aesthetic object (Ingarden 1973). One might say that the artistic object is akin to Kant's noumenon, while the aesthetic object, as a phenomenon, is the correlate of the reader's noetic act. This brings us close to Morton's ideas.[8] Contrary to Husserl, however, Ingarden believed that a work of art is an intentional being, designed to evoke specific affects and reflections in the audience. Here, we find ourselves near Marder's notion of non-conscious intentionality – it seems that the text's orientation towards concretization is not accompanied by any conscious, ideational, representational reflection (unless we assume that the text's intention is, in fact, the internalized intention of the author). Barad adds the final missing element in this puzzle.

As mentioned earlier, the proponent of agential realism does not draw from the phenomenological tradition. Their concept of phenomenon has been borrowed from the quantum physicist Niels Bohr, who argued that the entire experimental setup involved in the investigation is an inseparable element of any phenomenon. Phenomena always emerge in the interaction between the 'object' and the 'apparatus,' meaning that all measured values pertain to this interaction rather than to objects independent of observation (Bohr 1963b, 4). For Barad, this has significant ontological implications. Before employing a specific physical device and a particular conceptual apparatus – which themselves belong to the constituted phenomena – there is no distinction between subject and object; it is only the application of specific cognitive apparatuses that delineates the boundary between them.

Therefore, instead of 'interaction,' Barad introduces the neologism 'intra-action,' which they define as follows:

> The neologism 'intra-action' *signifies the mutual constitution of entangled agencies*. That is, in contrast to the usual 'interaction,' which assumes that there are separate individual agencies that precede their interaction, the notion of intra-action recognizes that distinct agencies do not precede, but rather emerge through, their intra-action. [...] *agencies are only distinct in relation to their mutual entanglement; they don't exist as individual elements*.
>
> (Barad 2007, 33)

In this perspective, phenomena do not precede intra-actions; rather, they emerge from them as an ontological inseparability of their components. As the author of *Meeting the Universe Halfway* writes, phenomena are 'relations without preexisting relata' (Barad 2007, 139), meaning that individual objects do not exist prior to entering into specific intra-actions. It is these intra-actions that bring them into existence and assign them specific meanings, and this process is dynamic: different intra-actions produce different phenomena. Thus, phenomena do not appear to us in pure consciousness through transcendental reduction; instead, they emerge from the very immanence of the world's becoming. Rather than a stable division between the knowing subject and the known object, or between separate, individual entities, intra-actions produce 'individuation-within-and-as-part-of-the-phenomenon' (Barad 2007,

321), and instead of an absolute external relationship between the 'observer' and the 'observed,' there is 'exteriority-within-phenomena' (Barad 2007, 140).

Consequently, in the agential realist view, Ingarden's 'artistic object' does not exist at all, or more precisely, before a specific intra-action, such as the reader's aesthetic experience (which is also only then brought into existence), it does not possess any stable ontological status. Contrary to appearances, this does not enable interpretative anything goes; rather, it means that the 'objective reference' of interpretation is constituted only subsequently as a result of intra-action.

Importantly, the focus here is not on epistemology (how I come to know a literary work), but on ontology (how the literary work – and I as its reader – are brought into existence). Therefore, Barad proposes replacing Heisenberg's uncertainty principle with the indeterminacy principle, which they attribute to Bohr. It is not that observation *disturbs* phenomena (Bohr 1963a, 63–64), but rather that their complementary properties can never be simultaneously determined. Prior to observation, they remain ontically indeterminate (lacking definite values), rather than epistemically unknowable (disturbed during observation). According to Barad, this principle operates not only at the quantum level but also in the macroscopic reality: 'there is no evidence to suggest that there are two separate "worlds" – the Newtonian (macro) world in which Newton's equations apply, and the quantum (micro) world in which Schrödinger's equation applies' (2007, 416). The fact that Newtonian physics 'works' in the macroscopic realm does not arise from it being a 'true' account of the world, but from its provision of sufficiently accurate approximations for quantum mechanical equations.[9]

Translating Barad's considerations into the agential realist ontology of a literary work, it seems necessary to emphasize two points. First, a poem is constituted through various intra-actions involving not only the agencies we retrospectively recognize as the author, the text, or the reader, but also a range of diverse material-discursive practices – such as the influences of capitalist economy – from which the poem has separated itself in modernity (and continues to separate itself) as a specific, de/pragmatized, non/utilitarian, non/functional type of communication (Bourdieu 1996). This emergence of the poem – or the broader poetic field – occurs through what Barad terms agential cuts: decisions that resolve indeterminacy *within* the phenomenon and stabilize its mattering (always at the cost of certain exclusions, such as depriving the poem of direct political influence). Second, within the poem itself, there is a constant play of intra-active becoming. The poem does not consist of 'form' and 'content'; rather, form and content are resolved through agential cuts within fundamental mattering entanglements. Nevertheless, the poem – ontologically dense, affectively intense, aesthetically open – always retains a certain quantum of indeterminacy. Let me refer to this quantum as 'nothingness.' The poem strives towards nothingness, much like Miłobędzka's run, which – note – gradually fades: successive lines become shorter, gravitating towards 'nothing' not only in 'content' but also in 'form.'

According to quantum field theory, nothingness is not an emptiness but 'quantized indeterminacies-in-action,' literally teeming with virtual particles (quantum fluctuations of the vacuum), establishing an 'infinite plentitude of openness' (Barad 2012, 11, 16). This concept is reminiscent of Derrida's chora: 'void [is not] nothing (while not being something), but it may in fact be the source of all that is, a womb that births existence' (Barad 2012, 9). Clearly, quantum nothingness does not fit within the agrilogistic metaphysics of presence – fluctuations of the vacuum do not imply particles stealthily appearing and disappearing, but rather the indeterminacy of being and non-being, existence and non-existence, presence and absence. Consequently, 'even the smallest bits of matter are an enormous multitude' (Barad 2012, 15).

In W.H. Auden's poem *In Memory of W.B. Yeats*, the famous statement appears: 'Poetry makes nothing happen.' In an agrilogistic interpretation, the meaning of these words is straightforward – poetry lacks agency, does not affect reality, and does not translate into any social change. However, there is a possibility for another reading, more faithful to the arche-lithic. According to this interpretation, poetry embodies the principle of indeterminacy, pure multiplicity of immanence, resembling a quantum field that literally brings forth nothing: nothing *does happen* in poetry. Within this 'nothing' pulses – quite literally – *something*: a no/thing that unravels the very fabric of the principle of non-contradiction, making nothing of its claim to coherence.

If this sounds weird, all the better. The etymology of the English word 'weird' contains, as Morton points out, a reference to 'worth.' The 'nothing' of poetry is weird and is worth something beyond both exchange and use value. It opposes the fetish of utility in agrilogistics with non/usefulness. It contrasts the seriousness of agrilogos with non/productive, dis/interested play. If intra-actions always rely on agential cuts that grant material significance only at the cost of constitutive exclusion of something else, then poetry would be a type of material-discursive practice that strives to sustain as many intra-actions as possible, deferring – through its semantic, imaginative, and aesthetic complexity – the agential cuts that resolve fundamental indeterminacy.

By Way of (Open-Ended) Conclusion

In Rich's oxymoronic phrase 'forgotten future', Morton's arche-lithic, Marder's plant thinking, and Barad's agential realism, we equally find the suggestion that the future does not emerge from transcendence but from immanence, from the interiority of the 'now,' from the residues of the past that have been erased. For can the 'past' truly be erased? According to literary scholar Pheng Cheah, literature, through its disruptive world-making power, allows us to see what we have been forbidden to know: that the world is never fully mapped out and controlled by capital. Importantly, Cheah argues that 'this openness is an *unerasable* normative resource for disrupting and resisting the calculations of globalization' (cited in: Vermeulen 2020, 73; italics PS).[10]

The stakes of poetry, like those of the becoming world, lie in this very unerasability. Against this backdrop, it is worth mentioning the quantum eraser experiments discussed by Barad. These experiments allow the recovery of interference patterns lost due to an atom taking a specific path through a slit. They enable this even '*after the atom has passed through the slits and registered its mark on the screen*' (Barad 2007, 311), thus disrupting the classical notion of causality. The quantum eraser does not erase the past but rather the distinction between past, present and future: 'the "past" and the "future" are iteratively reconfigured and enfolded through one another: phenomena cannot be located in space and time; rather, phenomena are material entanglements that "extend" across different spaces and times' (Barad 2007, 316–17). According to the ontology of agential realism, this is not limited to the quantum world, as evidenced also by Marder, who writes about the 'recovery of information' by plants (2013, 155–56), and poet J.H. Prynne, who develops his concept of poetry in dialogue with the phenomenon of reverse transcription performed by retroviruses (cited in: Solnick 2017, 150). Both agrilogistic and arche-lithic patterns extend across various spaces and times, virtually residing in the 'here and now.'

The run for nothing, to nothing in Miłobędzka's poem can be seen as a movement toward fundamental – though always already contaminated – indeterminacy. It is a run toward the nothingness from which something emerges, toward Life itself, which, as Sosnowski observes, 'is under true siege today,' yet still resists. It is also a run beyond the blind persistence of agrilogistics, toward that first step in Ostrychacz's poem, when people fled the volcano in Laetoli 'like other herds,' still believing that

> Antelopes are sacred for carrion, sacred carrion
> of buffaloes and hippos, sacred
> meat, brain, and marrow, sacred are mocking hyenas
> and saber-toothed cats, for they forced Paranthropus to fight.
> (Ostrychacz 2020, 33)

What is at stake here is a practice that Barad describes as re/membering – not just remembering, but rejoining, becoming members of a larger community once again.

Notes

1 All works of Polish poets are rendered here in my own translation.

2 When we are dreaming, we forget, but immediately forget that we have done so; since the gaps and lacunae in our memories are Photoshopped out, they do not trouble or torment us. What dreamwork does is to produce a confabulated consistency which covers over anomalies and contradictions … .

(Fisher 2009, 66)

3 Namely, as the dynamic of fluctuating affects within particular bodies that engenders the difference by sensations, which, on their part, destabilize petrified vision, thus constituting Body without Organs (Young, Genosko, and Watson 2013, 166–69).

4 Jason W. Moore, among others, argues that capitalism's 'externalities' or frontiers are becoming exhausted. According to him, this means the depletion of the 'Cheap Nature' strategy – a regime that has supported capitalism for the past six centuries (Patel and Moore 2018). Sosnowski's poem seems to suggest that in response to this situation, capital has discovered a new frontier by cheapening our cognitive-emotional labor, which aligns with Franco 'Bifo' Berardi's ideas on semiocapitalism (2009).

5 All these thinkers employ the concept of 'phenomenon,' although only Marder directly references the phenomenological tradition. Morton engages in a speculative-realist reinterpretation of Kant's distinction between phenomenon and noumenon, while Barad uses the term in a very specific sense attributed to Niels Bohr, explicitly stating that this conception of phenomenon should not be derived from Kant or phenomenology. Nevertheless, all three fit within the broader phenomenological tradition, particularly through their dialogue with Jacques Derrida's deconstruction. (see: Morton 2013; Fritsch, Lynes, and Wood 2018).

6 I use the slash notation following Barad to indicate a departure from binary thinking. A/telic is not the opposite of *telos*, just as the arche-lithic is not the reverse of agrilogistics. Strictly speaking, a/telic exists beyond the opposition of purposefulness and purposelessness. This also applies to a/nomy, an/archy, etc.

7 Which was pointed out by Derrida long ago (1972).

8 Unlike many other speculative realists, Morton identifies as a correlationist, although – as he puts it – he releases the anthropocentric copyright control on correlationism. This means he believes that all entities exist in a correlationist manner, as none have access to things-in-themselves.

9 Morton's arche-lithic and Marder's plant soul exhibit a formal similarity to quantum indeterminacy in this regard.

10 Or a non-resource, as Marder would have it.

References

Barad, Karen. 2007. *Meeting the Universe Halfway: Quantum Physics and the Entanglement of Matter and Meaning*. Durham: Duke University Press.

Barad, Karen. 2012. *What Is the Measure of Nothingness? Infinity; VIrtuality; Justice / Was Ist Das MaB Des Nichts? Unendlichkeit, Vlitualitat, Gerechtigkeit*. Ostfildern: Hatje Cantz Verlag.

Berardi, Franco "Bifo." 2009. *Precarious Rhapsody: Semiocapitalism and the Pathologies of the Post-Alpha Generation*. London: Minor Compositions.

Bohr, Niels. 1963a. *The Philosophical Writings of Niels Bohr. Vol. 2, Essays, 1933–1957, On Atomic Physics and Human Knowledge*. Woodbridge: Ox Bow Press.

Bohr, Niels. 1963b. *The Philosophical Writings of Niels Bohr. Vol. 3, Essays, 1958–1962, On Atomic Physics and Human Knowledge*. Woodbridge: Ox Bow Press.

Bourdieu, Pierre. 1996. *Rules of Art: Genesis and Structure of the Literary Field*. Trans. Susan Emanuel. Stanford: Stanford University Press.

Bristow, Tom. 2015. *The Anthropocene Lyric: An Affective Geography of Poetry, Person, Place*. Palgrave Macmillan.

Debord, Guy. 2005. *The Society of the Spectacle*. Trans. Ken Knabb. London: Rebel Press.

Deleuze, Gilles, and Felix Guattari. 1987. *A Thousand Plateaus: Capitalism and Schizofrenia*. Trans. Brian Massumi. Minneapolis-London: University of Minnesota Press.

Derrida, Jacques. 1972. "Structure, Sign and Play in the Discourse of the Human Sciences." In *The Structuralist Controversy*, edited by R. Macksey and E. Donato, 247–72. Baltimore: John Hopkins University Press.

Derrida, Jacques. 1998. *Of Grammatology*. Trans. Gayatri Chakravorty Spivak. Baltimore: John Hopkins University Press.

Fisher, Mark. 2009. *Capitalist Realism: Is There No Alternative?* London: Zero Books.

Fritsch, Matthias, Philippe Lynes, and David Wood, eds. 2018. *Eco-Deconstruction. Derrida and Environmental Philosophy*. New York: Fordham University Press.

Glosowitz, Monika. 2019. *Maszynerie afektywne. Literackie strategie emancypacji w najnowszej poezji kobiet*. Warszawa: Wydawnictwo IBL PAN.

Ingarden, Roman. 1973. *The Literary Work of Art. Investigations on the Borderlines of Ontology, Logic and the Theory of Literature*. Trans. Ruth Ann Crowley, Kenneth Olsen. Evanston: Northwestern University Press.

Marder, Michael. 2013. *Plant-Thinking: A Philosophy of Vegetal Life*. New York: Columbia University Press.

Miłobędzka, Krystyna. 2010. *Zbierane, Gubione: 1960-2010*. Wrocław: Biuro Literackie.

Morton, Timothy. 2013. *Hyperobjects: Philosophy and Ecology After the End of the World*. Minneapolis-London: University of Minnesota Press.

Morton, Timothy. 2016. *Dark Ecology: For a Logic of Future Coexistence*. New York: Columbia University Press.

Ostrychacz, Marcin. 2020. *Cielenie lodowca*. Poznań: Wydawnictwo WBPiCAK.

Patel, Raj, and Jason W. Moore. 2018. *A History of the World in Seven Cheap Things. A Guide to Capitalism, Nature, and the Future of the Planet*. Carlton: Black Inc.

Prynne, J.H. 2015. *Poems*. Hexham: Bloodaxe.

Rich, Adrienne. 2007. *Poetry and Commitment. An Essay*. London-New York: W. W. Norton & Company.

Solnick, Sam. 2017. *Poetry and the Anthropocene: Ecology, Biology and Technology in Contemporary British and Irish Poetry*. New York-London: Routledge.

Sosnowski, Andrzej. 2015. *Dom ran*. Wrocław: Biuro Literackie.

Steffen, Will, Wendy Broadgate, Lisa Deutsch, Owen Gaffney, and Cornelia Ludwig. 2015. "The Trajectory of the Anthropocene: The Great Acceleration." *Anthropocene Review* 2(1): 81–98.

Tsing, Anna Lowenhaupt. 2015. *The Mushroom at the End of the World: On the Possibility of Life in Capitalist Ruins*. Princeton and London: Princeton University Press.

Vermeulen, Pieter. 2020. *Literature and the Anthropocene*. New York-London: Routledge.

Williams, Raymond. 1977. *Marxism and Literature*. Oxford: Oxford University Press.

Young, Eugene B., Gary Genosko, and Janell Watson. 2013. *The Deleuze and Guattari Dictionary*. London-New York: Bloomsbury Academic.

Zapf, Hubert. 2016. *Literature as Cultural Ecology: Sustainable Texts*. London-New York-Oxford-New Delhi-Sydney: Bloomsbury Academic.

SECTION 7

Eco-Phenomenology in Dialogue

27
ENACTING AN ONTOLOGICAL SHIFT
Individuation and Passivity in Enactivism and Eco-Phenomenology

Maia Vige Helle

Introduction

News about heat records, natural catastrophes, and ecosystem collapses has become alarmingly routine. We are, in the words of António Guterres, "on a highway to climate hell" (COP 27 2022), and we need to do something about it. We are in this crisis because we have treated the more-than-human world as merely possessing instrumental value – as an inexhaustible source of resources or means for our desires. Like an overly ambitious horse trainer with their eyes on the prize, not the horse, we are pushing the natural world to its physical limits, threatening the very foundations of our way of life and the lives of those we share this planet with.

The combined forces of globalization, technological advances, and capitalism have entrenched unsustainable practices so deeply within human societies that any plausible solution seems to call for radical action.[1] Many now call for a new ethics of nature. But is a new set of rules on how we *should* behave enough? Or does something more fundamental need to change for us to truly shift our behavior toward sustainability? According to eco-phenomenologists Brown and Toadvine (2003), properly addressing the crisis means that we must look beyond mere technical solutions or moral guidelines. They argue that to take the problems at their roots, we must fundamentally change the view of nature that has led us into them. In other words, we need an "ontological shift" (Toadvine 2009, 134).

According to Varela (1999), everyday ethics is not primarily about figuring out the "right" thing to do but about acting based on established knowledge of how one should behave in given situations – that is, having good intuitions. These intuitions emerge from one's ontology, which in everyday life gives the passive, nonreflective background to one's activity. To create consistent change in action, which is what is called for in our times of environmental crisis, we need to change this background so that our intuitive actions are in line with the needs of more-than-human nature.

There are three key areas to explore when considering an ontological shift. The first is deconstructive. It focuses on the prevailing ontology, examining its consequences and causes in order to break it down. The second is constructive and is concerned with suggesting an alternative ontology. Significant work has already been done in enactivism and eco-phenomenology regarding these two.[2]

 DOI: 10.4324/9781003538097-34

But there is also a third area that is necessary to fully cover an ethics based on ontological shifts, which, to my knowledge, has not been written about yet. We need to explain how we transition from one ontology to another. Building on Gilbert Simondon, I will explore how knowledge might be conceived in a way that can capture the phenomenon of ontological shifts, which I argue can serve as common ground for enactive and eco-phenomenological approaches to an ethics of more-than-human nature.[3]

Eco-Phenomenology, Enactivism, and Simondon

Eco-phenomenology is the intersection of phenomenology and ecology (Brown & Toadvine 2003, xii). Like all phenomenology, it starts with experience, but it explicitly focuses on our relationship with more-than-human nature (Abrahm 1997). That is, the parts of nature or biosphere that relate to but are not reducible to humans – other species, ecosystems, the climate system, and so on. There are many ways to develop this cross-disciplinary inquiry, as demonstrated by the variety of projects represented in this volume. I build on the general idea that the ethics of nature is bound up in the experience of nature (Brown & Toadvine 2003; Toadvine 2009).

Enactivism, particularly the tradition following Francisco Varela, is a theory in cognitive science and philosophy of mind that emphasizes cognition as an active, embodied process arising from the interaction between an organism and its environment. According to enactivism, knowledge is not passively received but actively constructed through lived experience, perception, and action. In this view, meaning emerges from the dynamic, reciprocal relationships between living beings and their surroundings (Thompson 2010; Di Paolo 2023; Sepúlveda-Pedro 2023).

In addition to having a shared origin in phenomenology, Eco-phenomenology and enactivism have some complementary goals and positions. They both argue for a new view on nature – with a special emphasis on value. Both aim to develop new ways of understanding humans, more-than-human nature, and the relationship between them. Recently, some enactivists have shown increasing interest in issues of more-than-human ethics (Candiotto 2022; Werner og Kiełkowicz-Werner 2022). This interest brings enactivism closer to eco-phenomenology, which currently has more resources on the phenomenology of nature. In turn, eco-phenomenology could benefit from enactivism‘s (arguably) more complete theory of life, mind, and nature. I propose drawing on Simondon's work to connect these approaches.

Gilbert Simondon, student of phenomenologist Maurice Merleau-Ponty, has recently become a more prominent figure in enactivism.[4] In *Individuation in Light of Notions of Form and Information* (2020), Simondon constructs a system for understanding individuation – what makes something the specific thing that it is. Individuation is an ongoing process in which a being – whether a person, object, concept, or relation – comes into existence and continuously evolves in interaction with its environment. All being is a process of becoming, and individuation is this becoming. Things (in the broadest sense) emerge as individuals from a pre-individual state through constant interaction with their surroundings, which gives them a specific form, but never a final shape. This offers a comprehensive metaphysical framework that aligns well with and complements the wider project of enactivism.

Having an epistemological theory that takes care of the need for a strong belief in scientific knowledge is an advantage in our times of widespread climate skepticism. Simondon offers a view on knowledge that is compatible with the limitations to the human perspective that every theory based on phenomenology is committed to, without allowing it to collapse into a form of anti-realism that can be used as a tool for science skepticism.[5] This makes it well-suited to examine how we in these times of crisis can envision the transition from one way of knowing the world to another that is better in line with how the world is. That is, how we can undergo an ontological shift.

I have adopted the concept of "ontological shift" from Toadvine's examination of what is needed to change our ethical relationship with more-than-human nature (2009, 134). I define ontological shifts as fundamental transformations in how we know reality. "Know" must here be understood broadly as how we perceive, understand, feel, and act towards the known. Roughly, we can say that one's ontology consists of three aspects: the being, or "ontos," of the knower; the ontos of the known, and the structure of the history of interactions between them, which orders (gives logic to) the relationship. Put more simply, everyone's ontology depends on how they are, how the world they interact with is, and how that interaction is structured over time.[6] I use the word ontology to refer to the network of relations that we stand in with the world. Importantly, human knowing is fundamentally intersubjectively grounded, developed by communities of knowers within a shared framework. I will discuss the dynamic between the individual and societal ontology later.

Based on Simondon's theory of individuation, and especially his analysis of anxiety, I suggest the term "ontological strain" to depict the kind of tension that builds up over time when there is an imbalance in the relationship, which, if not alleviated, threatens to break down the being of one or all parts of the relationship. If the ontological strain grows too strong, it may cause a collapse, meaning a breakdown of the being of the individuated thing. In environmental terms, ontological strain occurs when more-than-human nature is known or treated primarily as a resource for economic exploitation, ignoring its ecological complexity. Over time, this strain can lead to a collapse in the relationship, creating conditions for an ontological shift, where the subject is forced to reconsider and reshape how they relate to the object. Ontological strain is a characteristic of a perpetually unbalanced relationship, and is present whether it is experienced or not. But as I will show, ontological strain can also be experienced as cognitive dissonance and anxiety by (human) individuals.

To illustrate the kind of strained ontology characterizing modern societies' relation to the more-than-human, I will build on Kym Maclaren (2002) and Hanne De Jaegher's (2019) example of an overly ambitious horse trainer, which they use to characterize a relation lacking in "letting be" – that is, maintaining a stably balanced knowledge-relationship over time.

Knowing a Horse Badly

Imagine a horse trainer owning a large stable where he trains horses for races.[7] The more his horses win, the more money the trainer earns. The trainer really likes money and is deeply committed to turning his horses into money-making machines. He puts them on a strict regimen of exercise and diet. To get the best results, he sometimes even uses tools like whips to push them to their limits. And it works! His strictly trained horses win race after race, confirming to the trainer that he has understood how horses work – as money-making machines.

But after a while, his stable starts to decline. The horses become stubborn or sick, and they no longer win races. Some horses even stop eating or become so ill that they die. If a horse-loving outsider were to visit the stable, they would be appalled by the trainer's treatment of the horses. It would be obvious to them that the horses are not getting what they need for a good horse-life, such as free time, fresh grass, and social interaction with other horses, humans, or cats. The trainer's hyperfocus on money has blinded him to these less race-related aspects of being a horse. He doesn't see that his horses are suffering under this regimen because, according to his standard of success, they were doing well – they were winning races.

When the horses stop winning, get sick, become stubborn, or die, the reason is perplexing to the trainer because, as far as he can tell, everything has been working well until now. He has been blind to the signs that the horses were not doing well. What is visible to the horse-lover is invisible to the trainer due to their fundamentally different ways of knowing horses.

It's not hard to see parallels between this trainer-horse relationship and the way modern societies relate to more-than-human nature. The most obvious parallel is the exploitative nature of the relationship, where the knower overdetermines the knowledge relationship to make it fit their interests, forcing their understanding of the other as purely a resource for them. Secondly, this creates an ontological strain that threatens the known. Thirdly, in both cases, the knower's reliance on the known means that by rejecting or neglecting the strain, they are indirectly threatening themselves: the trainer risks losing his means of income; societies risk destroying their foundation. This practice is self-refuting.

A fourth interesting parallel is that in both cases of the knowledge relation, the knower must be said to have been "proven right" – the horses *are* money-makers, and more-than-human nature *is* a source of resources. Treating them this way actually *makes them into what they are seen as*. The problem then isn't that the knower is wrong, but that the knowledge is limited, overdetermining, and forceful to such a degree that it creates a strain between the basic needs and form of the known and what the relation shapes it into.

We can see and treat horses as a means to make money. And doing so will likely, at least for a while, confirm this understanding because horses can be many things, including money-makers. But as Maclaren points out, the possibilities are not infinite (2002, 188). You cannot use a horse as a cup of tea, for example. The way a horse is refuses that way of being known completely. Likewise, over time, you cannot view horses *solely* as a means of making money, because horses are much more than that. Knowing horses in this way practically means denying them a good horse-life.[8]

Similarly, using one's environment to satisfy one's needs is perhaps the most basic kind of relationship living organisms have with their environments. There is nothing inherently wrong with that. However, using the environment solely as a means in the way humans of today can, due to our high level of social coordination and modern technology, without acknowledging the complexity and internal needs of the systems we interact with, creates a strain that, over time, becomes unbearable for the long-standing balances in nature. This leads to the collapse of ecosystems, climate patterns, and so on. We are too big and powerful to operate only with instrumental values without risking collapse, and we know too much to be excused.[9]

The alternative to this instrumental, exploitative way of knowing is "letting be." In this approach, one remains open and adaptive to the shifting nature of the known. Instead of aiming to determine the known, as the horse trainer does, the goal is to let it be on its own terms. This should not be understood as an appeal to disengage and not determine the other at all. Such underdetermination effectively hinders engagement altogether. To have a relation at all, one has to let the other be in a way where one actively engages with it on one's own terms, which always, to some degree, will change and thus determine the known. This balancing between over- and underdetermination must be open and adaptive to be able to deal with the constantly changing nature of both the knower, the known, and the relationship between them (De Jaegher 2019, 13). In other words, knowing horses in a way that lets them be means actively engaging with them without enforcing the way you want them to be. That is, exercise when they seem ready for and happy with that, and let them rest and do other things when they want that. Yet, if I am correct, letting be is not really an option for the knowers in either of the above examples, because they are too biased by their own interests to be able to engage in this way. Thus, for letting be to become an option, an ontological shift needs to take place. To understand how this happens, we first need to look closer at the concept of knowledge at work here.

Knowing

Toadvine is concerned with the unreachable, unknowable, wild aspect of nature and sees this as central to a phenomenological ethics of nature. He writes that the "basis for responsiveness is in the call for a more radical Outside. Nature in this radical sense [...] is the 'pocket of resistance' and the unpredictable par excellence" (2003, 150). For Toadvine, our ethics of nature cannot be based on how we (think we) know nature, but rather on its refusal to be fully known. I think this emphasis is important to prevent yet another form of over-determined knowledge relation. This view recognizes that there is more to nature than we can grasp. However, on its own, it can seem unsatisfying and, in some respects, even nonsensical. What does it mean to be responsive to something, to engage actively and knowingly with it, based on what is *not* available to us? What does it mean to know something as unknowable? It seems we need some kind of bridge between the (un)known and ourselves as knowers to be able to respond to it. To address this concern, we need a form of realism that is compatible with the indeterminacy and dynamic nature of the parts involved in a knowledge relation.

Di Paolo argues that the observation that in our attempts to know the world, the world "pushes back," resisting our efforts in some areas while yielding in others, is sufficient to avoid anti-realism. The world and everything in it cannot be purely a construction of our minds or our will because it resists our attempts to force it into specific forms. Again, you simply cannot use the horse as a cup of tea. Some absolute limitations exist: "The act is the joint result of my powers and sensibilities and the materiality of the world, which will not allow for just any kind of engagement" (Di Paolo 2023, 178).

Thus, we see that knowing something "other" means being aware of its otherness and our limited ability to fully grasp it. This otherness of the (un)known imposes both positive and negative constraints on our behavior, allowing us to know it within the limits of our possible interactions in a dynamic way. This is well and good; now we understand that there is something real outside of us and that we cannot simply know it however we would like. However, something is still missing to fully understand the specific nature of knowledge and how it can lead to an ontological shift. To explore this, I will use Simondon to examine the basic structure of knowledge and how it can collapse.

To understand how Simondon views knowledge, we first need to examine his ideas about *individuation*. As mentioned, individuation, according to Simondon, characterizes everything that exists.[10] "What exists" is for him a broad category encompassing everything from planets, solar systems, and galaxies, to crystals, bricks, bacteria, animals, humans, thoughts, feelings, friendships, societies, and so on. This means that he considers knowledge, for example, your specific knowledge of the piece of paper in front of you, to be a real individuating being. The horse trainer's knowledge of horses is a relatively stable, real structure shaped by the history of his interaction with horses. This knowledge continues to exist and evolve if the trainer maintains a relationship with horses that reinforces and builds upon the established relationship.

The idea that knowledge of the horse is just as real and existent as both the horse and the horse trainer might seem unintuitive and strange. But it makes sense if you accept that all of them – horse, trainer, and knowledge – are nothing more than processes of becoming, with no inherent or static features. It just so happens that the becoming of the horse and the horse trainer largely depends on structuring material/physical parts into one whole, which allows us to see them, while knowledge does not materialize in the same way. Instead, it structures itself as a pattern of interaction that achieves its boundaries from the way it is constrained by and constrains the modes of interaction between the two physical entities (Simondon 2020, 74–77).

The knower, the known, and the relationship between them – the knowledge – are all *vulnerable* because their existence depends on a metastable balance distinguishing them from what they are not. This idea of vulnerability, or in enactive terms, precariousness, reveals the processual character of the entities involved. For living beings, maintaining a certain metabolic balance is crucial for continued existence. Similarly, Simondon describes knowledge as a balance between the knower and the known (Simondon 2020, 76). Things cannot exist in any arbitrary way; there are practical limitations to how much something can be changed that must be respected, or else it will stop being what it is or become something else. This point is crucial for understanding how Simondon's notion of knowledge can account for how we make mistakes or know things badly, as in the examples above. Since knowledge is a process of individuation that takes shape through relatively stable patterns between knower and known, knowledge tends to move in the direction of stable laws. The tendency towards static generality is necessary for it to be knowledge rather than loose information (Simondon 2020, 292–293). But as the example shows, the relation can become too static and thus lose contact with the dynamic nature of the known.

True knowledge, according to Simondon, exists when the balance between the knower and the known is optimally stable. The more unstable the relationship, the "less true" or more uncertain and weaker the knowledge becomes, as evidenced by how easily it might collapse (Simondon 2020, 76). In the example of the horse trainer, there was a form of balance, at least for a while, between the trainer and the horses. This balance reassured the trainer that he was on the right track, encouraging him to continue his approach. However, because the knowledge was narrowly defined and restricted to the goal of making money, it constrained the trainer's perspective and effectively blinded him to the signs that the relationship was barely maintaining its balance and constantly at risk of collapse.

Thus, the form of the knowledge relation was what made the trainer-horse relationship possible in the first place – since one cannot relate to something without doing it in a specific way that creates boundaries and possibilities for further interaction. But given that the trainer's acts of knowing were constrained and overdetermined by his goal of making money, the knowledge relation became too static and narrow to recognize the different and changing needs of the horses. Under such strain, it is only a matter of time (and luck) before the horse collapses, causing a collapse of the knowledge-relation. But what exactly happens when individuated knowledge collapses?

What Happens When Knowledge Collapses?

If we were talking about living organisms, the question of what happens in a collapse would be much easier to answer: if the organism loses metabolic balance, it dies. But what happens when the individuated cognitive structures and habits that order our relationship with the world fall out of balance? First, let's note what does *not* happen: the knower does not physically perish together with their ontology. If your knowledge of the world crumbles, your body persists, and you live on. So, what changes?

Throughout life, individuals develop a wide repertoire of relations with the world. While constantly changing along with the individual and their environment, these relationships maintain a distinct shape given by the boundaries created by previous interactions. Most instances of individuation at this cognitive level are gradual and additive – usually evolving rather than dismantling and replacing structures entirely. Simondonian individuation can thus usually be understood as a purely positive process without a need for a negating element for change to take place. Shaking hands with an acquaintance might progress to giving them a hug as your relationship develops into a friendship. The shift is gradual and does not imply a negation or break with the previous state.

Being open-ended, individuated knowledge-relations allow you to actively shift details of how you act. If you want to be environmentally friendly, you might resist the urge to check flights and rather look for train options when going on a vacation. But you probably don't even think of the possibility of jumping to your destination on a bouncy ball. In this way, individuated knowledge passively constrains you. These constraints allow for active engagement, which creates new constraints, keeping one's practical knowledge of the world constantly open and changing, but also bounded. This forms your background intuitive know-how that structures everyday action and, as Varela (1999) suggests, are crucial to understanding everyday ethical behavior.

Simondon addresses the collapse of cognitive structures in his discussion of anxiety (Simondon 2020, 283–284). According to him, anxiety arises when passivity – internal and external processes of change outside volitional control (pre-individual sparks of new individuation) overwhelm the active subject, dismantling the established or individuated knowledge-relations. Anxiety is triggered by a tension between the active and the passive aspects of your being. You are being torn between your active continuation of an established way of being and something within and/or around you that refuses this way of being, like the horse refuses the bad treatment of the trainer. In an anxious breakdown, all the established ways of relating to the world, your ontology, break down due to the passive strain. This breakdown leaves you unable to act, because every pattern of relation you thought you knew and that was the baseline for your free actions has now broken down to nothing. This leaves you in a passive state where there is an unbearable painful indistinction between you and the world, and without any constraints to delineate your further individuation.

However, this state also holds the potential for transformation. Simondon describes anxiety as an opportunity for "restructuring," where the collapse of old individuations creates the possibility for new forms of being. "If the experience of anxiety could be supported and endured long enough, it would lead to a new individuation within the being itself [...]. To a veritable metamorphosis" (Simondon 2020, 284). This process can be understood as an ontological shift.[11]

The reason for this new individuation is exactly what I noted above, that having your mental structures collapse is not sufficient to kill you. And being a constant process of self-continuation, there is nothing else to do than to continue, which involves individualizing new relations with one's world. This highlights the transformative potential of knowledge collapse. The previously individuated structures dissolve, allowing for the formation of new knowledge structures that reflect a deeper integration of the subject's experiences. But for this transformative potential of anxiety to appear, Simondon argues that we need to accept the breakdown, allowing the experience of being dissolved and losing one's established relation with oneself and the world. As he writes,

> anxiety already contains the premonition of this new birth of the individuated being based on the chaos that spreads all throughout; the anxious being feels that it might be able to be reconcentrated within itself in an *ontological beyond* that supposes a change in all dimensions; but in order for this new birth to be possible, the dissolution of the previous structures and the reduction of the previous functions in potential must be complete, which is an acceptance of the annihilation of the individuated being.[12]
>
> (Simondon 2020, 284)

In this context, the "annihilation of the individuated being" represents the breaking down of static or overly rigid self-conceptions. Despite how it might feel, this is not a complete existential threat, but rather an opportunity for the individual to reconstitute themselves in a new form (Simondon 2020, 292, 299). Anxiety comes from a tension that emerges because something in your life is going against what you realize yourself as, making your way of life incompatible with itself. Accepting that what you have been until now can change, losing your rigid self-conception, opens you to new ways of being and knowing. These ways are not (yet) so clearly and rigidly individuated and therefore more able to adapt to the tensions in your life that triggered the breakdown in the first place. So how might we go through anxiety and open up to a new way of being?

Interestingly, Simondon writes that this transformative stage following anxiety is not something a person can achieve alone. We may break down in isolation, but we cannot just decide on our own who we want to become afterwards; this requires the collective from which we individuate and are part of (Simondon 2020, 284). So, the new ontology – the new way of relating to the world – needs to be individuated out of the ontology of the social system one is part of. This does not mean that every ontological shift on the individual level requires an ontological shift on the social level, since societies usually are plural enough to be able to capture different ways of being. But, as we shall see, some instances of anxiety might stem from a crisis in society, like, for example, the climate crisis, and then a shift in the social ontology is needed. In the rest of the chapter, I will look at conditions for going through with a shift both at the level of society and the individual, and how these are connected.

Moving from individual anxiety to social ontology, it is important to keep in mind that *if* anxiety at the individual level is structurally similar to an ontological shift at the social level, then such a shift cannot occur as a *gradual* transition from an old worldview to a new one. The reason an anxious person must accept the "annihilation of the individuated being" would apply equally to any similar shift at a societal level – this can be seen as anticipated in deconstructive approaches to ethics of social ontology. So, if my analysis is correct, social ontological shifts have radical political implications.

From Anxiety to Ontological Shift

We now have the resources to link Simondon's idea of anxiety to the phenomenon of climate anxiety. In some ways, it's misleading to talk about *climate* or any other specific kind of anxiety, since anxiety, by definition, lacks content. Anxiety is the breakdown of everything known and established in your mind. During the breakdown, there is no relevant distinction between the climate crisis and the pain in a toe; everything loses its order and structure. As Simondon writes, anxiety "unravels what has been woven" (Simondon 2020, 284), so nothing remains ordered and individualized. In the grip of anxiety, one's knowledge of climate is as disentangled as everything else.

However, the ontological strain leading to an anxious breakdown might still be the climate crisis. If, on any level, you recognize that your daily actions contribute to the destruction of the planet – a planet that you, your loved ones, and everyone else depend on – this can create a significant dissonance. If you cannot resolve this dissonance, it can threaten your mental well-being. Thus, while climate might not be the "object" of anxiety, it can still be a *cause*. Being the initial cause of tension, addressing it may also be essential to resolving the anxiety. Simply becoming aware that the climate crisis is causing the strain may not be enough; you must also have the means to alleviate this strain actively.

This brings us to a practical consequence: if the strain arises from a contradiction between a person's potential and actualized knowledge – such as when someone "knows" their lifestyle is incompatible with a stable climate, but lives as if there is no crisis – then to get through anxiety and achieve an ontological shift, practical conditions for behavioral change must be within reach. If you can't live your knowledge, you are forced back to the unstable relation of the pre-anxious state. To get out of this loop, we must change the physical, practical, and social conditions constraining our daily life so that living in a climate-sustainable way becomes a real, accessible option. Read in this way, it's clear that focusing narrowly on the life of the anxious individual is insufficient; we must examine the political and socio-material conditions in which they live.

To clarify, consider the horse trainer again. If practical or contractual limitations force him to operate his stable in a horse-unfriendly way, he will not change. If he is unable to let the horses go out to eat grass and hang around doing fun horse stuff, he will not do so, no matter how much he might wish it. When the horses get ill, he may understand why, but he still cannot, *in practice*, let his old way of knowing, his know-how of horses, break down and shift. Thus, he remains stuck: either accepting the horses' suffering but continuing as before, falling in and out of anxiety without ever going through it, or doubling down on his blindness, refusing to acknowledge their deterioration. In any case, the horses' situation remains unchanged.

This pattern mirrors what we often see in people's responses to climate issues (Norgaard 2011). Many accept that there is a problem, enduring silent pain while continuing to act as if none exists. Others refuse to see the problem at all and carry on as before. Only a privileged few – those with the right physical environment, time, money, cognitive and practical resources, and community support – can go through their climate anxiety and begin living in accordance with their knowledge.[13]

This analysis of the relation between individual anxiety and the socially determined affordances one encounters in daily life can help bridge the gap that has been underlying this discussion – the gap between the individual experience of anxiety and the ontology that exists at the level of society, that is, the ways of knowing more-than-human nature within a global, exploitative system. If we cannot practice or even envision how to change our lives to relieve the ontological strain manifesting as cognitive dissonance, we will not be able to address the root causes of the stress.[14] Feelings of hopelessness and despair when faced with news of the climate crisis and other catastrophes make more and more people disengage politically and avoid news about the state of the world (Newman 2023). Such backing away from uncomfortable knowledge is a way to avoid facing the threat of anxious breakdown. But if Simondon is right, accepting the breakdown might have a positive effect, but that requires a political rather than individual acceptance and response. Combined, this leaves us in a pickle where the factors that are creating the crisis are also facilitating the inaction to prevent it. But on the other hand, it also makes it very clear where we should put pressure if we want to make a real change.

If we hope to collectively respond to what the world is telling us about our unsustainable ways of living, we need political action to restructure the conditions that shape people's lives – not just for the privileged few, but for everyone. Achieving this ontological shift, on both individual and social levels, requires something analogous to anxiety at the political level – a dismantling of the societal structures that prevent us from re-relating to the more-than-human world. We must build new ways of living – not atop the old structures, but through embracing the breakdown and the new possibilities it opens.

This analysis suggests that we might need to abandon our belief in controlling the environmental crisis and accept the uncontrollable aspects of our nature-relationship. We must accept defeat and allow the more-than-human to reveal what is fundamentally unsustainable in our current ways – dismantling the instrumental exploitative ontology of modern societies. Only through this collapse – this annihilation of our individuated being – can we begin to construct a worldview based not on established socio-political-economic systems, but on the more-than-human world. Gradual approaches like "green growth" or "techno-optimism" might alleviate some issues short term, but fail to address the root causes. Our planet simply cannot sustain modern, affluent lifestyles. And believing it can only sustain inaction (Slaby 2024). We need to fundamentally rethink how we as a collective travel, eat, consume, and live all together. Understanding ontological shifts in light of anxiety underlines this positive potential for dealing with the crisis that lies in a radical approach to societal change.

Where Should the Shift Originate?

Any ethical theory implies a moral agent or target audience that one implores to enact the ethical change. From what I have said so far, it's not yet clear who or what should be the target for an ethics based on an ontological shift. Where should the shift originate? To answer this, I critically examine Brown and Toadvine's initial approach using enactive and Simondonian ideas to propose an answer I think is more in line with the radical potential of eco-phenomenology.

Brown and Toadvine both seem to suggest that ethical change starts with perception or experience – whether it is the experience of value or of otherness and unreachability in nature that works as the moral anchor (Brown and Toadvine 2003, xi–xiii, 13–16, 148–150). Brown states that for a proper ethical shift to occur, there needs to be a change in lifestyle that accompanies the shift in perception (2003, 3). However, stating it this way – combined with the fact that neither of them seems to focus on the practical constraints of daily life in their phenomenological analysis of nature – implies a certain separation between perception and action, at least theoretically. If such a separation exists, it appears to follow that this perspective prioritizes individual perception.

This interpretation might imply that if one seeks change, one should start with one's own relation to the more-than-human. I am not claiming that this is the intention of Brown or Toadvine, only that it can be read in a way that gives the impression that ethical change primarily needs to come "from within" – imploring the reader to focus on changing their experience and feelings towards nature (and from there, societal change will follow). Concluding that change must come "from within" seems problematic for several reasons. If global changes in attitude depend primarily on individual self-world relationships, we risk, as described earlier, limiting the ontological shift to only those already privileged enough to navigate the nature-destructive social structures. This means that very few can achieve this shift, and their impact will be limited as social-material conditions remain untouched. It is thus both elitist, ineffective, and prevents a shift at the social level.[15]

Focusing exclusively on experience is, thus, misguided if the goal is to help dismantle our current destructive system and reorient our relation to the natural world – a goal Brown and Toadvine set for eco-phenomenology (2003, ix–x). Given the role of the intersubjective, active body in Merleau-Ponty's work (2014) – and his importance to eco-phenomenology – it is possible to argue that, instead of prioritizing perception, and possibly being read as promoting an "inside first" view, eco-phenomenology could benefit from focusing on action-oriented views of perception, like those found in enactivism.

From an enactive standpoint, perception, action, and knowledge form an interconnected loop, where prioritizing one over the others as the source for ethics (or anything else) does not make sense. Knowing is doing and perceiving (Di Paolo 2023; Thompson 2010; Varela, Thompson, and Rosh 2017). Prioritizing perception as the locus of change risks overlooking the broader network of influences shaping behavior. Acknowledging that change in behavior relies on outer factors and not only one's own attitude or experience, entails a broader set of resources for going through an ontological shift at a societal level, opening this ethical work more towards the political realm. Here, one is not implored to develop a healthy personal relationship with nature (even though this can be a good thing to do in addition, if one has the opportunity and gets motivation from it), but to work politically to break down unsustainable and build up sustainable ways of living. Through this, one can start to change perception and knowledge patterns in people *en masse*, starting the process of individuating a new ontology of society.

Conclusion

There is much work to be done to fully flesh out the political potential in enactive thinking. Some important contributions already exist (see Slaby 2024, Maisie & Hanna 2019), but more is needed. Simondon's philosophy provides a robust framework for examining relationships between the various social levels and structures. The discussion of anxiety presented here illustrates how such an analysis might bridge the gap between the individual and the social:

Although "anxiety" here is primarily used as a tool for analyzing a collapse in one person's cognitive structures, I don't want to exclude the possibility that at least some instances of anxiety can play a role in creating the conditions for a societal ontological shift. Since the way we as individuals relate to the world around us is formed by the shared ontology of the societies we partake in, anxiety might emerge from a strained way of living that is socially determined. Analyzing anxiety from this perspective shifts the focus from individual responsibility to the structures that shape the available affordances in people's everyday lives, that is, politics. Slaby's (2024) work on how emotions can be shaped by political and economic systems that suppress rather than enable transformative action clarifies how emotions contribute to both maintaining and challenging established ontologies. From this perspective, at least some kinds of anxiety (like the ones caused by the climate crisis) are not something merely to "overcome" individually but rather *a call to dismantle the societal structures causing it*.

If transformation need not exclusively "come from within," attention can shift to leveraging external factors to encourage more sustainable relationships with the world. In other words, we can actively shape conditions for change by restructuring the socio-cultural-material milieu. By altering affordances in people's environments, we make it easier to engage in actions that are more in line with a well-balanced knowing of more-than-human nature. When sustainable ways of living are made accessible, people will be more inclined to adopt them, reducing the cognitive dissonance leading to anxiety – and, most importantly, move us away from a way of life that is destroying itself and everything around it.

By embracing the unity of knowledge, perception, and action in eco-phenomenology and opening the political potential in enactivism, we gain tools for a practically oriented and broad approach to more-than-human ethics that still preserves the valuable insights and methods already existing in eco-phenomenology and enactivism. My suggestion is that Simondon's theory of knowledge represents a promising way to make a deeper sense of these potentials in eco-phenomenology and enactivism, and at the same time connect them so they can stand stronger together.

Notes

1 These factors have impacted groups and societies differently. These varieties are important when discussing responsibility and solutions. But here it's of less relevance as I'm focused on the overall skewing of human societies towards unsustainable nature-relations.

2 For sources in eco-phenomenology, see chapters by my colleagues in this volume.

For sources on ontology of nature in enactivism, I recommend Thompson's *Mind in Life* (2010), and Sepúlveda-Pedro's *Enactive Cognition in Place* (2023). Slaby's "Habits of Affluence" (2024) provides a deconstructive account of modern affluent societies.

3 I use Abrahm's (1997) term "more-than-human" to underline that I reject dualistic thinking of humans as separate form nature. "More-than-human" is used to direct the focus of inquiry to the parts of nature that exists partly independently of us humans. It should in this context be read as an epistemological not an ontological or ethical term.

4 Simondon has influenced enactive thinking across psychology (Arandia & Di Paolo 2021; García 2023), tecnology (Dereclenne 2021), epistemology, and language (Di Paolo Cuffari & De Jaegher 2018; Di Paolo 2023), and ethics (Di Paolo & De Jaegher 2022). Thanks to Di Paolo for giudence.

5 Simondon's "realism of relation" (2020, 77) views reality as processes of becoming, where scientific knowledge arises from (meta)stable relations between the knower and the known. As I understand him, scientific knowledge is valid if it's able to withstand standard scientific scrutiny. Some readers might find this surprising given his influence on constructivists like Bruno Latour.

6 This use of the terms "ontos" and "ontology" is based on Di Paolo's "F/acts" (2023).

7 The example is based on Maclaren (2002) and De Jaegher (2019) with adjustments to fit my project.

8 I focus on two "simple" cases involving a kind of strain that goes in the direction of an undisputable collapse. The case is more complex the more flexible and interconnected the relationship is, like in human relations. See "Enactive Ethics" (2022) by Di Paolo and De Jaegher for a great introduction to enactive ethics.

9 This view on responsibility aligns with Hans Jonas: responsibility answers to knowledge, power, and freedom. In destroying a lake, unfree, unknowing plankton is unaccountable, while free, knowing human polluters are. (Jonas 1984, 129–130).

10 This might be an exaggeration as I don't grasp his view on quantum mechanics well enough to determine whether quantum phenomena exist as processes of individuation. But as quantum physics is less than relevant for the current project, I hope the reader will excuse this possible misstep. Also, for the sake of simplicity I have avoided most technical jargon of Simondons writing. This makes statements like this somewhat unprecise, although they should be in line with his main ideas. For a clear description of some central ides in Simondon, and their applicability enactivism see García's "Affectivity in Mental Disorders" (2023).

11 While I address shifts on the level of individuals here, my primary focus is on the ontology of growth-driven modern societies, discussed further below.

12 My emphasis.

13 There is then still a problem for them that they probably know that this is insufficient, which further promotes the focus on societal and political change that I argue for.

14 For collective cognitive dissonance regarding climate change, see Nordgaard's *Living in Denial* (2011).

15 Contemporary ethical theories in phenomenology and enactivism often emphasize what I will call an "inside first" approach, fostering individual virtues or attitudes like love of nature. For an example of this kind approach in enactive ethics of nature see Candiotto (2022). While valuable, I argue that such approaches need to be supplemented with a political and social focus to be realizable. For a critique of "inside first" approaches that do not take social factors into account see "The "Long Hope": An Unaffordable Luxury" in Swanigan (2024).

References

Abrahm, David. 1997. *The Spell of the Senuous: Perception and Language in a More-than-human World*. New York: Vintage.

Arandia, Iñigo R, and Di Paolo, Ezequiel A. 2021. "Placebo From an Enactive Perspective." *Frontiers in Psychology* 12: 1–20. https://doi.org/10.3389/fpsyg.2021.660118

Brown, Charles S, and Ted Toadvine. 2003. *Eco-Phenomenology: Back to Earth Itself*. New York: State University of New York Press.

Candiotto, Laura. 2022. "Loving the Earth by Loving a Place: A Situated Approach to the Love of Nature." *Constructivist Foundations* 17 (3): 179–189. https://constructivist.info/17/3/179

De Jaegher, Hanne. 2019. "Loving and Knowing: Reflections on an Engaged Epistemology." *Phenomenology and the Cognitive Sciences* 20 (5): 847–870.

De Jaegher, Hanne, and Di Paolo, Ezequiel A. 2007. "Participatory sense-making: An enactive approach to social cognition." *Phenomenology and the Cognitive Sciences* 6 (4): 485–507.

Dereclenne, Emilien. 2021. "Simondon and Enaction: The Articulation of Life, Subjectivity, and Tecnics." *Adaptive Behavior* 29 (5): 449–458. https://doi.org/10.1177/1059712319861915

Di Paolo, Ezequiel A. 2023. "F/acts; Ways of Enactive Worldmaking." *Journal of Conciousness Studies* 30 (11–12): 159–189.

Di Paolo, Ezequiel A., Elena Clare Cuffari, and Hanne De Jaegher. 2018. *Linguistic Bodies: The Continuity between Life and Language*. Camebridge: The MIT Press.

Di Paolo, Ezequiel A., and Hanne De Jaegher 2022. "Enactive Ethics: Difference Becoming Participation." *Topoi* 41, no. 2: 241–256.

Frank, Adam, Adam Gleiser, and Evan Thompson. 2024. *The Blind Spot: Why Science Cannot Ignore Human Experience*. Cambridge, MA: MIT Press.

García, Enara. 2023. "Affectivity in Mental Disorders: An Enactive-simondonian Approach." *Phenomenology and Cognitive Sciences* Vol. 23, no. 5: 747–765.

Guterres, António. 2022. *Secretary-General's remarks to High-Level opening of COP27 | United Nations Secretary-General*. https://www.un.org/sg/en/content/sg/speeches/2022-11-07/secretary-generals-remarks-high-level-opening-of-cop27

Jonas, Hans. 1984. *The Imperative of Responsibility: In Search of an Ethics for the Technological Age*. Chicago: The University of Chicago Press.

Maclaren, Kym. 2002. "Intercorperality, Intersubjectivity and the Problem of "Letting Others Be"." *Chiasmi International* 4. 187–208.

Maisie, Michelle, and Robert Hanna. 2019. *The Mind-Body Politic*. Palgrave Macmillan.

Merleau-Ponty, Maurice. 2014. *Phenomenology of Perception*. Oxon: Rutledge.

Newman, Nic. 2023. *Overview and key findings of the 2023 Digital News Report*. June 14. https://reutersinstitute.politics.ox.ac.uk/digital-news-report/2023/dnr-executive-summary

Norgaard, Kari Marie. 2011. *Living in Denial: Climate Change, Emotions, and Everyday Life*. Massachussetts: MIT Press.

Sepúlveda-Pedro, Miguel A. 2023. *Enactive Cognition in Place: Sense-Making as the Development of Ecological Norms*. Cham: Palgrave Macmillan.

Simondon, Gilbert. 2020. *Individuation in light of Notions of Form and Information*. Minnesota: University of Minnesota Press.

Slaby, Jan. 2024. "Habits of Affluence: Unfeeling, Enactivism and the Ecological Crisis of Capitalism." *Mind & Society*, 24, no. 1: 165–186.

Swanigan, Pamela. 2024. "It's Time To Give Up Hope For A Better Climate & Get Heroic." *Noema.* https://www.noemamag.com/its-time-to-give-up-hope-for-a-better-climate-get-heroic/

Thompson, Evan. 2010. *Mind and Life: Biology, Phenomenology, and the Sciences of Mind.* Camebridge: Belknap Press.

Toadvine, Ted. 2009. *Merleau-Ponty's Philosophy of Nature*. Evanston, IL: Northwestern University Press.

Varela, Francesco, Evan Thompson, and Elanor Rosh. 2017. *The Embodied Mind.* Cambridge: MIT Press.

Varela, Francisco J. 1999. *Ethical Know-How: Action, Wisdom, and Cognition*. Stanford: Stanford University Press.

Werner, Konrad, and Magdalena Kiełkowicz-Werner. 2022. "From Shared Enaction to Intrinsic Value: How Enactivism Contributes to Environmental Ethics." *Topoi* Vol. 41, no. 2: 409–423.

28

MAKING SENSE OF 'DESCRIPTION FROM WITHIN'

An Ontological Perspective

Svein Anders Noer Lie

Prefatory Note

It will become clear to the reader that this chapter does not proceed as a conventional eco-phenomenological analysis; its critique of the epistemological dogma – and the underlying notion of intrinsicness – should nevertheless be understood as operating in close conversation with eco-phenomenology. My aim is to extend the ontological implications of phenomenological insight by foregrounding a relational realist framework, which I see as complementary to, and in some cases sharpening, eco-phenomenology's central concern with embeddedness, interrelation and the limits of representational thinking.

Because the content of this work relies on an account developed in a previous book of mine, I need to – already now – excuse the extent of self-referencing.

28.1 The Trouble with Intrinsicness

Phenomenology represents an alternative to what is often called representationalism, namely the notion that the outside world is re-presented in our mind (See e.g. Taylor 1995). The idea is that if we can overcome representationalism, we will also overcome the subject/object dualism, which has created whole libraries of debate ever since Rene Descartes. If you want to understand what love is, you need to start from the experience of love. To describe the natural world, one must be receptive and allow the phenomenon to unfold.

For the eco-phenomenologist, this has an even deeper meaning, since the dualism is considered to be the main problem underlying our current detached relation to allo-human nature. Phenomenology is thus deemed as an avenue into a more primordial relation with nature. It is supposed to offer a perspective that is able to challenge or supplement what is considered to be the objective descriptions of science.

A major problem with this story is that the phenomenological stand *remains* an alternative to scientific descriptions, even if it attempts to go beyond and found science. Phenomenologists have not been able to convince the scientific community to change their position, at least not in the natural sciences. From their point of view, phenomenology comes across as subjective and unable to describe the world 'as it is in itself'. Its reputation as an alternative is that it has

 DOI: 10.4324/9781003538097-35

been one of the main contributors to the maturity of social constructionism, which is not a very popular position among scientists.[1]

What some environmental philosophers are typically worried about if phenomenology is to be accepted as an (or even the) alternative to the dualistic ontology underlying scientific descriptions, is that we are driven towards accepting the scientific stance and its ontology of nature, while environmental concerns effectively are looked up-on as not dealing with nature as such, but with projections of an idealized or romantic *position*, (see Toadvine, 139). This is also exactly how Steven Vogel interprets the situation, but his point is that the environmental philosopher should embrace this situation, adding however, that scientific descriptions do not have a default authority (Vogel 2015. See, also Lie 2021).

Martin Heidegger is an example of a notoriously ambivalent figure in this case. While he never really refrains from insisting that 'nature' shows itself through human activity (ready-to-hand, projects, care), he also insists that such disclosure is beyond human control and will (in the clearing *(Lichtung*), in *Ereignis*). Hence, humans can be more or less open to how 'nature' is phenomenally present, through authentic *care* (SuZ) and later through, what he called *Gelassenheit* (letting-be). His concern is being, not nature, and his overall framework rests on the assumption that we need to free ourselves from the historical route that the question of being has taken (*Seinsgeschichte*). Although I support Heidegger's efforts to rethink the philosophical tradition, I also believe that he is soaked in the very tradition he wants to depart from. The reason is simply that he concedes that any ontological idea of nature would exemplify a fall into the forgetting of being (*Seinsvergessenheit*).

The origin of this indebtedness to metaphysics, I contend, lies in the idea of describing the world as it is in itself. Not that Heidegger supports this claim, but that he fails to recognize this assumption (the 'in itself') as an ontological idea. An ontological idea that eventually lays down the foundation of what could be called the epistemological dogma, which Heidegger also accepts.[2]

The simple idea is this: Where you claim that beings actually *do* exist in themselves and that beings have intrinsic natures, you are bound to assess that we need to access these intrinsic natures to describe the world as it *is*. It is this *reality principle* – properties are real if they exist independently of everything else – that *prescribes* the subject/object-relation (which Heidegger wants to overcome) in the first place, *not* the seemingly uncontroversial notion that we need to know *how* to gain knowledge about the world *before* we can say anything about what the world is. (Which originally was posed as a knock-out argument against the naïve understanding of the relation between ontology and epistemology associated with Greek philosophy. C.f. the debate on realism (about universals) vs. nominalism in late medieval philosophy).

If it is the case that the epistemological dogma carries implicit ontological claims about the nature of the world, it would be misguided to indirectly accept the implications for our epistemological stance (subject vs. object), and attempt to refute it by for example, demonstrating that our primordial access to the world is grounded in being-in-the-world (Heidegger) or in bodily habituation (Merleau-Ponty). Such a refutation is only a refutation of how the 'traditional' subject is conceived, and thus an indirect confirmation of the idea that we should grasp the world as it is in itself.[3] What we need to do is to reject the ontological supposition in the first place *by* presenting an alternative and then, *eventually*, try to understand our epistemological situation. I will now present an alternative ontology and the way this ontology prescribes our epistemological situation, and come back to the question concerning Heidegger in particular and eco-phenomenology in general.

28.2 The Dispositional Ontology

To present an alternative ontology has two aims here. First of all, an alternative ontology may provide a different description of where 'the subject' is. If it is not up against external beings that exist independently of everything else, what is it up against? Secondly, if a relational ontology, which I will present here, can be viewed as a viable alternative, it will become clearer *why* it is pertinent to point out that the epistemological dogma is also a dogma that has influenced phenomenology. It is only in light of a relational ontology that we can properly assess whether phenomenology ever truly escaped the intrinsicness assumption of the epistemological dogma.

There are currently several relational ontologies on offer. Relational ontologies and process philosophies are growing increasingly popular, and we find them within new materialism, speculative realism, structural realism and other isms. What they have in common is the idea of the world being intrinsically active, where the main ingredient is not objects, but processes.[4]

For reasons that I will not substantiate here, I find the relational account of dispositionalism to be the best candidate.[5]

The dispositional ontology originated as an alternative to neo-Humean accounts of what properties are (Mumford 1998, 2004). Dispositionalists claim that properties are dispositions. (In this case, we need not worry about the many dispositionalists who argue that there are dispositional *and* categorical properties). This means that for something to exist at all, it needs to *do* or be able to do something. (The Eleatic Stranger principle). A dispositionalist argues that potencies are real.

Also within *this* field of discourse, there is a debate about whether dispositions are had in virtue of being intrinsic, or whether they are had in virtue of being relational.[6] While most dispositionalists would not even question the need to have intrinsicness in their theoretical inventory, I argue that it is crucial for the dispositional ontology that dispositions are accepted as being relational in nature (Lie 2016. Ch 4).[7]

A typical example would be that the dormitive effect of opium is routinely attributed to its intrinsic properties (i.e. its chemical structure), while a relational realist would argue that the disposition to manifest a dormitive effect could not be had without the particular receptors in our brains. The mutual manifestation of chemical bonding at a lower level is responsible for its typical chemical structure, but not for the dormitive effect. The dormitive effect is a (concurrent) mutual manifestation of these (+ other) partners. (On the term 'mutual manifestation, see Martin 2007, 3). Opium does not have a dormitive effect 'in itself'. Nothing has 'an effect in itself'. Every activity is a mutual manifestation between at least two dispositional partners (when *in* manifestation, I call them manifesting partners), and manifestations are what we usually would refer to as 'causal activity'.

28.3 Descriptions from within Nature

Where does this leave us with respect to our understanding of nature? It leaves us in the epistemological situation where we *need to* describe and understand nature from within. We need to understand the contribution of x to understand the contribution of y, to understand the whole – which is its manifestation.[8] And if humans are taking part in the manifestation, we are a part of this manifestation as well. We are an ingredient, as Graham Harman would say. We are on the inside. This does not mean that we construct the manifestation, as the social constructivist would have it. Color is a co-manifestation. The manifestation of a blue color is a 'co-production' between light, a surface and the eye. Does this co-production entail that color is a secondary quality? I would argue that it is a manifestation just like any other taking place in nature and cuts beneath the primary/secondary quality distinction. We don't say that a certain weight of an object is a secondary quality, even though weight (on this account) is a manifestation of several manifesting partners. The difference is just that humans are simply taking part in the first instance and not in the second (the weight of an object). The fact that we are taking part does not mean that the manifestation is less real. It is as real as any other manifestation where humans don't take part (i.e. a tree falling in the forest).

What we should discuss when we are worried about real and true descriptions is therefore not about describing the world independently of ourselves. The problem is rather about getting hold of the dispositions involved and *placing and attributing the right concurrent dispositions to the right bearers*. If color is attributed to the table alone, it would be a wrong attribution, not a secondary quality. The right attribution would be to say that the table has a particular structure reflecting light in a certain way. Color is a real property, despite the fact that it does not belong to the table alone. It is a manifestation of at least three manifesting partners that have further dispositions (That is: 'is able to do something').

The first problem with mutual manifestations (and for science in general) is that we don't see or sense (or observe) the manifestation of a single disposition. What we sense is always a joint manifestation. We never *positively* sense the solubility of sugar; we only see the solubility of sugar in water (or some other solvent). We can only *infer* that bearer x gives the contribution D when partnering with y, w, z, etc. The second problem is that we *cannot single out the different contributors without knowing the contribution from the other bearers*. One bearer may have a wide range of dispositions – even an indeterminate range (depending on partners).[9] Hence, to know the contribution from the other contributors, you need to see their joint manifestation and compare it with other sets of joint manifestations. It is when you 'see' the joint manifestation of what happens when you put a solid stone in water that you find it proper to *infer* (after the fact) that *sugar is* dissolvable (as opposed to the stone). This is inevitable. We cannot bypass this epistemological problem. All 'empirical data' comes with inference. Inference is a proxy for relations. Getting things right is a slow, untidy and deeply fallible process where you don't really know what x does before you know what all the other bearers do.

From what we have gathered this far, it may look like we are heading straight into relativism, where beings change according to their relational partnering. At this point, however, the idea of properties being dispositional becomes important. There is a difference between dispositions and manifestations. The bearer does not (necessarily) lose its disposition when manifesting, and a bearer can maintain its identity despite a seemingly 'dual nature'. A famous example would be that light can manifest itself as a wave with certain manifesting partners and as a particle (photon) with others and *still be ('the same') light*.[10] Another example can be drawn from the Human Genome Organisation (HUGO). Research involved in this project, initiated in the 90's, had to conclude (much to their own surprise) in 2000 that humans have fewer genes than rice. The fact that the same gene can code for (as it's wrongfully phrased) many thousands of proteins (i.e. protein isoforms: more than 38,000 in fruit flies. C.f. *Dscam*) does not change the identity of the gene. What explains this surprising result is that the properties are had in relation to a (large) set of partners. If you are an 'intrinsicalist' it will rather look as if the gene changes its *identity* (since this position would entail that its identity is determined by the properties it has independently of everything else).

I will not attempt to draw the possible conclusions that might follow from this view in terms of scientific *method* on this occasion. (My proposition would be to develop the hermeneutic method further). I will only point to the fact that this ontology necessarily puts the subject on the inside of the natural world. We don't have to state or justify that we only come to know the world by being-in-the-world or by bodily habituation. We are a dispositional and manifesting partner like any other, and we cannot choose not to be. We are not looking at the sky. The sky (as it looks) is a manifestation of many partners, and there is no reason to doubt the fact that this is a real manifestation just because we are partners. Even (non-recognized) hallucinations are 'real'. What is wrong with them (hallucinations) is the wrong attribution of the contribution from the different partners involved, i.e. what you see is what you get, but the attribution might be wrong.

This is why the classical belief in *objective* science is wrong. What is wrong with it is that the *stance itself* starts out with a preconceived misattribution of the partners involved.

All scientists are, from my proposed view, metaphorically *in* an ice hockey game playing part, but natural scientists (physics/chemistry/*in vitro* science) think that they are subjects observing objects from the outside, just because (most of) the other players don't move or are made rigidified in an experiment. But rigidified parts are not categorically different just because they don't (seem to) move. They are still dispositional, and they still take part on par with the others in the mutual manifestation. (An experiment is not 'screening off' the other contributors; they are just replaced with others, Lie 2016, 114).

28.4 Trapped in Phenomenology?

Based on the line of reasoning in Chapter 1, it should be obvious why we should not replace one epistemological stance (subject/object) with another (phenomenological) and call it a day. We should rather compare ontologies to see what an alternative to the prevailing ontology looks like for the scientist. There is, however, a common ground between relational realism and a central tenet in (especially later, such as Jean-Luc Marion) phenomenology. From a phenomenological point of view, the phenomena are deemed as inexhaustible, exceeding our grasp, saturating our intentions, etc. A central point for relational realism is that the bearers (objects, events, processes) are 'open'. A bearer *has* a wide and principally indeterminable range of dispositions, some of which are manifested, many of which are not.

Under normal conditions, there tends to be a regular set of dispositions that are manifest, while others are not. The cup that you use for coffee in the morning tends to have the same weight (when lifted), color (when looked at) and solidity (filled with coffee). This regularity is interpreted in many ways, but most importantly, the 'intrinsicalist' will try to pin down which of these properties belong to the cup itself, and which do not. Most often this account ends up with a reductionist understanding of the cup, and we look for microstructure 'responsible for' the properties that tend to manifest (Lie 2016, ch. 2.7). The phenomenologist does not agree with this move: The cup may have other properties, such as functional properties, affectional properties and properties like being beautiful, and Heidegger, especially, would say that when some of the properties are manifested, others are concealed (i.e. truth as *aletheia*). The intrinicalist will disagree and say that those properties are *only* relational (not real).

The theory presented here (relational realism) says that all properties are relational. Weight depends on the gravity of the Earth, and solidity depends on a certain temperature, for instance. If you try to reduce these properties all the way down to quarks and gluons, you will not find non-relational properties at the bottom. Relational realism based on dispositionalism will argue that properties are real as long as they do something, i.e. contribute to certain manifestations. The problem of projections is not a problem of realness but about attributing the right disposition to the right bearer in a relational co-production. Thus, beauty, for instance, is not in the eye of the beholder (solely). There has to be something about bearers that creates a sense of beauty in the beholder. Something needs to have mass for gravity to pull x downwards.

I have argued that the reason we see a fairly regular pattern of manifestations is not due to an intrinsic set of properties, but rather a historically created entrenchment of partners likely to be present. (Lie 2016, ch. 5.2, 5.3 and footnote 9). This entrenchment is what constitutes naturalness. (Hence the title of the book-title: 'Philosophy of Nature: *Rethinking Naturalness*'). What we may observe in a laboratory provides examples of the wider range of dispositions that bearers have. But those are not displaying the real and intrinsic properties of the being in question. Adding and subtracting partners is a deviation. It is neither a construction of truth nor a finding of the truth, but it is a construction of the truth *when* we infer that it is (by default) a finding of the truth. The contrived conditions in the lab give us an answer to the question: do these bearers also have other dispositions not displayed under normal (natural) conditions?

The manifestation we achieve in the lab is a result of a huge set of partners that dynamically contribute to the manifestation. This is why Heidegger's tool analysis is also interesting for the understanding of the situation of science. If the scientist could isolate each of the dispositional contributions, it would *not* be of interest, but since they contribute relationally (dynamically), the scientist needs an understanding from inside. A fairly similar understanding that Heidegger describes in his tool analysis (Heidegger 1993, ch. 3). This is far from being what the scientist *wants*. It's a result of how the world works. The subject/Dasein/scientist *needs* to have a circumspective (*umsichtich*) concern, and there needs to be a dynamic shift between what redraws and what is attended to. What is attended to is attended to on the basis of what is redraws. Singular (isolated) facts do not exist because facts are deeply relational.[11] It is the phenomenon itself that forces us to describe the facts about the world 'in context'. To describe something 'in context' should be attributed to the real relational nature of the phenomena, not our epistemological faculties. If we don't realize this, we get into all sorts of misunderstandings about the relationship between (natural) science and phenomenological experience.

The first misunderstanding is that we accept that science actually doesn't have to step into the muddled water and that phenomenology represents an alternative perspective of the world. Although it is not really clear what Heidegger means by 'present-at-hand,' it is safe to say that Heidegger, in important places, attributes the presence-at-hand to the scientific attitude (e.g. 1993, 54). Thus, he is making the question of relationality into a question of episteme.

The second problem is that, due to this schism, we get two different views on what it means to describe something in itself. The idea of intrinsicness prevails in both cases. *Zu dem sachen selbst* on one hand, and different versions of isolated beings having their properties independently of everything else on the other. 'What is being as such? By 'as such' Heidegger meant the conditions needed for beings 'to be', in the sense of manifesting themselves *in their intelligibility* or *in their availability* to Dasein' (Zimmerman 2003, 76). The impression is that science secures intrinsicness by describing the intrinsic *properties* of beings, while Heidegger insists the essential being of beings can only be disclosed by humans, i.e. humans adopting a particular phenomenological stance.[12] Neither of these positions would guarantee the right description of nature, which should rather be aimed at attributing the right relationally constituted dispositions to the right bearers in a manifestation.

The third problem that this schism creates is a mystification of the phenomenon. First of all, the manifestation is a *joint* manifestation of a large set of *simultaneously* manifesting partners. As previously claimed, we cannot see or observe what is going on directly, although we may infer – after the fact. There is something we see, and there is something that withdraws from conscious experience. Hence, there is *more* to the phenomenon *if* we use the scientific stance (emphasizing transparency, conceptual grasp and certainty) as a measure-stick. But it is only when the phenomenologist compares the description (of the lived experience of the phenomena) to the scientific stance that the experience becomes mystical. Compared with what relational realism says about the phenomena, we may see the epistemic situation we face as obvious – although – extremely difficult. We are always embedded in a situation where the feeling of shortcoming and humility would be the most adequate emotions. At the same time, we can 'see more than we think' through our bodily presence, our pre-conscious receptivity and circumspective concern, as Merleau-Ponty would agree. But what we see is not easily explicable. It is tacit.[13] In Ted Toadevines' reading, this creates a *desire* which is 'a response to the insensible' (2003, 149):

> Granted that *Il y a* [there is!] is not nature as we usually conceive of it: it is neither wilderness nor animals or plants, nor an ecosystem striving for diversification. But that nature is none of these things is precisely the point. The *Il y a* is a "wild" or "brute" being, being as it shows itself in the possible limit experience or which redraws from the opening of the world or of subjectivity. There is an opacity of my body as a desiring being which subjectivity cannot penetrate (...).
>
> (2003)

This realm beyond subjectivity has been described in many ways, even as the numinous (Rodolf Otto). But given the theme of this book, I note that it has also been named 'nature-as-miracle' (Neil Evernden: see Toadevine 2003, 141), and Toadevine goes on to describe it in the following way:

> The basis for responsiveness is in the call of a more radical Outside. Nature in this radical sense is, if anything, the refusal of the hegemony of perception, language and thought; it is the 'pocket of resistance' and the unpredictable par excellence.
>
> (Toadevine 2003, 150)

As most readers would recognize, Heidegger's obsession with being is a version of the same issue, often laid out as the distinction between being and beings.

Above, I called this a mystification. This does not mean that I don't find our existence a mystery and nature unfathomable. The mystification here originates from the epistemological dogma: The fact that we think that we cannot describe the world beyond what is here called subjectivity (Toadvine). To the extent that phenomenology tries to describe the world as it occurs to the subject, being a part among parts (i.e. from within), I believe it is on the right track, but the right track gets sidetracked when what is 'beyond subjectivity' gets mystified. The mystification originates from the subject/object deal: the idea that we should describe the world from the outside to describe the world *as it is in itself*, and that this should be done in a clear and transparent way. But this is, as emphasized many times, not possible when the world 'itself' is relational. As I said above: We are all forced into an ice hockey game where what every part does depends on what the other does, and where you need something close to a mystical ability to 'sense the whole' to know what to do next.

From this outlook, I would say that we are 'trapped in phenomenology' when phenomenology mystifies this relationality as otherness, instead of challenging the ontology of science. As long as phenomenology allows this mystification it will also pray on its counterpart (the shortcoming of 'subjectivity') and allow science to play its own game with isolated parts and prevail in their steady belief in a passivist view on nature (see below) – allowing for science to keep its false promise of transparent and apodictic knowledge, (or what David Lewis called 'a complete physics'). This is an unholy marriage where phenomenologists get to pat themselves on the shoulder for producing a seemingly attractive account that, in the shadow of the dominant role of scientific descriptions, allows us to keep some version of (radical) otherness in our hearts.

The attractiveness of phenomenology is its affection for 'descriptions from within'. This is attractive because the relational ontology will deem 'descriptions from outside', trying to describe the world as a world of distinct existences, as problematic by default. The reason would be that the outside stance already gets the attribution of dispositions wrong. ('We are not looking at the sky'). But this is not a question of perspective or stance, it is a question of the immensely difficult task of getting attributions right. It is a serious task that should not be washed out by finding confident satisfaction in 'saturated phenomena'.

28.5 Eco-Phenomenology

For the eco-phenomenologist, there is an extra dimension to how the scientific ontology puts strings on how nature is described. Eco-phenomenologists often describe how the scientific (or maybe techno-scientific) view of nature plays a deeper role in ecological destruction (e.g. Wood 2003). Thus, as previously conceded, what an eco-phenomenologist wants is to provide positions that give rise to descriptions that may challenge the way science reveals nature, and to warrant these descriptions as viable and important.

From this outlook, we should consider how the idea of intrinsicness (spurring the epistemological dogma) also plays a role in what can be referred to as a passivist view on nature (see, e.g. Ellis 2002, 2; Lie 2016, ch. 2.6. and 2.8). If the parts of the world are what they are *independently of everything else*, we may go with Aristotelian teleology (which emphasizes the independence of substances) and argue that every part has an inner urge towards actualizing an inner goal. But if Aristotelian teleology is not tenable anymore, what would be the common denominator of the parts inhabiting the world? This was the question raised in the wake of the scientific revolution by the philosopher that wanted to put science on a firm ground 3–400 years ago. The result was a passivist ontology of nature.

A passivist ontology is an ontology where the parts are seen as passive. The ontology can be construed in different ways. Lockean essentialism would argue that the being in question is doing what it does in virtue of its minute parts (c.f. the distinction between nominal and real essences), (Locke 1995 (1690), 212, 356). Hume argues that any being is a contingent arrangement of distinct properties (Hume 1978 (1739–1740), 173). The most common view would be to say the behavior of the being in question is determined by *external* forces. This is the view of law governance, where laws are external to the being in question.[14] Passivism and 'modern' intrinsicalism go hand in hand with the scientific enterprise of apodictic descriptions of an external world of fixed, quantifiable facts. *If* parts are active, however, they would behave differently depending on the activity of the researcher, for instance.[15] This is precisely why social sciences, humanities and some parts of biology (e.g., ethology) are deemed 'soft sciences,' and why the double-slit experiment caused such a shock within the physics community in the early 20th century (Heisenberg 1959). A world of active beings (a dispositional world) represents a problem for passivist ontologies as well as those who think that the world is controllable and at our disposal (what Heidegger dubs *Ge-stell*).[16]

If we accept that phenomenology and its relative attractiveness are a result of the epistemological dogma and that this dogma arises from the idea of intrinsicness, it should also be clear why eco-phenomenology is not providing the best position to challenge the dominant passivist understanding of nature that science generally holds.[17] This is why I think that eco-phenomenology should move in a direction where the importance of our stance is not left behind, but is supplemented by the curiosity (described by Toadevine) towards the immensely difficult task of correct ascriptions. If our method and philosophical outlook actually do have ontological precepts, we cannot just deny the role of ontology on the basis that it would be naïve or somewhat forbidden to enter into such questions (which, by the way, is a similar strategy that logical positivists had). Of course, we will never reach a point where a single, definitive ontology is universally agreed upon. But this does not mean that our philosophical orientation cannot rest on a mistaken ontology – or that a more compelling alternative might not be available.

Notes

1 A typical reference would be Berger and Luckmann (1966) who acknowledges phenomenology as one of their sources of inspiration. (Also, Alfred Schutz, a student of Heidegger, was Bergers supervisor.).

2 This is referred to as the 'philosophy of access' by Graham Harman. See, for example Harman (2013). With the qualifications given here (i.e. its relation to intrinsicness) I call it 'the epistemological dogma', and I will continue to use this phrase onwards (see e.g. Lie 2016, ch. 1.3.1; 2021).

3 Some 'realists' would then typically argue that this embodied *in-der-welt-sein* subject excludes the possibility of 'objective knowledge' and label e.g. Heidegger 'a forerunner to social constructivism'. Heidegger or Heideggerians would then say that this is a misunderstanding, but we need to ask ourselves why is this a debate at all, if the misunderstanding is so obvious? My answer would be: It's a debate because Heidegger never leaves the epistemological dogma in the first place. Heidegger identifies the problem as problem of how we relate to the world. To get out of the epistemological dogma we need to pose the problem of how we relate to the world as a problem that initially occurred as a result of how reality was described in the first place.

4 I consider Graham Harman to be a relationalist although his ontology is 'object oriented'. See his notions of 'overmining' and 'undermining', e.g. in The Quadruple Object 2011.

5 The reader may consult Lie (2016, ch. 3, 4 and 5) (ch. 4.6 in particular) and Lie forthcoming to get a fuller insight into what I call relational realism.

6 In truth I must admit that most relationalists (who are not so many) argue that *manifestations* are relational, while the *dispositions* are intrinsic. In Lie (2016) I argue that also the dispositions are relational (ch. 4.5), and I call this position relational realism. (ch. 4.6). The worry with such an account is that the relata disappears in the relation. I seek to solve this problem by substituting, to keep it short, (what you gain from) intrinsicness for *manifestations* on lower levels (ch. 3.6 and 4.6).

7 That dispositions are 'relational in nature' does not mean that they *are* relations. 'Relational in nature' means that they are not 'intrinsic in nature'. They are properties which means that they adhere to beings, (or *bearers* as the dispositionalist would call it).

8 Within a causal framework (causa efficiens) we say that e.g. a causes b. In this dispositional framework the effect is a manifestation, but it is (always) a manifestation of several mutually manifesting partners. This is why I call it a whole. Importantly however the manifestation instantly gains new dispositions (that manifests in different ways), depending on the partners on the higher level that the manifestation is taking place. Hydrogen and oxygen (+ other partners) come together in a way that

manifests something we call water, but water *is* not H2O since 'water' gains dispositions that are not computable from the parts that made the manifestation. This is my main argument against reductionism. The 'manifestation' is a hyleo-morphic (Aristotle) unity of the manifesting aspect and the disposition aspect. See Lie (2016, figure 4.3 and ch. 4.6).

9 If you are an essentialist or intrinsicalist you would assume that the bearer (with necessity) only contributes with a confined set of essential properties. One way to explain why essentialism looks viable is the idea that during the historical development of the universe, there are certain manifestation-habits that has become rigidified. When two or more partners come together to manifest something, they may continue to stay together (often called objects), creating a new bearer that has further (and different) manifestations on a higher level. If we think about our current universe in this way, it follows that lower levels are more fixed than higher levels (Lie 2016, ch. 4.6 and 5.2). The fact that something is fixed in this way does not mean that bearers do actually have essences however, which is evidenced by the new dispositions we discover under contrived conditions (i.e. laboratory research) (Lie 2016, ch. 5.4).

10 According to quantum field theory it is *the field* that is manifesting as both wave and particle, and which maintains its identity.

11 In Heidegger's analysis, zeug (tools) is also 'deeply relational', as many would recognize, but he reserves this relationality for 'useful things' which eventually refers to a goal (*werk*), c.f. telos. 'Nature' turns up to the degree that it is useful. In his later works he changes his view somewhat (e.g. *'Das Ding'*), but not in a direction similar to relational realism.

12 Which could be in the form of eidetic reduction (Husserl), circumspective concern (Heidegger) or just simply being receptive or passive, which might be a more contemporary version of the stance (e.g. Jean-Luc Marion: 'saturated phenomena').

13 Which was also something that Thomas Kuhn, following Michael Polanyi emphasized in his 'Structure of Scientific Discovery'.

14 If the laws of the universe are contingent, they would also be external (Mumford 2004, 147). I have argued elsewhere that passivism is what makes it hard to for environmental ethics to attribute intrinsic value to non-human beings. See Lie (2016, ch. 1.4.2).

15 Historically, when we left what I call the whole-limiting cosmology (Lie forthcoming) in the 1700'th century to harvest the seeds of nominalism it became very difficult to explain what parts (that are not a part of an ordered whole) really are. At that point in time one could choose to insist on the intrinsic nature of the parts (in line with Aristotle) that we need to access from outside, or one could choose a relational account where we need to understand the part as a part of a relational web, which looks awfully difficult and which, further down the line, carries little promise of a ordered scientific enterprise that can assure apodictic knowledge for further returns.

16 One would assume that *enactivism* would fit nicely to the account presented here since the relational realist account is emphasizing potency and the relational activity of the beings in question, but is this really the case? In one sense it seems to fit the idea of a manifestation being the result of all partners involved and that this manifestation is as real as any other. On the other hand, enactivism does not come with an ontology. Enactivism is absorbed in being an alternative to representationalism and the computational model of the mind. When this is the case I can only repeat the criticism presented throughout this chapter: If enactivism does not challenge representationalism on an ontological level, *it will only confirm the idea of intrinsicality by accepting that the best we can do is to challenge representationalism.*

In the context of eco-phenomenology it does not make sense to challenge the passivist ontology by just showing that reality is enacted by the organism in its environment. There is no 'outside' to this argument. Intrinsicalism says that reality is out there in terms of a pre-given reality of beings that have intrinsic properties. To counter *this* argument, you need to have an idea of a world outside enactment, a view of reality that shows why the enactivist position is offering the best epistemic stance. You need an ontology to support the idea that reality is not *found* out there. An example of such an ontology would be a dis-positional ontology ('without intrinsicality'). To establish the viability of such an ontology one has to go beyond the epistemological dogma.

17 A pertinent example would again be Wood (2003), who strictly speaking tries to describe an alternative to what he calls naturalism through epistemological arguments. I have given a more substantiated critique of this strategy in Lie (2021). As far as I can see, Kūle (2018) represents a different attempt to overcome this problem which seems to be more in line with the thinking presented here.

References

Berger, Peter & Luckmann, Thomas. 1966. *The Social Construction of Reality*: Anchor Books.

Ellis, Brian. 2002. *The Philosophy of Nature: The Guide to the New Essentialism*: Acumen Publishing.

Harman, Graham. 2011. *The Quadruple Object*: Zero Books

Harman, Graham. 2013. An Outline of Object-Oriented Philosophy. *Science Progress*, 96(2), 187–199.

Heidegger M. 1993 [1927]. *Sein und Zeit*: Max Niemeyer Verlag. Tubingen.

Heisenberg, Werner. 1971 [1959]. *Physics and Philosophy: The Revolution in Modern Science*. (Ed. R.N. Anshen), 3rd ed., Ruskin House.

Hume, David. 1978 [1739–1740]. *A Treatise on Human Nature* (Ed. L.A. Selby-Bigge & P.H. Nidditch): Oxford University Press.

Kūle, Maija. 2018. "Eco-Phenomenology: Philosophical Sources and Main Concepts." In *Eco-Phenomenology: Life, Human Life, Post-Human Life in the Harmony of the Cosmos*, edited by A.-T. Tymieniecka and Maija Kūle, Analecta Husserliana, Vol. CXXI. Dordrecht: Springer.

Lie, Svein Anders Noer. 2016. *Philosophy of Nature: Rethinking Naturalness*: Routledge.

Lie, Svein Anders Noer. 2021. A Critique of Steven Vogel's Social Constructionist Attempt to Overcome the Human/Nature Dichotomy. *Environmental Values*, 30(5), 593–612.

Locke, John. 1995 [1690]. *An Essay Concerning Human Understanding*: Prometheus Books.

Martin, Charlie. 2007. *The Mind in Nature*: Oxford University Press.

Mumford, Stephen. 1998. *Dispositions*: Oxford University Press.

Mumford, Stephen. 2004. *Laws in Nature*: Routledge.

Taylor, Charles. 1995. "Overcoming Epistemology." In *Philosophical Arguments*, 1–19. Cambridge, MA: Harvard University Press.

Toadevine, Ted. 2003. The Primacy of Desire and its Ecological Consequences. In *Ecophenomenology: Back to the Earth Itself* (Ed. Brown, C.S. & Toadvine, T.): State University of New York Press.

Vogel, Steven. 2015. *Thinking like a Mall: Environmental Philosophy after the End of Nature*. MIT Press.

Wood, David. 2003. What is Ecophenomenology? In *Ecophenomenology: Back to the Earth Itself* (Ed. Brown, C.S. & Toadvine, T.). State University of New York Press.

Zimmerman, Michael. 2003. Heidegger's Phenomenology and Contemporary Environmentalism In *Ecophenomenology: Back to the Earth Itself* (Ed. Brown, C.S. & Toadvine, T.). State University of New York Press.

29

MINDFULNESS AND ECO-PHENOMENOLOGY

Similarities, Differences and Critical Potentials

Jonas Jakobsen and Ida Solhaug

Introduction

Over the past few decades, mindfulness-based interventions have experienced a remarkable rise in influence, transcending their origins in clinical psychology to become a pervasive force across various fields, including psychological research, education, business and leadership, health care and popular culture. At the heart of this widespread adoption is the growing recognition of mindfulness as a powerful tool for reducing stress and emotional reactivity and promoting overall well-being across a variety of populations (Goldberg, Riordan, Sun & Davidson 2022; Goyal et al. 2014; Keng, Smoski & Robins 2011; Kohury et al. 2013). Mindfulness training has also been recognized as a potential means for fostering pro-environmental behavior through its influence on factors such as environmental attitudes and nature-connectedness, eudaimonic well-being, the alignment of values with behavior and the cultivation of compassion (Ericson, Kjønstad & Barstad 2014; Thiermann & Sheate 2021; Wamsler & Brink 2018). Mindfulness practices have been adopted by parts of the environmental movement and found their way into the so-called Inner Development Goals initiative, which aims at accelerating the UN's Global Sustainability Goals by way of changing human attitudes and relations to nature "from within".

A growing body of research focuses on the relationship between phenomenology and mindfulness (Bitbol 2019; Depraz 2019; Ferrarello & Hadjioannou 2024; Stone and Zahavi 2021; van den Bold 2025), yet the connection between mindfulness and *eco*-phenomenology remains underexplored. This raises important questions: What concerns do mindfulness practice and eco-phenomenology share, and how do they diverge? What critical insights do they offer in addressing the unsustainable thinking and behaviors promoted by capitalist consumer culture? This article argues that mindfulness and eco-phenomenology share essential concerns: epistemic, ontological and ethical. What distinguishes mindfulness practice from eco-phenomenology is its practically embodied approach to these three concerns and its emancipatory goal: to reduce suffering among humans and ultimately all.[1] The final section argues that mindfulness practice aligns well with a critical eco-phenomenology that interrogates the human–nature relationship as it appears in our contemporary lifeworld, colonized as it is by capitalism, consumerism and instrumental rationality. First, however, we introduce

DOI: 10.4324/9781003538097-36

mindfulness practice and its underlying premises and position our approach within the "mindfulness-Husserl debate".

Mindfulness

Mindfulness has been described as a psychological trait, a temporary state of awareness, a practice, a process, an outcome, a way of living and even a placeholder for the core Buddhist teachings on suffering and its alleviation, or the "dharma" (Vago & Silbersweig 2012; Williams & Kabat-Zinn 2011). The term mindfulness is the English equivalent of the Pali word "sati", which implies awareness, attention and remembering (Germer 2005). In conceptualizing mindfulness, some Buddhist scholars emphasize specific aspects of attention, while others highlight the dynamic interplay of various factors, such as cognitive, emotional, social and ethical (Grossman & Van Damm 2011). Despite great diversity in the Buddhist tradition regarding its philosophies, meditation techniques as well as the understanding of mindfulness, Buddhist scholars commonly view mindfulness as a practice or process implying (a) deliberate awareness of moment-to-moment experience; (b) qualities such as acceptance, kindness, tolerance and non-judgement; (c) non-discursive, non-analytical investigation of present, ongoing experience; and (d) the need for systematic practice to gradually refine the mindful process (Grossman & Van Dam 2011).

Many authors have attempted to extract the psychological aspects of Buddhist teachings into a secular "Buddhist psychological theory", and aspects of this theory inform many of the contemporary mindfulness-based interventions (Epstein 1995; Hayes et al. 2006a; Williams & Kabat-Zinn 2011; Teasdale & Chaskalson 2011). According to this extracted Buddhist psychology, much human suffering is due to the habitual tendency to want things to be different. Humans habitually crave for or cling to some experiences and attempt to push others away, resisting the fact that all experiences change continuously. Further, Buddhist psychology posits that much human suffering comes from the "delusion" of being overly attached to a self-centered perspective, a subtle obsession with our personal stories about who we are and ought to be, and constant judgment of ourselves and others (Solhaug 2016).

The classic method for cultivating mindfulness generally progresses from the development of attention and awareness (through concentration or *samatha* meditation) to practices aimed at fostering experiential inquiry and understanding (insight or *vipassanā* meditation). Thus, mindfulness is understood as a compound of awareness and "clear comprehension", where the aim is not simply to know what is happening in the present moment, but to know it in a certain way; "without grasping, aversion, or delusion` (Salzberg 2011, p. 177). Mindfulness is thus considered an antidote to resistance to change, allowing one to "hold" fluctuating experiences in awareness without reactivity. Repetition of this practice is thought to invite the following insights: (a) everything is impermanent; (b) suffering arises from greed/clinging, hatred/aversion and ignorance/delusion; and (c) everything is connected, nothing is separate (Teasdale & Chaskalson 2011). With mindfulness training, one is encouraged to step back and observe one's internal narratives at a distance, so to speak. Insight in this tradition is seen as the gradual recognition of maladaptive clinging to "me" and "mine", fostering a clearer, more profound seeing of the world and all that living beings have in common.[2] According to Buddhist psychology, such embodied insights promote "lovingkindness", which implies an awareness of the interconnectedness and interdependence of all beings. Ultimately, these insights are thought to alleviate suffering and promote ethical conduct toward all forms of life.

Whereas mindfulness in the Buddhist tradition is viewed as a component of a larger system of ethics and is practiced against the backdrop of contemplating key aspects of the Buddha's teachings, Western contemporary conceptualizations of mindfulness are largely independent of any specific philosophy or ethical code. Nevertheless, the above-mentioned notions of suffering, interconnectedness and compassion/lovingkindness have become widespread in the pedagogy of mindfulness, notably so in the widely implemented Mindfulness-Based Stress Reduction program developed by Jon Kabat-Zinn (Kabat-Zinn, 1982, 2005).

Despite diversity in the understanding of mindfulness and its intention across contemporary mindfulness-based interventions, mindfulness is typically defined as consisting of two components: attention to present experience from moment to moment (a) with an attitude of acceptance and non-judgment (b) (Bishop et al. 2004). These components are primarily trained through formal meditation practice, but also informally in daily life and interaction with others. For example, dedicating full awareness to the taste and feel of food while eating, or taking a minute to notice the flowers on your way to work, can be understood as elements of a non-formalized mindfulness practice.

In formal meditation practice, mindfulness is cultivated more systematically and structured. The practice includes a series of characteristic experiential states, which evolve in complexity over the course of training. Initial stages typically involve four key components: (1) selecting and sustaining attention on a particular object without judgment or conceptual elaboration, (2) maintaining alertness to potential distractions, (3) non-judgmentally noticing and disengaging from distractions when they arise and (4) reorienting attention back to the object of focus. In the early phases of practice, the objects of attention tend to be simple and concrete, such as the breath, bodily sensations, or sounds. As practitioners advance, the focus may shift to more complex objects, including discursive thoughts. In such cases, the practitioner is encouraged to observe thoughts as they arise and dissipate, perceiving them as "events" in the mind, without elaboration: "if possible not being drawn into thinking about them, or thinking the next thought, but just maintaining the 'frame' through which you are observing the process of thought" (Kabat-Zinn 2005, 73).

Once foundational mindfulness skills – such as sustained attention, awareness of distractions and reorientation – are developed, practitioners are often introduced to a form of "objectless" meditation, typically referred to as "choiceless awareness". This practice involves relinquishing the instruction to focus on a specific object and instead adopting an open, receptive state of awareness: "Don't hold on to anything, don't look for anything. Practice being completely open and receptive to whatever comes into the field of awareness, letting it all come and go, watching, witnessing in stillness" (Kabat-Zinn 2005, 74).

In accordance with Buddhist psychology, mindfulness-based approaches often highlight the role of discursive language practices in human suffering, focusing on how linguistic constructions and emotionally charged thought patterns shape and limit our perception of self, others and the world. Verbal constructions create symbolic divisions between "me" and "others", and the world is framed as a collection of "things" and "events" that are either "good" or "bad" in relation to this "self". Further, human beings tend to ruminate about the future and the past, thus missing out on the experiential richness, texture and detail of the present moment. Clinging mentally to negative experiences in the past (say, failures or insults) can prevent us from appreciating the beauty of the here and now, but so can clinging to past successes and positive experiences because they turn the present moment into a disappointment in comparison. Likewise, the human mind tends to be intensely preoccupied with the future and how it could be better than the present moment. This future-centeredness includes desires for products to consume, experiences to have, things to accumulate and people to impress. Mindfulness practice aims to "de-fuse" or decenter from the thinking process (Hayes, Strosahl & Wilson 2003); not to escape thoughts but to reduce their habitual dominance over us. While mindfulness practice is not opposed to acting on behalf of a better future, it asks us to temporarily suspend our fears and desires about the past and the future, turning curiously towards present experience without evaluating, analyzing, or trying to control it – also when present experience feels unpleasant, boring, painful, etc.

Thus, contrary to the tendencies of the wellness industry to focus on *avoiding the unpleasant* (Cederström & Spicer 2015), mindfulness practice *turns towards* it and embraces it. Mindfulness practices thus involve a dialectical, non-instrumental structure that differs from the causal thinking style central to modern medicine and psychology (Sauer, Lynch, Walach & Kohls 2011). Rather than striving for symptom alleviation, mindfulness is conceived of as a way of being in relation to all experience. One is invited to turn towards aversive feelings, sensations and thoughts, and befriend them as a common human experience, rather than "symptoms" to get rid of (Hayes et al. 2003; McCown 2013). However, because of such a change in the relationship with "unwanted" inner experience, its impact may change passively. For instance, physical pain could gradually be perceived as a fluctuating sensation when secondary cognitive-emotional reactions to pain are not fueled but held patiently within one's awareness. Similarly, when afflictive emotions and thoughts are allowed to fluctuate, with reduced attempts at control, modification, or avoidance, they could take on a less threatening and defining character. Mindfulness practices, therefore, combine active and passive components in a complementary, dialectic manner; for instance, actively and diligently attending to what happens, while simultaneously resisting reaction to unwanted psychological experience. A similar dialectic regards defining features such as acceptance and non-judgement. Rather than implying uncritical acceptance of all situations or experiences or refraining from all evaluative thoughts, mindfulness invites a suspension or attenuation of evaluative cognitive or emotional reactions in a given situation (Sauer, Lynch, Walach & Kohls 2011); a non-judgement of habitual judgement or acceptance of the struggle to accept current experience. This process has been referred to as a "second order", rather than "first order", change process (Hayes 2004) and inherently involves paradox (Solhaug et al. 2016).

Crucially, while formal mindfulness practice is emphasized in training, the aim is for mindfulness to be integrated into daily life, enriching ordinary experience with the distinctive qualities of attention and heart cultivated through practice (Kabat-Zinn 2005). Living our ordinary lives, we must analyze situations, evaluate claims, make choices and take action. The point of mindfulness is not to disqualify these actions and abilities, but to qualify them: regular withdrawal *from* life's activities can increase the ability to navigate and make conscious choices *in* life. Grossman (2015) thus defines mindfulness as *awareness informed by embodied ethics*, highlighting the attitudinal elements of mindfulness, such as openness, acceptance, curiosity, compassion, patience and even courage. Rather than remaining abstract virtuous ideals, mindfulness practice may serve as a tangible invitation to cultivate these virtuous attitudinal qualities in relation to immediate experience and explore the influence of their absence or presence on experience at any given moment.

The Husserl-Mindfulness Debate

Readers familiar with phenomenological philosophy will already have noticed the most obvious overlap between mindfulness and phenomenology, most notably the focus on first-person, subjective experience and the claim that unique insights can be gained from turning systematically towards experience as it appears in consciousness from moment to moment (Depraz 2019; Ferrarello 2024; Lundh 2020). Indeed, the relationship between mindfulness and phenomenology is a new and growing academic research field, as witnessed by the recent publication of *The Routledge Handbook of Phenomenology of Mindfulness* (Ferrarello & Hadjioannou 2024).

Like "mindfulness", "phenomenology" is a contested and multifaceted concept. Also, phenomenological philosophy includes a broad variety of authors, such as Edmund Husserl, Maurice Merly Merleau-Ponty, Jean-Paul Sartre, Simone de Beauvoir and Martin Heidegger. While comparisons between mindfulness and phenomenological thinkers such as Heidegger (i.e. Hadjioannou 2024) and Merleau-Ponty (i.e. van den Bold 2025) exist, most of those who compare mindfulness and phenomenology refer to Husserl's notion of the "epoché" as a method of suspending pre-given theoretical beliefs and habitual perceptions in order to return in a direct way to experiencing "things in themselves". In this section, we shall therefore focus on potential overlaps between mindfulness meditation and Husserl's epoché, leaving out other potentially relevant discussions and comparisons. Given that Husserl's philosophy is not our field of expertise, we will not make any judgments about what the "correct" interpretation of the epoché is. Rather, we discuss how some Husserl-scholars interpret this aspect of his work and what that means for the relationship between mindfulness and phenomenology.

There is a long list of psychologists, philosophers and Buddhist scholars who emphasize the overlaps, common presuppositions and shared goals of mindfulness and phenomenological theory, or who argue that the phenomenological method (or at least the epoché) is a kind of meditative practice (Bitbol 2019; Depraz 2019; Ferrarello and Hadjioannou 2024). For example, Depraz (2019) compares the practice of samatha-vipassanā sitting meditation (which overlaps with mindfulness meditation) to Husserl's epoché, emphasizing that both methods involve a suspension of prejudices and natural attachments regarding the world. The phenomenologist John Cogan (2025) describes Husser´s epoché as a "radical, rigorous, and transformative meditative technique". Susi Ferrarello contends that "what we want to achieve both in mindfulness and with Husserl's phenomenology is a state of 'bare attention'" (2024, 82). This state, she argues, allows us to abandon "mental constructions that prevent us from seeing reality as it is while hiding behind theories, interpretations and habitual patterns" (2024, 82). It also helps us to overcome our tendency to "see everything from the point of view of the ego and personality, rather than seeing things as they are" (2024, 82).

One of the most influential accounts of the mutual convergence between phenomenology and mindfulness is given in *The Embodied Mind* by Varela, Thompson and Rosch (1991). According to them, mindfulness and Husserlian phenomenology are united around the same goal, namely, to systematically and rigorously investigate first-person lived experience. However, they also argue that Husserl and later phenomenologists such as Heidegger and Merleau-Ponty got lost in abstract theorizing and failed to deliver on the promise of a more embodied philosophy that takes subjective experience seriously.[3] Notably, while Husserl has a lot to say about how to bracket the natural attitude through different steps, his approach remains reflective, that is, it reflects *on* experience, whereas Valera et al regard reflection as integral to experience, as it comes to pass in meditation (1991). In a similar vein, Depraz argues that Husserl's reflections on the epoché are "astonishingly theoretical and general" (Depraz 2003, 185). Chu, finally, states that Husserl seems to lack any interest in working out the "corporeal and spiritual techniques" required for carrying out his project (Chu, 2024, 40).

For these scholars, then, Buddhist meditation techniques and mindfulness practice in particular are not just compatible with Husserl's epoché but *necessary* in order to actually perform it.

Other scholars push back against the alleged happy marriage of (Husserlian) phenomenology and mindfulness. Stone and Zahavi (2021), notably, point to ambiguities and inconsistencies in the theoretical landscape associated with mindfulness as well as phenomenology and argue that existing comparisons (such as those mentioned above) suffer from various confusions. We agree with Zahavi and Stone that it would be a mistake to reduce phenomenology as a philosophical enterprise to mindful observation of subjective experience, or to expect philosophers to practice meditation in order to be good phenomenologists. We also agree that there are ambiguities in the theoretical landscape associated with mindfulness and welcome their discussion of these ambiguities. Importantly, we also agree that *if* their interpretation of the epoché is correct, then it has little to do with mindfulness practice. On their account, the whole point of epoché is to suspend the "natural attitude", that is, the assumption that "the world we encounter in experience exists independently of us" (Stone and Zahavi 2021, 171). In other words, we should examine how the world appears to us without assuming that it exists as an objective reality apart from our experience of it.

Stone and Zahavi are correct that this kind of highly abstract and counterintuitive theoretical suspension should not be confused with mindfulness practice. Mindfulness practitioners can safely assume that chairs and trees exist also when no one observes them. Mindfulness practice may lead to the experience that phenomena exist "empty of *independent* existence" (Kuiji 2001), as part of an interconnected web of conditions. However, the external world is assumed to exist beyond human perception. Furthermore, while mindfulness practitioners do pay attention to how and as what the world reveals itself in experience, it is too reductionistic to speak with Zahavi and Stone about "worldly objects" (2021, 172), as if the epoché is all about the experience of static and coherent physical entities. My emotional reactions when communicating with a difficult person, or the slowly changing colors of a sunset, are not "objects" like stones or houses, but they are certainly phenomena or processes worth paying attention to. Finally, as we shall elaborate further below, mindfulness practice includes an emancipatory dimension that is absent in Zahavi and Stone (2021) purely epistemic presentation of the epoché. While mindfulness does share the epistemic commitment to consult subjective experience without preconceived narratives and ideas, the goal is not to gain a correct theoretical understanding of worldly objects, but to reduce suffering. Thus, to the extent that there are overlaps between Husserl's epoché and mindfulness practice, it is the broader interpretations introduced in the beginning of this section that are relevant: those that emphasize a common interest in first person experience, and, in particular, how such experience tends to be limited by unexamined prejudices, emotionally charged mind-wandering, and habitual thought patterns and the like. The following section explores how this common interest is pursued in eco-phenomenology and so-called nature-based or "eco-mindfulness", respectively.

Mindfulness and Eco-Phenomenology

A longer story could be written about the theoretical development from Husserl to Heidegger, Merleau-Ponty and onwards, and its implications for what is today recognized as "eco-phenomenology": an interdisciplinary approach that draws on the philosophical tradition of phenomenology and combines it with ecological and environmental concerns. Whereas traditional phenomenology tends to focus on the subjective structures of experience in relation to objects, eco-phenomenology emphasizes the interconnectedness of humans and world, viewing perception not as a detached, internal activity but as something inherently embodied and situated within ecosystems. Eco-phenomenology, in other words, explores the relationship between human consciousness, lived experience and the non-human world, particularly the natural environment (Brown & Toadvine 2003; Booth 2025; Wood 2001).

We cannot go into depth with any particular eco-phenomenological thinker or position here. Instead, we present three common concerns shared by eco-phenomenology (understood in the broad sense indicated above) and mindfulness practice, respectively, namely the *epistemic*, *ontological* and *ethical* concerns. We then argue that the unique contribution of mindfulness practice lies in its *practical-embodied approach* to these three concerns, as well as its claim that human suffering tends to distort all three of them.

We have already addressed the epistemic concern shared by mindfulness and at least one possible interpretation of Husserl's epoché, namely the emphasis on direct first-person experience and the claim that such experience tends to be "clouded" by prejudices and mental presuppositions. In eco-phenomenology, this general concern is then applied to presuppositions that are seen as detrimental to nature and the environment:

> Eco-phenomenology's guiding idea, put simply, is that uprooting and replicating some of our deeply-entrenched but environmentally-destructive ethical and metaphysical presuppositions can help us heal the earth, combating environmental devastation at its conceptual roots, as it where.
>
> (Thompson 2004, 381)

This quote demonstrates how the epistemic concern with overcoming flawed mental presuppositions is intimately intertwined with an ethical concern: our duty to heal and protect the environment and the planet on which we live. As Brown and Toadvine state (2003, xi), "our conviction that nature has value, that it deserves or demands a certain proper treatment from us, must have its roots in an *experience* of nature". In other words, being able to experience nature without the influence of inadequate concepts and presuppositions is the foundation not only for an enhanced *understanding* of nature, but also for an ethically desirable change in conduct.

In mindfulness-based approaches, there is a similar interest in the relationship between experiencing nature (the epistemic concern) and treating it properly (the ethical concern). A theoretical link is established between mindful experience *of nature*, and *in nature*, at times referred to as "eco-mindfulness" (Mitchell et al. 2016) or "nature-based mindfulness" (Djernis et al. 2019), and sustainable attitudes and behaviors. In environmental psychology, individuals' feelings of *connectedness to nature* have been found to consistently predict pro-environmental behavior (Whitburn, Linklater & Abrahamse 2020). As mindfulness practice aims to dismantle mental distractions and self-centered ruminations, it can help practitioners to appreciate the sensory richness of the natural world more fully and intensely. Empirical studies support the link between mindfulness meditation (or disposition) and nature connectedness, which may be reciprocal (Djernis et al. 2019; Schutte & Malouffe 2018; Travis et al. 2020). A study found that mindfulness was linked to better psychological well-being, partly because it encouraged a sense of *selflessness*. This selflessness reflected one´s feeling of being interconnected with both the social and natural worlds, as well as the universe (Hanley et al. 2017). Thus, feelings of connectedness can reach beyond the natural world, encompassing a sense of unity with the universe and all living beings.

Overall, reviews and meta-analytic inquiries, such as those by Ericson, Kjønstad and Barstad (2014), Wamsler and Brink (2018) and Thiermann and Sheate (2021), indicate a positive relationship between mindfulness and pro-environmental behavior; yet, the intricate question of causality remains a largely uncharted territory, inviting deeper investigation. One discussion around causality goes along the lens of habits – deeply ingrained patterns of action that often serve as barriers to enduring transformation in personal conduct, particularly when propelled by unattended emotional impulses (Kollmuss & Agyeman, 2002). Mindful awareness offers a means to transcend the automaticity of our actions. Empirical findings give support to this claim, indicating that mindfulness practice enhances our capacity to react to emotions more consciously and adaptively (Hölzel et al., 2011; Tang, Hölzel & Posner 2015) and is linked to deliberate, values-driven behavior and eudaimonic well-being (Garland et al. 2015; Ryan & Deci, 2008).

We can therefore say that the epistemic concern of mindfulness practice is broader than that of eco-phenomenology: while the latter focuses on the "uprooting" of generalized metaphysical and ethical presuppositions, the former also focuses on mental stories and psychological presuppositions about ourselves, as individual agents, to the extent that these get in the way of pro-environmental behavior.

Finally, in mindfulness-based as well as eco-phenomenological approaches, the epistemic and ethical concerns (i.e. how we *experience* the world and how we *act*) rely on a deeper ontological claim about the human–nature relationship, namely that this relationship should not be conceived in terms of metaphysical dualism or separateness, but holistically, as mutually interdependent. It reflects a movement away from seeing the environment as a separate entity to be managed or controlled, towards experiencing it as an integral part of our being. Thus, both eco-phenomenology and mindfulness reject Cartesian dualism regarding subject and object, mind and body, human and nature (Kabat-Zinn 2005). In mindfulness, emphasizing awareness of the body and its sensations, often through practices like breath observation, body scans, or mindful movement, the mind and body are integrated and treated as inseparable aspects of experience. By fostering an intimate awareness of our sensory experiences and cognitive-emotional responses, mindfulness helps dismantle the habitual barriers that often disconnect us from the natural world. In eco-mindfulness, this awareness opens to "the subjective feeling of communion with nature as well as seeing oneself as a part of nature" (Richter & Hunecke 2022, 2); the felt experience of interconnectedness of all beings, in the natural world.

However, one thing that distinguishes mindfulness practice from eco-phenomenology is its *practical-embodied approach* to the three mentioned concerns (epistemic, ethical and ontological). In contemporary mindfulness-based interventions, these concerns are tied to concrete, bodily practices and exercises that require no familiarity with notoriously difficult authors such as Husserl, Merleau-Ponty or Heidegger. Instructions and explanations are generally kept in a simple, accessible language that can be shared across cultural, social and intellectual divides, and the focus is practical rather than theoretical. Eco-phenomenology, by contrast, mostly presents itself as a highly academic enterprise that requires mastery of technical and specialized theoretical vocabularies and traditions. According to Brown and Toadvine,

> Eco-phenomenology is based on a double claim: first, that an adequate account of our ecological situation requires the methods and insights of phenomenology; and, second, that phenomenology (…) becomes a philosophical ecology, that is, a study of the interrelationship between organism and world.
>
> (2003, xii)

Mindfulness practice can also study the relationship between the organism and the world, but its method is not philosophical reflection and theorizing. Mindfulness is not a philosophy *about* organism and world, as it were, but an embodied experience *of* the human organism *in* the world.

In addition, mindfulness practice offers an opportunity to look more closely at how human *suffering* tends to distort the epistemic, ethical and ontological concerns discussed above and, correspondingly, how we can emancipate ourselves from suffering (not completely, but gradually and partly). This emancipatory aim seems absent in the literature on eco-phenomenology. For example, in mindfulness practice, we are invited to pay attention to how reactive emotions, stress or anxiety tend to create a sense of ontological separation and a corresponding forgetting of our interconnectedness with others and the natural world, potentially leading to dualistic thinking and attitudes (*me* against *you*, us against them, human versus nature). On this account, relieving ourselves from reactivity is therefore crucial not just for our personal well-being but also for our ontological presuppositions and beliefs. The same goes for the epistemic and ethical concerns. It is typically not when we suffer the most that we are most receptive to the aesthetic qualities of nature or make the most climate-friendly decisions.

Mindfulness, Critical Eco-Phenomenology and Contemporary Consumer Culture

The final part of the article addresses a typical criticism of mindfulness and provides a response. The criticism is that mindfulness is an individualized and apolitical practice that takes our focus away from the structural conditions that produce suffering, injustice and environmental disaster today, and directs it towards our inner life and personal well-being instead. We agree that a subjective focus on personal experience should be combined with a critical focus on the broader lifeworld in which such experience takes place. However, we also argue that mindfulness practice can be understood as a form of existential resistance in a lifeworld colonized by capitalism, individualism and instrumental rationality. Thus understood, mindfulness practice aligns well with a *critical eco-phenomenology* that focuses on the structural and systemic roots of the ecological and climate crisis.

Given its seemingly apolitical and individualistic form, some critics and commentators are extremely skeptical of mindfulness meditation and regard it as part of the problem with capitalist modernity, not part of the solution. In his *McMindfulness: How Mindfulness Became the New Capitalist Spirituality*, Ronald Purser (2019) attacks mindfulness as a million-dollar self-help industry that teaches people how to cope with stress and anxiety instead of addressing the social and political mechanisms that produce stress, say, working conditions, cultural norms or political deregulation of the welfare state. Mindfulness, he argues, fits perfectly with the neoliberal strategy of individualizing stress, and it offers nothing more than superficial self-help in return. In a similar vein, Norwegian philosopher Arne Johan Vetlesen, in his book *The Denial of Nature*, compares "outer" cosmetic surgery with mindfulness as a form of "inner" cosmetic surgery: in the same way as the inhabitants of capitalist modernity fix their lips and noses or use Botox to remove their wrinkles, they also fix their "inner self" through mindfulness meditation (Vetlesen 2015). According to him, mindfulness is a way of instrumentalizing our own *inner nature*, much in the same way as we instrumentalize outer nature for strategic purposes.

We are no less skeptical of "McMindfulness", understood as a million-dollar self-help industry, or as an ideology that makes individuals responsible for socially produced stress, than Purser and Vetlesen. Nevertheless, some of Vetlesen and Purser's criticisms are misunderstood. For example, Vetlesen's claim that mindfulness aims to "overcome negative thinking and help us to think positively" (2015, 174) relies on a false understanding of the practice. As already discussed, in contrast to much mainstream clinical psychology, mindfulness does not aim to replace negative with positive thoughts but to *be with* negative as well as positive thoughts without aiming at controlling or "fixing" them (Hayes, Strosahl & Wilson 2003). Purser, on his side, confuses self-awareness with self-centeredness: Rather than boosting the personal self and its narratives, mindfulness practice – as understood across different Buddhist traditions as well as contemporary interventions[4] – aims to "decenter" our awareness from self-referential stories that too often preoccupy our minds and cloud our perception. By this, we learn to take our egocentric thoughts and feelings less seriously, allowing for a gradual loosening of habitual self-centeredness and, moment by moment, opening for deeper engagement in the world around us.

Purser and Vetlesen, that is, ignore the dialectical and paradoxical structure of mindfulness practice. Despite its seemingly passive method of observation and acceptance, it can be practiced and understood as a form of existential *resistance* against a capitalist consumer culture that celebrates efficiency, accumulation of goods and constant self-development. For example, by paying close attention to culturally produced *subjective* cravings and aversions, we can become more aware of these and the force they exercise on our thinking, feeling and behavior. Some of these cravings (say, the craving for more material goods) are clearly illusory because they imply a false promise that we will become happy once they are satisfied. Sometimes, they distort our ability to enjoy life as it is and act in accordance with our deeper values (say, ecological values). What mindfulness offers is a practical method of recognizing these cravings at the "micro-phenomenological" level and, by this, increasing our freedom to respond wisely and ethically to them.

It is true that mindfulness promises relief, joy or even happiness for the individual in a stressful, exploitative culture. But that does not necessarily turn mindfulness into the new "spirit of capitalism". On the contrary, given that the kind of happiness that is cultivated in mindfulness is based on radical acceptance rather than accumulation of goods and achievements, mindfulness can be seen as a countercultural act of resistance rather than a cultural pathology. As novelist Matt Haig puts it in his memoir *Reasons to Stay Alive* (2015, 181), dealing with his own history of anxiety and depression:

> The world is increasingly designed to depress us. Happiness isn't very good for the economy. If we were happy with what we had, why would we need more? How do you sell an anti-ageing moisturiser? You make someone worry about aging. How do you get people to vote for a political party? You make them worry about immigration. How do you get them to buy insurance? By making them worry about everything. How do you get them to have plastic surgery? By highlighting their physical flaws. How do you get them to watch a TV show? By making them worry about missing out. How do you get them to buy a new smartphone? By making them feel like they are being left behind. To be calm becomes a kind of revolutionary act. To be happy with your own non-upgraded existence. To be comfortable with our messy, human selves, would not be good for business.

On this background, we contend, mindfulness practice has a revolutionary potential if combined with a critical understanding of collective, socially produced suffering and how to alleviate it. It also aligns well with a *critical eco-phenomenology* that reflects on the human–nature relationship as it is formed and conditioned within our current, historically situated, and culturally formed lifeworld. Critical phenomenology has been defined as

> a practice of suspending hegemonic 'common sense' accounts of reality in order to reflect on the conditions of lived experience and the lifeworld in which it unfolds. It then describes, interrogates, and ultimately transforms the contingent, historical, yet quasi-transcendental structures that shape experience.
>
> (Guenther 2021, 6)

Building on this definition, we define critical *eco*-phenomenology as a practice of suspending hegemonic common-sense accounts of reality in order to reflect on the conditions of lived experience *in an unsustainable lifeworld*. Ultimately, critical eco-phenomenology aims not just to give a theoretical description but to transform the lifeworld in a more sustainable direction.

What are these "hegemonic common-sense accounts of reality"? As we have already addressed, there are many ways in which culturally hegemonic truth-claims, values and narratives can distort our sense of connectedness with the natural world as well as our sense of responsibility towards it. For example, the belief that happiness is a matter of accumulating goods or status can promote an unsustainable lifestyle of consumerism that leads to a sense of ontological disconnection. The same goes for economic and instrumental accounts of nature as a resource for human exploitation, including accounts of our "inner nature" as a resource for self-perfection and self-development. *Critical* eco-phenomenology, then, investigates how culturally hegemonic accounts of reality are defined and promoted in our lifeworld, including in politics, economy, education and the media, and how they perpetuate unsustainable beliefs and lifeforms. It also investigates how these accounts can be challenged and potentially transformed in a more sustainable direction through counter-accounts and phenomenological "suspensions" of beliefs taken for granted.

Mindfulness practice can be considered a close ally of critical-eco-phenomenology in the sense implied here. It also aims at emancipation from hegemonic "common sense accounts of reality" to the extent that these accounts produce suffering, reduce our experience of interconnectedness and promote unsustainable behavior. However, instead of r*eflecting discursively* upon these accounts and narratives, as in critical eco-phenomenology, mindfulness practice explores them *non-linguistically*, as embodied manifestations in the here and now. For example, as we have explored in this article, the culturally hegemonic belief that we need to be proactive, optimize ourselves and accumulate status or possessions, produces reactive experiences at the subjective level of our personal existence, from stress and fear of missing out to low self-esteem, social insecurity and the desire to own and consume things. These experiences are part and parcel of the unsustainable lifestyle and economy in capitalist modernity. Slowing down the metabolic rate individually and socially, therefore, can be seen as two sides of the same coin.

What we have been trying to argue is emphatically not that mindfulness meditation alone can liberate us from unwholesome subjective experiences and instantiate systemic and political change. Political activism and intellectual and moral criticism are much-needed responses to the ongoing climate crisis and ecological disaster. However, we suggest that politics, moral criticism and intellectual analysis may not be as likely to succeed in the long run without a corresponding change at the "inner" or subjective level of human experience. If humans continue to be caught up in a cycle of craving and desire, navigating mostly from a limited and limiting perception of the personal, separate self, then political programs and moral arguments may only do so much. Mindfulness practice is not the answer to the problem of subjective suffering and its negative consequences for nature and the environment, but it may be an important part of the answer.

Notes

1 Our understanding of suffering is not premised on a metaphysical claim in Buddhist tradition; that all life is suffering and consequently to be liberated implies to overcome the endless cycles of birth, death and rebirth. We argue that mindfulness practice may foster emancipation from constricting habits of perception, mind and heart, with implications for well-being, ethics and politics in this life.

2 Our understanding of subjectivity in the Buddhist tradition does not imply the rejection of any conception of selfhood or human agency. Non-attachment to the selfs' "contents", reducing "clinging" to the "illusion" of the self as a permanent and independent entity, is compatible in theory as well as in practice with for instance the notions of self as a fluid, everchanging process (Thomson 2017; Hayes et al. 2006b) or the self as context for conscious experience (Hayes et al. 2006b).

3 But see Thomsons' revision of this line of argumentation in the new introduction to the 2016-republication of The Embodied Mind (pp. xxii–xxv).

4 Such as Mindfulness-Based Stress Reduction, Mindfulness-Based Cognitive Therapy, Acceptance and Commitment Therapy and Dialectical Behaviour Therapy.

References

Bishop, Scott, Mark Lau, Shauna Shapiro, Linda Carlson, Nicole Anderson, Jon Carmody, et al. 2004. "Mindfulness: A Proposed Operational Definition." *Clinical Psychology: Science and Practice* 11 (3): 230–241.

Bitbol, Michel. 2019. "Consciousness, Being and Life: Phenomenological Approaches to Mindfulness." *Journal of Phenomenological Psychology* 50: 127–161.

Booth, Robert. 2025. "Environmental Philosophy and Phenomenology." In *Encyclopedia of Phenomenology*, edited by N. de Warren, and T. Toadvine. Springer Nature Link. https://doi.org/10.1007/978-3-030-47253-5

Brown, Charles S., and Toadvine, Ted. 2003. *Eco-phenomenology: Back to the Earth Itself*. New York: State University of New York Press.

Brown, Kirk W., and Richard M. Ryan. 2003. "The Benefits of Being Present: Mindfulness and Its Role in Psychological Well-Being." *Journal of Personality and Social Psychology* 84 (4): 822–848.

Cederström, Carl, and André Spicer. 2015. *The Wellness Syndrome*. Cambridge: Polity.

Chu, Ming-Hong. 2024. "Mindfulness as a Motivation for Phenomenological Reduction". In *The Routledge Handbook of Phenomenology of Mindfulness*, edited by Susi Ferrarello and Christos Hadjioannou. New York: Routledge.

Cogan, John. 2025. "The Phenomenological Reduction." *The Internet Encyclopedia of Philosophy*. Accessed on 23.2.2025. https://iep.utm.edu/phen-red/

Depraz, Natalie. 2019. "Epoché in Light of Samatha-Vipassana Meditation." *Journal of Consciousness Studies* 26 (7–8): 49–69.

Depraz, Natalie, Francisco J. Varela, and Pierre Vermersch, eds. 2003. *On Becoming Aware: A Pragmatics of Experiencing*. Amsterdam & Philadelphia: John Benjamins Publishing Company.

Djernis, Dorthe, Ida Lerstrup, Dorthe Poulsen, Ulrika Stigsdotter, Jette Dahlgaard, and Maureen O'Toole. 2019. "A Systematic Review and Meta-Analysis of Nature-Based Mindfulness: Effects of Moving Mindfulness Training into an Outdoor Natural Setting." *International Journal of Environmental Research and Public Health* 16 (17): 3202.

Epstein, Mark. 1995. *Thoughts Without a Thinker: Psychotherapy from a Buddhist Perspective*. New York: Basic Books.

Ericson, Torgeir, Bjørn Gunaketu Kjønstad, and Anders Barstad. 2014. "Mindfulness and Sustainability." *Ecological Economics* 104: 73–79.

Ferrarello, Susi. 2024. "Husserl and Mindfulness." In *The Routledge Handbook of Phenomenology of Mindfulness*, edited by Susi Ferrarello and Christos Hadjioannou. Routledge.
Ferrarello, Susi, and Christos Hadjioannou, eds. 2024. *The Routledge Handbook of Phenomenology of Mindfulness*. New York: Routledge.
Garland, Eric L., Norman A. Farb, Philippe R. Goldin, and Barbara L. Fredrickson. 2015. "Mindfulness Broadens Awareness and Builds Eudaimonic Meaning: A Process Model of Mindful Positive Emotion Regulation." *Psychological Inquiry* 26 (4): 293–314.
Germer, Christopher K. 2005. "Mindfulness: What Is It? What Does It Matter?" In *Mindfulness and Psychotherapy*, edited by Christopher K. Germer, Ronald D. Siegel, and Paul R. Fulton. New York: Guilford Press.
Goldberg, Simon B., Kelly M. Riordan, Shian-Ling Sun, and Richard J. Davidson. 2022. "The Empirical Status of Mindfulness-Based Interventions: A Systematic Review of 44 Meta-Analyses of Randomized Controlled Trials." *Perspectives on Psychological Science* 17 (1): 108–113.
Goyal, Madhav, Sonal Singh, Erica M. S. Sibinga, Neda F. Gould, Alicia Rowland-Seymour, Ritu Sharma, et al. 2014. "Meditation Programs for Psychological Stress and Well-Being: A Systematic Review and Meta-Analysis." *JAMA Internal Medicine* 174 (3): 357–368.
Grossman, Paul. 2015. "Mindfulness: Awareness Informed by an Embodied Ethic". *Mindfulness* 6:17–22.
Grossman, Paul, and Nicholas T. Van Dam. 2011. "Mindfulness, by Any Other Name…: Trials and Tribulations of Sati in Western Psychology and Science." *Contemporary Buddhism* 12 (1): 219–239.
Guenther, Lisa 2021. "Six Senses of Critique for Critical Phenomenology." *Puncta: Journal of Critical Phenomenology* 4 (2): 1–23.
Hadjioannou, Christos. 2024. "Heiderggerian and Stoic Mindfulness: Two Competing Models with Common Ground". In *The Routledge Handbook of Phenomenology of Mindfulness*, edited by Susi Ferrarello and Christos Hadjioannou. New York: Routledge.
Haig, Matt. 2015. *Reasons to Stay Alive*. London: Canongate.
Hanley, Adam W., Brian P. M. Nakamura, Andrew J. Garland, Kirk Warren Brown, and Eric L. Garland. 2017. "The Selflessness of Mindfulness: A Qualitative Investigation of the Experience of Self-Transcendence in Mindfulness Practice." *Mindfulness* 8 (5): 1302–1313.
Hayes, Steven C. 2004. "Acceptance and Commitment Therapy, Relational Frame Theory, and the Third Wave of Behavioral and Cognitive Therapies." *Behavior Therapy* 35 (4): 639–665.
Hayes, Steven C., Jason Luoma, Frank Bond, Akihika Masuda, and Jason Lillis. 2006a. "Acceptance and Commitment Therapy: Model, processes, and outcomes." *Behaviour Research and Therapy* 44: 1–25. https://doi.org/10.1016/j.brat.2005.06.006
Hayes, Steven C., Christine Shenk, Aiko Masuda, and Kristoffer Bunting. 2006b. "The Psychological Processes Underlying Attention: Exploring the Connection Between Buddhism and Contemporary Contextual Behavioral Psychology." In *Buddhist Thought and Applied Psychological Research*, edited by D. K. Nauriyal, Michael S. Drummond, and Y. B. Lai, 431–448. New York: Routledge.
Hayes, Steven C., Kirk D. Strosahl, and Kelly G. Wilson. 2003. *Acceptance and Commitment Therapy: An Experiential Approach to Behavior Change*. New York: Guilford Press.
Hölzel, Britta K., Sara W. Lazar, Tim Gard, Zev Schuman-Olivier, David R. Vago, and Ulrich Ott. 2011. How Does Mindfulness Meditation Work? Proposing Mechanisms of Action from a Conceptual and Neural Perspective. *Perspectives on Psychological Science* 6 (6): 537–559.
Kabat-Zinn, Jon. 1982. "An Outpatient Program in Behavioral Medicine for Chronic Pain Patients Based on the Practice of Mindfulness Meditation: Theoretical Considerations and Preliminary Results." *General Hospital Psychiatry* 4 (1): 33–47.
Kabat-Zinn, Jon. 2005. *Full Catastrophe Living: Using the Wisdom of Your Body and Mind to Face Stress, Pain, and Illness*. Delta Trade Paperbacks. Originally published 1990.
Keng, Shian-Ling, Moria J. Smoski, and Clive J. Robins. 2011. "The Effectiveness of Mindfulness-Based Therapies for Anxiety and Depression." *Journal of Affective Disorders* 137 (1–3): 1–9.
Khoury, Bassam, Tania Lecomte, Guillaume Fortin, Marjolaine Masse, Phillip Therien, Vanessa Bouchard, Marie-Andrée Chapleau, Karine Paquin, and Stefan G. Hofmann. 2013. "Mindfulness-Based Therapy: A Comprehensive Meta-Analysis." *Clinical Psychology Review* 33 (6): 763–771.
Kollmuss, Anja, and Julian Agyeman. 2002. "Mind the Gap: Why Do People Act Environmentally and What are the Barriers to Pro-environmental Behavior?" *Environmental Education Research* 8 (3): 239–260.

Kuiji. 2001. *A Comprehensive Commentary on the Heart Sutra (Prajñāpāramita- hr̥daya-sūtra)*. Berkeley, CA: Numata Center for Buddhist Translation and Research.
Lundh, Lars-Gunnar. 2020. "Experimental Phenomenology in Mindfulness Research." *Mindfulness* 11: 493–506.
McCown, Donald. 2013. *The Ethical Space of Mindfulness in Clinical Practice: An Exploratory Essay*. London and Philadelphia: Jessica Kingsley Publishers.
Mitchell, Debra, Deborah J. Tippins, Young Ae Kim, Gail D. Perkins, and Heather A. Rudolph. 2016. "Last Child in the Woods: An Analysis of Nature, Child, and Time through a Lens of Eco-Mindfulness." In *Mindfulness and Educating Citizens for Everyday Life. Bold Visions in Educational Research*, edited by Malgorzata Powietrzynska and Kenneth Tobin, Rotterdam: Brill.
Purser, Ronald. 2019. *McMindfulness: How Mindfulness Became the New Capitalist Spirituality*. London: Repeater.
Richter, Nadine, and Marcel Hunecke. 2022. "Mindfulness, Connectedness to Nature, Personal Ecological Norm and Pro-environmental Behavior: A Daily Diary Study." *Current Research in Ecological and Social Psychology* 3. https://doi.org/10.1016/j.cresp.2022.100038
Ryan, Richard M., and Edward L. Deci. 2008. "From Ego to Eco: Development and the Emergence of Values." *Social and Personality Psychology Compass* 2 (3): 1417–1442.
Salzberg, Sharon. 2011. "Mindfulness and Loving-Kindness." *Contemporary Buddhism* 12 (1): 177–182.
Sauer, Sebastian, Scott Lynch, Harald Walach, and Niko Kohls. 2011. "Dialectics of Mindfulness: Implications for Western Medicine." *Philosophy, Ethics, and Humanities in Medicine* 6: 10–17.
Schutte, Nicola S., and John M. Malouff. 2018. "Mindfulness and Connectedness to Nature: A Meta-analytic Investigation." *Personality and Individual Differences* 127: 10–14.
Shapiro, Shauna L., and Linda E. Carlson. 2009. *The Art and Science of Mindfulness: Integrating Mindfulness into Psychology and the Helping Professions*. 1st ed. Washington, DC: American Psychological Association.
Solhaug, Ida. 2016. *Cultivating Helper Qualities: Immediate and Long-Term Impacts of Mindfulness Training for Medical and Psychology Students. A Prospective Randomized Controlled Study and Qualitative Exploration*. Ph.D Dissertation, UiT The Arctic University of Norway.
Solhaug, Ida, Thor E. Eriksen, Hanne Haavind, Michael de Vibe, Oddgeir Friborg, Tore Sørlie, and Jan H. Rosenvinge. 2016. "Medical and Psychology Students' Experiences in Learning Mindfulness: Benefits, Paradoxes, and Pitfalls." *Mindfulness* 7: 838–850.
Stone, Oliver, and Dan Zahavi. 2021. "Phenomenology and Mindfulness." *Journal of Consciousness Studies* 28 (3–4): 158–185.
Tang, Y. Y., B. K. Hölzel, and M. I. Posner. 2015. "The Neuroscience of Mindfulness Meditation." *Nature Reviews Neuroscience* 16 (4): 213–225.
Teasdale, John D., and Michael Chaskalson. 2011. "How Does Mindfulness Transform Suffering? II: The Transformation of Dukkha." *Contemporary Buddhism* 12 (1): 103–124.
Thiermann, Ute B., and William R. Sheate. 2021. "The Way Forward in Mindfulness and Sustainability: A Critical Review and Research Agenda." *Journal of Cognitive Enhancement* 5: 118–139.
Thompson, Iain. 2004. "Ontology and Ethics at the Intersection of Phenomenology and Environmental Philosophy". *Inquiry* 47: 380–412.
Thomson, Evan. 2017. *Waking, Dreaming, Being: Self and Consciousness in Neuroscience, Meditation, and Philosophy*. New York: Columbia University Press.
Travis, N. Ray, Franz Scott, Nicole Jarrett, and Scott Pickett. 2020. "Nature Enhanced Meditation: Effects on Mindfulness, Connectedness to Nature, and Pro-environmental Behavior." *Environment and Behavior* 53 (8): 899–926.
Vago, David R., and David A. Silbersweig. 2012. "Self-Awareness, Self-Regulation, and Self-Transcendence (S-ART): A Framework for Understanding the Neurobiological Mechanisms of Mindfulness." *Frontiers in Human Neuroscience* 6. https://doi.org/10.3389/fnhum.2012.00296
van den Bold, Ingeborg. 2025. "Mindfulness and the Paradoxes in Merleau-Ponty's Phenomenology." *Journal of Humanistic Psychology* 65 (2): 351–369.
Varela, Francisco J., Evan Thompson, and Eleanor Rosch. 1991. *The Embodied Mind: Cognitive Science and Human Experience*. Cambridge, MA: MIT Press.
Vetlesen, Arne Johan. 2015. *Denial of Nature: Environmental Philosophy in the Era of Global Capitalism*. London: Routledge.

Wamsler, Christine, and Ebba Brink. 2018. "Mindsets for Sustainability: Exploring the Link Between Mindfulness and Sustainable Climate Adaptation." *Ecological Economics* 151: 55–61.
Whitburn, Julie, Warren Linklater, and Wokje Abrahamse. 2020. "Meta-Analysis of Human Connection to Nature and Proenvironmental Behavior." *Conservation Biology* 34: 180–193.
Williams, J. Mark G., and Jon Kabat-Zinn. 2011. "Mindfulness: Diverse perspectives on its meaning, origins, and multiple applications at the intersection of science and dharma." *Contemporary Buddhism*, 12(1): 1–18.
Wood, David. 2001. "What Is Ecophenomenology?" *Research in Phenomenology* 31: 78–95.

30

TIRED NATURE, TIRED OF NATURE

Eco-Phenomenology of Fatigue and Melancholy

Shannon Hayes and Kaja Jenssen Rathe

Introduction

"No more excuses for passivity" (Malm 2021, 1). So begins Andreas Malm's manifesto, *How To Blow Up a Pipeline: Learning to Fight in a World on Fire.* Writing during the COVID-19 pandemic, Malm argues that if governments can take emergency action for a health crisis, they should do the same for "a climate breakdown that threatens to kill off the very life-support systems of the planet do the same" (2). However, his critique targets not only world elites but also the climate movement, which he believes has remained ineffective by committing to absolute non-violence. For Malm, the movement's dedication to pacifism—which extends beyond avoiding harm to people to also prohibiting property destruction—has rendered it "gentle and mild in the extreme" (21) and ultimately futile against powerful interests. In a political climate where meaningful climate action needs to threaten "business as usual," Malm argues that peaceful appeals to reason are doomed to fail. Instead, he advocates moving "beyond peaceful protest" (25) into strategic property destruction and sabotage to force change through disruption rather than persuasion.

Beyond criticizing the climate movement's mildness, Malm reserves his sharpest critique for what he calls *climate fatalism*, exemplified by writers like Roy Scranton who suggest that "we're fucked" and that we should simply "learn to die" amid inevitable catastrophe (133). Crucially, Malm distinguishes between momentary despair as "an eminently understandable emotional response to the crisis" and climate fatalism's systematized cultivation of passivity that forecloses political action before it can begin. For Malm, Scranton's call to "cultivate detachment" (135) to climate crises represents not just passive resignation but a dangerous "*reification* of despair" that transforms a natural emotional response into a "foundation for a politics" (141). Moreover, Scranton's reified despair is not based on the fact that real climate action is technically or logically *impossible*, but that *given how the world is,* it is "a task that is *scarcely imaginable*, much less feasible" (141). In other words, this position is "based on a judgement of extreme improbability, hypostasized into impossibility" (141): a self-defeating, "anti-political stance."

For Malm, Scranton's claim that he cannot possibly make a difference simply means that he is "unwilling to make a difference" (143). He *chooses* this attitude. For Malm, there is no

 DOI: 10.4324/9781003538097-37

good reason to think that it is "too late" or that the fight for the climate is "lost." Not because global heating could be averted—Malm agrees that this would be naïve. Rather, because the task is not one of "all or nothing," but rather "a giant salvaging operation: safe-guarding as much space as possible on this scarred planet for human and other life to survive and maybe thrive and, in the best case, healing some of the wounds from the past centuries" (146). But *action* is required to maintain hope in this kind of salvaging.

Today, four years after the publication of Malm's book, we find ourselves in the situation that Malm feared. Over the past few years, the climate problem has only gotten worse, with all-time high temperatures and increasingly extreme weather across the planet. At the same time, paradoxically, the climate movement is weaker than ever and has lost a lot of impact in the past few years, as young people are reporting declining engagement despite increasing climate instability. "We can declare Generation Thunberg dead," Ola Svenneby, leader of Norway's conservative youth party, proclaimed in 2023 after the country's two rightwing parties together had won a large majority in the national "student elections" (VGTV 2023). The environmental cause is struggling to reach headlines in a political climate that has seen a massive shift towards the priority of other concerns and even outright climate denial. It indeed feels as though we are surrounded by a general, widespread case of "eco-fatigue." People are tired of nature, tired of talking about the environment and the climate. Tired of—not only in the sense of "being fed up," but actually as in the sense of being *exhausted* (Stine 2024).

Is such widespread weariness merely a sign of the climate fatalism that Malm condemns, or does it represent something more complex—a phenomenon that requires its own theoretical framework? In this paper, we introduce the concept of eco-fatigue as distinct from both climate fatalism and climate burnout. We argue that eco-fatigue constitutes a specific temporal experience that emerges from the contradictions of human relationships to nature under capitalism, particularly the strain of navigating multiple, often conflicting temporal frameworks simultaneously.

Eco-fatigue has been described as "an attitude characterized by a form of discouragement, or even an indifference to environmental issues" (The Brussels Times Newsroom 2022). Others describe it as "the idea that people have become weary of discussions about sustainability, pessimistic about the future of sustainability, and distrustful of business claims about sustainability" (Moscardo and Pearce 2019, 140), or as "a regression of eco-anxiety into the belief that climate and ecological breakdown have surpassed a point where our efforts will make a difference" (McCubbin 2023). While these definitions capture aspects of the phenomenon, we argue that they fail to recognize eco-fatigue's deeper temporal dimensions and potential political significance.

It is important to distinguish eco-fatigue from what has come to be understood as "climate burnout." As Lucy Bird et al. (2024) conceptualize it, climate burnout comprises "exhaustion, capturing experiences of feeling drained and weary" and "disengagement, which involves distancing oneself from the stressor," positioning it as both "detrimental to individuals" and "detrimental for society" (Cher Weixia & Gorski 2015) due to its association with withdrawal from climate activism. While climate burnout is framed as a psychological condition to be overcome—something that has "no place in a burning world" (Wood 2022)—eco-fatigue emerges as a complex temporal experience that cannot be reduced to individual psychological deficit. Rather than viewing it solely as an obstacle to action, we propose understanding eco-fatigue as a necessary dialectical moment in environmental consciousness that makes visible the contradictions of relating to nature under capitalism.

This distinction highlights the essential difference between eco-fatigue and the climate fatalism that Malm so vehemently rejects. Whereas for Malm, the attitude of climate fatalism is connected to a reification of despair that is cultivated and used as "foundation for a politics," eco-fatigue points to a politicizable affective experience. More crucially, eco-fatigue cannot be reduced to what Malm rejects as *passivity*, as something that is opposed in a binary way to "action." It is not simply an "unwillingness to act," even though it is certainly an experience that may affect or influence our willingness or ability to act. While Malm seems to acknowledge that there exists a kind of climate despair that is not yet reified, but simply *lived*, his one-dimensional rejection of passivity and call for action reflects that he, in fact, does not take this despair—or the fatigue it can lead to—seriously. In order to better understand the turn towards climate-hostile politics and increased climate denial, it is in fact necessary to acknowledge the *power* of eco-fatigue, rather than simply rejecting and dismissing it as despicable; as a lack of commitment, strength, or willingness to resist.

While we share Malm's concern about climate inaction, our critique centers on his reductive understanding of "passivity" through a linear, progressive view of history embedded in capitalist logics. His binary between action and inaction fails to recognize how what appears as passivity may constitute an embodied critique of capitalism's temporal logic that opens space for alternative forms of environmental relationship. The problem with this opposition is that thinking about passivity in general as entirely opposed to activity is both reductive and misleading. In short, this is not how either passivity or action works. The same goes for the experiences of fatigue, including eco-fatigue: it is not "entirely" passive, just as political "action"—even the one that works—will never be completely "active." And if fatigue is not "mere" passivity, and passivity itself not "merely passive," then eco-fatigue, too, may be ambiguous and politically generative.

Our thesis is that eco-fatigue represents not a failure of political will but an embodied response to the temporal contradictions of environmental crisis under capitalism. We argue that eco-fatigue emerges specifically from the strain of attempting to navigate multiple temporal relationships to nature without adequate spatial grounding—a condition particularly acute under capitalism's abstraction of time from place. By reconceptualizing eco-fatigue as a complex temporal experience rather than an individual psychological deficit, we can recognize its potential as a site of critical insight and collective transformation. This approach requires that we move beyond binaries of action and passivity to understand how fatigue itself contains active dimensions of resistance and refusal.

What Malm's perception of the climate crisis misses is how human experiences of nature are characterized by temporal multiplicity. By focusing on eco-fatigue, we reconceive passivity not as opposed to activity but as a dialectical moment within it—revealing how both fatigue and climate action contain active-passive elements. The central question then becomes: How might we work from within eco-fatigue, acknowledging it as an embodied experience not reducible to mere passivity, to develop a more constructive alternative to what Malm aptly describes as climate fatalism's "reification of despair?"

This temporal complexity becomes particularly evident in Indigenous-led environmental movements like the Standing Rock resistance to the Dakota Access Pipeline (DAPL), where Water Protectors simultaneously maintained traditional relationships to land while confronting immediate political realities—demonstrating an alternative to the linear progress narrative that underlies both climate fatalism and Malm's call for escalating action. Malm's characterization of the Standing Rock movement as part of a "cycle" that "came to a dead end" when Donald Trump came to power in 2017 reveals the limitations of his progressive temporal framework (2021, 98). This analysis misses how Water Protectors maintained relationships to land and water outside the logic of commodification, inhabiting an "outside-within" perspective that linear narratives cannot accommodate.

To understand this complexity, we develop a theoretical framework that brings together Emmanuel Levinas's phenomenological analysis of fatigue, Walter Benjamin's concept of the aura, and Ernst Bloch's theory of nonsynchronism. Together, these frameworks help us recognize eco-fatigue not as an individual psychological deficit but as a shared temporal condition reflecting contradictions in our relationship to nature under capitalism. Through this theoretical lens, we can begin to understand eco-fatigue as a mode of melancholic attention to ecological crisis—one that refuses both paralysis and premature resolution.

First, we explore Levinas's phenomenological analysis of fatigue as simultaneously passive and active, challenging the binary opposition underlying both climate fatalism and calls for intensified action. We analyze fatigue's four key dimensions: intentional slackening, intrinsic relationship to labor, distinctive temporality, and passive-active resistance. Second, we connect this analysis to Benjamin's melancholic concept of the aura, examining how the decay of auratic experience in modernity produces a distinctive form of attention that recognizes the constructedness of seemingly natural relationships. Third, we incorporate Bloch's theory of nonsynchronism to reveal how eco-fatigue emerges from and responds to capitalism's contradictory temporalities, particularly how it operates as both subjective emotional resistance and objective remnant of alternative ways of relating to nature.

Finally, we analyze the Standing Rock resistance to the DAPL as a concrete manifestation of temporal multiplicity in practice. While our theoretical framework draws primarily from European philosophical traditions, we acknowledge their limitations when confronted with Indigenous knowledge systems. Indigenous conceptions of time as spatially grounded, articulated by thinkers like Vine Deloria Jr. (1972) and Leanne Betasamosake Simpson (2017), offer crucial interventions that transform our understanding of environmental exhaustion. We examine how these spatial-temporal frameworks enabled Water Protectors to maintain multiple temporal relationships simultaneously, demonstrating an alternative to both climate fatalism and the burnout-inducing demand for constant action within capitalism's progressive narrative. This case study reveals how apparent passivity often contains complex forms of agency that Western philosophical paradigms struggle to recognize, suggesting that addressing eco-fatigue requires not overcoming it through intensified action but creating conditions where multiple temporal relationships to nature can be acknowledged and maintained through practices grounded in specific places.

Part I: Levinas and the Lived Experience of Fatigue

What does it mean to be tired of nature; to be exhausted from having to deal with anxiety over impending and present climate disaster, in combination with the seeming lack of international commitment to turn the situation around? Is this fatigue simply a failure of commitment that can be freely overcome, or does it represent something more complex that must be taken seriously as embodied experience? Understanding eco-fatigue requires navigating between these perspectives, acknowledging both our freedom and the weight of our embodied circumstances.

For Malm, experiences of fatigue or despair do not have the power to determine your actions or commitments regarding the climate cause. According to Malm, Scranton seems to be lacking precisely "an enduring commitment to resistance" and rather goes for "resignation before the inevitable" as his "main credo" (2021, 138). This lack of commitment is for Malm central to Scranton's conflicted and ultimately untenable position on the climate: if he were only more committed and would not give up so fast, his political position would be different. It is Scranton who has chosen to linger in despair and resignation, rather than acting in order to maintain hope.

Interestingly, this line of argumentation resonates with that of another European thinker, who in his magnum opus more than 80 years ago also emphasized our freedom to choose our attitude faced with experiences of despair or other affects and embodied motives, and who moreover—just like Malm—rejected probability as a motive for action. This thinker is, of course, Jean Paul Sartre. For Sartre, human beings retain their freedom in confrontation with the world's "facticity," i.e., everything that the subject encounters in the world as existing seemingly independently of itself, whether it concerns the physical environment, our body, affects, or even motives for one's own actions, such as feelings.[1] In his analysis of the "tired hiker," Sartre describes the following situation:

> I have gone on an excursion with some friends. After several hours of walking, I am growing tired, and eventually my fatigue becomes oppressive. At first I resist and then suddenly I let myself go: I give in; I throw my bag down on the side of the road, and I drop down beside it.
>
> (Sartre 2018, 594–595)

Sartre's question becomes: Was I free to do otherwise, and if so, then "at what cost" (595)? For Sartre, importantly, an experience like fatigue "cannot induce my decision. [The experience of fatigue] is only … the way in which I exist my body" (595). As a bodily motive, fatigue is thus part of facticity, part of what I, the subject, encounter in the world. More specifically, even though I suffer my fatigue and feel like I cannot continue, it is I who interprets my experience in this particular way—someone else might indeed experience the same level of fatigue and give it another meaning, for example, that it exists in order to be overcome. The crucial argument, however, is that even these ulterior or more "deep-seated" motives—someone's attitude to physical challenges as projects to be overcome vs. someone else's lack of interest in them and priority of physical comfort—are themselves never imposed or already given, but *also* freely chosen. In other words, the decision to surrender or not to one's fatigue is ultimately one's own, even though (or precisely because) such decisions are always bound up with more deep-seated motives that are central to how we see our "fundamental project"—projects that are themselves always already freely chosen by the subject.

Of course, this interpretation of the relation between experiences such as fatigue and freedom more generally has been contested, among others, by Sartre's contemporary, Maurice Merleau-Ponty. Referring back to Sartre's own example of the tired hiker, Merleau-Ponty points out that the question of succumbing to fatigue cannot be reduced entirely to a question of free choice, at least not in the sense that Sartre is interpreting this question. More specifically, he urges us to "recognize a sort of sedimentation of our life: when an attitude toward the world has been confirmed often enough, it becomes privileged for us" (Merleau-Ponty 2012, 466). We are, existentially speaking, embodied, "habitual beings"—not merely intellectual projects—and our freedom must be understood accordingly: as the freedom of embodied, habitual beings, not as the freedom of abstract intellects. Concrete freedom is, in fact, to be found in the large field that stretches out *between* absolute freedom and absolute determination—the exact same field where, as Michel Foucault (1982) would later show, relations of power always coexist with and influence concrete freedom without negating it.

More specifically, the problem with Sartre's and Malm's argumentation is that they, by focusing solely on the fact that our fatigue does not determine our actions, somehow seem to bypass the *phenomenon* of fatigue entirely. Sartre is right in pointing out that there is an element of freedom in how we choose to relate to the situations we find ourselves in, including lived embodied experiences like fright, despair, being in love, or fatigue—just like Malm has a lot of valid points in his critique in the specific case of Scranton. Yet Sartre's problem, which he seems to share with Malm, has always been—despite his commitment to always study freedom in concrete situations—to take too lightly the kind of weight or pull that embodied experiences actually do have upon our concrete experience of situated freedom, or in other words, our freedom to commit to a project, without taking that to mean that the freedom in question would be cancelled. Understanding eco-fatigue requires, on the contrary, taking this weight or pull *seriously*, while at the same time acknowledging that the experience of fatigue does not fully determine further actions.

So, what does it mean to take fatigue seriously? What pull and meaning does fatigue tend to have in concrete lived experience? What characterizes our experiences of fatigue? A thinker who can really help us think through the dynamic of fatigue, as well as its effect on the subject, is Emmanuel Levinas, who in his work *Existence and Existents* describes fatigue's "internal dialectics" (1978, 31). We will here point to four key dimensions of Levinas' account that are particularly relevant for understanding eco-fatigue: fatigue as intentional slackening, the relationship of fatigue and labor, the temporality of fatigue, and fatigue's passive-active dimension.[2]

The first thing to notice is that Levinas describes the experience of fatigue as characterized by "a stiffening, a numbness, a way of curling up into oneself," which takes the form of

> an impossibility of following through, a constant and increasing *lag* between being and what it remains attached to, like a hand little by little letting slip what it is trying to hold on to, letting go even while it tightens its grip. Fatigue is not just the cause of this letting go, it is the slackening itself.
>
> (Levinas 1978, 30)

What this means is that the dynamics of fatigue can first of all be said to be characterized by a certain detachment from worldly matters. As Katherine Morris puts it in her own phenomenological analysis of fatigue, "the intentional threads that bind me to the world are somewhat slackened" (Morris 2019, 138). In fact, "fatigue makes the world—including one's own projects—matter less. Fatigue makes us more indifferent; the world as well as our own projects lose their pull on us" (Rathe 2025, 13). This "intentional slackening" extends beyond my relationship to the world. Interestingly, it involves the relationship between me and myself: fatigue indeed "involves a *lag* or *gap* in the subject, which [Levinas] also describes as a lag 'between a being and itself'" (Rathe 2025, 13). Fatigue is thus characterized by a certain duality of the subject, which becomes increasingly manifest as fatigue increases.

The second critical aspect of fatigue concerns its dialectical movement, which involves an intrinsic relationship between fatigue and *labor*. The reason for this is that fatigue's "intentional slackening" happens while we are *trying to hold on* to whatever we are letting go of. In other words, letting go of something because you are tired is very different from letting go of something because you simply decide to. As Levinas himself puts it,

> [fatigue] does not occur simply in a hand that is letting slip the weight it finds tiring to lift, but in one that is holding on to what it is letting slip, even when it has let it drop but remains taut with the effort. For there is fatigue only in effort and labor.
>
> (Levinas 1978, 30)

This insight is crucial for our understanding of eco-fatigue. For Levinas, fatigue is a slackening *within effort*—not something that is opposed to effort. The fatigue we experience in relation to ecological crisis is not separate from our continued efforts to care about and address this crisis, but intrinsically bound up with these efforts.

The third important aspect of fatigue is its distinctive temporality, which, as Morris points out, is different from that of ordinary tiredness.[3] Ordinary tiredness has a cyclical temporality characterized by the fact that rest or sleep tends to reset it and makes the sense of being tired disappear. In fatigue, on the other hand, which can be generally understood as cases of *extreme tiredness*, the temporality is rather *cumulative*: fatigue is defined precisely by the way in which rest or sleep does not seem to help. Rather, fatigue increases until it reaches a tipping point. At the tipping point, one's accumulated tiredness "moves resolutely *from the background to the foreground* of … experience, and with that comes an increased *sensation of effort*" (Rathe 2025, 15). The tipping point is hence characterized by an increased awareness of one's experience of fatigue, which has the potential to change or influence one's strategies for dealing with it.

Crucially, however, the moment where fatigue shifts from becoming bearable to seemingly unbearable reveals how "giving in to fatigue" has both an active and a passive dimension. In a certain sense, it is a choice to eventually "let go," yet at the same time it is not: the fatigue is something we eventually *succumb* to, i.e., a kind of internal pressure that we can no longer resist. The temporal dimension helps explain why eco-fatigue cannot be overcome simply through individual willpower or occasional rest: even if we could have held out a little longer, fatigue continues to accumulate and will result in a tipping point where it seemingly "takes us over."

Finally, and perhaps most importantly for our argument, we can conclude that fatigue is nevertheless not simply a passive experience or event, but has a clear active dimension in which Levinas in fact sees the potential for revolt: for him, fatigue is precisely a passive-active form of "impotent nonacceptance" (1978, 24) This passive-active dimension of fatigue is related to the close connection between fatigue and labor, which—crucially—cannot be reduced to a causal relation where labor is said to "cause" fatigue. Rather, for Levinas, "effort as it were lunges forward out of fatigue and falls back upon it" (31). While not synonymous, effort and fatigue are, in other words, intimately linked and interdependent. Interestingly, this challenges the very binary between activity and passivity that underlies Malm's critique, and reveals it as a false opposition. There exists a large field of possibilities between "active action" and "passive eco-fatalism," which is precisely where fatigue and effort dynamically co-exist. This is why fatigue is not only the passive "limit" of action, but carries within itself a potential for revolt and resistance.

Moreover, Levinas's refusal to understand action as "entirely active" and as something opposed to passivity is connected to his more general critique of the Western liberal concept of action as the (free) subject's "struggle with matter." It involves a parallel critique of how the body has been understood as opposed to action, as a "chain." By contrast, Levinas understands action and the body, and more generally, action and "being," as profoundly related, not antagonistic forces.

It is within this framework that fatigue must be understood, not as a bodily "enchainment" or mere limitation of action, but as the very condition of existence from which any effort originates. More specifically, fatigue for Levinas is a pre-reflective event of refusal (of the continuous labor of existence), from which any effort has to "lunge out." Importantly, this refusal is not just a mental state, but a pre-reflective, embodied event.

We can extend Levinas's suggestion that the refusal embedded in fatigue seems to exist in either "latent" or "explicit" forms. In the "latent" form, the resistance of fatigue is pre-reflective, happening *merely through one being tired*. Already at this level, however, it is passive-active: not entirely active, not entirely passive.

The more explicit level of fatigue's resistance happens, however, after the tipping point, when the experience of fatigue can no longer be neglected.

> [A]t the "tipping-point" things change. The refusal which was before "latent"—or, in Levinas' words, "impotent"—now becomes more explicit, in so far as the fatigue comes to the forefront of our experience and thereby forces us to take a more conscious stance toward our fatigue: We cannot go on as before. We "give in" to fatigue ... Again, this event is passive-active, but now it has a more articulated sense of "activity" than the "mere" experience of fatigue before the tipping-point. Indeed ... giving in to fatigue is not only "impotent nonacceptance," but has, at the tipping-point, the momentum to institute something *new*.
>
> (Rathe 2025, 17)

This nuance is crucial for understanding the political potential of eco-fatigue. Rather than seeing eco-fatigue as mere resignation or withdrawal that must be overcome through sheer will to action, Levinas's analysis reveals how it might contain within itself the seeds of a different kind of engagement with ecological crisis—one that emerges precisely at the moment when continuing "business as usual" environmental activism becomes impossible.

Levinas's analysis of fatigue as a passive-active "impotent nonacceptance" provides a crucial foundation for understanding eco-fatigue beyond simplistic binaries of action and resignation. This phenomenological account reveals how fatigue embodies a complex temporal experience that both withdraws from and engages with the world. To further develop this temporal complexity in the context of ecological crisis, we turn to Walter Benjamin's concept of the aura and its relationship to melancholic attention. Where Levinas illuminates the embodied experience of fatigue as a slackening of intentional threads that paradoxically contains the seeds of refusal, Benjamin's auratic theory helps us understand how this experience relates specifically to our changing relationship with nature under capitalism. Both thinkers reveal how apparent "passivity" contains active dimensions that challenge linear narratives of progress and action.

Benjamin's Aura and the Melancholy of Eco-Fatigue

German Jewish philosopher Walter Benjamin (1892–1940), whose critical theory of modernity examines how technological and social changes transform human experience, provides a valuable framework for rethinking eco-fatigue. Benjamin's concept of the aura helps us understand eco-fatigue not as passive resignation, but as a complex temporal experience that challenges Malm's binary between action and passivity. By revealing the historically contingent nature of our perception of ecological crisis, Benjamin's theory helps us reconceptualize eco-fatigue as a potentially critical stance.

Benjamin defines the aura as "a strange tissue of space and time: the unique apparition of a distance, however near it may be" (2008, 23). This formulation captures a historically embedded mode of perception that invests objects with temporal depth. As Miriam Bratu Hansen notes, "the aura is not an inherent property of objects but pertains to the medium of perception" (2008, 342), recognizing how both objects and our perception of them are shaped by historical and social relations. Through this lens, eco-fatigue emerges not simply as emotional exhaustion but as a response to changing conditions of ecological perception within late capitalism.

Central to Benjamin's analysis is the decay of auratic experience in modernity—a process accelerated by technological reproduction and capitalist commodification. In ecological contexts, this decay manifests itself as nature becoming simultaneously hyper-visible as a resource and impossibly distant as something with intrinsic value. As ecosystems are increasingly mediated through satellite imagery, climate models, and digital interfaces, the possibility of an immediate, reciprocal relationship with nature diminishes. The result is a distinctive form of melancholic attention that characterizes eco-fatigue: a painful awareness of both proximity to and distance from the natural world.

Yet, this auratic decay carries productive ambivalence. Hansen writes that "to perceive the aura of an object we look at means to invest it with the 'ability to look at us' in turn" (2008, 339). This reciprocal gaze establishes a relationship that capitalism's commodification systematically undermines. When Benjamin writes that "[a]ll nature would begin to lament if it were endowed with language" (2004, 72), he attributes to nature a latent expressivity and autonomy that capital silences. When ancient forests are reduced to board-feet of lumber or carbon sequestration assets, the possibility of that forest "looking back"—of our recognizing its autonomous existence beyond human utility—gradually recedes. Eco-fatigue responds to this loss of reciprocity, acknowledging the impossibility of returning to pre-capitalist nature while refusing to accept its complete commodification.

Unlike the "reified despair" that Malm critiques, Benjamin's melancholy maintains a critically productive relationship to loss. As we've argued elsewhere (Hayes 2020), the melancholic subject refuses to smooth over the fractures of capitalist alienation, instead dwelling in the contradictions that these fractures reveal. Similarly, the eco-fatigued subject refuses to assimilate nature fully into either commodification or nostalgia, preserving critical distance from both technological optimism and romanticism while maintaining awareness of ecological relationships as historically contingent and subject to transformation.

Georges Didi-Huberman describes the decay of the aura as an "'uncompleted' and 'always open' phenomenon" (2005, 5). Eco-fatigue similarly exists in this liminal space—registering nature as simultaneously present and fading. This critical attention that recognizes both what has been lost and what might yet be possible connects directly to Levinas's understanding of fatigue as "impotent nonacceptance." The slackening of intentional threads that Levinas identifies in fatigue parallels Benjamin's account of how auratic experience loosens fixed meanings, revealing the constructedness of seemingly natural relationships. Both perspectives challenge the notion that apparent passivity lacks political potential, suggesting instead that the withdrawal characteristic of fatigue and melancholy might constitute a form of critique that precedes transformative action.

Our analysis suggests that eco-fatigue's value lies in how it disrupts capitalism's homogenizing temporality, creating openings for different kinds of environmental politics. This disruption becomes particularly evident when viewed through Bloch's theory of nonsynchronism, which provides a framework for understanding how multiple temporal relationships to nature can coexist within our present moment. While Benjamin illuminates the dialectical potential within auratic decay, Bloch enables us to see how this melancholic stance manifests differently across communities and contexts. This movement from Benjamin to Bloch transforms eco-fatigue from an individual experience of melancholy into a collective resource for environmental politics.

Ernst Bloch and the Dialectics of Eco-Fatigue

The utopian philosophy of the German Marxist philosopher Ernst Bloch (1885–1977), with its theory of nonsynchronous temporalities and "educated hope," provides tools for understanding the collective political potential of eco-fatigue. While Benjamin's analysis of auratic decay already reveals our alienated relationship to nature and Levinas's phenomenology illuminates the passive-active dynamics of fatigue, Bloch's dialectical theory situates this insight within a temporally complex framework, which helps us mobilize eco-fatigue politically without reducing it to either pure passivity or instrumental action. As Bloch observes: "Not all people exist in the same Now. They do so only externally, by virtue of the fact that they may all be seen today" (1977, 22). This insight is critical for understanding how different relationships to environmental crisis—and thus distinct experiences of eco-fatigue—can coexist within our present moment.

Central to Bloch's thought is the distinction between alienation (*Entfremdung*) and estrangement (*Verfremdung*). While alienation signifies a deadening disconnection that separates us from our potentialities, estrangement—a form of "distanced nearness" (*Fernahe*)—functions as a productive disorientation that opens up space for critical recognition. As he writes: "all nearness makes matters difficult, and if it is too close, then one is blinded" (1998, 120). This estrangement serves a critical function by offering "insight into what lies nearest, drawn from astonishment at what lies farthest" (245), creating "a mirror image that reflects in striking and unfamiliar fashion the all too familiar aspects of our world" (245). Estrangement thus operates as a dialectical pivot between alienation and recognition: "a form of exoticism that looks homeward, acting against alienation in favor of self-encounter" (246).

This dialectical tension allows us to move beyond a purely negative reading of eco-fatigue that would align with Malm's (2021) critique of climate fatalism. Research by Panu Pihkala (2020) and Sarah Jaquette Ray (2020) on climate anxiety shows that the constant pressure to maintain hope amid ecological devastation often leads to a uniquely depleting form of emotional labor. This tension between hope and exhaustion exemplifies what Bloch calls the "educated hope" that must wrestle with concrete conditions rather than escape into false optimism. Unlike wishful thinking that "builds castles in the air … without foundation" (339), genuine hope for Bloch is inherently tied to disappointment:

> the hope that consists only of dreams always can and will be disappointed … and yet does well-founded hope, mediated, guiding hope, have any better prospects? It too can be, and will be, disappointed; indeed, it must be so, as a matter of honor, or else it would not be hope.
>
> (340)

This paradoxical formulation illuminates why eco-fatigue serves as a necessary counterpart to hope rather than its negation. As Bloch insists, "hope must be unconditionally disappointable

… because it is open in a forward direction … it does not address itself to that which already exists" (341). Applied to eco-fatigue, environmental exhaustion becomes not hope's failure but the necessary condition for hope's authentic emergence—"hope that travels onward with knowledge in its backpack" (196).

Eco-fatigue expresses our weariness with the perpetual task of repairing our relationship to nature within capitalism's relentless logic. However, this weariness emerges differently across various temporal and social contexts, pointing toward what Bloch identifies as nonsynchronous contradictions. In his framework, nonsynchronism operates both "subjectively" as emotional resistance and "objectively" as the persistence of earlier ways of relating to nature that remain as remnants within the present (1977, 31). This creates a complex landscape of temporal relationships to the environmental crisis.

This temporal complexity manifests through distinct but interconnected experiences across different groups. Indigenous communities experience eco-fatigue not as exhaustion with environmental solutions, but as the weariness of continually defending traditional ecological relationships against systemic erosion (Whyte 2017). Urban environmental activists confront fatigue through their perpetual struggle against capitalism's extractive logic, which positions development and progress as inevitable and non-negotiable (Hoggett & Randall 2018). Climate scientists, inhabiting yet another temporal position, experience professional exhaustion as they grapple with the discord between the urgency of their findings and the glacial pace of political action (Head & Harada 2017).

This reconceptualization directly challenges Malm's strict binary between activity and passivity. From a Blochian perspective, what appears as passive fatigue actually constitutes an embodied critique that actively holds multiple temporal orientations in tension. Michelle Murphy's (2017) concept of "alterlife" illuminates how marginalized communities experience environmental harm through unique temporal registers, while Kyle Whyte (2017) demonstrates how Indigenous communities navigate this complexity by simultaneously defending traditional ecological relationships and contending with capitalism's accelerating destruction. As Nick Estes (2019) argues, these struggles often involve defending alternative temporalities while forging new forms of solidarity.

In this light, eco-fatigue emerges not as passive resignation, but as an embodied critique and potential catalyst for change: a mode of attention that refuses the endless drive for technological solutions, insisting instead on reflective engagement with multiple temporalities of ecological crisis. While Malm might interpret certain manifestations of eco-fatigue as politically ineffective, Bloch's concepts of nonsynchronous contradictions and educated hope help us recognize how these apparent forms of passivity might constitute sophisticated resistance to capitalism's temporal logic—forms that maintain the possibility of genuine transformation through their refusal of false optimism. This theoretical framework finds concrete expression in the Indigenous-led resistance at Standing Rock, where multiple temporal relationships to environmental crisis operate not as abstract concepts but as living practices.

The Standing Rock Sioux's Resistance to the Dakota Access Pipeline

The Standing Rock resistance to the DAPL[4] offers a concrete manifestation of the temporal multiplicity that our theoretical framework illuminates. Here, we see how Levinas's "passive-active" resistance, Benjamin's melancholic attention to what is being lost, and Bloch's nonsynchronous contradictions operate not as abstract concepts but as living practices within Indigenous-led environmental movements.

Our critique of Malm centers on his reductive view of history as progressive and linear—a framework embedded in capitalist logics of productivity that fails to recognize how alternative temporal relationships can constitute forms of resistance rather than withdrawal. Malm's characterization of the Standing Rock movement as part of a "cycle" that "came to a dead end" when Donald Trump came to power in 2017 exemplifies these limitations (2021, 98). This framework, which measures success through immediate political outcomes, cannot account for how the Water Protectors' resistance operated simultaneously across multiple temporal registers. While the pipeline was ultimately completed—representing a defeat in conventional political terms—this narrow analysis misses how Water Protectors maintained relationships to land and water outside the logic of commodification.

The concept of Mni Wiconi ("water is life") embodies the relational temporality underlying the Indigenous-led resistance. As Dhillon and Estes (2016) explain: "Mni Sose (the Missouri River) is not a thing that is quantifiable according to possessive logics. Mni Sosé is a relative: the Mni Oyate, the Water Nation. She is alive. Nothing owns her." This understanding of water as a living relative rather than a resource transcends capitalist frameworks that reduce nature to a commodity, capturing the still-living auratic relationship that Benjamin describes, where nature "looks back" at the observer. Mni Sose, Estes adds, translates to "roiling water," indicating "not merely static, inanimate form, but also action" (2017, 13). Neither strictly active nor passive, Water is Life.

> Mni Wiconi simultaneously speaks to the past, present, and future—catapulting us into a moment of critical, radical reflection about the colonial wounds and wounding in the spaces between calls to save planet Earth and the everyday sociopolitical realities facing Indigenous peoples.
>
> (Dhillon and Estes 2016)

Indigenous temporal frameworks offer an alternative to linear progressive narratives. As Vine Deloria Jr. (1972) demonstrates, Indigenous peoples often operate through a spatial conception of history, where time is understood through relationship to place rather than linear progression. Leanne Betasamosake Simpson builds on this understanding, noting that the spatial construction of time means that the present is "a colliding of the past and future" (2017, 193). She writes:

> My Ancestors are not in the past. The spiritual world does not exist in some mystical realm. These forces and beings are right here beside me—inspiring, loving, and caring for me in each moment and compelling me to do the same. It is my responsibility with them and those yet unborn to continuously give birth to my Indigenous present.
>
> (192–193)

This spatialization of time affords a rich temporal multiplicity that was embodied in daily practices at Standing Rock. Water ceremonies performed each morning enacted an ongoing relationship to water that predated colonial contact while simultaneously addressing immediate political realities. These ceremonies weren't merely symbolic gestures alongside "real" political resistance but constituted what Simpson describes as "constellations of coresistance with other humans and nonhumans" that generate "ethical engagements and practices" outside extractive logics (2017, 233). The camp's organization combined traditional governance structures with contemporary tactical approaches, creating what Whyte (2018) has called "spiraling temporalities" where past, present, and future coexist rather than following linear progression.

Rather than experiencing the exhaustion that comes from forcing multiple temporalities into a single progressive narrative—what we've identified as eco-fatigue—the Water Protectors' practice of "Wotakuye (kinship)" embraces temporal multiplicity as a source of resilience (Dhillon & Estes 2016). This stands in contrast to Malm's emphasis on "organized upscaling" as necessary for effective resistance (2021, 76). While physically confronting the extractive economy, Water Protectors simultaneously maintained relationships to land and water outside the logic of commodification. Their practices reveal how a critical stance toward capitalist temporality need not result in paralysis or detachment but can generate forms of collective action that maintain awareness of multiple, sometimes contradictory, temporal relationships to environmental crisis.

Moreover, the resistance continues in forms that transcend the specific site of protest and have fostered enduring solidarities. As Nick Estes writes:

> [T]his was not just about Standing Rock water on the Missouri River … [The pipeline] cut under the Mississippi River at the Iowa-Illinois border, where a coalition of Indigenous peoples and white farmers, ranchers, and environmentalists in Iowa opposed it.
>
> (2019, 9)

While Indigenous-led, the camps attracted Water Protectors from across the nation and fostered solidarities that endure beyond the specific site of resistance. "[A] very indigenous thing happened [in 2016]: we made and remade relations" (Estes 2020). On these enduring connections, he observes:

> I saw Water Protectors fight fascists in the streets of Seattle and burn down the Third Precinct in Minneapolis as payback for the killing of George Floyd and the daily terror police inflicted on Black life. To this day, I walk the streets with Water Protectors to feed and clothe our relatives, not for charity but solidarity. The spirit of Mni Wiconi — Water is Life — and the #NoDAPL movement lives on. In many ways, it has grown.
>
> (Estes 2020)

The Water Protectors' conception of Mni Wiconi illuminates the dual aspect of our title: "Tired Nature, Tired of Nature." Their understanding of water as a living relative highlights how nature itself becomes "tired" through extractive exploitation. Their experience also helps us understand how maintaining alternative relationships to nature within hostile political and economic systems produces different forms of eco-fatigue. The strain of continuously translating between competing temporal frameworks represents a form of labor that helps explain why eco-fatigue does not necessarily indicate disengagement but often reflects the costs of maintaining complex ecological relationships.

The difference between experiences of eco-fatigue and collective movements like Standing Rock suggests that embodying shared practices of temporal multiplicity may offer pathways for transforming fatigue from a limitation into a source of critical insight and collective action. Standing Rock thus offers not just a case study but a methodological reorientation, suggesting that addressing eco-fatigue requires not overcoming it through intensified action within existing temporal frameworks, but rather creating conditions where multiple temporal relationships to nature can be acknowledged and maintained as the foundation for environmental politics.

Conclusion

Our analysis has traced eco-fatigue as a complex temporal experience that transcends the "climate fatalism" that Malm rejects. Through Levinas's phenomenology of fatigue, Benjamin's concept of auratic decay, and Bloch's nonsynchronous contradictions, we've revealed eco-fatigue not as mere passivity but as an embodied critique of capitalism's linear temporal logic.

What emerges most powerfully from our examination of Standing Rock is the recognition that European philosophical frameworks, however sophisticated, remain limited when confronted with Indigenous resistance practices. Eco-fatigue can be understood as the strain of attempting to navigate multiple temporal relationships without spatial grounding, a condition particularly acute under capitalism's abstraction of time from place. The spatial conception of time articulated by Indigenous thinkers offers a crucial intervention that transforms our understanding of environmental exhaustion, which emerges from the contradictions of inhabiting multiple temporal relationships to nature without adequate grounding in place. The concept of "Indigenous presents" articulated by Simpson—where the present is understood as "a colliding of the past and future" (2017, 193), anchored in specific places—offers an alternative to both climate fatalism and the burnout-inducing demand for climate action within capitalism's progressive narratives.

This understanding points toward a different environmental politics, one that recognizes how apparent passivity often contains within it complex forms of agency that Western frameworks struggle to recognize. By bringing European philosophical traditions into conversation with Indigenous knowledge systems, we begin to see how eco-fatigue might be transformed through practices that spatially ground temporal multiplicity. In practical terms, this might mean developing environmental movements that incorporate ceremonies acknowledging relationships to specific places, governance structures that make space for multiple temporal frameworks, and forms of resistance that refuse to separate cultural practice from political action.

By learning from Indigenous-led movements, we can understand eco-fatigue not as an obstacle but as a necessary dialectical moment in environmental consciousness that makes visible the contradictions in our relationship to nature under capitalism and points toward transformative practices. This approach won't eliminate the real exhaustion of environmental struggle in a world of increasing climate instability. But it might allow us to experience this exhaustion differently, not as evidence of failure, but as a shared condition connecting us across multiple temporal frames. In this sense, eco-fatigue becomes not an endpoint but a beginning—a critical moment that reveals the need for new temporal-spatial practices that can sustain environmental consciousness beyond the limitations of linear progress narratives.

Notes

1 This does not mean that the outer world does not impose any actual limits on our situation—of course it does, and we can never do exactly "what we want." Sartre knows this very well, and this is of course not what his concept of freedom entails, contrary to certain simplistic misrepresentations of his view. His point is rather, roughly put, that it is we who *give meaning* to our experiences of

facticity: they can never be said to determine our actions directly, as if they were simply acting "upon" us. Rather, they only have meaning through us: we choose what signification we give them, and for this reason we remain free in relation to them.

2 In what follows, we build on Rathe's (2025) reading of Levinas and phenomenological analysis of fatigue as passive-active resistance, which we here connect more specifically to the question of eco-fatigue.

3 Note that on this point, we are moving slightly away from Levinas' analysis.

4 The Standing Rock resistance (April 2016–February 2017) was led by the Standing Rock Sioux Tribe and joined by representatives from over 300 tribes and thousands of non-Indigenous allies who established camps to physically block construction of the Dakota Access Pipeline (DAPL) in North Dakota. The resistance sought to protect the Missouri River watershed from potential oil spills and to defend sacred sites threatened by the pipeline's route. See Nick Estes, *Our History Is the Future* (2019) for a comprehensive account.

References

Benjamin, Walter. 2004. "On Language as Such and on the Language of Man." In *Selected Writings, Volume 1: 1913–1926*, edited by Marcus Bullock and Michael W. Jennings. Harvard University Press.

Benjamin, Walter. 2008. *The Work of Art in the Age of Its Technological Reproducibility and Other Writings on Media*, edited by Michael W. Jennings, Brigid Doherty, and Thomas Y. Levin. Harvard University Press.

Bird, Lucy H., Emma F. Thomas, Michael Wenzel, and Morgana Lizzio-Wilson. 2024. "Thinking about the Future: Examining the Exacerbating and Attenuating Factors of Despair-Induced Climate Burnout." *Journal of Environmental Psychology* 98: 102382. https://doi.org/10.1016/j.jenvp.2024.102382

Bloch, Ernst. 1977. "Nonsynchronism and the Obligation to Its Dialectics." Translated by Mark Ritter. *New German Critique* 11: 22–38.

Bloch, Ernst. 1998. *Literary Essays*. Stanford University Press.

Cher Weixia, Chen and Paul C. Gorski. 2015. "Burnout in Social Justice and Human Rights Activists: Symptoms, Causes and Implications." *Journal of Human Rights Practice* 7 (3): 366–90. https://doi.org/10.1093/jhuman/huv011

Deloria Jr., Vine. 1972. *God is Red: A Native View of Religion*. Fulcrum Publishing.

Dhillon, Jaskiran and Nick Estes. 2016. "Introduction: Standing Rock, #NoDAPL, and Mni Wiconi." *Hot Spots, Fieldsights*, December 22. Accessed March 3, 2025. https://culanth.org/fieldsights/introduction-standing-rock-no-dapl-and-mni-wiconi

Didi-Huberman, George. 2005. "The Supposition of the Aura: The Now, the Then, and Modernity," in *Walter Benjamin and History*, edited by Andrew Benjamin. Routledge.

Estes, Nick. 2019. *Our History is the Future*. Verso.

Estes, Nick. 2020. "Water Protectors Gave Us a Vision of the Future." *Society and Space*, October 12. Accessed March 13, 2025. https://www.societyandspace.org/articles/water-protectors-gave-us-a-vision-of-the-future

Foucault, Michel. 1982. "The Subject and Power." *Critical Inquiry* 8 (4): 777–95. http://www.jstor.org/stable/1343197

Hansen, Miriam Bratu. 2008. "Benjamin's Aura." *Critical Inquiry* 34 (2): 336–75. https://doi.org/10.1086/529060

Hayes, Shannon. 2020. "Can Melancholy Be Heroic? Walter Benjamin and the Vicissitudes of Melancholy," In *Faces of Depression in Literature*, edited by Josefa Ros Velasco. Peter Lang.

Head, Lesley and Theresa Harada. 2017. "Keeping the Heart a Long Way from the Brain: The Emotional Labour of Climate Scientists." *Emotion, Space, and Society* 24: 34–41. https://doi.org/10.1016/j.emospa.2017.07.005

Hoggett, Paul and Rosemary Randall. 2018. "Engaging with Climate Change: Comparing the Cultures of Science and Activism." *Environmental Values* 27 (3): 223–43. https://doi.org/10.3197/096327118X15217309300813

Levinas, Emmanuel. 1978. *Existence and Existents*. Translated by Alphonso Lingis. Martinus Nijhoff.

Malm, Andreas. 2021. *How to Blow Up a Pipeline*. Verso.

McCubbin, Megan. 2023. "From Eco-Fatigue to Eco-Action: How Local Conservation Can Re-ignite the Fight." Interview by Kieran Murnane. *CIWEM*. Accessed March 17, 2025. https://www.ciwem.org/news/from-eco-fatigue-to-eco-action-how-local-conservation-can-re-ignite-the-fight

Merleau-Ponty, Maurice. 2012. *Phenomenology of Perception*. Translated by Donald A. Landes. Routledge.

Morris, Katherine J. 2019. "Toward a Phenomenology of Fatigue," In *Phenomenology of the Broken Body*, edited by Espen Dahl, Cassandra Falke, and Thor Eirik Eriksen. Routledge.

Moscardo, Gianna and John Pearce. 2019. "Eco-Fatigue and its Potential Impact on Sustainable Tourist Experiences." *BEST EN Think Tank XIX: Creating Sustainable Tourism Experiences*, 140–64. https://researchonline.jcu.edu.au/61316/1/61316_Moscardo_and_Pearce_2019.pdf

Murphy, Michelle. 2017. "Alterlife and Decolonial Chemical Relations." *Cultural Anthropology* 32 (4): 494–503. https://doi.org/10.14506/ca32.4.02

Pihkala, Panu. 2020. "Anxiety and the Ecological Crisis: An Analysis of Eco-Anxiety and Climate Anxiety" *Sustainability* 12 (19): 7836. https://doi.org/10.3390/su12197836

Rathe, Kaja Jenssen. 2025. "Tired of Trying to Belong: On Affective Justice and Fatigue." *Philosophy and Social Criticism* 0 (0), 1–24. https://doi.org/10.1177/01914537251334948

Ray, Sarah Jaquette. 2020. *A Field Guide to Climate Anxiety*. University of California Press.

Sartre, Jean-Paul. 2018. *Being and Nothingness. An Essay in Phenomenological Ontology*. Translated by Sarah Richmond. Atria: Washington Square Press.

Simpson, Leanna Betasamosake. 2017. *As We Have Always Done: Indigenous Freedom through Radical Resistance*. University of Minnesota Press.

Stine, Alison. 2024. "How Does Climate Fatigue Hurt Our Efforts?" *Nonprofit Quarterly*, March 21. Accessed March 17, 2025. https://nonprofitquarterly.org/how-does-climate-fatigue-hurt-our-efforts/

The Brussels Times Newsroom. 2022. "Frustration and Disillusionment See Some Suffer from 'Eco-Fatigue'," December 11. Accessed March 17, 2025. https://www.brusselstimes.com/334792/frustration-and-disillusionment-see-some-suffer-from-eco-fatigue

VGTV. 2023. "SEE: Reaching Out to 'Generation Thunberg,'" September 6. Video, 33 sec. Accessed February 13, 2025. https://tv.vg.no/nyheter/se-langer-ut-mot-generasjon-thunberg?id=273323

Whyte, Kyle. 2017. "Indigenous Climate Change Studies: Indigenizing Futures, Decolonizing the Anthropocene." *English Language Notes* 55 (1–2): 153–62. https://doi.org/10.1215/00138282-55.1-2.153

Whyte, Kyle. 2018. "Indigenous Science (Fiction) for the Anthropocene: Ancestral Dystopias and Fantasies of Climate Change Crises." *Environment and Planning E: Nature and Space* 1 (1–2): 224–42. https://doi.org/10.1177/2514848618777621

Wood, Charlie. 2022. "There's No Place for Burnout in a Burning World." *Waging Nonviolence*, August 29. Accessed March 3, 2025. https://wagingnonviolence.org/2022/08/no-place-for-burnout-in-a-burning-world/

INDEX

Page numbers followed by 'n' refer to notes.